T0190125

Harnessing Synthetic Nanotechnology-Based Methodologies for Sustainable Green Applications

Nanotechnology is at the forefront of many of the latest developments across science and technology, but to generate and deploy these applications, macroscopic levels of nanoscale materials have to be carefully generated whilst remaining cost effective. These materials need to be reliable, consistent, and safe, and as a general principle, industries should consider green sustainable methods in the synthesis of these material and their applications as well. This book introduces readers to the field of green nanotechnologies and their possible applications to create a safer world. This accessible and practical guide will be a useful resource for material scientists, engineers, chemists, biotechnologists, and scientists working in the space of nanomaterials, in addition to graduate students in physics, chemistry, biomedical sciences and engineering.

THIS BOOK

- Presents an accessible introduction to the topic in addition to more advanced material for specialists in the field.
- Covers a broad spectrum of topics in this new field.
- Contains exciting case studies and examples, such as quantum dots, bionanomaterials, and future perspectives.

Dr Gérrard E.J. Poinern holds a Ph.D. in Physics from Murdoch University, Western Australia and a Double Major in Physics and Chemistry. Currently he is an Associate Professor in Physics and Nanotechnology in the School of Engineering and Information Technology at Murdoch University. He is the director of Murdoch Applied Innovation and Nanotechnology Research Group, Murdoch University. In 2003, he discovered and pioneered the use of an inorganic nanomembrane for potential skin tissue engineering applications. He is the recipient of a Gates Foundation Global Health Grand Challenge Exploration Award for his work in the development of biosynthetic materials and their subsequent application in the manufacture of biomedical devices. He is also the author of the 2014 CRC Press experimental textbook "A Laboratory Course in Nanoscience and Nanotechnology".

Associate Professor Suraj Kumar Tripathy is Associate Dean of the School of Chemical Technology at Kalinga Institute of Industrial Technology, Bhubaneswar, India. He currently leads the Chemical & Bioprocess Engineering Lab (CBEL) at KIIT which focuses on achieving sustainability in materials processing and utilization. CBEL explores opportunities in valorization of waste materials (secondary resources) and investigate their applications in catalysis, water treatment, and biomedical systems. CBEL also works closely with industries to develop suitable waste management and resource recycling strategies to optimize the potential of circular economy model.

Dr. Derek Fawcett is the Defence Science Centre research fellow at Murdoch University, Australia. His research involves the investigation and development of new advanced materials and their use in innovative engineering systems. He has published over seventy peer-reviewed research papers in international journals and is the co-author of four book chapters on applied nanotechnology.

Harnessing Synthetic Nanotechnology-Based Methodologies for Sustainable Green Applications

Edited by
Gérrard Eddy Jai Poinern, Suraj Kumar Tripathy
and Derek Fawcett

CRC Press is an imprint of the
Taylor & Francis Group, an informa business

First edition published 2023
by CRC Press
4 Park Square, Milton Park, Abingdon, Oxon, OX14 4RN

and by CRC Press
6000 Broken Sound Parkway NW, Suite 300, Boca Raton, FL 33487-2742

British Library Cataloguing-in-Publication Data
A catalogue record for this book is available from the British Library

Library of Congress Cataloging-in-Publication Data

Names: Poinern, Gérrard Eddy Jai, editor. | Tripathy, Suraj, editor. | Fawcett, Derek, editor.
Title: Harnessing synthetic nanotechnology-based methodologies for sustainable green applications / edited by Gérrard Eddy Jai Poinern, Suraj Tripathy and Derek Fawcett.
Description: First edition. | Boca Raton : CRC Press, 2023. | Includes bibliographical references. | Summary: "Nanotechnology is at the forefront of many of the latest developments across science and technology, but to generate and deploy these applications, macroscopic levels of nanoscale materials have to be carefully generated whilst remaining cost effective. These materials need to be reliable, consistent and safe, and considering green sustainable methods at the same time. This book introduces readers to the field of green nanotechnologies and their possible applications to create a safer world. This accessible and practical guide will be a useful resource for material scientists, engineers, chemists, and biotechnologists and scientists working in the space of nanomaterials, in addition to graduate students in physics, chemistry, and engineering"-- Provided by publisher.
Identifiers: LCCN 2022060689 | ISBN 9780367764920 (paperback) | ISBN 9781032020082 (hardcover) | ISBN 9781003181422 (ebook)
Subjects: LCSH: Nanotechnology--Industrial applications. | Nanostructured materials. | Synthetic products. | Sustainable engineering.
Classification: LCC T174.7 .H3835 2023 | DDC 620.1/150286--dc23/eng/20230323
LC record available at https://lccn.loc.gov/2022060689

ISBN: 978-1-032-02008-2 (hbk)
ISBN: 978-0-367-76492-0 (pbk)
ISBN: 978-1-003-18142-2 (ebk)

DOI: 10.1201/9781003181422

Typeset in Times
by Deanta Global Publishing Services, Chennai, India

Contents

THEME 1 Sustainable Energy Technologies

THEME 2 Developing Advanced Materials for Medicine

THEME 3 Advanced Materials for Agriculture and Environmental Concerns

Contributors

Shirsendu Banerjee
School of Chemical Technology, Kalinga Institute of Industrial Technology,
Bhubaneswar, India

Sarmistha Basu
Department of Physics, Indian Institute of Technology Kharagpur,
Kharagpur, India

Susanta Kumar Behera
School of Biotechnology, Kalinga Institute of Industrial Technology,
Bhubaneswar, India
and
IMGENIX India Pvt. Ltd.,
Bhubaneswar, India

Archana Bhaw-Luximon
Biomaterials, Drug Delivery and Nanotechnology Unit, Center for Biomedical and Biomaterials Research, University of Mauritius,
Reduit, Mauritius

Donny Bhuana
Murdoch Applied Nanotechnology Research Group, Department of Physics, Energy Studies and Nanotechnology, Murdoch University,
Murdoch, Western Australia, Australia

Sankha Chakrabortty
School of Chemical Technology, Kalinga Institute of Industrial Technology,
Bhubaneswar, India

Wisut Chamsa-ard
Murdoch Applied Nanotechnology Research Group, Department of Physics, Energy Studies and Nanotechnology, Murdoch University,
Murdoch, Western Australia, Australia

Huu Dang
Innovation Fuels, Curtin University,
Bentley, Western Australia, Australia

Debasish Das
School of Nanoscience and Technology, Indian Institute of Technology Kharagpur, Kharagpur, India

Suprem Ranjan Das
Department of Industrial and Manufacturing Systems Engineering, Kansas State University, Manhattan, KS, USA
and
Department of Electrical and Computer Engineering, Kansas State University,
Manhattan, KS, USA

Derek Fawcett
Murdoch Applied Nanotechnology Research Group, Department of Physics, Energy Studies and Nanotechnology, Murdoch University,
Murdoch, Western Australia, Australia

Yuanyuan Feng
Murdoch Applied Nanotechnology Research Group, Department of Physics, Energy Studies and Nanotechnology, Murdoch University,
Murdoch, Western Australia, Australia
and
Key Laboratory of Agro-Environment in Downstream of Yangtze Plain, Ministry of Agriculture and Rural Affairs, Jiangsu Key Laboratory for Food Quality and Safety/ State Key Laboratory Cultivation Base of Ministry of Science and Technology, Institute of Agricultural Resources and Environment, Jiangsu Academy of Agricultural Sciences, Nanjing, China

Cormac Fitzgerald
Murdoch Applied Nanotechnology Research Group, Department of Physics, Energy Studies and Nanotechnology, Murdoch University,
Murdoch, Western Australia, Australia

Chun Che Fung
School of Engineering and Information Technology, Murdoch University,
Murdoch, Western Australia, Australia

A.F.M. Fahad Halim
Murdoch Applied Nanotechnology Research Group, Department of Physics, Energy Studies and Nanotechnology, Murdoch University,
Murdoch, Western Australia, Australia

David Henry
School of Mathematics, Statistics, Chemistry and Physics and Centre for Sustainable Farming Systems, Food Futures Institute, Murdoch University,
Murdoch, Western Australia, Australia

Owen Horoch
School of Mathematics, Statistics, Chemistry and Physics and Centre for Sustainable Farming Systems, Food Futures Institute, Murdoch University,
Murdoch, Western Australia, Australia

Md. Imran Khan
School of Biotechnology, Kalinga Institute of Industrial Technology,
Bhubaneswar, India

Cecilia Stalsby Lundborg
Department of Public Health Sciences, Karolinska Institutet,
Stockholm, Sweden

Subhasish Basu Majumder
Materials Science Centre, Indian Institute of Technology Kharagpur,
Kharagpur, India

Amrita Mishra
School of Biotechnology, Kalinga Institute of Industrial Technology,
Bhubaneswar, India

Ananyo Jyoti Misra
School of Chemical Technology, Kalinga Institute of Industrial Technology,
Bhubaneswar, India
and
School of Biotechnology, Kalinga Institute of Industrial Technology,
Bhubaneswar, India

Marjan Nasseh
Murdoch Applied Nanotechnology Research Group. Department of Physics, Energy Studies and Nanotechnology, Murdoch University,
Murdoch, Western Australia, Australia

Jayato Nayak
Centre for Life Science, Mahindra University,
Hyderabad, India

Itisha Chummun Phul
Biomaterials, Drug Delivery and Nanotechnology Unit, Center for Biomedical and Biomaterials Research, University of Mauritius,
Reduit, Mauritius

Gérrard Eddy Jai Poinern
Murdoch Applied Nanotechnology Research Group, Department of Physics, Energy Studies and Nanotechnology, Murdoch University,
Murdoch, Western Australia, Australia

Kh M Asif Raihan
Department of Industrial and Manufacturing Systems Engineering, Kansas State University, Manhattan, KS, USA

Supriya Rattan
Murdoch Applied Nanotechnology Research Group. Department of Physics, Energy Studies and Nanotechnology, Murdoch University,
Murdoch, Western Australia, Australia

Julio Sanchez-Nieva
Academy, School of Veterinary and Life Sciences, Environmental & Conservation Sciences
Murdoch University,
Murdoch, Western Australia, Australia

Rupam Sharma
ICAR-Central Institute of Fisheries Education, Panch Marg, Yari Road,
Versova, Mumbai, India

Suraj Kumar Tripathy
School of Chemical Technology, Kalinga Institute of Industrial Technology,
Bhubaneswar, India

Rajeev Kumar Varshney
State Agricultural Biotechnology Centre & Centre for Crop & Food Innovation
International Chair in Agriculture & Food Security
Food Futures Institute, Murdoch University
Perth, Australia

Jennifer Verduin
Academy, School of Veterinary and Life Sciences, Environmental & Conservation Sciences
Murdoch University, Murdoch, Western Australia, Australia

Triana Wulandari
Murdoch Applied Nanotechnology Research Group, Department of Physics, Energy Studies and Nanotechnology, Murdoch University,
Murdoch, Western Australia, Australia

1 Introduction

Derek Fawcett and Gérrard Eddy Jai Poinern

CONTENTS

1.1 OVERVIEW

In this book, we summarize and discuss current research efforts aimed at mitigating the effects of three global challenges facing humanity. The increasing rapid growth of the global population and industrialization is not only creating serious problems for humanity today, but it is also creating challenges for the future [1]. The result of increasing global population growth and industrialization is greater demand for both energy and water. Thus, industrialization, economic growth, and improving human health are all dependent on energy and water. The sustainable management of both resources must be achieved to overcome the challenges of climate change and deliver food security. Crucially, both challenges are exacerbated by population growth, industrial development, and diminishing natural resources [2]. Fossil fuels are predominantly used to address the current energy demand. These fuels supply around 86% of the energy needed to power the world's economy. Significantly, over the next 20 years, the global population is expected to increase by 1.5 billion and will result in a total global population of around 8.8 billion people [3]. And because of the reliance on fossil fuels, the population increase will also translate into larger quantities of harmful greenhouse gases being discharged into the atmosphere. The detrimental impact of accumulating greenhouse gases in the atmosphere is a very serious problem. Recent studies have highlighted the need to limit global temperature rises to no more than 2 °C of pre-industrial levels [4, 5]. Consequently, global decarbonization of electricity generation, industrial and transportation sectors must immediately take place to reduce the global warming trend [6].

Recent research efforts have focused on developing alternative, clean, and viable renewable energy sources [7]. However, transforming the world's energy supply from fossil fuels to renewable energy sources is not going to happen overnight. Current estimates indicate fossil fuel consumption will continue to increase at a rate of about 1.3% per year until around 2030 [8]. During this period, greenhouse gas emissions will continue to increase and exacerbate global climate change. Also, during this period, agricultural, industrial, and transport sectors that are currently dependent on fossil fuels will need to implement change management strategies to promote the use of sustainable renewable energy sources. In the case of marine and aviation sectors that are heavily dependent on fossil fuels, new propulsion and power generation systems will need to be developed. So, during this translational period, harm reduction methods, such as reducing sulfur dioxide, nitrogen oxides, and other pollutants from marine vessel exhausts and the production of clean synthetic aviation fuels, will be needed to alleviate the effects of greenhouse gas emissions.

Water is an indispensable resource that is essential for all life on Earth. Most people believe water is plentiful because around 1,338,000,000 km^3 is stored in oceans and covers 71% of Earth's

DOI: 10.1201/9781003181422-1

surface. This equates to about 97% of all Earth's water and is too salty to drink [9]. The remaining 3% is freshwater, but of this, 90% is frozen, and only the remaining 0.3% is accessible. Of this, only 1% of the freshwater is accessible from rivers, streams, and lakes, with the remainder being underground. Overall, this equates to around 0.003% being available for human usage. However, freshwater is not evenly distributed across Earth's land masses [10]. In addition, according to the World Health Organization, around 2 billion people have only gained access to improved and sustainable water since 1990 [11]. Moreover, there are currently around 800 million people who still live without safe clean fresh water. The health problem is further exacerbated by almost two-fifths of the world's population not having access to proper waste disposal systems [12]. Crucially, current global efforts are focused on finding new and alternative freshwater sources that will be needed to service the predicted global population of 12.3 billion in 2100 [3, 13]. The annual global demand for water is increasing annually by 2% and is predicted to reach 6,900 billion m^3 by 2030 [14]. However, Earth's annual natural water cycle of 4,200 billion m^3 will not be able to meet this demand [15]. At present, global water demand exceeds Earth's natural sustainable water cycle and is being alleviated by the use of high energy-intensive, fossil fuel-driven desalination plants.

The successful deployment of new technologies and renewable energy resources, combined with improved wastewater treatments and mitigation of water pollution, together with advanced biotechnological solutions for agriculture, can deliver sustainable, cost-effective energy and clean water supplies and food security. Ultimately, these technologies, treatments, and processes can decrease greenhouse gas emissions and treat wastewater, hence reducing environmental impacts on the global economy and promoting public health and well-being. Nanotechnology is well-placed to achieve these outcomes, where engineered nanometer-scale materials for a wide range of applications can be manufactured. Compared to conventional bulk materials, nanomaterials with a larger proportion of atoms located at the material's surface have unique surface area-related properties. For instance, nanoparticles have a greater tendency to absorb, interact, and react with other chemical species to attain charge stabilization. During the past few decades, nanotechnology has evolved exponentially to address some of the challenges facing humanity. In the energy sector, nanotechnology-based methods can be used to reduce the impact of conventional fuels, generate new synthetic fuels, and produce nanostructured electrodes in new advanced electrolysis cells, fuel cells, and lithium-ion batteries. In the health sector, nanotechnology-based systems can be applied to applications such as biosensors, cancer nanomedicine, tissue regeneration of wounds, and vaccine development. In the agriculture and environmental sectors, nanotechnology-based methods can be used to deliver advanced biotechnological solutions to improve crop productivity, treat wastewater, and reduce environmental degradation. This book addresses the abovementioned issues facing humanity arising from population growth, industrial development, and diminishing resources, and the opportunities resulting from recent nanotechnology-based advances.

1.2 SCOPE OF THE BOOK

The aim of this book is to provide an overview of three significant areas of concern to humanity today. As mentioned above, the three main concerns are to develop sustainable energy technologies that are not dependent on fossil fuels; to develop advanced medical materials for the well-being of humanity; and to develop advanced materials for agriculture, new methods of treating wastewater, and to address the current problems associated with microplastic-based water pollution. Each of these concerns is addressed by a theme. Each theme contains several chapters that address specific aspects relating to the respective field. The three themes are:

1. Sustainable energy technologies
2. Advanced materials for medicine
3. Advanced materials for agriculture and environmental concerns

1.2.1 Theme 1: Sustainable Energy Technologies

Chapters 2 to 6 consider Theme 1. Each chapter provides an overview and extensive discussion relating to specific key sustainable energy technologies currently being developed for use in the near future.

Chapter 2 discusses the current use of marine fuels in shipping and the resulting enormous levels of exhaust emissions. These emissions contain pollutants such as carbon dioxide, nitrogen oxides, fine particles, and sulfur dioxide which is responsible for acid rain. The sulfur content in fuels, and its subsequent combustion, has been identified as the most serious pollutant in exhaust emissions. Therefore, no meaningful strategy for reducing air pollution will be successful until the sulfur levels in marine fuels are significantly reduced to near zero. This chapter discusses the importance of reducing the sulfur content in marine fuels and the subsequent reduction in harmful exhaust emissions. Currently, the oil refining industry is producing low-sulfur fuels, and their use is expected to reduce greenhouse gas emissions by at least 50% by 2050. However, there is still an urgent need to completely remove sulfur from marine fuels. To this end, Chapter 2 examines the synthesis and use of active molybdenum-based catalysts such as MoO_2 and MoS_2 nanoparticles as the basis for novel sulfur-removal technology.

Chapter 3 reviews the pivotal role the aviation industry plays in the modern world by facilitating communication, trade, and marketing on a global scale. However, the aviation industry is also a major producer of greenhouse gases such as CO_2. Alternatives to conventional fossil fuels used in the aviation industry are synthetic fuels, biologically derived fuels, or biofuels. Biofuels can be thought of as any energy-rich chemical compound generated from biomass. Both synthetic and biofuels are special because they combine many desirable qualities. For instance, they are renewable, since they are made from readily available feedstock. In particular, biofuels are biodegradable, have lower toxicity, are diverse, and are composed of easily available biomass. Notably, Chapter 3 examines the use of catalytic nanomaterials for producing renewable synthetic fuels and biofuels. The chapter also discusses the Fischer-Tropsch process as the downstream technology for sustainably producing aviation fuels.

Chapter 4 examines the current research into inorganic membranes for gas separation processes and their use in high-temperature solid oxide cells. Solid oxide cells can be used to produce hydrogen, synthetic fuels, and electrical power. Microporous membranes made from silica, carbon, and zeolite are initially discussed, before dense metallic and ceramic membranes are considered. Progress made toward developing fluorites and Perovskite-based materials for membrane technologies is also discussed. This is followed by an overview of the recent progress made toward developing high-temperature solid oxide cells for gas separation processes, and electrolysis applications for producing synthetic fuels and their use as fuel cells for generating electrical power. The basic operating principles of both electrolysis and fuel cells are examined. In addition, the challenges involved in designing high-performance electrodes, electrolytes, and issues relating to degradability and long-term durability are also reviewed. Notably, the chapter considers the recent interest in CO_2 capture as a method of reducing greenhouse gas emissions.

Once captured, CO_2 can either be sequestered for long-term storage or utilized as an important resource for the production of value-added products. Crucially, renewable energy sources are not suitable fuel sources for aircraft, shipping, and land-based transport such as trucks and trains. Thus, there is a need to change our philosophical mindset toward CO_2 from being a global menace to an important renewable resource. A CO_2 conversion technology based on inorganic membranes can be used to produce initial gases such as carbon monoxide (CO) and oxygen (O_2). Then from these gases, mixtures of H_2 and CO can be used to produce a variety of carbon-based fuels such as methane (CH_4), synthetic natural gas (SNG), and methanol (CH_3OH). Further processing means these products can be converted into fuels for specific transport applications while the pure O_2 generated can be directly used in many existing industrial and manufacturing processes. Furthermore, when inorganic oxide membranes are used in the fuel cell mode, they become promising electrochemical conversion devices with high electrical efficiencies and fuel flexibility. Typically, gases such as H_2 and O_2 (usually from air), are supplied, and through the electrochemical oxidation of H_2 (as fuel), electrical energy is generated. The chapter concludes with a discussion of new and improved nanometer-scale materials, combined with advanced fabrication processes, which have the potential to deliver commercially viable inorganic membranes for gas separation processes and high-temperature solid oxide cells. These inorganic membranes and devices have the potential to produce synthetic fuels and electrical power in the near future.

Chapter 5 explores the use of advance materials such as graphene, oxide and reduced graphene oxide additives to improve basin water evaporation rates. Basin water-based systems are the main component in solar-thermal still desalination systems. This type of research is very important because the demand for clean drinkable water is rapidly increasing due to global population growth, increasing agricultural activities, and growing industrial development. At present, fossil fuels provide the energy needed to drive energy-intensive desalination processes around the world. Unfortunately, fossil fuel-driven desalination processes have significantly contributed to high greenhouse gas emissions, environmental degradation, and global warming. On the other hand, solar-thermal desalination is a low-cost, sustainable, and eco-friendly alternative for generating high-quality freshwater from brackish water without using fossil fuel-driven systems. However, improving solar-thermal desalination performance levels still remain elusive. But new nanometer-scale materials have the potential to considerably improve solar-thermal still performance and promote greater water production rates. This chapter evaluates the use of graphene oxide and reduced graphene oxide additives to improve the photothermal response and evaporation rates of basin water used in stills.

Chapter 6 discusses electrophoretic deposition and inkjet printing methods as promising fabrication routes for manufacturing flexible rechargeable cells and supercapacitors. In recent years, there has been considerable interest shown in developing flexible electronic devices, such as wearable/implantable sensors and monitoring devices, e-textiles, rollable displays, and even robotic suits. The advent and growth of these types of flexible electronic devices have also necessitated the development of flexible electrochemical energy storage devices. The two frontrunners among such energy storage devices are lithium-ion batteries and supercapacitors. Rechargeable lithium-ion batteries are considered the most promising power source for such flexible devices due to their higher nominal voltage, and higher energy and power densities. Lithium-ion batteries also have longer life spans compared to other battery systems while flexible supercapacitors, with features such as an intrinsic high-power density, faster charge/discharge capability, and extended cycle life, are also considered to be a potential technology to power up various flexible devices. Furthermore, many of these flexible devices could use a supercapacitor in conjunction with a flexible lithium-ion rechargeable battery for added electrical performance. The chapter discusses the salient features of bendable and stretchable electrochemical energy storage devices. The chapter also reviews various materials used in the manufacture of flexible energy storage devices. The chapter then discusses two promising fabrication techniques that can be used to produce flexible energy storage devices. The first method discussed is electrophoretic deposition and focuses on the electrochemical performance of a 3D carbon cloth graphite anode. The second method discusses the inkjet printing method and focuses on the fabrication and electrochemical performance of an all-solid graphene supercapacitor.

1.2.2 Theme 2: Developing Advanced Materials for Medicine

Chapters 7 to 10 consider Theme 2. Each chapter summarizes and discusses a specific key area of medicine where new advanced nanometer-scale materials can improve, prolong, and promote human health and well-being.

Chapter 7 examines the use of sustainable nanotechnology-based strategies for developing targeted therapies using cell-encapsulated hydrogels. Nanotechnology contains all the technologies involved with science, engineering, and medicine operating at the nanometer scale. The fields of nanomedicine, tissue regeneration, and materials engineering have allowed the emergence of applicable nanotechnology-based strategies aimed at sustainable and targeted human medical therapies. Nanotechnology and these fields have evolved exponentially during the past decades to address new worldwide challenges such as such as novel cancer treatments and vaccine development. In the health sector, nanotechnology-based systems have very diverse applications ranging from nanofiltration systems to nanomedicine. In particular, nanotechnology-based methods have been able to take advantage of the versatile physiochemical properties of nanometer-scale materials and structures, and apply these unique properties to fields such as cancer nanomedicine, tissue regeneration of wounds, and viral vaccines. Similarly, hydrogels with nanostructures made from sustainable, accessible, and renewable materials can mimic the properties of living systems. These nanostructured hydrogels can respond to environmental changes such as pH, temperature, light, pressure, electric field, or a combination of different stimuli. Because of their versatility, hydrogels are used as scaffolds for tissue engineering, drug delivery systems, biosensors, pre-clinical cancer models, and cancer cell capture. In addition, hydrogels can embed exogenous cells to attract native cells and accelerate a number of therapies. In addition, hydrogels offer the advantage of being injectable and compressible, allowing liquid retention and sustaining cell encapsulation. Chapter 7 summarizes and discusses several targeted therapies using cell-encapsulated hydrogels engineered from sustainable materials. The chapter discusses the challenges facing hydrogel development in this medical field and future perspectives regarding the translation of cell-encapsulated hydrogels to several beneficial medical applications.

Chapter 8 considers wound healing and infection control, and the use of nanometer-scale materials to assist in the healing process. Skin wounds and infections will be experienced by all of us during our lifetime. The skin has the vital role of protecting the internal environment of the body. Normal healthy skin forms a defensive barrier to prevent the invasion of harmful microorganisms and other pathogenic agents into the body from the surrounding environment. When the skin is damaged, a wound forms. The subsequent healing process may be disturbed by both internal and external factors. The most important factor is the invading microorganisms that contaminate, infect, and delay wound healing. The traditional approach for treating wounds is by using dressings and topical therapies. However, chronic wounds are becoming a serious problem due to the increasing number of multidrug-resistant microorganisms. Nanomaterial-based technologies offer a new approach to fighting infection and promoting wound healing. Currently, nanoparticles and encapsulated nanoparticle-based dressings are being used to treat bacterial infections and enhance wound healing. The chapter summarizes and discusses recent studies regarding wound healing and the application of new nanoparticle-based therapies specifically developed for wound management.

Chapter 9 reviews the most recent progress made toward using calcium carbonate ($CaCO_3$) micro/nanoparticles in therapeutic applications such as the controlled release of pharmaceuticals for cancer treatment, tumor imaging, and gene therapy. The advantages of using $CaCO_3$ micro/nanoparticles in these medical applications arise from their unique material properties. Some of these advantageous properties include pH sensitivity, biocompatibility, safety, and biodegradability. Crucially, $CaCO_3$ micro/nanoparticles are stable at normal physiological pH (~7.3) in the blood circulation system, whereas in more acidic pH tumor environments, they readily decompose. It is this slow degradation of their porous core structures that allows these particles to be employed as carriers for contrast agents used in current biomedical imaging procedures or utilized as sustained-release carriers for the targeted delivery of anticancer drugs and gene therapies. The chapter provides an overview of recent research into developing $CaCO_3$ delivery carriers. It also discusses cancer, the advantages of using a nanomedicine approach for treating cancer, and various methods for producing $CaCO_3$ micro/nanoparticles designed for delivering anticancer pharmaceutical agents. Also considered in the chapter is the current state of research into using $CaCO_3$-based particle carriers for biomedical imaging and gene therapy. Notably, it highlights the potential use of $CaCO_3$ micro/nanoparticles as a safe and efficient drug delivery platform for cancer treatment.

Chapter 10 evaluates an ultrasonically engineered silicon-doped nanometer-scale hydroxyapatite powder designed for use in a number of dental and bone restorative procedures. Globally, millions of people each year require a bone transplant and many more millions need dental restorative procedures. In particular, oral caries is a serious global issue, and poor dental health can seriously affect the well-being of a patient. Traditional restorative methods use the patient's own hard tissue grafts, which have the advantages of excellent biocompatibility and osteogenic properties, and deliver hard tissue-forming cells to the implant site. However, problems such as donor site morbidity and limited donor sites have resulted in the search for alternative sources of hard tissues. An attractive alternative to natural hard tissue grafts is to create synthetic materials that can be formed into scaffolds for specific tissue regeneration procedures. Importantly, the fabricated scaffold needs to replicate the various physical, chemical, and mechanical properties found in natural hard tissues. Synthetically manufactured hydroxyapatite, which has a close chemical similarity to the natural form found in hard tissues, is a suitable scaffold candidate. But synthetic hydroxyapatite does not possess all the desirable properties of natural hydroxyapatite. Therefore, there is a need to improve the properties of synthetic hydroxyapatite. The addition of small quantities of silicon to synthetic hydroxyapatite can beneficially promote biological mineralization and hard tissue formation. The chapter examines the synthesis of silicon-doped nano-hydroxyapatite powders produced by a combined ultrasonic and microwave heating-based method. Advanced characterization techniques were carried out on annealed nano-powders. X-ray peak broadening analysis revealed that increasing silicon content inhibited grain growth and reduced crystallinity. Furthermore, increasing silicon content generated a distortion in the lattice structure and increased lattice constants. It also resulted in changes in the unit cell volume and increased lattice strain. Importantly, the analysis revealed that $(SiO_4)^{4-}$ ions were being substituted for $(PO_4)^{3-}$ ions in the nano-hydroxyapatite lattice structure.

1.2.3 Theme 3: Advanced Materials for Agriculture and Environmental Challenges

Chapters 11 to 16 consider Theme 3. The chapters summarize and discuss several aspects of advanced nanometer-scale materials and processes for treating wastewater and improving agricultural productivity. Chapters 12 and 13 focus on novel nanotechnology-based methods for treating wastewater. This is followed by Chapters 14 and 15, which outline and discuss the global impact of microplastic pollution and its detrimental effect on coastal ecosystems. Chapter 16 then discusses the role of genes, nanogenomics, and nanotechnology-based genetic engineering procedures in delivering advanced biotechnological solutions for many of the challenges facing agriculture and horticulture today.

Chapter 11 assesses the increasing use of nanoparticles within the agricultural sector in areas such as fertilizers, pesticides, biosensors, and soil amendments. In particular, the chapter focuses on the many advantages offered by nanomaterials, including the ability to achieve the targeted delivery of nutrients or pesticides, which in turn can reduce application rates. These advantages have benefits in terms of improved efficacy and lower costs. However, the longer-term effects of nanoparticles in the environment are only just being explored. In this regard, a challenge arises from the diversity of available nanoparticles and the massive variation in their properties. These properties not only depend on the composition but also the size, shape, porosity, and functionalization of the nanoparticles. Therefore, understanding how nanomaterials interact with plants and soils is imperative to increasing their efficacy and minimizing their environmental impact. In addition, determining the fate of nanoparticles in the agricultural environment must also be considered for their safe and effective use. The chapter discusses these issues and specifically focuses on the effective application of nanoparticles to crops.

Chapter 12 investigates the design of composite nano-systems for photocatalytic water treatments and opportunities for using these systems in off-grid applications. Photocatalysis is an advanced sustainable oxidation technology for the removal of microorganisms and organic compounds from wastewater. Photocatalysis has the potential to be a low-cost, eco-friendly, and sustainable treatment process capable of delivering a zero-wastewater strategy from wastewater handling facilities. Currently, the world needs effective water purification and wastewater treatment processes. Rapid population growth and industrialization have put large demands on limited freshwater supplies. At the same time, population growth and industrialization have produced extremely large quantities of pollutants. These pollutants are found in ground and surface water sources, and in wastewater. Therefore, to preserve food security, human health, and economic development, it is necessary to develop innovative, eco-friendly, low-cost, and highly effective wastewater treatment processes. In recent years, there has been significant progress made in developing advanced photocatalytic materials for the disinfection of wastewater in treatment facilities. A number of potential and commercially available photocatalytic materials are discussed in the chapter. The chapter goes on to identify several challenges that need to be addressed before this technology can gain wider acceptance. The chapter also discusses the potential use of decentralized, off-grid, and point-of-use wastewater disinfection systems. Crucially, the use of these types of systems can offer a cost-effective method for disinfecting water and wastewater in many remote and vulnerable rural populations.

Chapter 13 summarizes recent research and current applications of carbon-based heterogeneous nanomaterials for industrial wastewater treatment. Globally, the generation of wastewater has dramatically increased with increasing industrialization and population growth. The problem is further exacerbated by the decreasing supply of natural sources of clean freshwater. These factors necessitate the rapid development of effective wastewater treatment processes for delivering usable water supplies. In recent years, a great deal of interest has focused on using carbon-based nanomaterials as one strategy for treating wastewater. Due to their large surface area, mesoporous structure, tunable surface characteristics, and chemical stability, carbon nanomaterials have attracted considerable interest. In particular, carbon nanotubes and graphene-based composites have enormous potential to be used as adsorbents in wastewater treatment processes. Their unique properties make them resistant to high temperatures and a wide range of wastewater conditions, including acidic, basic, and salty environments. Even though a lot of research has been done to establish how carbon-based nanomaterials operate in wastewater systems, there are still engineering, toxicological, and environmental hurdles that must be overcome before they can be fully commercialized. The chapter explores the use of carbon nanotubes, graphene, and related nanomaterials for treating wastewater. The chapter also summarizes and discusses the various types of carbon nanomaterials, their structures and morphologies, their effectiveness in treating wastewater, and their ability to be used in water purification systems.

Chapter 14 examines and discusses the global microplastic pollution problem and its detrimental impact on coastal ecosystems and mid-ocean gyres. The presence of microplastic materials in the environment is widespread, and their largest concentrations can be found in coastal ecosystems and within mid-ocean gyres. Since the inception of mass plastic product manufacturing in the middle of the twentieth century, these durable, lightweight, and inexpensive materials have been, and continue to be, produced in large numbers. However, the presence of large numbers of microplastics in marine ecosystems in recent years has become a serious environmental issue that has attracted considerable widespread interest, both in the scientific and the broader community. In particular, the ingestion and subsequent detrimental health effects on many marine species are the most noticeable and alarming impacts of microplastics. Furthermore, recent studies have also shown that microplastics can accumulate, concentrate, and act as vectors for conveying toxic pollutants within the food chain. Another feature of microplastics is their ability to transport marine species from one ecosystem to another ecosystem. The invading species then become a threat to local indigenous marine species. Because of the serious nature of microplastic pollution, it is important to understand its impact on coastal ecosystems and ocean gyres. Chapter 14 reviews and discusses four aspects of microplastic pollution: 1) sources of both primary and secondary microplastics; 2) their physical and chemical behavioral properties; 3) bioavailability and behavioral properties toward marine organisms, and 4) future perspectives, which highlight key areas of research needed to elucidate the effects of microplastic pollution in the marine environment. Therefore, understanding these four aspects will assist in directing future marine pollution research and assist policymakers to develop appropriate management strategies.

Chapter 15 presents the results of a case study that determined the concentration levels of microplastic fibers present in sediments in South Beach, located on the southwest coast of Western Australia. The ever-increasing use of plastic materials has also led to high levels of waste entering the environment and becoming a serious pollutant. The chapter considers both the abundance and spatial distribution of microplastic fibers in the beach sediments. The analysis technique used density separation, *via* a new elutriation system using hyper-saline solutions, to separate microplastic fibers from the sandy sediments. Overall, the study found a mean fiber density of 43.3 fibers kg^{-1} and lengths ranging from 800 to 1824 microns. In addition, the chapter discusses the use of a novel elutriation system for collecting fibers and its suitability for routine monitoring programs.

Chapter 16 examines and discusses the role of genes, nanogenomics, and nanotechnology-based genetic engineering procedures as well as novel Dip-Pen Nanolithographic techniques in delivering advanced biotechnological solutions for many of the challenges facing agriculture today. Eco-friendly nanoparticle-based carrier systems can deliver bioengineered sustainable agrochemicals to improve crop productivity and reduce environmental degradation. The chapter discusses the positive effects produced by several types of nano-fertilizers and nano-pesticides on crop production. Also examined is the use of nanomaterial-based carrier systems designed to deliver genes directly into plant cells and enhance genome editing efficiency. The chapter also summarizes the current efforts being made to develop nanomaterial-based carrier systems for the delivery of the next generation of agrochemicals and genetic materials. Further discussed are pathogen control through nano-pesticides and nutrient delivery through nano-fertilizers for five major global crops, namely rice, wheat, maize, potato, and chickpeas. The chapter also considers the use of nanomaterial-based technologies to overcome barriers associated with gene therapy and conventional genome editing and the novel avenues afforded by Dip-Pen Lithographic technique of DNA bases manipulation, control and engineering. It is believed that advances in this field will greatly assist in improving crop quality and productivity, thereby providing environmentally friendly and sustainable food security to the ever-increasing global population.

The collected chapters demonstrate how nanotechnology-based techniques used in global studies can help solve problems resulting from population growth, industrial development, and diminishing natural resources. The development of new nanotechnology-based techniques and their effective application, combined with sustainable management of energy and wastewater, and eco-friendly agricultural practices, can deliver effective long-term solutions to many of the challenges currently facing humanity. Therefore, this book provides new outlooks for scientists, researchers, engineers, and postgraduate students working in this rapidly evolving field.

REFERENCES

1. Mekonnen MM, Hoekstra AY. Four billion people facing severe water scarcity. *Science Advances* 2016; 2(2): e1500323.
2. Vorosmarty CJ, Green P, Salisbury J, Lammers RB. Global water resources: Vulnerability from climate change and population growth. *Science* 2000; 289: 284–288.
3. Gerland P, Raftery AE, Sevcikova H, Li N, Gu D, Spoorenberg T, Alkema L, Fosdick BK, Chunn J, Lalic N, Bay G. World population stabilization unlikely this century. *Science* 2014; 346(6206): 234–237.
4. Ellabban O, Abu-Rub H, Blaabjerg F. Renewable energy resources: Current status, future prospects and their enabling technology. *Renewable and Sustainable Energy Reviews* 2014; 39: 748–764.
5. Panwar NL, Kaushik SC, Kothari S. Role of renewable energy sources in environmental protection: A review. *Renewable and Sustainable Energy Reviews* 2011; 15: 1513–1524.
6. Fawzy S, Osman AI, Doran J, Rooney DW. Strategies for mitigation of climate change: A review. *Environmental Chemistry Letters* 2020; 18: 2069–2094.
7. Dincer I, Bicer Y. *Enhanced Dimensions of Integrated Energy Systems for Environment and Sustainability: Integrated Energy Systems for Multi-Generation*. Hoboken: Elsevier, 2020.
8. BP Statistical Review of World Energy. 2020. www.bp.com/en/global/corporate/energy. Last accessed 28 September 2021.

9. U.S. Geological Survey. The water cycle: Water storage in oceans. https://www.usgs.gov/special-topics/water-science-school/science/water-cycle. Last accessed 28 October 2022.
10. Conard BR. Some challenges to sustainability. *Sustainability* 2013; 5: 3368–3381.
11. World Health Organisation (WHO). https://www.who.int/teams/environment-climate-change-and-health/water-sanitation-and-health/water-safety-and-quality. Last accessed 21 October 2022.
12. World Vision. 2022. https://www.worldvision.com.au/global-issues/work-we-do/climate-change/clean-water-sanitation. Last accessed 24 October 2022.
13. Watkins K. *Human Development Report 2006 – Beyond Scarcity: Power, Poverty and the Global Water Crisis* 2006; 28.
14. Ng KC, Thu K, Kim Y, Chakraborty A, Amy G. Adsorption desalination: An emerging low-cost thermal desalination method. *Desalination* 2013; 308: 161–179.
15. Wallace JS, Gregory PJ. Water resources and their use in food production systems. *Aquatic Sciences* 2002; 64(4): 363–375.

Theme 1

Sustainable Energy Technologies

2 Synthesis of Molybdenum-Based Nanomaterial Additives for the Sustainable Use of High-Sulfur Marine Fuels

Triana Wulandari, A.F.M. Fahad Halim, and Yuanyuan Feng

CONTENTS

2.1 THE USE OF SULFUR FUELS IN MARINE TRANSPORTATION

Today, fossil fuels and their fractions are the main sources of energy, and they are also used as feedstock for the chemical industry. Sulfur is the most common non-hydrocarbon atom. It can exist in its pure form or be complexed with hydrocarbons in the form of mercaptans, sulfides, and cyclic compounds such as thiophene and dibenzothiophene (DBT). When hydrocarbon-based fuels containing sulfur compounds are combusted to generate energy, they not only produce CO_2 and H_2O, but they also produce noxious gas emissions. Since ancient times, the maritime industry was the most commonly used mode for transporting both freight and humans. Historically, ships have always relied on human and wind power to transport goods until the early nineteenth century. During the second half of the nineteenth century, steam-powered ships and later internal combustion engine-powered ships replaced sailing ships. In both cases, these technologies used heavy fuel oil (HFO) as the source of energy to generate their power. Today, the vast majority of the maritime industry uses either HFO or marine diesel oil (MDO) as fuel, as shown in Figure 2.1 [1]. For more than a century, HFO, or high-sulfur fuel oil (HSFO), was the fuel of choice, since it was both economical and energy-intensive, i.e., a small quantity can propel a ship a long distance. The fuel, also known as bunker, is a sticky, tar-like residue that remains after crude oil catalytic cracking. The lighter, more expensive fuels like gasoline and automotive vehicle diesel are removed. The viscous residue is a mixture of paraffin, olefins, aromatics, and asphaltenes, as well as sulfur, nitrogen, and some metal

DOI: 10.1201/9781003181422-3

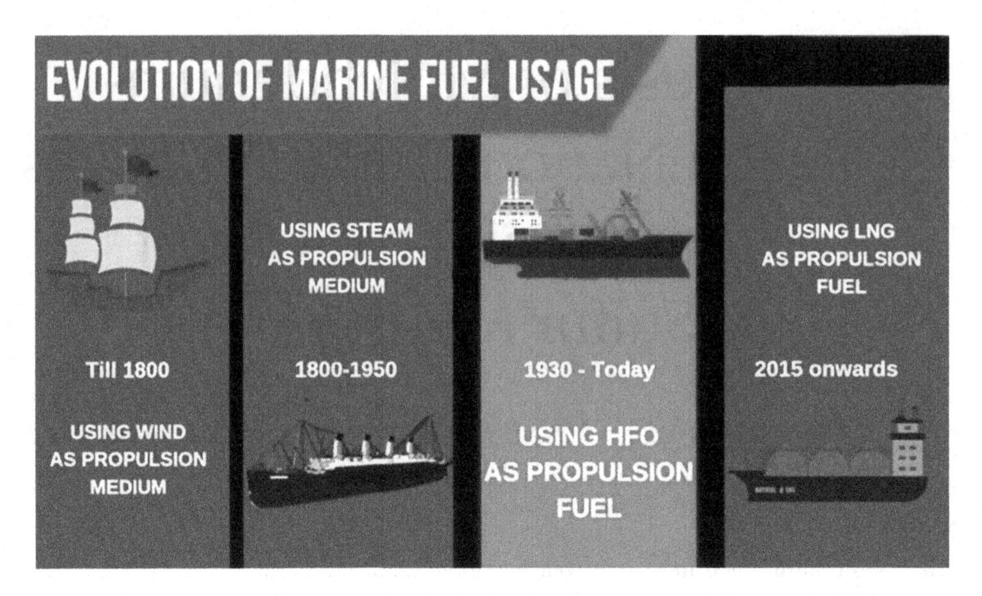

FIGURE 2.1 History of marine fuel usage. (Reprinted from Reference [2], open access source under Creative Commons Attribution 4.0 Licence, Copyright 2021, Atlantis Press.)

compounds. Generally, this fuel must be preheated so that it can be pumped through marine engines prior to combustion. Currently, HFO is used by around 60% (~60,000 ocean-going ships) of the world's marine transport industry, which also makes it a major component of the global economy. The main reason for its widespread use is its low cost. It is roughly 30% less expensive than alternative low-sulfur fuels on the market. In spite of the International Convention for the Prevention of Pollution from Ships (MARPOL) that establishes emission standards for nitrogen dioxide and sulfur dioxide, HFO fuels are still being used globally.

Marine diesel engines are generally low-speed, two-stroke compression ignition internal combustion engines used for propulsion. Residual heavy fuel (RHF), which contains sulfur, asphaltenes, and ash, is generally used to power marine diesel engines. The main reason for using this kind of fuel is cost, since this type of marine diesel has a very high fuel consumption rate. The problem arises when the fuel is combusted. The resultant exhaust emissions contain particulate matter, NO_x, or SO_x. The scale of the pollution problem was highlighted by both the International Maritime Organization's (IMO's) 3rd Greenhouse Gas Study and British Petroleum's (BP's) Statistical Review of World Energy, which identified that around 300 million tons of marine fuel oil were being burned each year, as of June 2015 [3]. Thus, the large amounts of low-cost high-sulfur marine fuel oils, once combusted, generate pollutants (including CO_2) that significantly contribute to global air pollution, as seen in Table 2.1.

The marine transport sector is essential for the global economy, since it transports large tonnages over long distances. It accounts for around 80% of global trade, with a total fleet of 94,171 vessels with a gross weight of 1.9 billion deadweight tonnage (DWT) [4]. The downside of marine operations is the large greenhouse gas (GHG) emissions and other hazardous compounds [5]. These detrimental factors have a significant impact on the global ecosystem and contribute to climate change [6, 7]. For example, global shipping generated 796 million tons of carbon dioxide in 2012 alone, which equated to around 2.2% of global GHG emissions, and according to the IMO's 3rd GHG study, this proportion could rise by a further 50% to 250% by 2050 due to increased global shipping volumes. Furthermore, maritime activities also contribute around 13% of SO_X and 12% of NO_X emissions globally.

Maritime shipping using HFO has the capacity to emit more than 2,700 times SO_2 than automotive fuels. In addition, HFOs also contribute significantly larger amounts of particulate emissions

TABLE 2.1
Marine fuels emission parameters [8]

Emissions (g/g of fuel)	LNG	MDO	HFO
Sulfur oxides (SO_x)	Trace	0.003	0.10
Carbon dioxide	2.80	3.21	3.11
Methane	0.10	Trace	Trace
Nitrogen oxide (NO_x)	0.01	0.10	0.10
Particulate matter (PM)	Trace	0.001	0.007

that are harmful to humans, animals, plants, and the environment. The health impacts of these toxic emissions go beyond geographical borders. For instance, shipping-related air pollution is estimated to have caused more than 50,000 premature deaths in Europe. Moreover, exhausted SO_2 molecules will ultimately combine with atmospheric water molecules to form acid rain. This acid rain then interacts and directly affects soils, soil biome, and surface waters, and overall damages the global biodiversity of the planet. However, in recent years, international environmental standards addressing SO_X emissions have focused on promoting the use of ultra-low (1–10 ppm) or sulfur-free fuels. Thus, there is an urgent need to reduce exhaust emissions from ships and also reduce the sulfur content of marine fuels [9, 10].

Fuel costs are constantly fluctuating as a result of market conditions and the price of crude oil. Importantly, the cost of MF follows crude oil prices more closely than fuel for diesel trucks. Significantly, HFO accounts for about 80% of the total marine fuel use globally and is typically made up of leftover refinery distillation or cracking streams [12]. Generally, most marine fuels are blends, as seen in Figure 2.2. For example, intermediate fuel oil (IFO) 380 is composed of 98% residue and 2% distillate, whereas IFO 180 contains 88% residue and 12% distillate. Other marine fuels include MDO, which is primarily made up of distillates, while marine gas oil (MGO) is the most refined distillate and has the lowest sulfur level [13]. However, the cost difference between the various fuels is a major contributor to the economics of ship management. For instance, MDO is around 87% more expensive on average than LS (low sulfur) 380. Moreover, because of the growing demand and expense of de-sulfured operations, marine distillate fuels are roughly twice as expensive as residual fuels. In particular, to operate in either the North American Emission Control Area or the US Caribbean Sea Emission Control Area (ECAs), ships must use low-sulfur fuels (LSF) with a sulfur content of less than 0.1%. Accordingly, these fuels are more expensive and increase operational costs for shipping businesses.

Several fuels are used as bunker fuel or could have the potential to be used in the future. The typical costs of these fuels are presented in Table 2.2. LNG is preferable to HFO since it produces fewer CO_2 emissions and costs about the same. On the other hand, exhaust gas emissions from LNG combustion have higher methane levels than those of HFO. However, life cycle assessment studies have shown that LNG is overall better for the environment than HFO in the long term. New fuels like hydrogen have the potential to drastically reduce global emissions of greenhouse gas emissions. However, the environmental advances of hydrogen come with a cost, since hydrogen has higher production costs. The difficulty for ship owners in selecting which fuel to use is also influenced by market dynamics, national strategic directions, and global events. Studies have shown there are generally four factors that need to be considered: 1) it is difficult to predict the costs of fuel (i.e. IFOs, MDOs, and MGOs) over time. Importantly, the price of crude oil, as well as the supply and demand for bunker fuels, will dictate their ultimate cost to users. Currently, several oil sector reports indicate oil producers will be capable of producing enough low-sulfur fuel to fulfill the ECA's shipping criteria in the near future [16]. Furthermore, studies state that LS fuels are plentiful, but the costs of producing the fuel have risen due to blending processes and the disposal of residues resulting

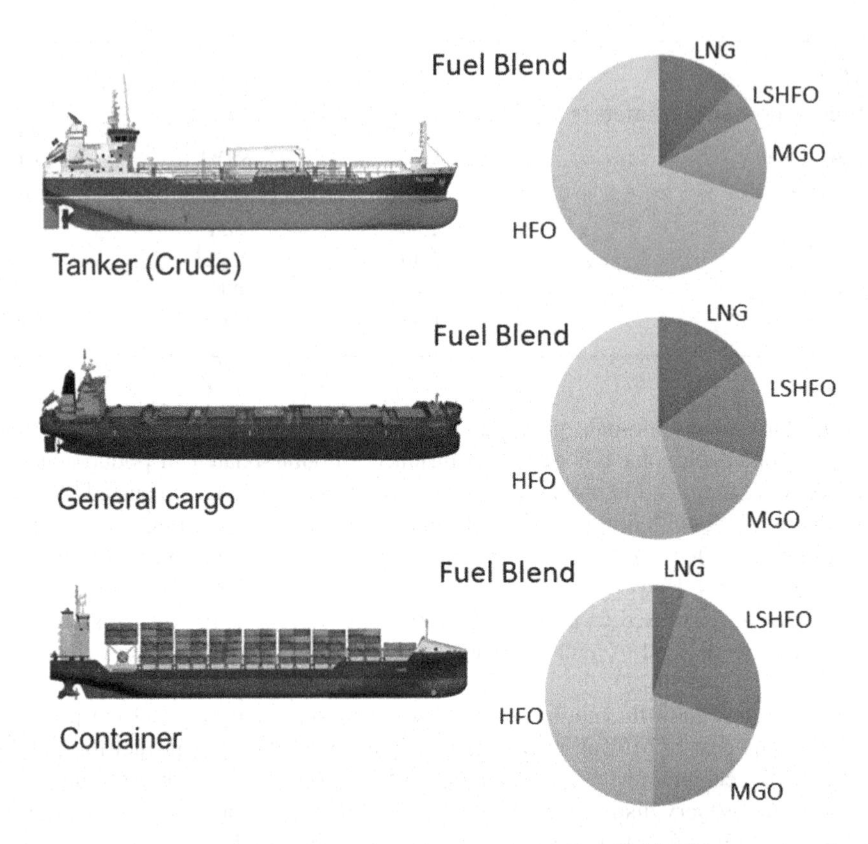

FIGURE 2.2 Fuel blend for different types of marine vessels. HFO; MDO/MGO = marine diesel/gas oil; LSHFO = low-sulfur heavy fuel oil; LNG = liquefied natural gas. The chart forecasts 2030 levels. (Reprinted from Reference [11] with permission, Copyright 2019, Elsevier.)

TABLE 2.2
Comparison between different marine fuel types and their costs [14, 15]

Fuel name	Cost $/gigajoule
HFO	9.41–14.11
LNG	8.23–14.11
MGO	14.11–22.35
Hydrocarbon	12.00
Ammonia	28.20
Methanol	16.30
Dimethyl ether	15.06

from the desulfurization process [17]. However, the refining industry in Europe needs to commit more resources to develop more desulfurization capacity to produce fuels with lower sulfur levels. 2) Because a major proportion of the cost of diesel for trucks involves taxes, rising oil prices have a more direct influence on MF costs than on trucking fuel costs. 3) The trucking industry has demonstrated greater adaptability to shifting pollution regulations. One explanation for this adaptability is due to trucks generally having a three- to four-year life span, while marine ships have a 20- to 25-year life span. To put it another way, the trucking industry can renew its fleet in just a few years and take advantage of new technologies. Thus, improvements in the land-based freight industry appear swiftly but take much longer to deploy in the shipping industry. 4) Importantly, ship owners can also gain by adopting LSF from a technological standpoint. For instance, distillate fuels have higher calorific values, which equates to reduced engine wear, lower levels of fuel consumption, and lower greenhouse emissions. Furthermore, higher-quality distillate fuels minimize the production of onboard sludge, which is beneficial to operators who are finding it increasingly harder to deal with disposing of sludge on land. Overall, increased fuel costs are likely to be offset by improvements in ship engine maintenance and operation [16].

2.2 FUTURE SUSTAINABILITY OF MARINE FUELS

2.2.1 Sustainability of Marine Fuels

Several studies have suggested that the shipping industry should cut greenhouse CO_2 emissions by at least 80% by 2050, as part of reaching the climate target to restrict global warming to less than 2 °C [11, 18]. To meet this goal, several goal entities have imposed regulations on the shipping industry to reduce fuel-related emissions. For instance, the amended form of the MARPOL is one of the conventions of the Marine Environment Protection Committee (MEPC) [19, 20]. This important global agreement is designed to reduce and prevent pollution of marine ecosystems by ships. Currently, MARPOL has six technical annexes, these are [19–21]:

I. Rules for preventing oil contamination.
II. Directions for stopping contamination from toxic liquids in bulk.
III. The avoidance of contamination by hazardous compounds conveyed in packaging.
IV. The preclusion of sewage pollution from ships.
V. The avoidance of pollution from ship waste.
VI. The mitigation of ship-related air pollution.

In 2020; MARPOL recommended the reduction of sulfur in MFs to 0.5% [21]. The subsequent upheaval in the marine industry resulted in changes in fuel mixes and significant interest in alternative transportation fuels. Thus, the demand for MGO or low-sulfur fuel oil (LSFO) has placed considerable pressure on refiners to meet future production levels [22]. The IMO is responsible for the safety and security of shipping, as well as for promoting the reduction of pollution from marine shipping. It also develops policies and protocols specifically designed for the global marine industry. These policies include how ships are designed, built, staffed, run, and discarded [19]. The IMO also established emission control areas (ECAs) to mitigate the negative impact of detrimental emissions from ship exhausts [23]. For instance, in comparison to other parts of the world, ECAs also set emission limits on SO_x, NO_x, and other pollutants from vessels operating within the controlled areas. Globally, the sulfur emission limit was reduced from 3.5% to 0.5% in 2015 [24], while the sulfur emission limit was reduced to 0.1% in ECAs. Thus, the demand for LSF and alternatives like scrubbers has been steadily increasing since 2015 [22]. Currently, the marine industry has four options available for operating in the low-sulfur emission environment. These include 1) moving to LSFO; 2) moving to MGO; 3) using scrubbers while burning HSFO, and 4) moving to LNG.

The IMO also utilizes predictive models to track expected future emission rates as part of its efforts to reduce greenhouse gas emissions (GHG). These models take into account factors such as transportation demand estimates, fleet productivity development, and independent monitoring or efficiency enhancements. The emissions are then categorized by vessel type and tonnage [21]. The IMO has also created a set of operative and technical guidelines for marine vessel operations. Importantly, governments, industry associations, and civil society organizations have all expressed interest in the IMO's Energy Efficiency Design Index (EEDI) and Ship Energy Efficiency Management Plan (SEEMP) as methods of enhancing energy efficiency and reducing GHG emissions [25, 26]. Importantly, by 2050, the European Union (EU) wants to cut CO_2 emissions from marine vessels entering their ports by 40% to 50%. Furthermore, in the near future, Sweden will calculate its fairway dues using the Clean Shipping Index (CSI), a protocol that incorporates CO_2 emissions generated by the vessel [18].

2.2.2 Mitigation Measures

Technological innovation, market-based intervention, and operational change are all methods for minimizing energy consumption, pollution levels, and costs in marine transportation [27]. Examples of technological improvements include the utilization of cleaner MFs, or the implementation of more effective ship and machine technologies, while also reducing pollution levels. Operational change examples include adopting kites to help with ship maneuvering and adopting more efficient shore-to-ship power transfers. In addition, from a market perspective, adopting particular methods can provide leverage in the fuel or energy supply market or influence push or pull carbon-reduction initiatives; for example, implementing cap-and-trade carbon regimes for shipping businesses or levying general tariffs on emissions levels. In recent years, some wealthy countries have become interested in using cleaner fuels as a possible alternative to traditional fossil fuels. This alternative has attracted considerable interest in recent years since many countries are now compelled by international conventions to reduce emissions in order to combat global warming. For instance, many countries developed and signed an agreement during the Climate Change Conference in 2015 (COP 21). The agreement's goal was to keep the average global temperature increase below 2 °C [28].

There are three types of energy consumption on marine vessels. These include energy for the propulsion system, energy for secondary engines associated with auxiliary systems, and energy for onboard thermal (furnace) power production. This means a typical ship has three different power systems. In most cases, the propulsion system accounts for around 68% of the vessel's annual energy requirement [29]. However, in recent years, there has been an interest in incorporating these systems to improve the overall energy efficiency of the ship. Recently, a study indicated that combining these energy systems could lead to a 2% reduction in the overall consumption of fuel annually [18]. A number of integration methods have been investigated and implemented to improve ship performance while at the same time still providing flexibility to satisfy the energy needs of all onboard operations. As seen in Figure 2.3, single-fuel/dual-fuel/hybrid-electric propulsion engines are all approaches that have been developed to meet the energy requirements for ship propulsion. For instance, high-capacity batteries can be utilized to store electricity for future use in hybrid-electric propulsion systems. Importantly, the use of hybrid-electric systems has the potential to reduce the amount of fossil fuel use and, in turn, reduce the levels of harmful emissions discharged to the atmosphere over the course of the journey between ports. Interestingly, the use of hybrid-electric/fossil fuel systems is one approach used to reduce pollution levels emitted from gasoline-powered automobiles [30]. However, ultimately, the use of any fossil fuel-derived product to generate power through combustion will produce exhaust pollutants.

From a global shipping perspective, the main pollutants from the high-temperature combustion of MFs are NO_x, SO_x, and CO_2. These pollutants have both local and international consequences. For instance, nitrogen oxides (NO_x) and sulfur oxides (SO_x) have the greatest impact on local air quality, while carbon dioxide (CO_2) emissions have a long-term negative impact on global warming [23]. To mitigate the impact of these pollutants, several methods have been investigated and used in marine vessels, for instance, the recirculation of exhaust gases incorporating particular catalytic reduction, scavenging moisture control (humidification), and the injection of water are methods for reducing NO_x emissions [31]. Catalytic reduction is the most effective method since it can reduce NO_x levels by up to 95% [32]. Moreover, Goldsworthy has investigated several methods for lowering NO_x emissions from marine engines. These include boosting compression pressure while manipulating fuel injection rates, enhancing injection configurations, and improving the shape of combustion chambers [33]. In addition, scrubbers can be used to reduce SO_x emissions. Saltwater scrubbing is an effective method and is capable of reducing SO_x emissions by up to 95%. Recently, Zwoliska

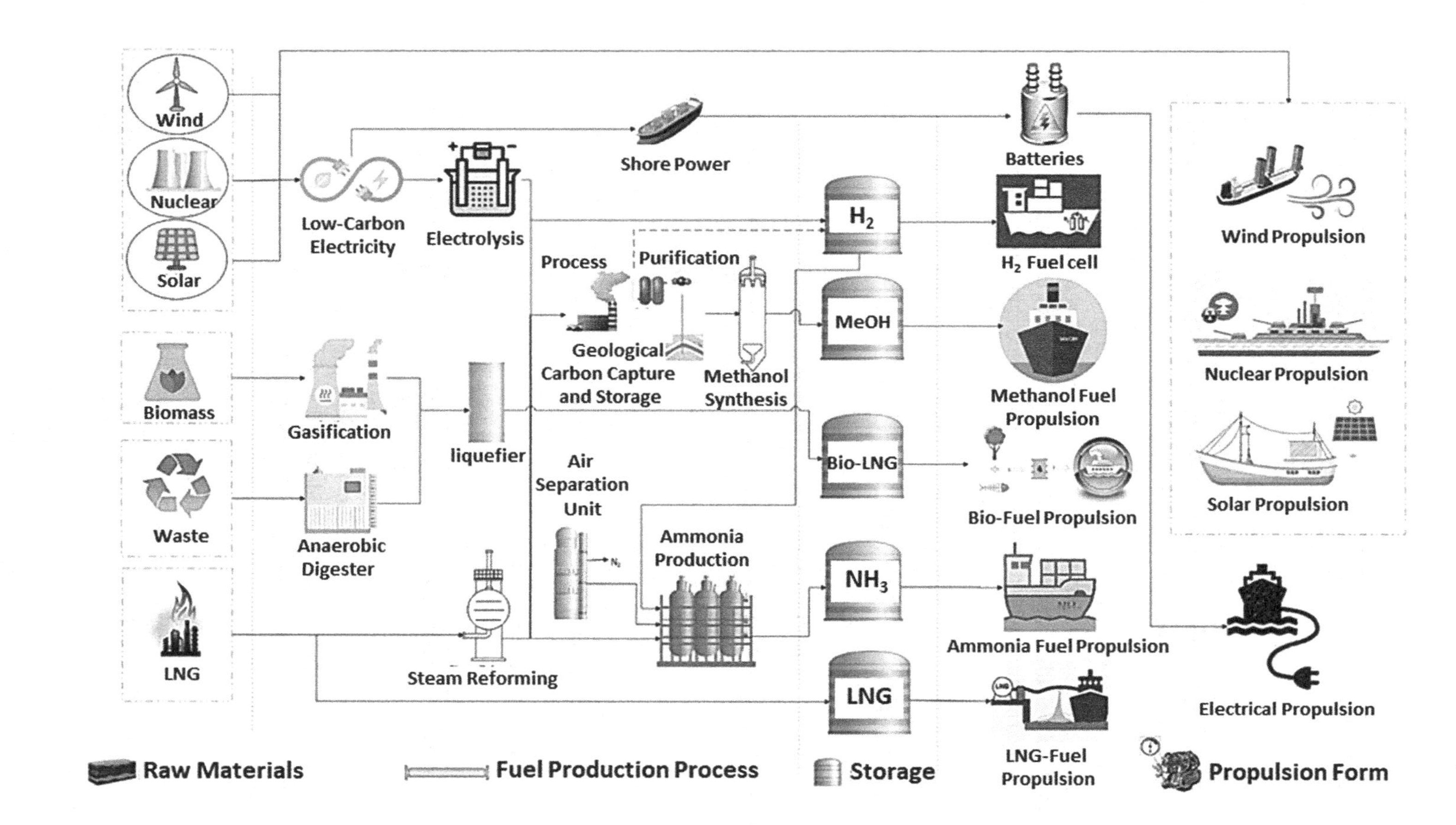

FIGURE 2.3 Different technologies and MFs leading to emissions-free shipping. (Reprinted from Reference [25], open access source under Creative Commons Attribution 3.0 Licence, Copyright 2021, Elsevier.)

et al. tested a method that utilizes a wet scrubber and a hybrid electron beam system, which reduced NO_x levels by 89.6% and SO_2 levels by 100% [34]. In reducing CO_2 gas emissions, there has been a focus put on developing specialized antifouling coatings to reduce ship friction while underway and the use of ultra-long-stroke engines to improve the combustion process [31].

Currently, HFs are traditionally and extensively used since they are considered well-established and economically viable due to existing processing facilities being in place, hence the widespread use of HFs by the transportation industry. However, because of price volatility, dwindling crude oil reserves, and concern over pollutant emissions as discussed above, the marine transportation industry is now considering alternative methods of improving the properties of HFs and MFs [35]. For instance, research has focused on adding hydrogen or ammonia (NH_3) to MFs to improve their flame consistency during combustion. The addition of either hydrogen or ammonia has been found to result in lower levels of NO_x being generated. However, concerns over the storage and delivery of hydrogen are a hindrance to its widespread use as a fuel additive present. As a result, both methanol and ammonia have been identified as promising indirect energy storage media for hydrogen [36]. Alternatively, the use of catalysts during the refinement process in oil refineries is another method for reducing the levels of sulfur and nitrogen compounds present in HFs and MFs. In particular, the desulfurization of marine fuels with the use of nanometer-scale catalysts has attracted considerable interest since it allows for the continued use of current HFs and MFs until new single-fuel/dual-fuel/hybrid-electric systems become available in the future.

2.3 THE APPLICATION OF NANO-SCALE CATALYSTS FOR THE DESULFURIZATION OF MARINE FUELS

2.3.1 Nanotechnology Background

Nanotechnology encompasses the study and creation of materials at the nanometer scale and the subsequent use of these materials in a wide range of applications. In Greek, nano means dwarf and denotes a unit of 10^{-9}. So nano-gram is equivalent to 10^{-9} of a gram, and a nanometer is 10^{-9} of a meter. The nano range is very small and nanomaterials have at least one dimension in the size range from 1 to 100 nanometers (nm). At the nanometer scale, materials have unique properties not found in their bulk counterpart. Importantly, at the nanometer scale, the properties of the materials are inherently linked to their size, shape, and composition. Because of the unique size, shape, and compositional properties, many nanomaterials have the potential to be patented for specific applications. For instance, nanometer-scale gold appears red in color (particles are around 20 nanometers in diameter) while particles around 80 nanometers in diameter appear orange in color. This is unlike bulk gold, which is a metal that appears yellow (gold) in color. Since its inception, nanotechnology has gone through a long process of exploration. From studying naturally existing nanometer-scale materials found in nature (viruses, DNA, RNA, etc.) to artificially created nanometer-scale inorganic structures (carbon nanotubes and graphene), nanotechnology translates the initial concept to ultimately manufacturing a product for a specific application. Thus, nanotechnology has been identified as one of the three frontier high-tech fields in the world. Importantly, nanotechnology has been identified as an important field and driving force for economic development [37, 38].

The earliest study of nanomaterials occurred around 1861 when Thomas Graham identified gelatin as a class of substances that did not form crystals. His study also found that gelatin diffused very slowly after dissolution and formed a colloid. Later, in 1905, Albert Einstein calculated the size of a sugar molecule from experimental data of sugar diffusion in water. Analysis of the data found that the diameter of each sugar molecule was around 1 nm and, for the first time, gave mankind an empirical sense of the molecular scale, and in particular the nanometer scale. However, it was not until 1935 that Max Knoll and N. Ruska developed the world's first transmission electron microscope (TEM). This device, with a resolution of 50 nm, enabled researchers for the first time to explore the nanometer world [39]. And in recent decades, the unique physical and chemical properties of nanomaterials such as high reactivity, large surface area, tenable size, and morphology have attracted considerable scientific and industrial interest. For instance, when a material is subjected to surface patterning at the nanometer scale, material properties like melting point, magnetic effect, and electrical charge can be modified without changing the material's chemical composition. Also, the wave properties of electrons are greatly influenced by surface changes in the nanometer scale. Thus, materials produced at the nanometer scale have unique material properties that can be tailored to a wide variety of scientific and industrial applications. Typical applications include agriculture (nano-fertilizers, nano-pesticides); chemicals (paints, coatings) and cosmetics; electronics (carbon nanotubes, biopolymers); energy (semiconductor chips, memory storage, optoelectronics, photonics, solar cell, fuel cells) and environment remediation (water and air purifying filters); food science (nano-capsules, food processing); military equipment (biosensors, nano-weapons, sensory enhancement technology); nanomedicine (nano drugs, medical devices, tissue engineering), and information and communication technologies and textiles [40, 41].

To date, many processes based on chemical, physical, and biological methods are routinely used to produce nanomaterials. Essentially two strategies are used to produce nanomaterials. The first is the bottom-up approach which involves a homogeneous system that typically uses reducing agents or enzymes to synthesize the nanostructures. This strategy generally yields pristine nanoparticles with controllable size and morphology. The second is the top-down approach which begins by taking a material in its bulk form and then uses a size reduction method to produce nanometer-scale materials. Typical methods used in size reduction include thermal decomposition, mechanical grinding, etching, cutting, and sputtering [42]. Importantly, both these methods have the potential to produce nanometer-scale catalytic materials that can be used to remove sulfur from fuel oils.

2.3.2 Nanocatalysts and Their Impact on Sulfur Removal from Fuels

Nanoparticles are very small in size but have large specific surface area-to-volume ratios. These unique properties also create different bonding states on the surface and just inside the particles not seen in their bulk counterparts. At this scale, nanoparticles also have incomplete atomic coordination at their surface which leads to an increase in the number of active surface sites. Furthermore, at this small scale, the particles' surface is less smooth and forms uneven atomic steps, which further increases the contact surface for chemical reactions to take place. These unique surface features make nanoparticles useful as catalysts.

Nanocatalysts have a very thin uniform surface layer (typically between two and five atomic layers) and a special crystal structure. Their favorable surface electronic states and unique surface properties make them ideal for promoting adsorption and surface chemical reactions. Thus, because of their high activity and selectivity, they are suitable for various types of chemical reactions like catalytic oxidation, and reduction and cracking reactions. In addition, nanocatalysts do not need to be attached to a carrier and can be added directly into the liquid phase. Because nanocatalysts are extremely small, they can be considered to promote a homogeneous catalytic reaction. The advantages of a homogeneous catalytic reaction are 1) high activity; 2) good selectivity; 3) low catalyst dosage; 4) low reaction temperature and pressure; and 5) reusable catalyst. Research has shown catalytic activity is dependent on surface defects, such as edges, corners, ribs, dislocations, layer errors, and crystal errors. In addition, when the nanocatalysts are at the quantum level, the catalytic activity dramatically increases due to the much larger surface areas. Many developed countries, because of the unique properties of nanocatalysts, have invested significant human and financial resources into investigating and developing high-performance catalysts. Today's nanocatalysts are known as fourth-generation catalysts and are extensively studied for a variety of applications. For example, the US Department of Energy's Industrial Technologies Program in the United States has made nanocatalyst research a key development project, and some of their research is already being translated into pilot plant scale. Moreover, according to Fior Markets, the total market for nanotechnology-based products will be about USD 5.96 billion by 2027. This prediction includes products in diverse fields like medicine, defense, agriculture, transportation, nanocatalysts, and nanosensors [43]. In particular, nanocatalysis has the potential to deliver large economic and social benefits. At present, nanocatalysts have been used in several chemical and environmental protection applications [44]. The reasons why nanocatalysts are currently being used in these applications are due to the following advantages: 1) they have higher catalytic activity due to the large surface area of the individual nanoparticles when compared to conventional bulk equivalents [44, 45]; 2) they have high selectivity [46, 47]; 3) they have high catalytic stability [48, 49]; and 4) they enable reactions to take place under moderate operational conditions.

2.3.3 Molybdenum-Based Catalysts for Sulfur Removal

The reduction of sulfur, nitrogen, and aromatics in oil fractions is usually achieved industrially by using hydro-treating processes. For many years, the most important hydro-treating reaction carried out was for the removal of sulfur from fuel fractions. Thus hydro-treating catalysts are commonly referred to as hydro-desulfurization catalysts. Importantly, the failure to remove sulfur during the petroleum refining process results in the formation of SO_x during combustion in automobile engines.

Transition metal sulfides, like molybdenum-based sulfides, are widely used in the oil refining industry as hydro-treating catalysts [50]. Typically, industrial catalysts are composed of nickel (or cobalt) and molybdenum supported on γ-Al_2O_3. They are first prepared in an oxidic state and then converted to a sulfidic state before use [51]. During the hydro-desulfurization (HDS) process and other types of petrochemical desulfurization processes, MoS_2 is utilized as the nanocatalyst. In addition, doping MoS_2 catalysts with trace amounts of cobalt or nickel has also been found to improve their efficiency and make them more useful. Importantly, in recent years sulfur content in motor fuels has been reduced in response to the adverse effects of SO_x exhaust emissions from motor vehicles on the environment. Thus, in recent years, many methods have been investigated to remove sulfur-containing compounds in petroleum fuels. While a number of these methods are new, most have focused on improving current hydro-treatment technologies (HDT). The latter option is the more appealing option for fuel refiners. Thus, the utilization of highly active catalyst materials is important if the performance of traditional HDT is to be improved. Because of this realization, there has been a recent increase in the number of HDT catalyst-related research projects undertaken [52–55]. The results of these studies suggest the performance of MoS_2-based catalysts could be improved by enhancing the active phase loading or making use of bulk sulfides during the reaction. Currently, the commercially available HDS catalyst is MoS_2 enhanced with either nickel or cobalt promoter atoms [56–59]. It is believed that the promoter atoms locate themselves at the margins of the MoS_2 sheets in the so-called Ni (Co)-Mo-S structure. Furthermore, the dispersion of the MoS_2 phase, the ratio of edge position to the basal plane area, and the stacking of MoS_2 sheets all play a significant role in determining the catalytic activity. Therefore, in order to create a highly active HDS catalyst, it is necessary to exercise control over the shape of the nanometer-scale MoS_2 material [60–62].

Thus, current research has shown MoS_2 is a very promising material for the oxidative desulfurization of fuels [63, 64]. The mechanism for the thermal-driven oxidative desulfurization process is presented in Figure 2.4 [65]. Al_2O_3, TiO_2, and SiO_2 are typical catalyst support materials and have been shown to improve the performance of oxidative desulfurization catalysts. Therefore, the selection of appropriate support is also a key parameter in improving the oxidative removal of sulfur contaminants in fuels [66].

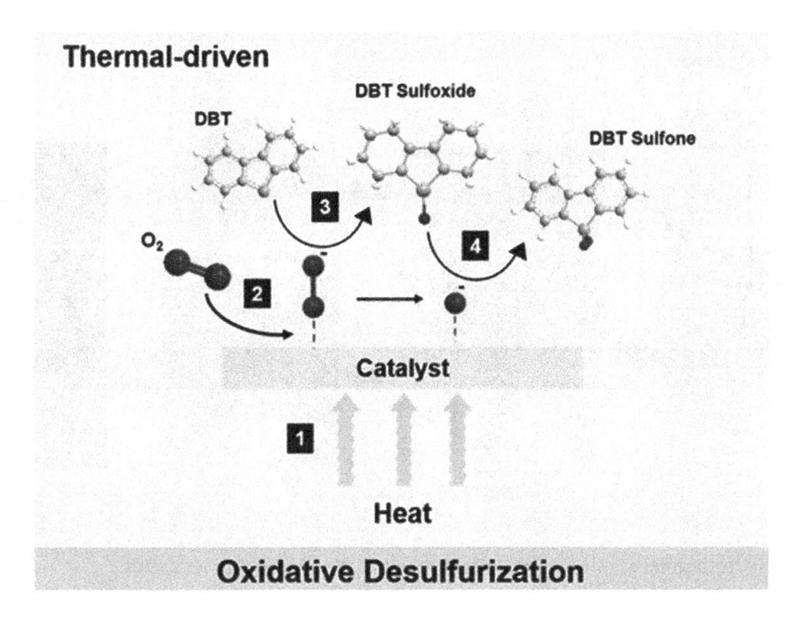

FIGURE 2.4 Mechanism for thermal-driven oxidative desulfurization using metal oxide catalyst. (Image is reproduced from Reference [65] with permission, Copyright 2021, Royal Society of Chemistry.)

2.4 PRODUCING MoO_2 AND MoS_2 NANOPARTICLES VIA HYDROTHERMAL SYNTHESIS AND THEIR CHARACTERIZATION BY X-RAY DIFFRACTION SPECTROSCOPY AND RIETVELD REFINEMENT

2.4.1 Hydrothermal Synthesis of MoO_2 and MoS_2 Nanoparticles

The MoO_2 and MoS_2 nanoparticles produced in this study were synthesized by a hydrothermal method. Hydrothermal synthesis is generally defined as crystal synthesis or crystal growth at temperatures above 100 °C and high-pressure water conditions above 1 atmosphere in an enclosed system [67]. Above 100 °C (water boiling point) and up to 374 °C (critical point) water is said to be superheated. Hydrothermal synthesis is usually carried out below 300 °C. Thus, the technique offers a single-step method for producing anhydrous oxide powders. Hydrothermal synthesis also has several advantages over other methods such as high-temperature reduction. The key advantage is that crystalline nanoparticles can be synthesized at relatively low temperatures that are typically below 300 °C. A number of studies have identified some parameters that influence the resultant crystal structures and morphologies produced by hydrothermal synthesis [68]. These parameters include 1) type and concentration of precipitant; 2) solvent medium; 3) reaction time; and 4) temperature and pressure conditions during the reaction. Each of these parameters plays a significant role in defining the resulting crystal structure, crystallite size, phase, and morphologies [69].

The present hydrothermal method involves the chemistry of a molybdenum species in solution. During the synthesis process, acidification of the molybdate solution will result in condensed molybdate species of various polymolybdates. In brief, the synthesis process begins with the molybdenum precursor, in this case, 0.173 grams of ammonium heptamolybdate tetrahydrate, $(NH_4)_6Mo_7O_{24}.4H_2O$, (AHM) powder. AHM is commonly used in solution chemistry because it readily dissolves in water to form MoO_4^{2-}(aq) species [70]. AHM was then dissolved in 4 ml of distilled water before 0.5 ml of 1 M HNO_3 and 0.5 ml of the reducing agent (ethylene glycol) were added to the solution. The acid was used for controlling the pH of the solution. This is because MoO_2 tends to crystallize better in acidic conditions [71]. The precursor solution was then transferred to a 50 ml Teflon-lined container in the hydrothermal reactor, as seen in Figure 2.5. The

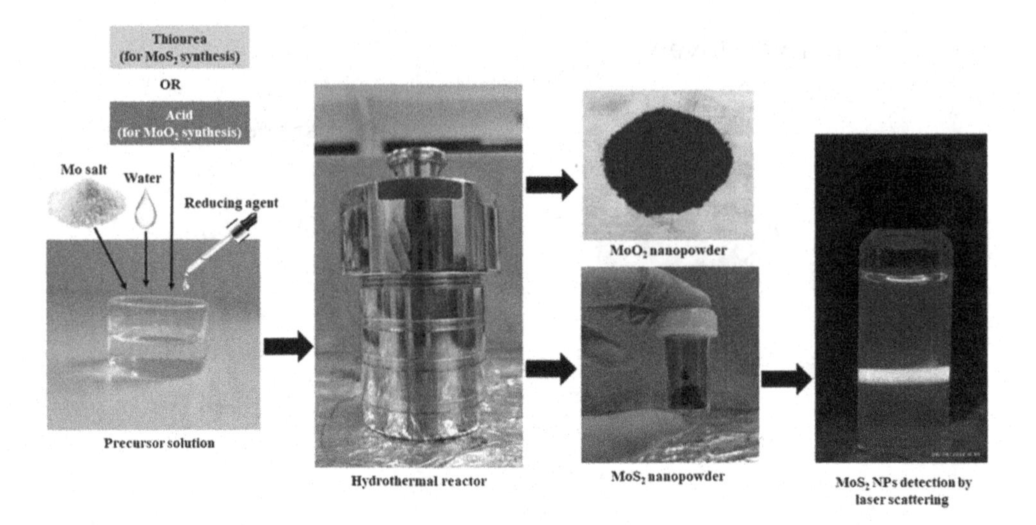

FIGURE 2.5 Hydrothermal synthesis illustration of MoO_2 and MoS_2 nanoparticles.

reactor was then sealed and then placed inside an oven at 220 °C for 24 hours. After this period, the reactor was allowed to cool down to room temperature. The resulting black product was removed from the reactor and washed twice using distilled water and then allowed to dry at room temperature overnight. In addition to producing MoO_2, the hydrothermal synthesis process was also used to produce MoS_2. The procedure begins with adding 1.21 g of sodium molybdate, 1.56 g of thiourea, and 0.14 g of polyethylene glycol to a 30 ml solution of distilled water. The ingredients are then thoroughly dissolved into the solution by agitation. The solution was then poured into a 50 ml Teflon-lined container in the hydrothermal reactor. The reactor was then sealed and then placed inside an oven at 220 °C for 11 hours. At the end of this period, the reactor was allowed to cool down to room temperature. The resulting black product was removed from the reactor and washed with distilled water and then ethanol, and then dried at 80 °C for 8 hours.

The synthesized powders were investigated using X-ray powder diffraction spectroscopy (XRD). XRD measurement was carried out using a Bruker D8 Discover between 10° and 80° at room temperature, using Bragg-Brentano geometry. The diffractometer operated with a Cu Kα radiation (λ = 1.54187 Å) source, with a stepping angle of 0.02° and speed of 1° per minute. Powder diffraction pattern data, simulations, and Rietveld refinements were carried out using the Fullprof program [72]. Starting values for the refinements were scale factor and lattice parameters for MoO_2 derived from the ICSD-152316 model. Profiles were fitted with a pseudo-Voigt function [72]. In addition, the following parameters were used in the Thompson-Cox-Hasting peak shape: one Gaussian half-width (IG) parameter and one Lorentzian (Y) parameter. Three Gaussian (U, V, W) parameters and two Lorentzian (X and Z) parameters were corrected by measuring the standard material (LaB_6). The background was analyzed using WinPLOTR [72].

2.4.2 X-ray Structural Characterization and Rietveld Refinement of Nanoparticles

Molybdenum is a transition metal and has oxidation states ranging from Mo^{2+} to Mo^{6+}. The stable oxide forms are MoO_2 and MoO_3. All of the molybdenum oxide structures can be built by [MoO_6] octahedra, [MoO_4] tetrahedra, or [MoO_7] pentagonal bipyramid [73]. Each structure can consist of either one or more types of these polyhedral, each sharing corners and edges. Molybdenum dioxide (MoO_2) mainly exists in three polymorphic forms. These forms are hexagonal ($P6_3/mmc$) [74], tetragonal ($P4_2/mnm$) [75], and monoclinic ($P2_1$) [76]. The tetragonal MoO_2 with a distorted rutile structure and the monoclinic structure are common forms, as seen in Figure 2.6, while the hexagonal form is unstable.

The crystal structure of MoO_2 can be described as a deformed rutile type [77]. The Mo atom coordinates with six oxygen atoms to form [MoO_6] octahedra. Each octahedron is joined by sharing

FIGURE 2.6 (a) Structure of monoclinic MoO_2 along the *a*-axis; (b) the distorted rutile structure chain is elongated on the *a*-axis; (c) by rotating the *a*-axis 90° to the left, the chain which consists of $[MoO_6]$ sharing edges can be seen. Images are reproduced from ICSD-152316.

edges to form chains. These chains are mutually connected to a three-dimensional structure by the octahedra having corners in common. In the ideal rutile structure type, the Mo atoms are arranged equally distant within the chain. However, in the MoO_2 structure type, the Mo atoms are alternately nearer to and farther away from each other. The Mo–Mo distances alternate along the rutile *c*-axis to give two distinct metal-to-metal bond lengths of 2.51 and 3.02 Å [78]. This results in a distortion of the $[MoO_6]$ octahedra. The symmetry is also correspondingly lowered from tetragonal rutile type to monoclinic MoO_2. Therefore, the monoclinic MoO_2 lattice can be derived from the tetragonal rutile structure through a small distortion [79]. The lattice constants are a = 0.56109 nm, b = 0.48562 nm, and c = 0.56285 nm, with the monoclinic angle β = 120.95° [80].

MoO_2 commonly crystallizes in either the monoclinic or tetragonal phase. However, in 2017, Ludtke *et al.* discovered the orthorhombic phase of MoO_2 that was only stable at high pressure [81]. The diffraction patterns of all three phases are presented in Figure 2.7a. Inspecting the XRD patterns presented in Figure 2.7a reveals that the bottom pattern is for MoO_2 nanoparticles synthesized in the present work. The other three XRD patterns taken from the ICSD database are orthorhombic (ICSD-243549), tetragonal (ICSD-99714), and monoclinic (ICSD-152316). The patterns for tetragonal and monoclinic MoO_2 are similar, but the monoclinic phase has some distinct peaks indicated by asterisks.

Figure 2.7b illustrates some small peaks associated with monoclinic MoO_2 that do not appear in the tetragonal pattern. Figure 2.7b reveals that the MoO_2 nanoparticles synthesized in the present study are in agreement with the monoclinic phase. Diffraction peaks of MoS_2, presented in Figure 2.7c, match the XRD pattern for MoS_2 nano-sheets studied by Quilty *et al.* [82]. However, to extract more useful information from the generated XRD patterns, a refinement process is needed. In the present work, the least square Rietveld refinement procedure was performed by using Fullprof software [72]. Figure 2.8 represents the refinement result for MoO_2 nanoparticles generated in the present work. Inspection of Figure 2.8 reveals a very good fit with R_{wp} = 10.8%. The particle size was calculated by a combination of both the Scherrer formula and Voigt function [72]. In addition, the size calculation was also based on a study by Carvajal [72]. The angular dependence of the profile width due to size broadening is described by the Scherrer equation:

$$D = \frac{K\lambda}{\beta \cos\theta} \qquad 2.1$$

where D is the crystallite size in nanometers, K is the shape factor or also known as the Scherrer constant [83], λ is the wavelength of the diffraction beam in nanometers, θ is the Bragg angle, and β is the peak width of the diffraction peak profile at half maximum height resulting from the small crystallite size. From a historical perspective, Scherrer, in 1918, defined β as "breadth". In 1936, Von Laue introduced β as the integral breadth [83]. The value of β must be in radians, while θ can be in degrees or radians since $\cos\theta$ corresponds to the same number [84]. In addition, Scherrer defined K as a dimensionless number of the order of unity [83]. However, a more detailed review of the shape factor published 60 years later by Langford and Wilson revealed there were systematic variations of the Scherrer constant. [83]. More recently, K has been defined as a constant related to crystallite

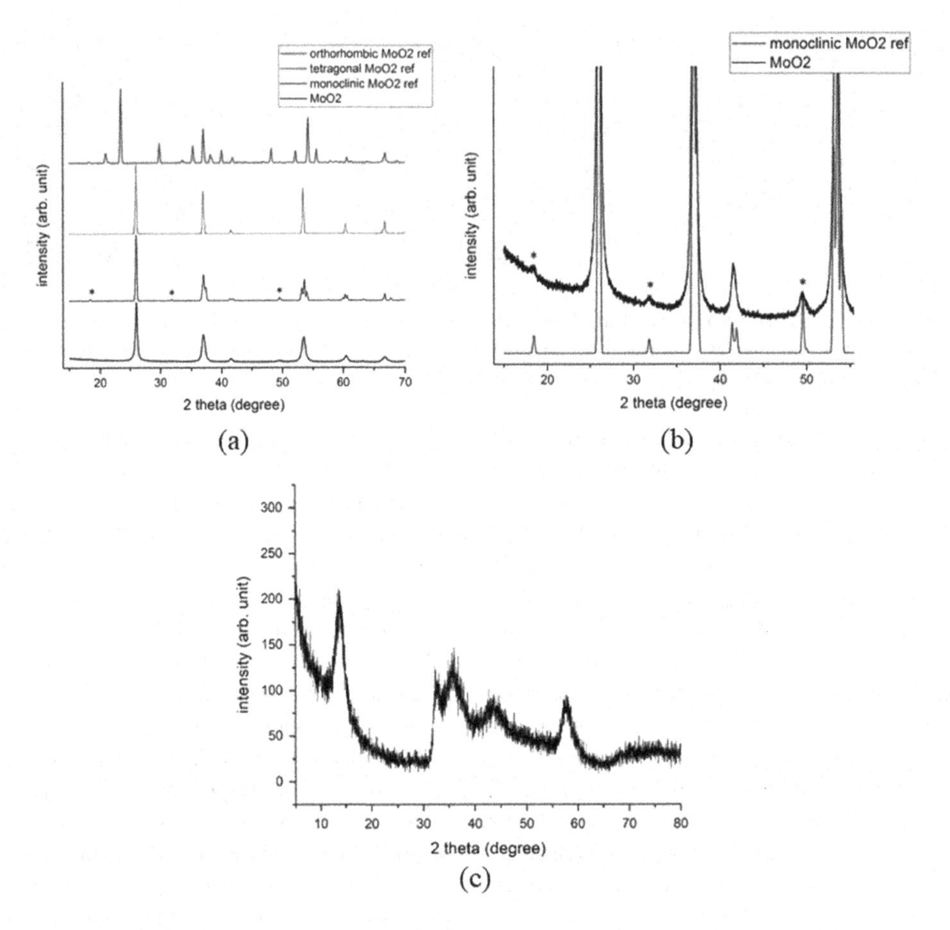

FIGURE 2.7 (a) Different phases of MoO_2 diffraction patterns; (b) monoclinic MoO_2 from the database compared to experimental data; and (c) MoS_2 diffraction pattern.

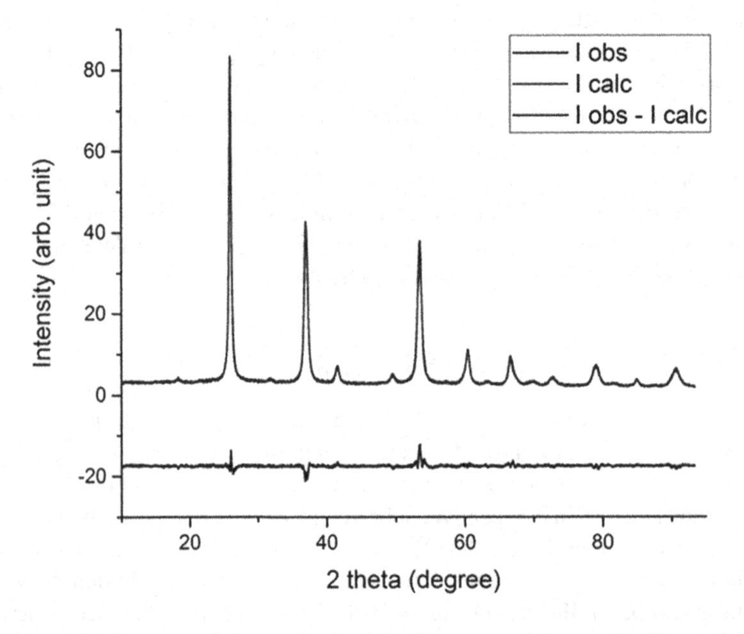

FIGURE 2.8 Rietveld refinement of the synthesized MoO_2 nanoparticle sample. The experimental data (I obs) and the calculated data (I calc) coincide (top), and the resultant differences are shown below.

shape and is normally taken as 0.9 [84]. The integral breadth (β) is calculated by using a pseudo-Voigt function provided by the Thompson-Cox-Hastings numerical approximation [72]:

$$H^5 = H_G^5 + 2.69269 H_G^4 H_L + 2.42843 H_G^3 H_L^2 + 4.47163 H_G^2 H_L^3 + 0.07842 H_G H_L^4 + H_L^5 \quad 2.2$$

where H is the full width at half maximum height (FWHM) for the whole pattern which has an angular dependence for Gaussian (H_G) and Lorentzian (H_L) components, given by:

$$H_G^2 = \left(U + (1-\xi)^2 D_{ST}^2(\alpha_D)\right)\tan^2\theta + V\tan\theta + W + \frac{I_G}{\cos^2\theta} \quad 2.3$$

and

$$H_L = \left(X + \xi D_{ST}(\alpha_D)\right)\tan\theta + \frac{\left[Y + F(\alpha_Z)\right]}{\cos\theta} \quad 2.4$$

The instrumental resolution function (IRF) file has been created to fix the value of U, V, W, and X, leaving the rest of the variables as:

$$H_G^2 = \frac{I_G}{\cos^2\theta} \quad 2.5$$

and

$$H_L = \frac{Y}{\cos\theta} \quad 2.6$$

Thus, by inserting expression (2.5) and (2.6) into equation (2.2), the FWHM can be obtained:

$$H^5 = \frac{1}{\cos^5\theta}\left(I_G^{\frac{5}{2}} + 2.69269 I_G^2 Y + 2.42843 I_G^{\frac{3}{2}} Y^2 + 4.47163 I_G Y^3 + 0.07842 I_G^{1/2} Y^4 + Y^5\right. \quad 2.7$$

Therefore, by substituting β in expression (2.1) with H in expression (2.7), the particle size can be calculated. The results of the particle size calculation and Rietveld refinement procedure are summarized and presented in Table 2.3.

TABLE 2.3
Results of particle size determination and the Rietveld refinement procedure for MoO_2 nanoparticles generated in the present work

Cell parameters	
a	5.622877 Å
b	4.849396 Å
c	5.615647 Å
β	120.652°
Shape parameter (Y)	0.322002
FWHM parameter (IG)	0.003204
B-factor (Mo atom)	1.4907
Bragg R-factor	3.49
R_f factor	2.23%
R_p	11.1%
R_{wp}	10.8%
Calculated particle size	26 nm

Analysis of the MoO_2 nanoparticle data reveals that the unit cell parameters are in agreement with the space group $P2_1/c$ with a single-phase monoclinic symmetry. In addition, the temperature factor, or so-called B-factor or Debye-Waller factor, was used to take into account the thermal vibration of the atoms and refined along with atomic coordinates. The value of the B-factor tends to be between 0 and 1, and lighter elements have larger values [85]. However, a study by McCusker *et al.* suggested that the B-factor be constrained for similar atoms and thus reduce the temperature factor value required [86]. In the present work, the B-factor is refined only for Mo atoms (heavy atoms), whiles the lighter O atoms are set to a constant value of 0.5. After refining the parameters, a whole pattern error (R_{wp}) of 10.8% was obtained. Thus, the pattern error value was found to be reliable for determining the nanoparticle size. In addition, using the Scherrer equation, the MoO_2 nanoparticle size was found to be 26 nm. For most laboratories, X-ray diffraction is the initial method for identifying metal oxide samples. Detection limits for crystalline to moderately crystalline materials are influenced by many factors. The identification of poorly crystalline phases by XRD is difficult and needs complementary characterization methods. In the case of nanometer-scale MoS_2 generated in the present study, nanoparticle size is complicated by the structural disorder which gave rise to XRD patterns consisting of disordered broad peaks. This amorphous characteristic has also been identified and confirmed by a study by Cao *et al.* [87].

2.5 CONCLUSION

Due to the new IMO sulfur limit regulation, which lowers the allowable sulfur content of MFs to 0.50% m/m, marine transport companies are undergoing a dramatic fuel translational change. Therefore, to remove sulfur from exhaust emissions, shipping companies can either pay for more compliant eco-friendly fuel mixes or install onboard sulfur scrubbers. Sulfur scrubbing before the sulfur cap went into force was not an ideal solution to the problem. Open-loop scrubbers emit hazardous wastewater, which has led to an increasing number of ports around the world banning or restricting their usage. However, nanotechnology appears to offer a more refined and effective alternative. Molybdenum-based nanocatalysts (MoO_2) can be utilized for fuel desulfurization. Their use can reduce sulfur emissions and improve fuel economy in marine diesel engines. To this end, recent research and development efforts have focused on improving the performance of molybdenum-based nanocatalysts for the removal of sulfur during the fuel refining process.

REFERENCES

1. Anish. Marine heavy fuel oil (HFO) for ships - Properties, challenges, and treatment methods. 2019 [Cited 2022 March 9]. Available from: https://www.marineinsight.com/tech/marine-heavy-fuel-oil-hfo-for-ships-properties- challenges-and-treatment-methods/.
2. Egemen S, *et al.* A decision-making tool based analysis of onboard electricity storage. In: *Proceedings of the 14th International Renewable Energy Storage Conference 2020 (IRES 2020).* 2021. Atlantis Press.
3. Concawe. Marine fuel facts. 2016 [cited 2022 July 26]. Available from: https://www.concawe.eu/wp-content/uploads/2017/01/marine_factsheet_web.pdf.
4. U.N.C.O. and Development Trade. *World Investment Report 2018.* United Nations, 2018.
5. Li L, *et al.* Ship's response strategy to emission control areas: From the perspective of sailing pattern optimization and evasion strategy selection. *Transportation Research Part E: Logistics and Transportation Review*, 2020; 133: 101835.
6. Russo MA, *et al.* Shipping emissions over Europe: A state-of-the-art and comparative analysis. *Atmospheric Environment*, 2018; 177: 187–194.
7. Zhen L, *et al.* The effects of emission control area regulations on cruise shipping. *Transportation Research Part D: Transport and Environment*, 2018; 62: 47–63.
8. I.M. Organization. Prevention of air pollution from ships. [Cited 2022 July 27]. Available from: https://www.imo.org/en/OurWork/Environment/Pages/Air-Pollution.aspx.

9. Vedachalam S, Baquerizo N, Dalai AK. Review on impacts of low sulfur regulations on marine fuels and compliance options. *Fuel*, 2022; 310: 122243.
10. Seddiek IS, Elgohary MM. Eco-friendly selection of ship emissions reduction strategies with emphasis on SOx and NOx emissions. *International Journal of Naval Architecture and Ocean Engineering*, 2014; 6(3): 737–748.
11. Schnurr RE, Walker TR. Reference module in earth systems and environmental sciences. In *Marine Transportation and Energy Use*. Elsevier, 2019.
12. Sui C, de Vos P, Stapersma D, Visser K, Ding Y. Fuel consumption and emissions of ocean-going cargo ship with hybrid propulsion and different fuels over voyage. *Journal of Marine Science and Engineering*, 2020; 8(8): 588.
13. Houda S, Lancelot C, Blanchard P, Poinel L, Lamonier C. Oxidative desulfurization of heavy oils with high sulfur content: A review. *Catalysts*, 2018; 8(9): 344.
14. Bengtsson SK, Fridell E, Andersson KE. Fuels for short sea shipping: A comparative assessment with focus on environmental impact. *Proceedings of the Institution of Mechanical Engineers, Part M: Journal of Engineering for the Maritime Environment*, 2013; 228(1): 44–54.
15. Al-Breiki M, Bicer Y. Comparative cost assessment of sustainable energy carriers produced from natural gas accounting for boil-off gas and social cost of carbon. *Energy Reports*, 2020; 6: 1897–1909.
16. Vierth I, Karlsson R, Mellin A. Effects of more stringent sulphur requirements for sea transports. *Transportation Research Procedia*, 2015; 8: 125–135.
17. Notteboom T. The impact of low sulphur fuel requirements in shipping on the competitiveness of roro shipping in Northern Europe. *WMU Journal of Maritime Affairs*, 2011; 10(1): 63–95.
18. Smith TWP, *et al. Third IMO Greenhouse Gas Study 2014*. London, UK: International Maritime Organization, 2015.
19. Structure of IMO. [Cited 2022 July 20]. Available from: http://www.imo.org/en/About/Pages/Structure .aspx.
20. Rule Finder. MARPOL - International convention for the prevention of pollution from ships. 2005 [Cited 2022 July 27]. Available from: http://www.mar.ist.utl.pt/mventura/Projecto-Navios-I/IMO-Conventions %20%28copies%29/MARPOL.pdf.
21. Company, M.A. MARPOL 2020: An opportunity for OPEC to reclaim market share. 2020 [Cited 2022 July 20]. Available from: https://www.mckinsey.com/industries/oil-and- gas/our-insights/petroleum-blo g/marpol-2020-an-opportunity-for-opec-to-reclaim- market-share.
22. Baldi F, *et al.* Optimal load allocation of complex ship power plants. *Energy Conversion and Management*, 2016; 124: 344–356.
23. Bengtsson S, Andersson K, Fridell E. A comparative life cycle assessment of marine fuels: Liquefied natural gas and three other fossil fuels. *Proceedings of the Institution of Mechanical Engineers, Part M: Journal of Engineering for the Maritime Environment*, 2011; 225(2): 97–110.
24. Bengtsson S, Andersson K, Fridell E. An environmental life cycle assessment of LNG and HFO as marine fuels, an environ. life cycle assess. LNG HFO as Mar. *Fuels*, 2013; 225(2): 97–110.
25. Al-Enazi A, *et al.* A review of cleaner alternative fuels for maritime transportation. *Energy Reports*, 2021; 7: 1962–1985.
26. IMO. Work on carbon capture and storage and ocean fertilization under the London Convention and Protocol. 2011 [Cited 2022 July 27]. Available from: https://unfccc.int/resource/docs/2011/smsn/igo/140 .pdf.
27. Walker TR, *et al.* Chapter 27: Environmental effects of marine transportation. In: *World Seas: An Environmental Evaluation* (Second Edition). Edited by C. Sheppard. Academic Press, 2019, pp. 505–530.
28. UNFCCC. Report of the Conference of the Parties on Its Twenty-First Session. 2015 [Cited 2022 July 27]. Available from: https://unfccc.int/resource/docs/2015/cop21/eng/10.pdf.
29. Baldi F, *et al.* Energy and energy analysis of ship energy systems - The case study of a chemical tanker. *International Journal of Thermodynamics*, 2015; 18(2): 82–93.
30. Jianyun Z, *et al.* Bi-objective optimal design of plug-in hybrid electric propulsion system for ships. *Energy*, 2019; 177: 247–261.
31. Zincir B, Deniz C. An investigation of hydrogen blend fuels applicability on ships, 2014.
32. Elgohary MM, Seddiek IS, Salem AM. Overview of alternative fuels with emphasis on the potential of liquefied natural gas as future marine fuel. *Proceedings of the Institution of Mechanical Engineers, Part M: Journal of Engineering for the Maritime Environment*, 2014; 229(4): 365–375.
33. Goldsworthy L. Design of ship engines for reduced emissions of oxides of nitrogen, 2002.

34. Zwolinska E, *et al.* Removal of high concentrations of NO_x and SO_2 from diesel off-gases using a hybrid electron beam technology. *Energy Reports*, 2020; 6: 952–964.
35. Bicer Y, Dincer I. Life cycle evaluation of hydrogen and other potential fuels for aircrafts. *International Journal of Hydrogen Energy*, 2017; 42(16): 10722–10738.
36. Valera-Medina A, *et al.* Ammonia for power. *Progress in Energy and Combustion Science*, 2018; 69: 63–102.
37. Hansen SF, *et al.* Nanotechnology meets circular economy. *Nature Nanotechnology*, 2022; 17(7): 682–685.
38. Salamanca-Buentello F, Daar AS. Nanotechnology, equity and global health. *Nature Nanotechnology* 2021; 16(4): 358–361.
39. Mulvey T. The electron microscope: The British contribution. *Journal of Microscopy*, 1989; 155(3): 327–338.
40. Jiang J, Chen DR, Biswas P. Synthesis of nanoparticles in a flame aerosol reactor with independent and strict control of their size, crystal phase and morphology. *Nanotechnology*, 2007; 18(28): 285603.
41. Husen A, Siddiqi KS. Carbon and fullerene nanomaterials in plant system. *Journal of Nanobiotechnology*, 2014; 12(1): 16.
42. Ali A, *et al.* Synthesis, characterization, applications, and challenges of iron oxide nanoparticles. *Nanotechnology, Science, and Applications*, 2016; 9: 49–67.
43. Fior Markets. Nanotechnology market is globally expected to drive growth of USD 5.96 billion by 2027. 2022 [Cited 2022 August 17]. Available from: https://www.globenewswire.com/news-release/2022/05/18/2445638/0/en/Nanotechnology-Market-is-Globally-Expected-to-Drive-Growth-of-USD-5-96-billion- by-2027-Fior-Markets.html.
44. Carnes CL, Klabunde KJ. The catalytic methanol synthesis over nanoparticle metal oxide catalysts. *Journal of Molecular Catalysis A: Chemical*, 2003; 194(1): 227–236.
45. Hwang CB, *et al.* Synthesis, characterization, and highly efficient catalytic reactivity of suspended palladium nanoparticles. *Journal of Catalysis*, 2000; 195(2): 336–341.
46. Balint I, Miyazaki A, Aika KI, Methane reaction with NO over alumina-supported Ru nanoparticles. *Journal of Catalysis*, 2002; 207(1): 66–75.
47. Sulman E, *et al.* Hydrogenation of acetylene alcohols with novel Pd colloidal catalysts prepared in block copolymers micelles. *Journal of Molecular Catalysis A: Chemical*, 1999; 146(1): 265–269.
48. Keshavaraja A, She X, Flytzani-Stephanopoulos M. Selective catalytic reduction of NO with methane over Ag-alumina catalysts. *Applied Catalysis B: Environmental*, 2000; 27(1): L1–L9.
49. Malyala RV, *et al.* Activity, selectivity and stability of Ni and bimetallic Ni–Pt supported on zeolite Y catalysts for hydrogenation of acetophenone and its substituted derivatives. *Applied Catalysis A: General*, 2000; 193(1): 71–86.
50. Eijsbouts S, Mayo SW, Fujita K. Unsupported transition metal sulfide catalysts: From fundamentals to industrial application. *Applied Catalysis A: General*, 2007; 322: 58–66.
51. Hong ST, *et al.* Characterization of the active phase of $NiMo/Al_2O_3$ hydrodesulfurization catalysts. *Research on Chemical Intermediates*, 2006; 32(9): 857–870.
52. Babich IV, Moulijn JA. Science and technology of novel processes for deep desulfurization of oil refinery streams: A review. *Fuel*, 2003; 82(6): 607–631.
53. Brunet S, *et al.* On the hydrodesulfurization of FCC gasoline: A review. *Applied Catalysis A: General*, 2005; 278(2): 143–172.
54. Song C. An overview of new approaches to deep desulfurization for ultra-clean gasoline, diesel fuel and jet fuel. *Catalysis Today*, 2003; 86(1): 211–263.
55. Song C, Ma X. New design approaches to ultra-clean diesel fuels by deep desulfurization and deep dearomatization. *Applied Catalysis B: Environmental*, 2003; 41(1): 207–238.
56. Candia R, *et al.* Effect of sulfiding temperature on activity and structures of $CO\text{–}MO/AL_2O_3$ Catalysts. II. *Bulletin des Sociétés Chimiques Belges*, 1984; 93(8–9): 763–774.
57. van Veen JAR, *et al.* A real support effect on the activity of fully sulphided CoMoS for the hydrodesulphurization of thiophene. *Journal of the Chemical Society, Chemical Communications*, 1987; 22: 1684–1686.
58. van Veen JAR, *et al.* On the formation of type I and type II NiMoS phases in $NiMo/Al_2O_3$ hydrotreating catalysts and its catalytic implications. *Fuel Processing Technology*, 1993; 35(1): 137–157.
59. Bouwens SMAM, *et al.* On the structural differences between alumina-supported comos type I and Alumina-, Silica-, and Carbon-supported comos type II phases studied by XAFS, MES, and XPS. *Journal of Catalysis*, 1994; 146(2): 375–393.

60. Topsoe H, Clausen BS. Importance of Co-Mo-S type structures in hydrodesulfurization. *Catalysis Reviews*, 1984; 26(3–4): 395–420.
61. Chianelli RR, Daage M, Ledoux MJ. Fundamental studies of transition-metal sulfide catalytic materials. In: *Advances in Catalysis*. Edited by DD Eley, H Pines, WO Haag. Academic Press, 1994, pp. 177–232.
62. Daage M, Chianelli RR. Structure-function relations in molybdenum sulfide catalysts: the "Rim-Edge" model. *Journal of Catalysis*, 1994; 149(2): 414–427.
63. Qiu L, *et al.* Oxidative desulfurization of dibenzothiophene using a catalyst of molybdenum supported on modified medicinal stone. *RSC Advances*, 2016; 6(21): 17036–17045.
64. Safa MA, *et al.* Oxidative desulfurization kinetics of refractory sulfur compounds in hydrotreated middle distillates. *Fuel*, 2019; 239: 24–31.
65. Lim XB, Ong WJ. A current overview of the oxidative desulfurization of fuels utilizing heat and solar light: from materials design to catalysis for clean energy. *Nanoscale Horizons*, 2021; 6(8): 588–633.
66. Rajendran A, *et al.* A comprehensive review on oxidative desulfurization catalysts targeting clean energy and environment. *Journal of Materials Chemistry A*, 2020; 8(5): 2246–2285.
67. Somiya S, Roy R. Hydrothermal synthesis of fine oxide powders. *Bulletin of Materials Science*, 2000; 23(6): 453–460.
68. Wu W, *et al.* Multilayer MoOx/Ag/MoOx emitters in dopant-free silicon solar cells. *Materials Letters*, 2017; 189: 86–88.
69. Chithambararaj A, Rajeswari Yogamalar N, Bose AC. Hydrothermally synthesized h-MoO_3 and α-MoO_3 nanocrystals: New findings on crystal-structure-dependent charge transport. *Crystal Growth & Design*, 2016; 16(4): 1984–1995.
70. Gumerova NI, Rompel A. Polyoxometalates in solution: Speciation under spotlight. *Chemical Society Reviews*, 2020; 49(21): 7568–7601.
71. Krishnan CV, *et al.* Formation of molybdenum oxide nanostructures controlled by poly (ethylene oxide). *Chinese Journal of Polymer Science*, 2009; 27(1): 11–22.
72. Carvajal J. Study of micro-structural effects by powder diffraction using the program FULLPROF. *Laboratoire Leon Brillouin (CEA-CNRS)*, CEA/Saclay. 91191.
73. Inzani K, *et al.* Electronic properties of reduced molybdenum oxides. *Physical Chemistry Chemical Physics*, 2017; 19(13): 9232–9245.
74. Yang LC, *et al.* Tremella-like molybdenum dioxide consisting of nanosheets as an anode material for lithium ion battery. *Electrochemistry Communications*, 2008; 10(1): 118–122.
75. Shi Y, *et al.* Ordered mesoporous metallic MoO_2 materials with highly reversible lithium storage capacity. *Nano Letters*, 2009; 9(12): 4215–4220.
76. Yang LC, *et al.* MoO_2 synthesized by reduction of MoO_3 with ethanol vapor as an anode material with good rate capability for the lithium ion battery. *Journal of Power Sources*, 2008; 179(1): 357–360.
77. Eyert V, *et al.* Embedded Peierls instability and the electronic structure of MoO_2. *Journal of Physics: Condensed Matter*, 2000; 12(23): 4923–4946.
78. Tokarz-Sobieraj R, Grybos R, Witko M. Electronic structure of MoO_2. DFT periodic and cluster model studies. *Applied Catalysis A: General*, 2011; 391(1): 137–143.
79. Schroeder T, *et al.* Formation of a faceted MoO_2 epilayer on Mo (112) studied by XPS, UPS and STM. *Surface Science*, 2004; 552(1): 85–97.
80. Brandt BG, Skapski AC. A refinement of the crystal structure of molybdenum dioxide. *Acta Chemica Scandinavica*, 1967; 21: 661–672.
81. Ludtke T, *et al.* HP-MoO_2: A high-pressure polymorph of molybdenum dioxide. *Inorganic Chemistry*, 2017; 56(4): 2321–2327.
82. Quilty CD, *et al.* Ex situ and Operando XRD and XAS analysis of MoS_2: A lithiation study of bulk and nanosheet materials. *ACS Applied Energy Materials*, 2019; 2(10): 7635–7646.
83. Langford JI, Wilson AJC. Scherrer after sixty years: A survey and some new results in the determination of crystallite size. *Journal of Applied Crystallography*, 1978; 11: 102–113.
84. Als-Nielsen J, McMorrow D. Kinematical scattering II: Crystalline order. In: *Elements of Modern X-Ray Physics*, 2011, pp. 147–205.
85. Lipkin HJ. Physics of Debye-Waller factors. 2004. Available from: cond-mat/0405023.
86. McCusker LB, *et al.* Rietveld refinement guidelines. *Journal of Applied Crystallography*, 1999; 32(1): 36–50.
87. Cao P, *et al.* Highly conductive carbon black supported amorphous molybdenum disulfide for efficient hydrogen evolution reaction. *Journal of Power Sources*, 2017; 347: 210–219.

3 Designing the Next Generation of Nanocatalysts for Sustainably Produced Aviation Fuels

Donny Bhuana, Derek Fawcett, A.F.M. Fahad Halim, and Gérrard Eddy Jai Poinern

CONTENTS

3.1 TOWARD THE SUSTAINABLE PRODUCTION OF AVIATION FUELS

The idea of flying has always captivated the human imagination: from Icarus's wings to Da Vinci's flying machines, and ultimately to the Wright Flyer. Many people with creative ideas have worked to make flying possible. Unfortunately, the early pioneers had little success in their endeavors due to the unavailability of high-power-to-weight propulsion systems. The advent of the internal combustion engine in the early twentieth century made flying a reality at last. The inability to overcome the force of gravity was the major hindrance that could not be overcome until portable and compact power sources became available. In the beginning, airplane engines were very much like automotive engines and even utilized the same fossil fuel (gasoline). However, it soon became evident that automotive engines and their fuel had limitations. Accordingly, experts in the field created aviation gasoline (avgas) and developed engine technology specifically designed for airplanes. By the 1940s, the gas turbine engine had been developed to meet the demand for even more power and speed. Similar to gasoline and avgas, kerosene, the original fuel for aircraft turbine engines, was later supplanted by special jet fuels (JF). Over the past 90 years, aviation has evolved from a revolutionary concept to an integral part of today's society. The

DOI: 10.1201/9781003181422-4

aviation industry helps propel a country forward economically and socially. Everything that involves the airborne transport of goods and people can be included in this industry. It also encourages international trade and tourism by making a market accessible to companies all over the world.

As a whole, air travel is on the rise, and this trend is expected to continue well into the future. According to the International Air Transport Association (IATA), the number of air travelers will have nearly doubled by 2035. Likewise, the International Civil Aviation Organization (ICAO) predicts a notable increase in total air travelers carried, from 3.5 billion in 2015 to around 10.5 billion by 2040. In addition, a significant increase in aircraft departures from 34 million to 95 million is expected during this period [1]. Studies have shown increasing air travel has increased fuel consumption and, in turn, has greatly increased exhaust emissions. Many of these emissions, such as CO_2, NO_x, and SO_x, cause serious environmental issues such as contributing to the global greenhouse gas load [2]. Recently, Google Flights was found to significantly understate the global impact of aviation on the climate [3]. In addition, other current environmental impact modeling only takes into account just over half of the real impact that aviation has on the climate, according to some experts. The real concern over the environmental impact of flights on global warming has been highlighted by the example of Google Flights, where its assessment of environmental impact did not take into account several pollution parameters, thereby making it difficult for airline customers to determine their real climate impact. Theoretically, for every kilogram of aviation fuel burned, 3.16 kg of carbon dioxide (CO_2) and 1.23 kg of water vapor (H_2O) are emitted with other combustion products to the atmosphere by typical aircraft engines, [4]. For instance, the estimated CO_2 emission from a Boeing 787-9 flight carrying 257 passengers, traveling from Zurich ZRH airport to San Francisco SFO airport can be estimated from several parameters. These flight parameters include total distance (nautical mileage), the landing and take-off (LTO) cycle, the climb, cruise, and descent (CCD) cycle, as well as the seating class multiplying factor. When calculated, the flight produces 418.71 kg of CO_2 per economy passenger [5]. Likewise, if we consider a much longer flight route, such as Sydney (Australia) to London (United Kingdom) that has a travel time of 19 hours and a distance of nearly 10,600 miles, with a similar aircraft and passenger numbers, the flight would generate 860 kg of CO_2 per economy passenger. However, traveling by aircraft is not just about the generation of CO_2; around two-thirds of detrimental environmental effects come from other types of pollutants, such as contrails (droplets of ice), fuel particles, and water vapor [6]. These pollutants lead to the formation of long thin clouds high up in the atmosphere that trap radiated heat from the earth, thereby contributing to a net warming effect on the planet.

The aviation industry has played a pivotal role in people's life since its inception in the 1900s. Its most important role was to cater to the rapid growth of the global mobility needs of humanity. The total number of global passenger flights has steadily increased since the 1970s, as seen in Figure 3.1. Inspection of Figure 3.1 reveals a dramatic rise in numbers occurring just after 2010, with numbers reaching 4 billion in 2017 and 4.5 billion in 2019 [7]. In spite of the catastrophic drop in numbers due to the recent COVID-19 pandemic, a positive trend in passenger numbers is expected to increase in the coming decades. Current estimates forecast an annual growth rate of about 4% per year, including the post-pandemic era [8]. For instance, the International Civil Aviation Organization has reported a significant recovery of passenger numbers in late 2022 [9].

In addition, to growing passenger numbers, the air freight industry has also continued to increase in recent decades. Figure 3.1 shows that the volume of freight transported by the air freight industry has steadily increased, and despite the shock of the COVID-19 pandemic, air freight is once again increasing. Furthermore, international air shipping services, such as FedEx, UPS, and DHL have shown a positive trend in daily package volume and freight pounds over the last couple of years. In particular, American-based freight companies, such as FedEx and UPS, recorded total shipping levels of 12.133 and 11.307 million tons respectively during the 2019 calendar year, while the European cargo giant DHL exceeded 24 million tons over the same period [10–12]. Regardless of COVID-19, the core business of air shipping services does not seem to be as severe as passenger services. As a matter of fact, most of the large air shipping companies registered unprecedented demands for residential delivery

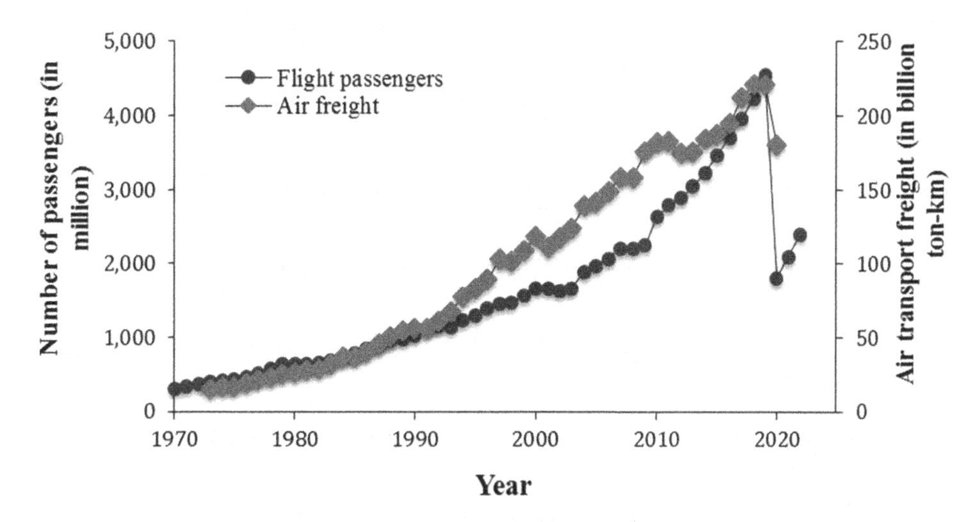

FIGURE 3.1 Number of worldwide aircraft passengers carried and air freight transported over time (Data from worldbank.org) [7]

services in 2021 due to the rapid increase in demand for online shopping deliveries during the pandemic period. The rapid demand equated to the busy peak normally seen during the holiday season [11, 12]. To put it another way, the rapid and consistent growth of both passenger and freight services highlights the importance of the aviation industry to people around the world today.

However, the downside the aviation industry faces is the concern over its high CO_2 and non-CO_2 emissions to the atmosphere and the subsequent detrimental effects on environmental climate change. Currently, around 5% of the total global greenhouse gases are emitted by the aviation industry. These emissions are predominantly in the form of CO_2, NO_x, SO_x, and other particulate materials [13, 14]. Additionally, aircraft produce contrails, and that equates to the aviation industry actually contributing almost 10% to anthropogenic global warming. The current emission levels are expected to increase with the increasing numbers of passengers and air shipping services [8, 11]. Thus, the aviation industry is facing a huge challenge in terms of dealing with the ambitious goal of the United Nation's Paris Agreement for the global reduction of CO_2 levels. Reducing carbon emissions and capping the global warming rate to 2.0 °C will be highly demanding for the aviation sector at large [15].

Comprehensive strategies are needed to mitigate the impact of global anthropogenic emissions by all countries in order to meet the United Nation's Paris Agreement target. Thus, long-term strategies in a number of areas, such as policy, regulations, and scientific research, are needed to abate climate change. Importantly, aviation fuels or jet fuels are predominantly based on kerosene, which is derived from the fractional distillation of crude oil, as shown in Figure 3.2. It is the high energy density of JF that has made commercial aviation possible. However, with the aviation sector expecting passenger numbers to double to more than 8 billion by 2050, alternative fuels are needed to reduce carbon emissions. One alternative to conventional JF is to develop "sustainable aviation fuel" (SAF) that is made from renewable feedstocks with a similar chemical composition.

Cooking oil and other non-palm waste oils from plants or animals are examples of potential feedstock materials that could be used to produce SAFs. Other examples include solid waste from households and workplaces, for instance, packing materials, paper, clothing, and leftover food that might otherwise be sent to landfill or incinerated. Forestry by-products like scrap wood and energy crops like fast-growing plants and algae are other possibilities. The use of SAFs throughout the fuel's entire lifecycle would potentially result in lower carbon emissions. Typically, eco-friendly feedstock, when combined with effective environmentally friendly processing of the feedstock and efficient delivery of the fuel to the airport, could lower carbon emissions by as much as 80% when compared to conventional JFs.

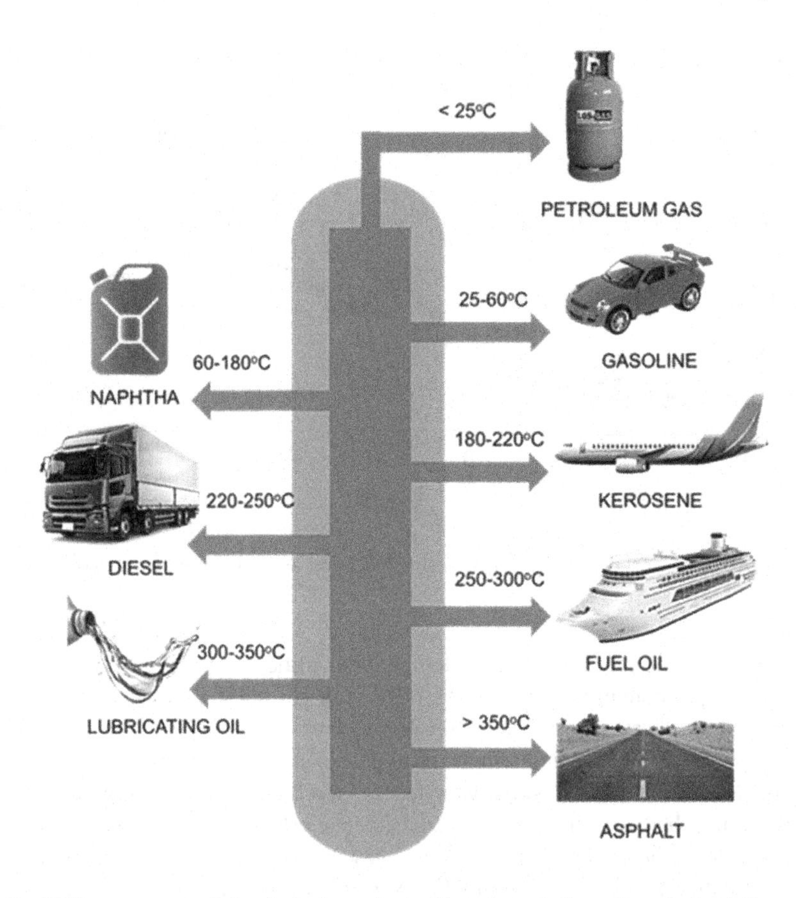

FIGURE 3.2 Different types of fossil fuel products (Based on information derived from Geoscience, University of Chicago) [16].

Another alternative is to utilize carbon dioxide (CO_2), which is a major contributor to greenhouse gas emissions. In this strategy, CO_2 is converted into synthetic kerosene, which makes this strategy an important pathway toward the sustainable production of eco-friendly aviation fuels. It also has the advantage of reducing the CO_2 load in the atmosphere and helping reduce global warming. The strategy is a two-step process. The first step involves the initial conversion of CO_2 into carbon monoxide (CO) and combining it with hydrogen to form syngas. The second step involves converting the syngas into liquid fuels. A number of methods can be used to convert CO_2 into syngas. These methods include reversed water gas shift reaction (also known as hydrogenation of CO_2), CO_2 reforming of natural gas, and auto-thermal reforming. Conversely, these methods are energy-intensive due to the highly endothermic processes involved and the elevated operational temperatures needed during the conversion processes. In addition, these methods also require a continuous supply of other reactants, such as methane and hydrogen [17]. However, there are alternative CO_2, conversion methods that can be utilized, such as photochemical, electrochemical, and photo-electrochemical processes. However, in spite of some favorable results from these alternative CO_2, conversion methods, a number of challenges still remain to be resolved. For instance, problems related to maintaining consistent operational performance during the process arise. Typical problems include non-durable electrochemical yields due to the deactivation of electrodes and the deposition of contaminants in the electrolyte [18]. Thus, the shortcomings of these methods have highlighted the need for feasible and sustainable processes that integrate multifunctional catalysts capable of economically converting CO_2 into liquid fuels.

In recent years, new technology has emerged in the form of solar thermochemical (STC) conversion. The STC method utilizes solar energy and a catalyst to convert CO_2 and water (H_2O) into synthetic kerosene. The technology involves a two-stage thermochemical redox cycle that utilizes a non-stoichiometric metal oxide, in this case, cerium oxide (ceria), catalyst, and solar radiation as the heat source [19]. During the first stage, there is high-temperature endothermic reduction of metal oxide, which is followed by the second stage which involves the lower temperature oxidation of the metal oxide with H_2O and CO_2 to produce liquid fuels. The STC technology has the potential to outperform several other current technologies such as solar-driven electricity generation, water electrolysis, and reverse water gas shift reaction (RWGS) [20]. Importantly, STC technology relies on two main factors. The first is having a high-performance catalyst material, and the second is using concentrated solar energy to provide adequate thermal energy for the process to be thermodynamically favorable [21]. STC technology has been translated from the laboratory to an industrial-scale pilot plant. The plant incorporated a solar tower to supply the thermal energy, a continuous supply of CO_2 and H_2O to generate a continuous stream of syngas. The facility operates at 1500 °C during the reduction stage and 900 °C during the oxidation stage, with all reactions occurring at a maximum operating pressure of 1 bar. An attractive feature of the STC method is the ability to control the quality of the syngas being produced, thus eliminating the need for further refining processes [20]. Because of the promising results of STC technology, attention has once again focused on the Fischer-Tropsch (FT) process for converting syngas into liquid fuels. The FT process is of particular importance since maximizing the syngas-to-fuel conversion efficiency will ultimately determine the viability of this new technology for sustainably producing aviation fuels.

3.2 FISCHER-TROPSCH METHOD: AN IMPORTANT METHOD FOR PRODUCING SYNTHETIC LIQUID FUELS

An ever-increasing demand for the supply of eco-friendly and sustainable aviation fuels has resulted in new technologies for converting CO_2 into aviation fuel. However, these new technologies can only convert CO_2 into syngas. Therefore, there is a need for converting downstream syngas into synthetic liquid aviation fuels. The Fischer-Tropsch method, a technology discovered in the 1920s, has re-emerged as a viable method for converting syngas into liquid aviation fuels. The original process, which used a highly exothermic catalytic reaction, converted syngas into long-chained hydrocarbons [22, 23]. However, the basic FT method produces some waxes and long-chained linear hydrocarbons, which are unsuitable for aviation fuels [23, 24]. Therefore, further refining is needed to shift the product distribution into the intermediate-chained isoparaffin region. The isoparaffin region is where kerosene is normally found [25]. In order to shift the product distribution into the intermediate-chained isoparaffin region, it is important to use multifunctional nanocatalysts that can efficiently enhance the conversion process and assist in directing the production of hydrocarbons in the isoparaffin region. The overall process for converting CO_2 gas into syngas by the solar thermochemical conversion process and then the downstream FT method for converting syngas into aviation fuel is schematically presented in Figure 3.3.

The FT method is a technology specifically designed for catalytically (metal catalysts) converting carbon monoxide and hydrogen (syngas) into clean liquid fuels. The process was discovered and developed in 1923 by Franz Fischer and Hans Tropsch at the Kaiser Wilhelm Institute for Coal Research [22]. In their original study, Fischer and Tropsch reported that CO hydrogenation occurred between 180 and 250 °C in the presence of catalysts at atmospheric pressure. The resulting hydrogenation process resulted in the formation of hydrocarbon products. The products typically consisted of straight-chain hydrocarbons, which can be in the form of paraffins ranging from CH_4 to waxes C_nH_{2n+2}, where n ranges from 1 to over 100, and olefins ranging from ethylene to much longer molecules (C_nH_{2n}, with n>2), and to a lesser extent oxygenated products such as alcohols [24]. Due to the highly exothermic chemical reaction taking place during the FT process, large amounts of heat are generated, as seen in a simplified reaction presented in Equation 3.1. For comparison purposes, the water gas shift reaction is present in Equation 3.2 and clearly shows the much larger amount of heat generated by the FT method.

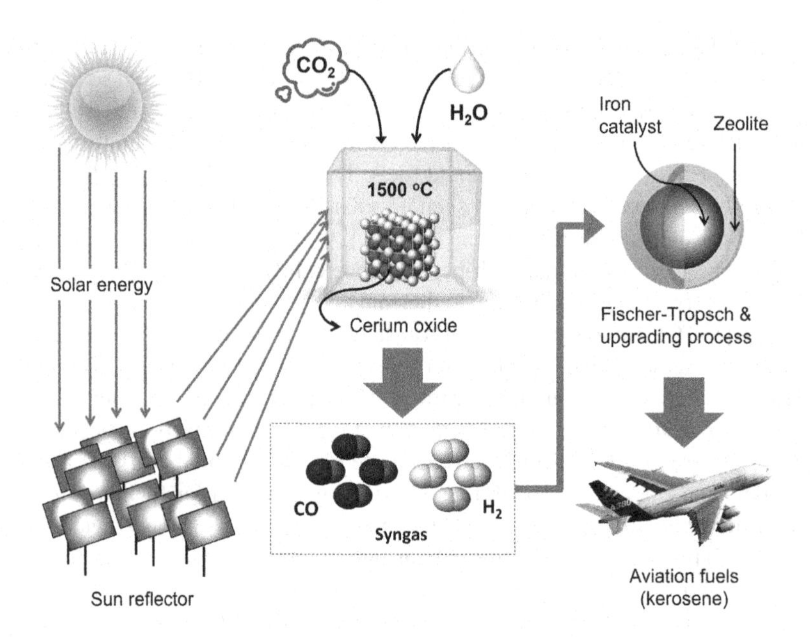

FIGURE 3.3 Schematic diagram of the solar thermochemical and Fischer-Tropsch methods for converting CO_2 into aviation fuels.

Fischer-Tropsch:	$CO + 2H_2 \rightarrow -CH_2- + H_2O$	$\Delta H^C = -165$ KJ/Mol	(3.1)
Water gas shift:	$CO + H_2O \leftrightarrow H_2 + CO_2$	$\Delta H^C = -42$ KJ/Mol	(3.2)

The FT technology is considered to be the most effective method for converting syngas into synthetic liquid fuels and since its inception has been continuously improved.

3.2.1 Syngas Production

The FT method is a heterogeneous catalytic process designed to produce clean liquid fuels from syngas. The syngas can be derived from non-petroleum sources such as coal, biomass, and natural gas by methods like steam reforming, partial or auto-thermal oxidation, or gasification [19]. Several methods for producing syngas, their chemical equations, and the heat of the reactions are presented in Table 3.1. As mentioned above, the thermochemical conversion of CO_2 into liquid fuels involves two reactions. The first reaction is called the reversed water gas shift reaction, and the second reaction is called the Fischer-Tropsch reaction. The two-reaction process represents an environmentally friendly method for producing liquid fuel from carbon dioxide and hydrogen [26–28].

In spite of the advantages of the combined FT-CO_2 hydrogenation process, supplying the hydrogen for the upstream stage of the process could be technically demanding [29]. However, other methods of producing liquid fuels also have issues that need to be resolved. For instance, CO_2/H_2O splitting (see Table 3.1), another catalytic-based process that involves highly endothermic reactions, needs to improve its overall process performance by implementing more effective heat recovery strategies [20, 30, 31].

3.2.2 Fischer-Tropsch Product Distribution

At a molecular level, the FT method involves CO hydrogenation and then polymerization. The hydrogenation reaction results from the need to break the C-O bond so that new C-H bonds can be formed. In addition, C-C bonds are needed in order to promote hydrocarbon chain growth.

TABLE 3.1
Methods for producing syngas, their chemical equations, and heat of reaction

Reaction	Chemical equations	Heat of reaction ΔH^0 kJ/mol	Reference
Steam reforming	$CH_4 + H_2O \rightarrow CO + 3H_2$	+ 206	[32]
CO_2 reforming	$CH_4 + CO_2 \rightarrow 2CO + 2H_2$	+ 247	[32]
	$C_2H_6 + 2CO_2 \leftrightarrow 4CO+3H_2$	+ 428.1	
	$C_3H_8 + 3CO_2 \leftrightarrow 6CO + 4H_2$	+ 644.1	
	$C_4H_{10} + 4CO_2 \leftrightarrow 8CO + 5H_2$	+ 817.1	
Auto-thermal reforming	$CH_4 + 0.25\ O_2 + 0.5\ H_2O \rightarrow CO + 0.25\ H_2$	+ 85.3	[33]
Coal gasification	$C + O_2 \rightarrow CO_2$	- 393.4	[34]
	$C + 0.5\ O_2 \rightarrow CO$	- 111.4	
	$C + H_2O \rightarrow CO + H_2$	+ 130.5	
	$C + CO_2 \leftrightarrow 2CO$	+ 170.7	
	$CO + H_2O \leftrightarrow CO_2 + H_2$	- 40.2	
	$C + 2H \rightarrow CH_4$	- 74.7	
Water gas shift	$H_2O + CO \leftrightarrow H_2 + CO_2$	- 42	[34]
Partial oxidation	$CH_4 + 0.5\ O_2 \rightarrow CO + 2H_2$	- 35.6	[35]
CO_2/H_2O splitting	$MO \rightarrow MO_{1-\delta} + \frac{\delta}{2} O_2$	Catalyst: Ceria (CeO_2)	[30]
	$MO_{1-\delta} + \delta H_2O \rightarrow MO + \delta H_2$		
	$MO_{1-\delta} + \delta CO_2 \rightarrow MO + \delta CO$		

The carbon number distribution of the FT products tends to follow a statistical function called the Anderson-Schulz-Flory (ASF) distribution. The competition between mechanisms like chain growth and chain termination is characterized by a chain growth probability factor, known as the α-value. The higher the α-value, the longer the hydrocarbon chain and the heavier the hydrocarbon mixture produced. According to Anderson, the FT product distribution can be formulated by defining the ASF distribution [36]. The mass fraction of a particular hydrocarbon with a chain length (carbon number) of n is given by Equation 3.3.

$$W_n = n(1-\alpha)^2 \alpha^{n-1} \tag{3.3}$$

Thus, by using Equation (3.3), the hydrocarbon product distribution resulting from the FT process can be constructed, as seen in Figure 3.4. Studies have shown that the statistical distribution of FT products is inherently non-selective for a specific range of hydrocarbons [24, 37, 38–42]. For example, the maximum selectivity between C_5–C_{11} (gasoline range) and C_{12}–C_{20} (diesel-range) hydrocarbons are roughly 45 and 30%, respectively. In this context, current FT processing generally aims to produce long-chain alkanes (waxes, C_{21+}). Thus, the waxes need to be transformed into liquid fuels by hydrocracking over metal-acid dual functional catalysts. Therefore, to take full advantage of the FT method, it should be operated at high α values (> 0.9) to minimize the formation of undesired light products, especially methane [37, 42–46].

There are two operational temperature ranges for the FT process, as seen in Figure 3.4. High-temperature FT (HTFT) operations are normally carried out at temperatures between 320 to 350 °C with the reaction products being essentially in the gas phase. Low-temperature FT (LTFT) processes are normally carried at temperatures between 200 and 250°C. HTFT operations are carried out in either fluidized beds or fixed-bed reactors. Generally, alkali-promoted iron catalysts appear to be more suited to this process, since cobalt-based catalysts would only produce methane at higher

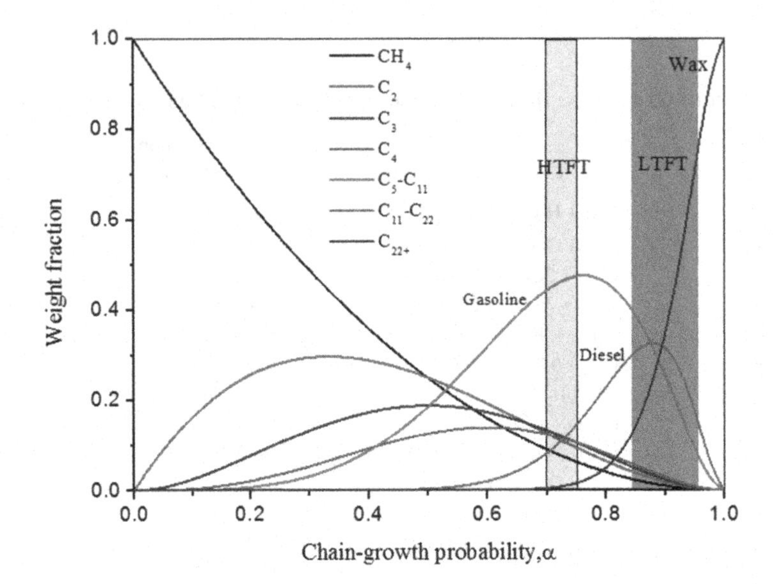

FIGURE 3.4 Spectrum of hydrocarbons produced by the Fischer-Tropsch process as a function of chain growth probability (adapted from Reference 37).

temperatures. LTFT processes can use both cobalt and iron-based catalysts. The cobalt-based catalyst is more suited to the lower end of the temperature. The LTFT produces high-quality middle distillates, such as diesel and jet fuels after the hydrocracking of long-chain waxes. The naphtha fraction produced by the process is used as feedstock for naphtha steam crackers, which typically produces ethylene and propylene [37].

3.3 A BRIEF HISTORY OF THE NANOCATALYST: THE JOURNEY OF A SMALL PARTICLE IN A BIG WORLD

Nanotechnology has promoted the development of new nanometer-scale materials with unique material properties. Some of these new nanomaterials have properties that make them useful as nanocatalysts [47, 48]. Several of these new nanocatalysts have the potential to be used in solar thermochemical conversion of greenhouse gases and in the FT process. While conventional catalysts are capable of promoting reaction rates by lowering the activation energy, nanocatalysts are capable of encouraging the process hundreds or even thousands of times more effectively due to their enhanced catalytic properties [49].

3.3.1 Fischer-Tropsch Nanocatalysts

Considering the reaction process involved in converting CO_2 into liquid fuels, it is essential to have effective catalysts that are capable of enhancing the performance of the FT process. The incorporation of nanocatalysts into the FT process is essential to promote both hydrocracking and isomerization. Studies have shown that combining catalysts with dopants, promoters, and catalyst supports can greatly improve process reactions. This is important because the increased catalytic activity will promote selectivity toward longer-chained hydrocarbons. Accordingly, this will promote higher yields of the desired liquid fuel [48]. For instance, coating iron (Fe)-based nanocatalysts with zeolite to form a core-shell structure has been found to produce hydrocarbons in the middle of the isoparaffin range. The shift toward hydrocarbons produced in the isoparaffin range is believed to be the result of easier mobility of reactant gases. Since the initial discovery of the FT process, different supported and unsupported metallic catalysts such as Fe, Co, Ni, Ru, and Rh have been investigated as a possible method of improving syngas conversion rates. However, the FT process follows the ASF product distribution, which is not favorable for producing liquid hydrocarbons over other products, such as waxes and methane [50]. The most common catalysts used for the FT process are Group VIII metals. Vannice investigated the activity of Group VIII metals for use in the FT process and found the catalytic activity of the metals could be ranked in the following order [51]:

$$Ru > Fe > Ni > Co > Rh > Pd > Pt$$

Among Group VIII metals, the most common catalysts are Fe, Co, and Ru, followed by Ni and Rh [50]. Iron and cobalt are more widely used as catalyst materials for the FT process. Cobalt-based catalysts possess a number of advantages in terms of high activity and long life. They also have a higher tendency to produce long-chained hydrocarbons as well as lowering the tendency to produce undesired carbon dioxide. The downside to using cobalt-based catalysts in the FT process includes low tolerance toward syngas with low H_2/CO ratios and their susceptibility to contaminants. However, the major disadvantage of using cobalt-based catalysts is their high cost compared to Fe. The most commonly used catalyst for the FT process is Fe. This is because Fe catalysts have a number of advantages over other catalytic materials. These advantages include 1) their lower cost compared to other metal catalysts; 2) they have higher water gas shift activity, which makes them effective in CO-rich syngas; 3) they have high selectivity toward olefins; and 4) they have high stability when exposed to H_2-rich syngas [52].

3.3.2 Dopants and Promoters

In order to enhance the performance of Fe- based catalysts, promoters are usually added to improve their selectivity, activity, and lifetime. Common promoters for Fe-based catalysts include potassium, manganese, copper, and magnesium. Potassium (K) is known to improve Fe-based catalyst activity and selectivity toward producing olefins and long-chain hydrocarbon products. Potassium doping in Fe-based catalysts acts as an electron promoter that improves the dissociative adsorption of CO and at the same time reduces the adsorption ability of hydrogen, thus leading to an increase in the olefin selectivity [43, 53]. The addition of manganese (Mn) to a Fe-based based catalyst is effective in reducing the selectivity for methane and increasing the olefin-to-paraffin ratio in both CO and CO_2 hydrogenation. The presence of Mn in Fe-based catalysts promotes the dispersion and carburization of the Fe_2O_3 precursor and increases in surface basicity of the catalyst by incorporating the Mn dopant. In addition, copper (Cu) can be used as a replacement dopant for Mn since it has a similar effect on Fe-based catalysts. In addition, Cu has been reported to enhance the reduction of hematite during carburization as well as improve catalytic dispersion [53]. However, care must be exercised, since excessive amounts of a promoter can result in decreased catalytic activity and shift the product distribution toward the formation of unwanted products.

3.3.3 Effect of Catalyst Supports

Catalyst supports have several important roles that include 1) dispersing the active phase; 2) providing a high surface area for the phase; 3) stabilizing the active phase; and 4) providing mechanical support to the catalyst [50]. Oxides such as Al_2O_3, SiO_2, and TiO_2 are the most widely used supports for cobalt (Co) catalysts. A number of studies have investigated the influence of these supports on the catalytic properties of Co catalysts. For instance, Davis et al. examined the influence of various supports such as Al_2O_3, SiO_2, and TiO_2 on the catalytic performances of Co catalysts [54]. The authors found the supports played a significant role in the reduction of Co oxide species. The study was also able to rank the performance of the supports as $Al_2O_3 > TiO_2 > SiO_2$. For Fe-based catalysts, the dispersion of the active phase on several types of supports tended to promote higher activity and selectivity. In general, studies have shown that the use of supports is beneficial in avoiding sintering or Oswald ripening of the catalytically active particles during reaction processes [55]. In particular, Dorner et al. have reported alumina as the best support for hydrogenation processes due to its ability to hinder sintering by providing strong metal-support interactions. The authors also report that when a K dopant is incorporated into the catalyst, the formation of potassium alanate ($KAlH_4$) is possible. Importantly, this compound is known for its reversible hydrogen sorption at moderate temperatures (250–350 °C). Thus, the presence of potassium alanate would lead to a reduction in the hydrogenation of surface-bound carbonaceous species that acts as a hydrogen reservoir. This in turn would result in higher olefin and lower methane products being produced [56].

3.3.4 The Need for Multifunctional Nanocatalysts

Hydrocarbons produced from the FT process follow the ASF distribution. Consequently, products have a wide distribution of carbon number base hydrocarbons. In addition, even after hydrocracking, these hydrocarbons, still contain normal aliphatic hydrocarbons with a high cetane number but a low octane number [57]. Thus, FT-derived products are naturally equivalent to diesel fuel but not to gasoline or kerosene [58]. Amid the stringent requirements of the Paris Agreement to limit carbon emissions and the fluctuating global oil market, there is also a drive to produce aviation fuels from syngas. But to achieve this goal, the FT process requires further refining processes such as isomerization and hydrocracking to produce synthetic aviation fuels. The Shell Company has commercially produced branched hydrocarbons from syngas using a two-step process. The first step involves converting syngas into linear paraffins through the FT process. Then during the second step, hydrocracking and isomerization reactions are carried out to produce an isoparaffin that is similar to kerosene or gasoline products [59]. However, these types of processes are complicated, with the initial FT process being prone to catalyst deactivation issues resulting from wax deposition on the surface of the catalysts. Thus, designing multifunctional catalysts for the direct conversion of kerosene-range isoparaffin from syngas would be advantageous. This would avoid the problem of catalyst deactivation.

3.3.5 Dry-Mixed Nanocatalysts

Dry mixing catalysts and zeolite for the FT process has been studied by several researchers. For instance, Yoneyama et al. investigated the physical mixing of varying amounts of zeolite and Fe-based catalysts to promote the synthesis of isoparaffin from syngas. Their study investigated four different mixture compositions and found no significant activity differences in terms of converting CO. However, in terms of product selectivity, the study found that higher concentrations of zeolite in the mixture resulted in increased CH_4 and CO_2 selectivity [58]. This is attributed to the formation of α-olefins as an F-T product in addition to n-paraffins, which could get re-adsorbed on the zeolite pores and eventually decompose as methane and, thus, increase the methane selectivity [60]. Generally, the addition of zeolite tended to reduce the presence of the heavy normal paraffins, and the resulting product distribution shifted toward the formation of lighter chained hydrocarbons (*i.e.* C_1-C_{10}) that are rich in isoparaffins. Thus, increasing the amount of zeolite in the mix increased selectivity toward lighter hydrocarbons. A study by Li et al. reported dry mixtures containing Co catalysts tended to produce heavier products compared to the Fe-based catalysts [61]. In addition, Fe-based mixed catalysts have been found to have a lower methane selectivity compared to Co-based mixed catalysts operating under the same conditions [58]. Studies have also reported the use of carbon monoxide or syngas, as the reducing gas, can give higher conversion rates than hydrogen gas. This difference was attributed to the formation of iron carbide that actively assisted in the FT process [62]. In addition, the effect of adding copper to Fe-based catalysts has been investigated and found to promote CO conversion rates in the FT process [53]. Increasing reaction temperatures during the FT process has been found to 1) increase olefin production; 2) decrease CO_2 production and; thus, suppress the water gas shift reaction; and 3) have little to no effect on methane selectivity [58].

3.3.6 Core-Shell Nanocatalysts

Synthetically produced fuels have several beneficial advantages over conventional petroleum-derived fuels. These advantages include 1) being sulfur-free; 2) being aromatic-free; and 3) being nitrogen-free. This makes synthetic fuels ideal for use as clean transportation fuels. As mentioned previously, FT-produced fuels contain normal paraffin hydrocarbons that make them suitable as synthetic diesel fuels, but additional processes such as hydrocracking and isomerization are needed to produce synthetic gasoline or kerosene [23, 63, 64]. While physically mixing FT catalysts with hydrocracking and isomerization is able to suppress the formation of longer-chained hydrocarbon and increase the selectivity for branched paraffins to a certain degree, the catalytic performance can still be improved. For Fe-zeolite dry-mixed catalysts, the active sites are distributed in random locations on the catalyst surface. This results in unrestricted reaction sites that allow FT, hydrocracking, and isomerization reactions to occur randomly and independently. This behavior suggests that it is possible for FT reaction products to leave the catalyst surface directly without further reacting with zeolite active sites. Apart from physically mixing metal-based FT and zeolite catalysts, a number of other studies have attempted to produce bi-functional metal/zeolite catalysts in different formats by utilizing an FT catalyst supported on zeolite for the direct production of isoparaffin [65, 66]. However, this method resulted in significantly lower CO conversion rates and also displayed a lower degree of reduction due to the extremely strong interaction between the metal and the zeolite support. In addition, the ease of migration and mobility of FT products appears to play an important role in determining the successful formation of middle isoparaffin hydrocarbons. However, in an attempt to overcome some of these shortcomings, several studies have investigated using novel multifunctional catalysts consisting of a core-shell structure [67, 68]. The core consists of conventional Fe-based FT catalysts, and the shell comprises a layer of H-ZSM5 zeolite, as seen in Figure 3.5. In this configuration, the syngas passes through the shell to reach the catalyst core where it undergoes FT processing to form linear paraffins. After processing, the linear paraffins pass through the zeolite shell where they undergo hydrocracking and isomerization. During this stage, due to high collision probability, the linear paraffins have a high likelihood of being converted into isoparaffins on the acidic zeolite sites. Thus, the core-shell structure of the Fe/zeolite catalysts has the ability

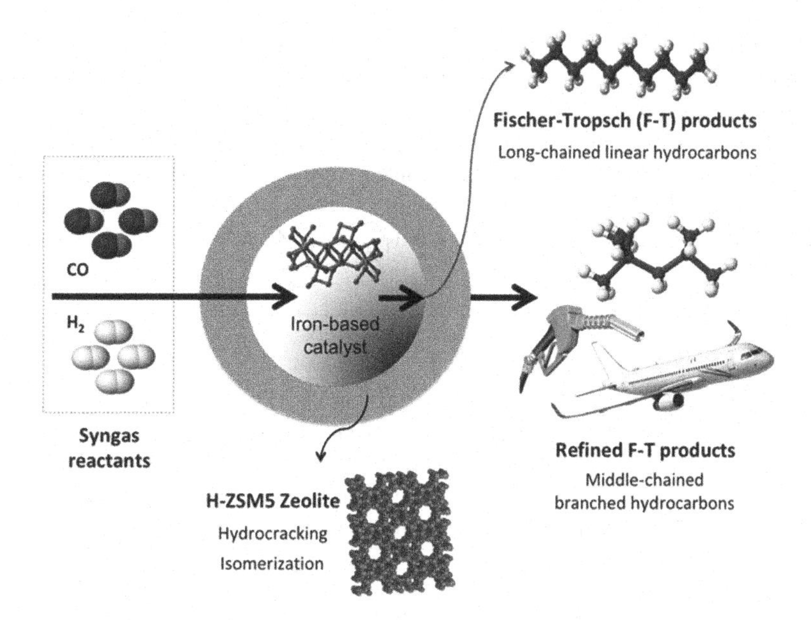

FIGURE 3.5 Schematic representation of the core-shell Fischer-Tropsch catalyst structure used for producing aviation fuel (kerosene).

to completely suppress the formation of C_{10}+ hydrocarbons and shifts the resulting products to the middle isoparaffins' range. Moreover, longer hydrocarbon chains take longer to diffuse through the zeolite layer, which means there is a greater probability they will be converted into isoparaffins *via* hydrocracking and isomerization processes. In terms of catalyst activity, the presence of the zeolite shell surrounding the core appears to have little effect on migration reactant gases. However, the core-shell structured catalyst does exhibit lower selectivity toward methane and carbon dioxide due to lower water gas shift (WGS) reaction activity. Consequently, as a result of the lower WGS activity, more unreacted H_2O and CO are expected to get trapped between the core and shell region. This phenomenon contributes to a lower H_2/CO ratio inside the catalyst core and leads to lower levels of methane formation [69]. However, in spite of the advantages of the core-shell structure pointed out in the abovementioned studies, more studies are necessary to examine the behavior of other forms of Fe-based catalysts, other catalyst materials, and dopants.

3.4 FISCHER-TROPSCH NANOCATALYSTS: A CASE STUDY – FROM CONCEPT TO CREATION

To manufacture FT nanocatalyst materials that will serve as well as downstream process catalysts for the overall conversion of CO_2 into synthetic aviation fuels consist of three types that include 1) synthesis of core catalysts; 2) synthesis of dry-mixed catalysts; and 3) synthesis of core-shell catalysts. Typically, dry-mixed catalysts were used as references to monitor the quality and performance of the core-shell type catalysts. A number of characterization studies have been carried out to investigate the material properties of the various catalyst types. The characterization studies include 1) X-Ray diffraction (XRD) analysis to investigate the structure of the crystalline materials present; 2) temperature-programmed reduction (TPR) testing to analyze the reducibility of the synthesized catalysts in gaseous reactants; and 3) scanning electron microscopy (SEM) analysis to explore the size and morphology of the engineered catalysts.

3.4.1 Synthesis and Characterization of Core Catalysts

Several studies have investigated synthesizing core catalysts and these are listed and referenced in Table 3.2. In the present case study, synthesizing core catalysts involved the co-precipitation method. In this method, an ammonium hydroxide solution was used as an alkali reactant and reacted with nitrate precursors to form metal hydroxide precipitates. Details of all the precursors and reactants used in these studies are presented in Table 3.2. Typically, the mixing of the metal precursors and the alkali solution was conducted by utilizing syringe pumps which allowed the reactants to be kept at a constant flow rate during mixing. The resulting solution of alkali and metal precursors was then heated and stirred throughout the process. After aging, the solution would turn to a deep blue

TABLE 3.2
Synthesis of core catalysts

Preparation technique	Base/promoter/support/dopant	Precursors and reactants	Remarks	Reference
Co-precipitation	$Fe/Cu/Al_2O_3/K$	$Fe(NO_3)_3 \cdot 9H_2O$	Source of Fe	[70–72]
	$Fe/Cu/SiO_2/K$	$Cu(NO_3)_2 \cdot 3H_2O$	Source of Cu	
		K_2SiO_3	Source of SiO_2	
		$Al(NO_3)_3 \cdot 9H_2O$	Source of Al_2O_3	
		K_2CO_3	Source of K	
		NH_4OH	Alkali reactant	

color in all cases, as shown in Figure 3.6a. The presence of the deep blue solution was caused by the formation of the "Cupramine" complex ion, $[Cu(NH_3)_4]^{2+}$, a by-product resulting from the reaction between copper hydroxide and ammonium hydroxide.

By monitoring the deep blue appearance of the Cupramine complex ion solution formed following the precipitation step, it was evident that a large amount of Cu had been dissolved in the solution rather than precipitating out of the solution with the Fe. In order to check the decrease in copper content due to the formation of the Cupramine complex and its possible effect on the reducibility of the Cu-promoted catalyst, a number of temperature-programmed reduction analyses using hydrogen were carried out. Inspecting the temperature programmed reduction (TPR) profile presented in Figure 3.6b, it can be seen the respective reduction peak locations peaks are consistent. The first peak is typically located between 340 °C and 350 °C, and the second peak is located between 560 °C and 580 °C. Generally, the copper content present in the catalyst is identified by the presence of a shoulder before the first reduction peak. The first peak in Figure 3.6b shows no shoulder present, thus indicating some degree of copper loss occurred during the co-precipitation stage of core catalyst preparation, therefore, indicating that some modification was needed to improve the preparation procedure. In the modified procedure, copper nitrate was added to the dried metal hydroxide precipitates. The new procedure was aimed at reducing the precipitated copper during the filtration and washing steps [71].

In order to study the effect of co-precipitation conditions, XRD studies were performed on the core catalysts produced with different precipitation pHs, i.e. FeAlCu-K-01 (pH = 4.75) and FeAlCu-K-04 (pH = 6.98), as shown in Figure 3.7b. Inspection of the XRD pattern reveals the crystalline structure of catalyst core materials is weakly affected by the co-precipitation conditions. In addition, further TPR analysis was performed for each catalyst core material synthesized using the modified procedure. In general, the precipitated iron precursors were converted to magnetite during activation, irrespective of the gas used. When hydrogen is used as the activation gas, α-Fe is the final iron phase, but when the activation gas is CO, Hagg-carbide (χ-Fe_5C_2) is the final iron species [37]. Under FT working conditions, several iron species like α-Fe and Fe_3O_4 can coexist with χ-Fe_5C_2. Furthermore, oxidic iron species are also responsible for the WGS reaction [47]. The results of the TPR analysis are shown in Figure 3.7a. The analysis indicates that the iron oxides reduce to magnetite (Fe_3O_4) and α-Fe, which correspond to the two main peaks, one located between 300 and 500 °C and the other centered on 590 °C. The results of the study suggest the pH at which the precipitate was aged had a significant effect on reduction behavior. For instance, as the pH increased, more copper was precipitate from the solution and resulted in higher levels of Cu in the catalyst, thus indicating that alkaline catalysts are more catalytically favorable in FT reactions than acidic catalysts [73]. In addition, the abovementioned studies have shown Cu is capable of promoting the reduction of iron compounds during pre-treatments. And this most likely results from hydrogen spilling over from the Cu surface to the iron oxide surface. Furthermore, Figure 3.7a shows the

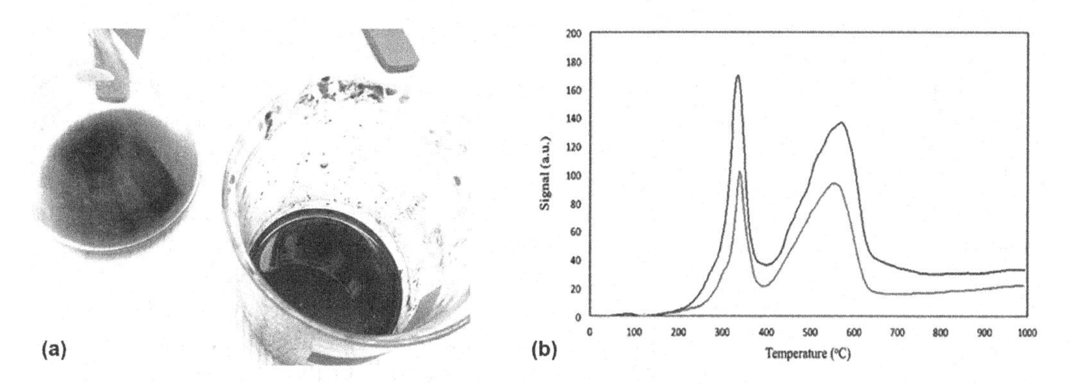

FIGURE 3.6 The formation of blue Cupramine complex ion in the resulting solution and the subsequent H_2 reduction profile of the core catalyst.

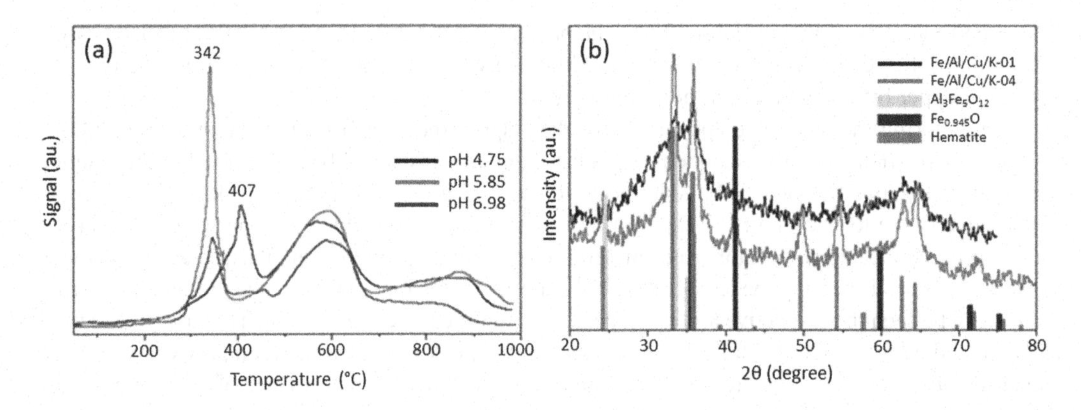

FIGURE 3.7 (a) The influence of pH on core catalysts and (b) the reducibility of XRD profile patterns for core catalysts.

TABLE 3.3
Precursors and reactants used for the synthesis of core-shell catalysts

Preparation technique	Core catalysts	Precursors and reactants	Remarks	Reference
Hydrothermal synthesis	$Fe/Cu/Al_2O_3/K$	Deionized H_2O	Solvent	[67–69]
	$Fe/Cu/SiO_2/K$	C_2H_5OH	Solvent	[74–81]
		TPAOH	Structure template	[67]
		TEOS	Source of silica	
		$Al(NO_3)_3 \cdot 9H_2O$	Source of alumina	

presence of a smaller peak appearing on the shoulder of the first reduction peak of almost all the modified catalysts, thus indicating that copper has been successfully incorporated into the catalysts.

3.4.2 Synthesis and Characterization of Core-Shell Catalysts

In the present work, to coat the zeolite onto a Fe-based catalyst in the form of a core-shell structure that we adapted, and we optimized methods currently available in the literature [67–69, 74–81]. The core-shell catalysts in this study were synthesized *via* a hydrothermal method. All the precursors and reactants used are presented in Table 3.3.

Once all precursors were added, the solution was then thoroughly stirred until a gel-based solution formed. The iron-based catalyst was added as a "core" to the gel solution and mixed. The mixture was then subjected to hydrothermal synthesis which was carried out in a hydrothermal reactor at 180 °C for 48 hours. During this period, the hydrothermal reactor was periodically rotated at a rotational speed of 2 rpm every two hours. At the end of the hydrothermal synthesis period, the product underwent washing, filtration, and drying/heat treatment.

Characterization of the samples was carried out by scanning electron microscopy (SEM), XRD, and TPR. Figure 3.8 presents a selection of SEM images of core-shell structured Fe-based zeolite-coated catalysts produced using the abovementioned procedure. The images shown in Figure 3.8 reveal that zeolites have been successfully grown on the surface of the Fe-based catalyst cores. However, the images also show that Fe-based cores are not completely coated with zeolites. To address this issue an improved procedure was developed.

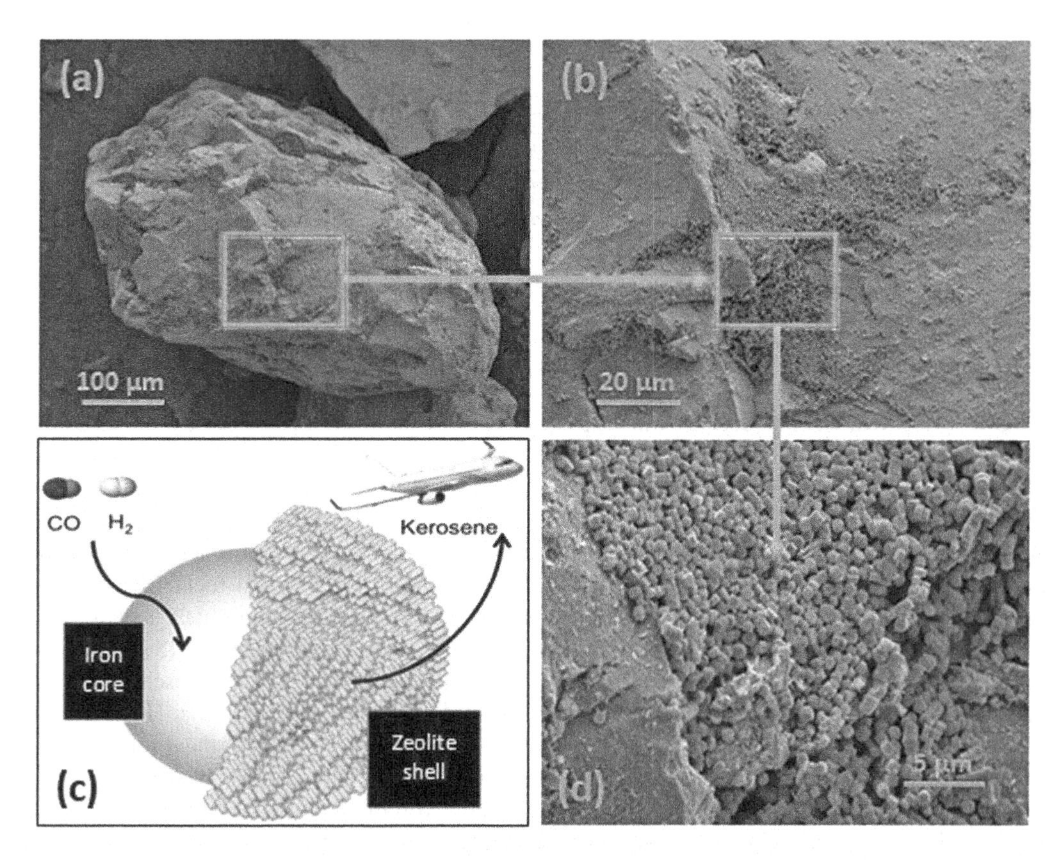

FIGURE 3.8 SEM images (a, b, d) showing the zeolite coating on a Fe-based catalyst (core-shell structure) from low to higher magnification and a schematic diagram of the core-shell structure (c).

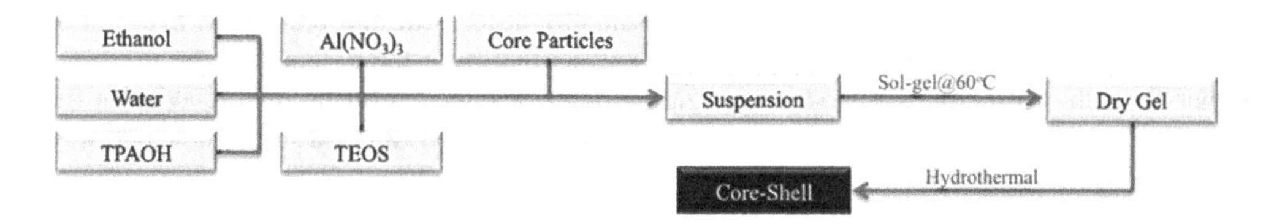

FIGURE 3.9 Schematic of the improved procedure for producing zeolite coating on Fe-based catalysts.

In the improved procedure, the Fe-based core was added at an earlier step in the previous procedure. Thus, instead of being added to the final solution when the gel had been formed, the new procedure added the Fe-based cores to the initial solution containing water, ethanol, and TPAOH. The mixture was then thoroughly mixed for one hour. This was followed by adding TEOS to the mixture and then thoroughly stirring for a further three hours. At the end of this period, the aluminum nitrate was added and thoroughly mixed. The improved procedure was designed to promote a longer contact period between the core and the template agent (TPAOH) and facilitate a more uniform distribution of zeolite coating on the core. The mixture was then subjected to hydrothermal synthesis following the abovementioned procedure. The improved procedure is schematically presented in Figure 3.9.

Both XRD spectroscopy and TPR were used to confirm the presence of the zeolite coating on Fe-based cores, as seen in Figure 3.10. The TPR profiles presented in Figure 3.10b indicate only a slight increase in reduction temperature for the zeolite-coated Fe-based catalysts produced by the improved procedure. Thus, the improved procedure was found to produce core-shell zeolite/Fe-based catalysts with the potential to be used in the FT process.

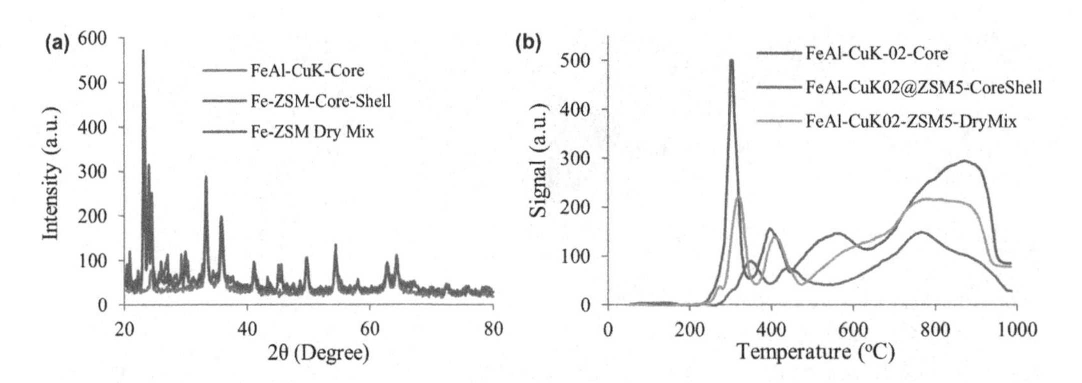

FIGURE 3.10 XRD and TPR data confirm the presence of the zeolite coating on Fe-based catalyst cores.

3.5 CONCLUSION

Because of global warming and ever-increasing levels of greenhouse gas emissions, the aviation industry needs to find an eco-friendly fuel to replace its current use of fossil fuels. Significantly, the strategy of converting CO_2, a major greenhouse gas, into synthetic aviation fuels is considered an important pathway for the aviation industry. While the solar thermochemical (STC) process is capable of catalytically converting CO_2 and H_2O into syngas, the Fischer-Tropsch process is needed to convert syngas into synthetic aviation fuels like kerosene and gasoline. However, the Fischer-Tropsch typically produces diesel-range hydrocarbons. Therefore, additional refining processes like hydrocracking and isomerization are required to convert diesel-range hydrocarbons into kerosene-range hydrocarbons. To undertake these additional refining processes, nanocatalysts are needed. Their nanometer-scale size gives them much higher and distinctive catalytic activities that are unachievable by similar materials at the bulk scale. This chapter has summarized and discussed the reactions and mechanisms involved in the Fischer-Tropsch process and the importance of nanocatalysts. The chapter has also presented a case study that discusses producing core-shell structured Fe-based/zeolite-coated catalysts. The study presented a successful and optimized hydrothermal method for producing Fe-based/zeolite-coated catalysts that have the potential to be used in the Fischer-Tropsch, hydrocracking, and isomerization processes needed for producing aviation fuels. Thus, these nanocatalytic technologies coupled with solar radiation can be a sustainable pathway in the future to create a greener aviation fuel for the benefit of everyone.

REFERENCES

1. ICAO. Annual report 2019. Accessed 22 July 2022. http://icao.int/annual-report-2019/Pages/the-world-of air-transport-in-2109.aspx.
2. Lee DS, Fahey DW, Forster PM, et al. Aviation and global climate change in the 21st century. *Atmos. Environ.* 2009; 43: 3520–3537.
3. BBC. Google 'airbrushes' out emissions from flying. *BBC Reveals*, 2022. Accessed 25 August 2022. https://www.bbc.com/news/science-environment-62664981.
4. Lee DS, Fahey DW, Skowron A, et al. The contribution of global aviation to anthropogenic climate forcing for 2000 to 2018. *Atmos. Environ.* 2012; 244: 2021.
5. Google. Travel impact model, 2022. Accessed 25 August 2022. https://github.com/google/travel-impact-model/background.
6. Karcher B, Burkhardt U, Bier A, et al. The microphysical pathway to contrail formation. *J. Geophys. Res. Atmos.* 2015; 120: 7893–7929.
7. The World Bank. Air transport passengers carried, 2020. Accessed 25 July 2022. https://data.worldbank.org/indicator/IS.AIR.PSGR.
8. Grewe V, Rao AG, Gronstedt T, et al. Evaluating the climate impact of aviation emission scenarios towards the Paris Agreement including COVID-19 effects. *Nat. Commun.* 2021; 12: 3841.

9. The International Civil Aviation Organization. The 2021 global air passenger totals show improvement from 2020, but still only half pre-pandemic levels. Accessed 21 July 2022. https://www.icao.int/Newsroom/Pages/2021-global-air-passenger-totals-show-improvement.aspx.
10. FedEx Express. FedEx annual reports, 2022. Accessed 22 July 2022. https://investors.fedex.com/financial-information/annual-reports/default.aspx.
11. United Parcel Service (UPS). UPS annual reports, 2022. Accessed 22 July 2022. https://investors.ups.com/company-profile/annual-reports.
12. Deutsche Post DHL Group. DHL annual reports, 2022. Accessed 22 July 2022. https://reporting-hub.dpdhl.com/downloads/2021/4/en/DPDHL-2021-Annual-Report.pdf.
13. Lee DS, Pitari G, Grewe V, et al. Transport impacts on atmosphere and climate: Aviation. *Atmos. Environ.* 2010; 44: 4678–4734.
14. Skeie RB, Fuglestvedt J, Berntsen T, et al. Global temperature change from the transport sectors: Historical development and future scenarios. *Atmos. Environ.* 2009; 43: 6260–6270.
15. United Nations Climate Change Conference (COP21). The Paris agreement, 2015. https://www.un.org/en/ climatechange/paris-agreement.
16. Energy Policy Institute at the University of Chicago. http://epic.uchicago.edu/area-of focus/energy-markets/fossil-fuels.
17. Wu J, Zhou XD. Catalytic conversion of CO_2 to value-added fuels: Current status, challenges, and future directions. *Chin. J. Catal.* 2016; 37: 999–1015.
18. Brundavanam S, Poinern GEJ, Fawcett D. Electrochemical synthesis of micrometre amorphous calcium phosphate tubes and their transformation to hydroxyapatite tubes. *Int. J. Sci.* 2015; 4: 7–15.
19. Carrillo RJ, Scheffe JR. Advances and trends in redox materials for solar thermochemical fuel production. *Solar Energy.* 2017; 156: 3–20.
20. Zoller S, Koepf E, Nizamian D, et al. A solar tower fuel plant for the thermochemical production of kerosene from CO_2 and H_2O. *Joule.* 2022; 6: 1606–1616.
21. Steinfeld A. Solar thermochemical production of hydrogen - A review. *Solar Energy.* 2005; 78(5): 603–615.
22. Fischer F, Tropsch H. Uber die Herstellung synthetischer ¨olgemische (Synthol) durch Aufbau aus Kohlenoxyd und Wasserstoff, Brennst. *Chem.* 1923; 4: 276–285.
23. Van der Laan. Kinetics, Selectivity and Scale Up of the Fischer-Tropsch Synthesis. PhD Thesis, Rijksuniversiteit Groningen, 1999.
24. Davis BH. Fischer-Tropsch synthesis: Current mechanism and futuristic needs. *Fuel Process. Technol.* 2011; 71: 157–166.
25. Koide S, Komatsu Y, Shibuya M. Kerosene composition. International Patent. World International Property Organization, 2004.
26. Wang W, WangSP, Ma XB, et al. Recent advances in catalytic hydrogenation of carbon dioxide. *Chem. Soc. Rev.* 2011; 40(7): 3703–3727.
27. Dorner RW, Hardy DR, Williams FW. Influence of gas feed composition and pressure on the catalytic conversion of CO_2 to hydrocarbons using a traditional cobalt-based Fischer-Tropsch catalyst. *Energy Fuels.* 2009; 23: 4190–4195.
28. Landau MV, Vidruk R, Herskowitz M. Sustainable production of green feed from carbon dioxide and hydrogen. *ChemSusChem.* 2014; 7: 785–794.
29. Steynberg A, Dry ME. *Studies in Surface Science and Catalysis: Fischer-Tropsch Technology*, Vol. 152. Elsevier, 2004.
30. Brendelberger S, Sattler C. Concept analysis of an indirect particle-based redox process for solar-driven H_2O/CO_2 splitting. *Solar Energy.* 2015; 113: 158–170.
31. Romero M, Steinfeld A. Concentrating solar thermal power and thermochemical fuels. *Energy Environ. Sci.* 2012; 5(11): 9234–9245.
32. Maitlis PM, de Klerk A. *Greener Fischer-Tropsch Processes for Fuels and Feedstocks.* Wiley-VCH Verlag GmbH & Co. KGaA, 2013.
33. Ayabe S, Omoto H, Utaka T, et al. Catalytic autothermal reforming of methane and propane over supported metal catalysts. *Appl. Catal. A Gen.* 2003; 241: 261–269.
34. Luque, R, Speight JG. *Woodhead Publishing Series in Energy: Gasification for Synthetic Fuel Production - Fundamentals, Processes, and Applications.* Elsevier, 2015.
35. Osazuwa OU, Cheng CK. Catalytic conversion of methane and carbon dioxide (greenhouse gases) into syngas over samarium-cobalt-trioxides perovskite catalyst. *J. Clean. Prod.* 2017; 148: 202–211.
36. Anderson RB. *Catalysts for the Fischer-Tropsch Synthesis*, Vol. 4. Van Nostrand Reinhold, New York, 1956.

37. van de Loosdrecht FGJ, Botes IM, Ciobica A, et al. *Comprehensive Inorganic Chemistry II.* Elsevier, Amsterdam, 2013; 7: 525–557.
38. Rostrup-Nielsen JR. Fuels and energy for the future: The role of catalysis. *Catal. Rev.* 2004; 46(3–4): 247–270.
39. Davis BH. Fischer-Tropsch synthesis: Overview of reactor development and future potentialities. *Top. Catal.* 2005; 32(3–4): 143–168.
40. Khodakov AY, Chu W, Fongarland P. Advances in the development of novel Cobalt Fischer-Tropsch synthesis of long chain hydrocarbons and clean fuels. *Chem. Rev.* 2007; 107: 1692–1744.
41. Hao X, Dong G, Yang Y, et al. Coal-to-liquid (CTL): Commercialization prospects in China. *Chem. Eng. Technol.* 2007; 30: 1157–1165.
42. Davis BH. Fischer-Tropsch synthesis: Comparison of performances of iron and cobalt catalysts. *Ind. Eng. Chem. Res.* 2007; 46: 8938–8945.
43. Dry ME. The Fischer-Tropsch synthesis processes. In *Handbook of Heterogeneous Catalysis* 2008; 6: 2965–2944.
44. Van Steen E, Claeys M. Fischer-Tropsch catalysts for the biomass-to-liquid (BTL) process. *Chem. Eng. Technol.* 2008; 31: 655–666.
45. De Smit E, Weckhuysen BM. The renaissance of iron-based Fischer-Tropsch synthesis: On the multifaceted catalyst deactivation behavior. *Chem. Soc. Rev.* 2008; 37: 2758–2781.
46. Guettel R, Kunz U, Turek T. Reactors for Fischer-Tropsch synthesis. *Chem. Eng. Technol.* 2008; 31: 746–754.
47. Zhang Q, Kang J, Wang Y. Development of novel catalysts for Fischer-Tropsch synthesis: Tuning the product selectivity. *ChemCatChem.* 2010; 2: 1030–1058.
48. Chaturvedi S, Dave PN, Shah NK. Applications of nano-catalyst in new era. *J. Saudi Chem. Soc.* 2012; 16: 307–325.
49. Singh SB, Tandon PK. Catalysis: A brief review on nano-catalyst. *J. Energy Chem. Eng.* 2014; 2(3): 106–115.
50. Weatherbee GD, Bartholomew CH. Hydrogenation of CO_2 on group VIII metals. *J. Catal.* 1984; 87: 352–362.
51. Vannice MA. The catalytic synthesis of hydrocarbons from H_2/CO mixtures over the group VIII metals. *J. Catal.* 1975; 37: 462–473.
52. Jager B, Espinoza R. Advances in low temperature Fischer-Tropsch synthesis. *Catal. Today.* 1995; 23: 17–28.
53. Dorner RW, Hardy DR, Williams FW, et al. Heterogeneous catalytic CO_2 conversion to value-added hydrocarbons. *Energy Environ. Sci.* 2010; 3: 884–890.
54. Davis BH, Jacobs G, Das TK, et al. Fischer-Tropsch synthesis: Support, loading, and promotor effects on the reducibility of cobalt catalysts. *Appl. Catal. A Gen.* 2002; 233: 263–281.
55. Prasad PSS, Bae JW, Jun KW, et al. Fischer-Tropsch synthesis by carbon dioxide hydrogenation on Fe-based catalysts. *Catal. Surv. Asia.* 2008; 12: 170–183.
56. Dorner RW, Hardy DR, Williams FW, et al. K and Mn doped iron-based CO_2 hydrogenation catalysts: Detection of KAIH4 as part of the catalyst's active phase. *Appl. Catal.* 2009; 373: 112–121.
57. Schulz H. Short history and present trends of Fischer-Tropsch synthesis. *Appl. Catal. A Gen.* 1999; 186: 3–12.
58. Yoneyama Y, He J, Morii Y, et al. Direct synthesis of iso-paraffin by modified Fischer-Tropsch synthesis using hybrid catalyst of iron catalyst and zeolite. Catal. *Today* 2005; 104: 37–40.
59. Anderson RB. *The Fischer-Tropsch Synthesis.* New York, 1984.
60. Fan L, Yoshii K, Yan S, et al. Supercritical-phase process for selective synthesis of wax from syngas: Catalyst and process development. *Catal. Today.* 1997; 36(3): 295–304.
61. Li X, Asami K, Luo M, et al. Direct synthesis of middle iso-paraffins from synthesis gas. *Catal. Today.* 2003; 84: 59–65.
62. Bukur DB, Lang X, Ding Y. Pretreatment effect studies with a precipitated iron Fischer-Tropsch catalyst in a slurry reactor. *Appl. Catal. A Gen.* 1999; 186: 255–275.
63. Li S, Krishnamoorty S, Li A, et al. Promoted iron-based catalysts for the Fischer-Tropsch synthesis: Design, synthesis, site densities, and catalytic properties. *J. Catal.* 2002; 206: 202–217.
64. Iglesia E. Design, synthesis, and use of cobalt-based Fischer-Tropsch synthesis catalysts. *Appl. Catal. A Gen.* 1997; 161(1–2): 59–78.
65. Chen YW, Tang HT, Goodwin JG. Effect of preparation methods on the catalytic properties of zeolite-supported ruthenium in the Fischer-Tropsch synthesis. *J. Catal.* 1983; 83: 415–427.

66. Stencel JM, Rao VUS, Diehl RJ, et al. Dual cobalt speciation in CoZSM-5 catalysts. *J. Catal.* 1983; 84: 109–118.
67. Bao J, He J, Zhang Y, et al. A core/shell catalyst produces a spatially confined effect and shape selectivity in a consecutive reaction. *Angew. Chem.* 2008; 120: 359–362.
68. Yang G, He J, Yoneyama Y, et al. Preparation, characterization and reaction performance of H-ZSM-5/cobalt/silica capsule catalysts with different sizes for direct synthesis of isoparaffins. *Appl. Catal. A Gen.* 2007; 329: 99–105.
69. Bao J, Yang G, Okada C, et al. H-type zeolite coated iron-based multiple-functional catalyst for direct synthesis of middle isoparaffins from syngas. *Appl. Catal. A Gen.* 2011; 394: 195–200.
70. Bukur DB, Lang X, Rossin JA, et al. Activation studies with a promoted precipitated iron Fischer-Tropsch catalyst. *Ind. Eng. Chem. Res.* 1989; 28: 1130–1140.
71. Bukur DB, Lang X, Mukesh D, et al. Binder/support effects on the activity and selectivity of iron catalysts in the Fischer-Tropsch synthesis. *Ind. Eng. Chem. Res.* 1990; 29: 1588–1599.
72. Bukur DB, Ma WP, Carreto-Vasquez V. Attrition studies with precipitated iron Fischer-Tropsch catalysts under reaction conditions. *Top. Catal.* 2005; 32(3–4): 135–141.
73. Akbarzadeh O, Mohd Zabidi NA, Hamizi NA, et al. Effect of pH, acid and thermal treatment conditions on Co/CNT catalyst performance in Fischer-Tropsch reaction. *Symmetry.* 2019; 11(50): 1–19.
74. He J, Yoneyama Y, Xu B, et al. Designing a capsule catalyst and its application for direct synthesis of middle isoparaffins. *Langmuir.* 2005; 21: 1699–1702.
75. He J, Liu Z, Yoneyama Y, et al. Multiple-functional capsule catalysts: A tailor-made confined reaction environment for the direct synthesis of middle isoparaffins from syngas. *Chem. A Eur. J.* 2006; 12: 8296–8304.
76. Li Y, Wang T, Wu C, et al. Gasoline-range hydrocarbons over cobalt-based Fischer-Tropsch catalysts supported on SiO_2/HZSM-5. *Energy Fuels.* 2008; 22: 1897–1901.
77. Li X, He J, Meng M, et al. One-step synthesis of H-β zeolite-enwrapped Co/Al_2O_3 Fischer-Tropsch catalyst with high spatial selectivity. *J. Catal.* 2009; 265: 26–34.
78. Li C, Xu H, Kido Y, et al. A capsule catalyst with a zeolite membrane prepared by direct liquid membrane crystallization. *ChemSusChem.* 2012; 5: 862–866.
79. Yang G, Thongkam M, Vitidsant T, et al. A double-shell capsule catalyst with core-shell-like structure for one step exactly controlled synthesis of dimethyl ether from CO_2 containing syngas. *Catal. Today.* 2011; 171: 229–235.
80. Yoneyama Y, He J, Morii Y, et al. Direct synthesis of iso-paraffin using hybrid catalyst of iron and zeolite. *Catal. Today.* 2005; 104: 37–40.
81. Yoneyama Y, San X, Iwai T, et al. One-step synthesis of isoparaffin from synthesis gas using hybrid catalyst with supercritical butane. *Energy Fuels.* 2008; 22(5): 2873–2876.

4 Inorganic Membranes for Gas Separation and High-Temperature Solid Oxide Cells for Producing Synthetic Fuels and Electrical Power Generation

A Review

Huu Dang, Triana Wulandari, A.F.M. Fahad Halim, Derek Fawcett, and Gérrard Eddy Jai Poinern

CONTENTS

DOI: 10.1201/9781003181422-5

4.1 INTRODUCTION

Global challenges facing the world today include escalating energy demands, increasing greenhouse gas emissions, global warming, and environmental degradation. Importantly, the voracious global demand for energy has never been satisfied and continues to intensify. Current consumption relies heavily on fossil fuel resources to supply around 86% of the energy needed to power the world economy. Furthermore, over the next 20 years, the global population is expected to increase by 1.5 billion, resulting in a total global population of around 8.8 billion people [1]. This increasing global population will also drive an annual increase in energy demand of around 1.4%, with most of this expected to meet the increasing demand for electrical power generation [2]. Because of this heavy reliance on fossil fuels, large quantities of harmful greenhouse gases are discharged daily into the environment. Because of the detrimental impact of accumulating greenhouse gases, recent studies have highlighted the need to limit global temperature rises (up to 2 °C compared to pre-industrial levels) [3, 4]. Thus, recent research efforts have focused on developing alternative, clean, and viable renewable energy sources. At present, renewable energy sources are only minor contributors to the global energy mix. For instance, solar energy only contributes around 0.33% to the current total energy mix. Current renewable energy source modeling data indicates an annual growth rate of around 6.6% for the next 20 years, which will ultimately lead to an annual contribution of 9% to the total global energy mix [2]. Accordingly, increasing the contribution of renewable energy sources to the energy mix is an important global objective. On the other hand, increasing the contribution of renewable energy sources, like solar and wind power, to the energy mix also presents challenges such as electrical load distribution and load management. This is because both solar and wind power are intermittent by nature and may not be available when needed. This is also true for renewable energy sources such as hydropower and biomass. For example, if a country has a natural landscape cable supporting hydropower facilities, the necessary rainfall needed throughout the year may not always eventuate due to climatic changes. Moreover, biomass has the potential to be a sustainable energy source, but it needs fertile land, and this is in direct competition with global food production. Therefore, there need to be economically viable energy-conversion and storage systems developed to fully utilize intermittent renewable energy sources.

In addition to developing viable energy-conversion and storage systems, considerable global interest has also been expressed in developing viable carbon capture and sequestration technologies to mediate the detrimental impact of carbon dioxide (CO_2). Not only is CO_2 a problematic greenhouse gas, but its presence during natural gas liquefaction can cause pipeline plugging and reduced plant equipment performance. Consequently, to be a successful carbon capture process, CO_2 must be efficiently removed from either exhaust gas streams or chemical processing streams [5, 6]. Once captured, CO_2 can then be sequestered for storage or utilized. Thus, there is a need to change our philosophical mind-set toward CO_2 from being a global menace to an important renewable resource. Importantly, none of the abovementioned renewable energy sources are suitable fuel sources for aircraft, shipping, and land-based transport like trucks and trains. Therefore, conversion technologies are needed to generate suitable fuels for transportation from energy supplied by renewable and sustainable energy sources. One way to achieve this goal is to separate CO_2 to produce value-added products like carbon monoxide (CO), oxygen (O_2), and hydrogen (H_2) [7]. Then from these gases, mixtures of H_2 and CO can be used to produce a variety of carbon-based fuels like methane (CH_4), synthetic natural gas (SNG), and methanol (CH_3OH), while the pure O_2 generated can be directly used. Importantly, O_2 is one of the five most produced and consumed industrial chemicals in the world [8]. In particular, O_2 is extensively used in large-scale metallurgical-metal manufacturing, chemical industries, and pharmaceuticals. While on a smaller scale, O_2 is used in medical applications, welding, and wastewater treatment. Global estimates indicate that hundreds of millions of tons of O_2 are consumed annually, and the demand is increasing [9, 10]. O_2 is currently supplied by two separating technologies, namely, cryogenic distillation and non-cryogenic distillation. The more expensive, energy intensive, and technically complicated cryogenic distillation process is preferred when large quantities (~100 tons per day) are required [11]. The liquefaction of air down to extremely low temperatures (~ –185 °C) and subsequent separation of constituent molecular/atomic components can produce high-purity (> 99 % vol.) O_2 streams [12, 13]. Non-cryogenic distillation processes are generally employed when smaller O_2 quantities are required. Non-cryogenic processes consist of pressure swing adsorption (PSA) and air separation using membranes. In PSA, sorbents forming a molecular sieve (typically composed of zeolites) are exposed to high-pressure air streams. The sorbent at ambient temperatures adsorbs nitrogen (N_2) from the air stream, resulting in an enriched O_2 stream. When exhausted, the molecular sieves are regenerated by reducing air pressure. The reduced air pressure decreases N_2 equilibrium over the surface of the sorbent, which in turn releases nitrogen into the atmosphere. Continuous production requires parallel processing, with each side of the production line alternatingly from adsorption or desorption mode. While both cryogenic distillation and PSA systems are currently in commercial operation, they tend to be energy intensive and operationally expensive to maintain.

Membrane separation technologies are an attractive alternative since they are considered simpler systems, versatile, eco-friendly, and have lower energy demands compared to both cryogenic distillation and PSA [14]. In addition, membrane-based separation technologies offer the possibility of continuous operation and lower investment costs [15]. Because of these advantages, membrane separation technologies have been extensively used in chemical, pharmaceutical, food processing, and biotechnological industries [16]. Importantly, membranes have the potential to be used as the principal component in conversion technologies designed to produce very pure gases from gas mixtures. Then the pure gases can be directly used or further processed to produce high-value chemicals or fuels. Fundamentally, a membrane is a selectively permeable barrier that allows the flow of desirable substances, under the influence of a driving force, to pass through it. At the same time, it prevents undesirable substances to pass through, as seen in Figure 4.1. The driving force is a consequence of a difference in concentration, electrical potential, or partial pressure [17]. Factors like composition, morphology, or structure give the membrane the ability to separate gases or liquids [18].

Membranes used for purification or separation can be broadly categorized as either biological (not discussed in this review) or artificial (human-made). Artificial membranes can be classified as adapted natural materials or synthetic materials. Synthetic materials can be further classified as either organic or inorganic membranes. Organic polymers of varying molecular weight and varying degrees of cross-linking are currently used to produce a wide variety of membranes. Commonly used polymers include aromatic polyamides, cellulose acetate, and fluorocarbons [19]. The lower material and fabrication costs involved in producing polymeric membranes make them more economically viable than other materials. These two factors make them extremely popular and have resulted in them being extensively used in purification and separation processes [20]. However, polymeric membranes are not suited for several applications, for example, separating O_2 from mixed gas streams, where low purities (~40% vol.) are achieved [13, 21]. In addition to lower separation performance, their efficiency decreases with time due to compaction, plasticization, and thermal instability when exposed to harsh chemical environments [22]. For example, when polymeric membranes are exposed to high-pressure CO_2 environments, they steadily become plasticized [23]. The resulting plasticization process leads to membrane swelling and increased permeability [24]. So, despite their remarkable transport properties and lower costs, their susceptibility to abrasion, chemical attack, pressure vulnerability, and limited thermal stability (maximum operating temperatures less than 250 °C), they face increasing competition from inorganic membrane technologies

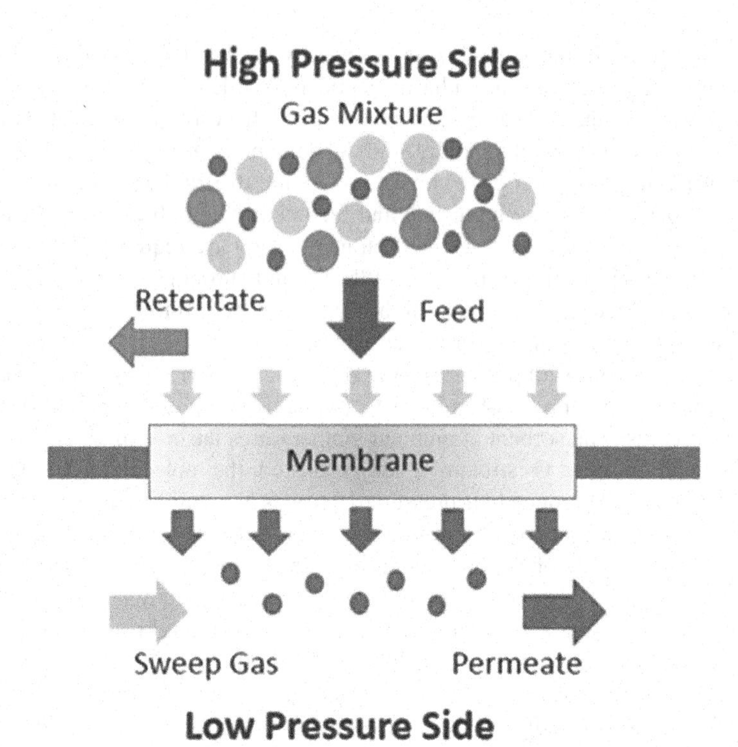

FIGURE 4.1 Schematic illustration of gas flow through a typical membrane.

[25, 26]. The manufacture of inorganic membrane technologies is well established. These alternative membranes can deliver high selectivity and superior permeability. They are also mechanically robust, chemically stable, and capable of operating at high temperatures and pressures [27].

As mentioned above, gas mixtures of H_2 and CO can be used to produce a variety of carbon-based fuels. In fact, pure H_2 gas alone is considered an eco-friendly alternative to conventional fossil fuels and is also judged to be an effective renewable energy carrier. However, producing large quantities of pure H_2 gas using convention gas processing technologies is expensive and energy intensive. Utilizing inorganic membranes to separate H_2 from mixed gas streams is one method for producing large amounts of this valuable gas fuel. Alternatively, a similar inorganic membrane-based technology has been used to produce solid oxide electrolysis cells (SOEC). These cells operate at high temperatures (around 800 °C) and can efficiently electrolyze water to produce $H_{2.}$ Advantageously, the electrolysis process also produces O_2 which can be collected, condensed, and used in other industrial processes. Interestingly, SOEC can operate in reverse mode and become a solid oxide fuel cell (SOFC). In the reverse configuration, H_2 and O_2 (usually air) are supplied, and the fuel cell generates electrical energy. In brief, the advantages of using SOFCs as an alternative energy source include 1) higher efficiencies compared to traditional power generation facilities; 2) they do not consume fossil fuels; 3) very low to negligible CO_2 emissions and reduced air pollution; 4) a variety of fuels sources can be used; and 5) can be considered as a renewable energy source and can be used with renewable energy sources. Importantly, both cell configurations can easily be scaled up to the industrial scale.

This chapter presents an overview of current research into inorganic membranes and solid oxide electrolysis cells for gas separation, and solid oxide fuel cells for generating electrical energy. The review begins by outlining the pore diameter-based method for classifying membranes and then discusses gas flow/diffusion mechanisms involved in the transport of desirable gases through the porous membrane. This is followed by an overview and discussion of the three main types of inorganic microporous membrane materials currently in use. These membrane types include silica, carbon, and zeolite. Then the dense inorganic membranes are discussed, and unlike other membrane types, they rely on solution diffusion for gas separation. Both metallic and ceramic-dense inorganic membranes are discussed at length. Recent research into developing fluorites and perovskite-based ceramic membranes is discussed. The review then examines recent research into high-temperature solid oxide cells for gas separation processes and fuel cell applications. The basic operating principles of solid oxide cells are discussed. In addition, challenges in designing fuel and air (oxygen) electrodes and electrolytes are addressed. Also discussed are issues relating to the degradability and long-term durability of solid oxide cells. Following on the review discusses the high electrical efficiencies and fuel flexibility of SOFCs. Typically, gases like hydrogen (H_2) and oxygen (O_2, usually from air) are supplied, and through the electrochemical oxidation of H_2 (as fuel) electrical energy is generated. The basic operating principles of fuel cells and challenges in designing cell components (anodes, cathodes, and electrolytes) are discussed. The review concludes with a discussion of the challenges and future perspectives. It is believed that future advances in producing new nanomaterials and improving the properties of existing materials, combined with advanced fabrication processes will deliver commercially viable and large-scale manufacturing of inorganic membranes, SOECs, and SOFCs soon.

4.2 CLASSIFYING INORGANIC MEMBRANES

In the past two decades, there has been considerable research into developing new and improving the performance of existing inorganic membrane materials [28, 29]. Although inorganic membranes cost more than polymeric membranes, they offer higher permeability, improved selectivity, and superior resistance to aggressive environments at higher temperatures and pressures [30]. Because of these advantageous properties, there has been extensive research and industrial effort into developing inorganic membranes for a variety of applications. There are two different types of inorganic membranes, with the type being based on its structure being either microporous or dense. Generally, an inorganic membrane assembly consists of an underlining strengthening macroporous structure with an overlying thin separation membrane layer. The overlying membrane layer can be either porous or dense (non-porous) [31]. An important parameter of a porous material is the size of pores present in its structure. Accordingly, the pore size range is used to classify the type of porosity. The IUPAC classification of porosity based on pore size is presented in Table 4.1.

Microporous inorganic membranes, as seen in Table 1, are defined by pore diameters smaller than 2 nm. Membranes with pore sizes around 1 nm and above are commercially available and currently used in a variety of applications such as purification, water treatment, chemical processing, and the medical and food industries [33–36]. Membranes with pore sizes of 0.5 nm and smaller are currently being investigated for future development. Furthermore, microporous membranes can be further classified by having either symmetric or asymmetric properties. Symmetric membranes have a uniform structure, while asymmetric membranes have a structure consisting of varying chemical and physical properties across their thickness [37]. Thus, a wide range of amorphous and

TABLE 4.1
IUPAC classification of porosity based on pore diameter [32]

Classification	Pore diameter range (nm)
Macroporous	$d > 50$
Mesoporous	$2 < d < 50$
Microporous	$d < 2$

crystalline materials that include oxides (alumina, titania, zirconia), metals, carbons, zeolites, silica, and organic-inorganic hybrids like metallic organic frameworks (MOFs) have been investigated and found to offer improved chemical and physical properties that allow these materials to perform under difficult and severe operating conditions [38, 39]. At the same time, the membranes' microporous structure resists the flow of molecules traveling through them [40]. This resistive behavior is quantified by applying a specific gas diffusion mechanism or a combination of different diffusion mechanisms [41, 42]. These gas flow/diffusion mechanisms include: 1) Knudson diffusion, 2) surface diffusion, 3) capillary condensation, 4) molecular diffusion or sieving, and 5) solution diffusion, as schematically presented in Figure 4.2 [43]. Thus, the performance and efficiency of a microporous membrane not only depends on its composition and material properties but also on the type or types of gas diffusion mechanisms occurring during operation.

Dense membranes are primarily used for the selective separation of hydrogen and oxygen. They are mainly made of metals like palladium and its alloys, while silver, nickel, and stabilized zirconia have also been used. In addition, polycrystalline ceramic materials like perovskites have also been used [44, 45]. Both H_2 and O_2 can permeate selectively through various types of dense membranes via the transport of charged particles, while electron transport may also take place in the lattice structure. Studies have shown that Group V metals (vanadium, niobium, and tantalum) have high hydrogen permeability suitable for dense membranes, but they also form oxide layers that can impede hydrogen dissociation and solubility [46, 47]. Because of oxidation issues, non-oxidizing palladium (Pd), with its high hydrogen permeability, is currently being studied as a dense metal membrane [48]. O_2 can permeate selectively through perovskites, thus making them a suitable dense membrane for oxygen separation processes [44, 49]. Importantly, for diffusion to take place in dense membranes, an activation and operating temperature of several hundred degrees Celsius is required. However, at present, the commercial inorganic membrane market is dominated by microporous ceramic membranes.

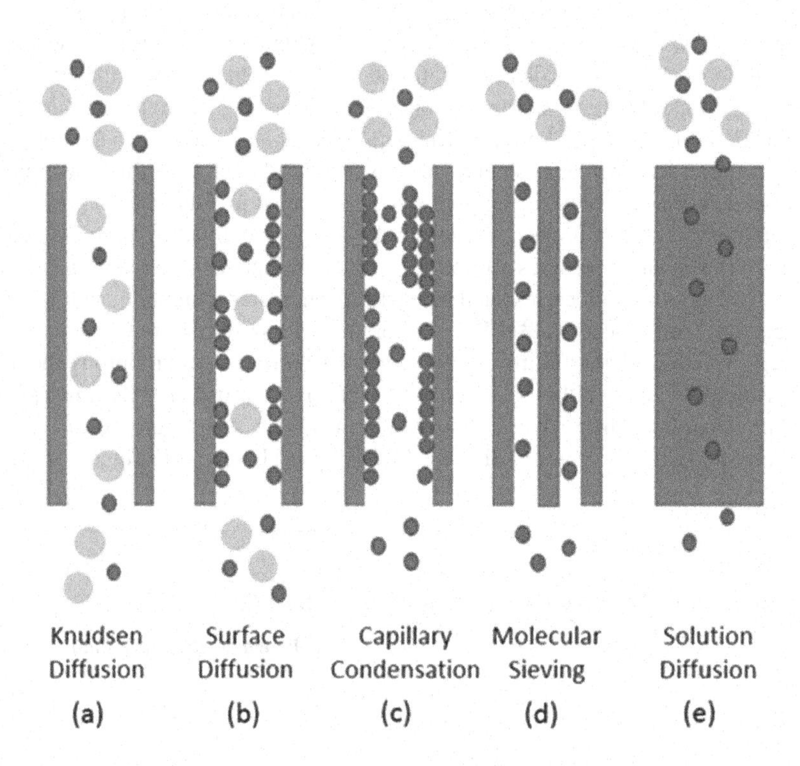

FIGURE 4.2 Schematic illustration of the four diffusion mechanisms through a porous membrane (a → d) and (e) solution diffusion through a dense membrane [43, 44].

4.3 MICROPOROUS INORGANIC MEMBRANES

Gas diffusion methods and porous inorganic membranes were initially used for the separation of uranium isotopes during the 1940s. However, it was not until the early 1980s that inorganic membranes were manufactured on a large scale for industrial applications [50]. In recent years, considerable research effort has been made to develop new and improve existing inorganic membranes to replace conventional polymer membranes for use in aggressive and operationally demanding environments [51]. Currently, inorganic membranes are used globally in microfiltration, separation of liquids, food and dairy processing, gas separation, and filtration processes [52–54]. The main membrane materials used in these applications include oxides, silica, zeolite, and several types of carbon materials. These membranes are usually prepared as a thin microporous film on porous multi-layered inorganic self-supporting structures composed of mesoporous and macroporous layers that provide mechanical strength. The topmost thin microporous film with the finest pores can selectively let liquid water, water vapor, or smaller gas molecules pass through [55]. Thus, the overall transport properties of the multi-layered membrane structure are governed by the properties of the topmost thin layer with the smaller pore sizes. This regulating effect is due to the greater resistance offered to molecular flow by the smaller pore dimensions present in the top layer compared to the lower resistance offered by the larger pores of the underlining layers [56]. Commercial inorganic membranes consist of three configurations: disks, tubes, and multi-channels. Multi-tubular configurations are preferable to plates or disks because the bundle configuration of multi-tube configurations offers higher packing density and greater strength. From an industrial perspective, it is much easier to scale up multi-tube configurations into modules that can be further aligned into series and/or parallel arrangements for great throughput [57]. The remainder of this section focuses on recent advances in the use of silica, zeolites, and carbon materials for the manufacture of microporous membranes designed for separation and filtration processes [58, 59].

4.3.1 Silica Membranes

Silica-based microporous membranes have been extensively studied and commercially available for a few decades. These membranes typically have a three-layer structure consisting of a thin microporous SiO_2 top layer, which can be either crystalline or amorphous, an intermediate mesoporous γ-alumina layer, and a macroporous α-alumina support layer [60]. These membranes can be manufactured using either sol-gel techniques or chemical vapor deposition (CVD) methods [61–63]. Microporous silica membranes have tunable pore sizes that give them favorable molecular sieve characteristics. This property enables these membranes to be used in a variety of liquid separations, per-evaporations, and gas separations over a wide range of temperatures and aggressive chemical environments [64]. In addition, these membranes have been found to be highly selective toward the H_2 and CO_2 present in methane gas mixtures, which makes them particularly useful for the separation and purification of methane [44, 65]. These membranes have also been evaluated for CO_2 capture and sequestration [66]. However, improving membrane performance (permeation and selectivity) and stability remain major challenges. A major challenge facing these membranes is to improve their stability by overcoming their water sorption sensitivity that results from their hydrophilic nature [67, 68]. In practice, the sorption of water vapor from gas mixtures produces pore blocking that results in a serious reduction in separation performance. Industrial separation processes involving high temperatures and steam promote the condensation of silanol groups in the silica structure, which results in the structure becoming brittle and losing its permeability [69]. Because of this problem, recent research has focused on developing hydrophobic and hydrothermally stable silica membranes. Several studies have revealed the addition of methyl groups into microporous silica, making it more hydrophobic and enhancing its operational lifetime [70, 71]. Similar studies have shown the addition of small quantities of alumina, zirconia, and titania can also improve the long-term stability of silica membranes exposed to high-humidity environments

[72–74]. Another option that has been adopted by several researchers is to add small quantities of metal and metal oxide nanoparticles into the silica matrix during synthesis. Their studies suggest the addition of nanoparticles into the silica matrix reduces thermally induced molecular motion of silica networks within the microporous structure when exposed to wet gas separation processes at elevated temperatures [75, 76]. For instance, cobalt oxide has been added to silica membranes to improve hydrothermal stability and enhance both H_2 and CO_2 separation from gas mixtures [77, 78]. In addition, nickel oxide-doped silica membranes have shown good thermal stability when used to separate H_2 from gas mixtures at elevated temperatures (~500 °C) [79, 80], while the addition of combinations of Iron (Fe) and Cobalt (Co) oxide to silica membranes during synthesis have shown selective separation properties between some gases. For example, the permeability of smaller gas molecules such as hydrogen and helium increased with temperature (max. 500 °C), while the permeability decreased for larger gas molecules like N_2 and CO_2 [81].

4.3.2 Carbon Membranes

Carbon membranes are generated from the pyrolysis of polymeric precursors at elevated temperatures (typically between 500 and 900 °C) in a closely controlled atmosphere. Carbonization under these conditions causes the decomposition of polymeric chains in the precursor material (e.g., polyimide, polyfurfuryl alcohol, polyvinylidene chloride, or phenolic resin) to form an amorphous carbon skeleton containing a vast array of interconnected pores [82, 83]. The pores produced have relatively wide diameters (similar in size to molecular dimensions) with narrow constrictive channels that hinder the passage of larger molecules. These microporous carbon membranes are classified as either activated carbons or ultra-microporous carbons. Activated carbon has pore sizes ranging from 0.8 to 2 nm, and ultra-microporous carbons have pore sizes ranging from 0.3 to 0.5 nm [84]. The small pore size range of ultra-microporous carbon membranes is responsible for their selective molecular sieving properties [85]. Compared to polymeric membranes, carbon membranes are more expensive to fabricate and tend to be brittle [86]. However, careful selection of precursors and optimizing fabrication processes can alleviate material property problems. Despite this disadvantage, carbon membranes are used in large-scale applications for the separation of hydrogen from multi-component hydrocarbon mixtures and for the purification of methane. Carbon membranes are currently used in the large-scale production of low-cost and high-purity nitrogen from the air [87]. Hollow carbon fiber membranes are also used to separate a variety of industrial gases from gas mixture feedstock [88]. In recent years, both graphene and graphene oxide films have been investigated for possible use as separation membranes. Both are extremely thin, consisting of a single layer of carbon atoms, and both are impermeable to all gases and liquids [89]. However, selective permeability can be achieved by either etching an extremely small opening into the layer or by stacking a few layers to form lateral nano-channels. These nano-channels would give the graphene/graphene oxide-based membranes molecular sieve properties, thus allowing small molecules like He, H_2, and H_2O to traverse but preventing the passage of larger molecules [59, 90].

4.3.3 Zeolite Membranes

Zeolites are crystalline metal oxides composed of aluminum, silicon, and oxygen which form a uniform crystal structure. Their structure consists of tetrahedral orientations composed of SiO_4 and AlO_4. The tetrahedral structures then assemble to form secondary polyhedral building units like cubes, hexagonal prisms, and octahedral and truncated octahedral configurations [91, 92]. These building units then form uniform three-dimensional ring crystalline frameworks. The rings form the micro-pores with molecular dimensions. Studies have shown that both pore size and molecular affinity can be modified *via* ion exchange with cations [93, 94]. Thus, the advantage of zeolite membranes is that their properties can be designed for applications [95]. Pore size can be modified to produce three sizes, small (0.3 to 0.4 nm), medium (0.5 to 0.6 nm), and large (0.7 to 0.8 nm). These pore size ranges are extremely narrow and reflect separation mechanisms associated with zeolite membranes. The separation mechanisms include adsorption-controlled permeation, diffusion-controlled permeation, and molecular sieving [96]. Thus, zeolite membranes offer good separation properties for gas molecules, and they are used in vapor separation applications [97, 98]. However, the presence of any microscopic or post-synthesis macroscopic defects in the microporous structure can have a detrimental impact on the selectivity of the membrane [99]. Recent studies have shown that several methods can be applied to improve performance, minimize defects, and reduce membrane manufacturing costs. For example, defects can be significantly reduced by a rapid calcination procedure that promotes the condensation of surface hydroxyl groups at zeolite grain boundaries [100]. Post-treatments like counter-diffusion chemical vapor diffusion, filling non-zeolite pores using wet impregnation, and hydrolysis of organics and silica precursors can significantly improve overall membrane performance [101–103]. However, these post-treatments can significantly add to the costs of membrane manufacturing [104].

4.4 DENSE INORGANIC MEMBRANES

4.4.1 Introduction

Dense inorganic membranes are made of metallic materials or polycrystalline ceramics that are gas-tight and only allow certain gas species to permeate through their crystal lattices. For the membrane to be effective in purifying and producing specific gas species from suitable feedstock, it must have superior permeability and selectivity performance, as well as be practical for the specific application. However, it must be pointed out that both membrane permeability and selectivity performance can vary significantly under fluctuating operational conditions such as a changing gas composition of the feedstock, as well as varying humidity, pressure, and temperature [105, 106]. Factors like membrane fabrication, production costs, robustness, and durability are important and must be considered when the membrane technology is being applied to a specific industrial application [107]. For example, dense metallic membranes can deliver very high-purity hydrogen (H_2) because of their selectivity toward H_2. This selectivity results from the membrane's dense structure which prevents the passage of larger molecules like CO, CO_2, and N_2 [108]. Ceramic-based membranes composed of fluorite and perovskite structures have also been extensively studied for oxygen (O_2) separation from various gas mixtures [109]. The following subsections provide a summary of recent literature and summarize the current state of research in the field of dense inorganic membranes.

4.4.2 Dense Metallic Membrane Technology

The effect of absorption and desorption of metal-hydrogen systems like metallic hydrides has attracted considerable multidisciplinary interest due to their electrical, optical, and material properties, and in particular their hydrogen storage capability [110–112]. Metal-hydrogen systems using dense metal membranes to generate high-purity H_2 gas from hydrocarbon sources have attracted considerable scientific and industrial interest in recent years [113]. Dense metal membranes are made from a range of metals in the element groups III to V. Of the several metals present in these groups, the most frequently studied are palladium (Pd), nickel (Ni), and platinum (Pt), and their alloys [114, 115]. Tantalum, niobium, and vanadium, and their respective alloys have also been studied as a cheaper alternative to the more expensive palladium and palladium-based alloys. An important characteristic of these membrane materials is their very high hydrogen selectivity, which enables them to produce ultra-pure hydrogen [116]. The high selectivity is due to the dense structure of metals and metal alloys that blocks the passage of larger gas molecules. Because of this dense and non-porous structure, hydrogen permeability is relatively low compared to other types of membrane materials [117]. The hydrogen separation process is believed to be a solution-diffusion mechanism that consists of 1) feed gas (gas mixture containing H_2 and unwanted gases) being delivered to the inlet side of the membrane; 2) dissociative chemisorption of H_2 into hydrogen ions (H^+) and electrons (e^-) at the surface of the membrane; adsorption of H^+ ions into the membrane structure; 4) diffusion of H^+ ions and electrons through the membrane (diffusion through the structure is driven by the partial pressure drop to the outlet side of the membrane); 5) desorption of H^+ ions at membrane surface; 6) re-association of the H^+ ions and electrons into discrete molecular H_2; and finally 7) the diffusion of the H_2 from membrane surface [118–120].

Importantly, the hydrogen flux through the membrane is dominated by the diffusion rate. Therefore, to achieve higher and more advantageous permeation rates, an ultrathin metal membrane incorporating an integrated porous substrate support is used [121, 122]. In addition, the separation process is also heavily dependent on permeability, which is influenced by the purity of the metal's surface, its surface roughness, and the metal's lattice structure [123]. For instance, permeation can be reduced by the formation of surface oxides on metals like niobium and vanadium. Oxide formations can significantly reduce H_2 molecule dissociation, dissolution, and subsequently hydrogen absorption by the underlining lattice structure. Furthermore, the presence of chemical contaminants in feed gases can also inhibit hydrogen dissociation, recombination, and permeation [124]. For example, one of the most studied dense metal membrane materials, palladium, is susceptible to harmful contamination by carbon monoxide, hydrocarbons, and sulfur compounds that form a variety of palladium-based sulfur and surface carbon compounds [125]. Membranes can be severely damaged by the generation of large structural expansions in the lattice when palladium-based sulfur compounds form [126, 127]. Further challenges facing dense metal membranes are the formation of hydrides and hydrogen embrittlement [128]. Hydrogen embrittlement results from changes in the chemical structure of the lattice that results in dimensional changes. These changes lead to volumetric expansion and result in increases in internal stresses throughout the lattice structure and ultimately lead to material failure and the selectivity loss of the membrane [129]. For example, in the case of palladium, hydrogen embrittlement occurs because of a phase transition (α to β), when membranes are operating below 300 °C and pressures lower than 2 MPa in hydrogen environments [125, 128].

To date, the most effective method for improving a pure metal's physical properties such as corrosion resistance, durability, and strength is by alloying [130]. Alloying is a very well-established method and involves adding metallic elements to the base metal to improve its physical properties and enhance its performance and stability [115, 131]. Studies have shown that there are numerous combinations of metallic alloying elements that can prevent hydride formation and resist hydrogen embrittlement caused by hydride formation pathways [132–134]. Palladium-based alloys have displayed high selectivity toward hydrogen and display improved mechanical properties and membrane stability [135]. For example, the addition of silver (Ag) not only improves mechanical strength and stability in the presence of hydrogen compared to pure palladium, but it can also enhance permeability (up to five times), and suppress hydrogen embrittlement [117, 136]. However, despite Pd-Ag being the most used alloy for hydrogen purification and separation applications, they are still sensitive to sulfur and chloride poisoning [115]. Two other thoroughly investigated palladium-based alloys for hydrogen separation membranes are palladium-copper (Pd-Cu) and palladium-gold (Pd-Au). The addition of Cu to produce the Pd-Cu alloy is believed to suppress hydrogen embrittlement and reduce poisoning by chemical contaminants like CO, CH_4, and sulfur [137, 138]. Importantly, Cu alloying significantly reduces the cost of the membrane compared to pure Pd membranes [139]. Unfortunately, studies have also shown that the presence of Cu can significantly reduce hydrogen permeation through the membrane [140]. However, unlike Pd-Cu alloys, the addition of Au to produce Pd-Au alloys has not only been found to suppress contaminant poisoning, but also increase hydrogen permeation by around 30% [141]. Further details of dense metal membranes can be found elsewhere in the literature [47, 114–116].

4.4.3 Dense Ceramic Membrane Technologies

In recent years considerable scientific and industrial interest has focused on ionic transport membranes, also known as dense ceramic membranes, for gas separation and carbon capture and storage

[142, 143]. In principle, these dense membranes or oxygen transport membranes (OTM) are based on solid oxide materials that can conduct gases like oxygen *via* ion diffusion through their crystal lattice. The process includes oxygen reduction and oxidation steps. Thus, to undertake these steps the design of oxygen transport membrane materials is critical. Considerable research has gone into the design of solid oxides from primary materials like fluorites and perovskites [144, 145]. Perovskites are a very attractive material because they not only conduct oxygen ions but also conduct electrons [146]. These materials are often referred to as mixed (oxygen) ionic-electronic conductors (MIECs) and are candidate materials for OTMs. Typical applications currently under study for these materials include pure oxygen separation, purification of hydrogen, and incorporation into power generation systems for CO_2 capture *via* oxy-fuel combustion enhancement [147, 148].

4.4.3.1 Fluorites in Electrical-Driven Membrane Technologies

Oxide compounds with a fluorite structure are pure ion-conducting (electrolyte) materials. However, they lack the electronic conductivity necessary for oxygen production and therefore need to be electrically driven by an external current source at elevated temperatures. Operationally, an electric potential is applied across the separating membrane. On the cathode side of the cell, a gas mixture containing oxygen is introduced. O_2 molecules in the gas mixture are reduced to oxygen ions at the electrolyte surface and then migrate through the electrolyte layer. Arriving at the anode (permeate) side, oxygen ions undergo electrochemical oxidation to recombine to form O_2 gas and release electrons. The released electrons then travel back to the cathode side to make up for electrons consumed during oxygen reduction and thus, achieving a complete electric circuit [149]. Since an external electric potential is applied across the cell, its operation is like a solid oxide electrolysis cell. Consequently, similar oxide compounds are used in both types of electrically driven cells [150]. Fluorite compounds were among the earliest oxygen-ion conductors investigated for potential use as a solid electrolyte in solid oxide fuel cells (SOFCs) [150]. Several different types of fluorite oxides, each with different dopants have been investigated for electrically driven oxygen separation cells over the past two decades [151, 152]. Typically, fluorite compounds have a general formula of AO_2, where the A cations form a cubic close packing and occupy the face-centered cubic positions, while the O anions occupy the tetrahedral interstices [153]. For example, ceria (CeO_2) ceramics with the typical fluorite structure, both in the pure form or acceptor-doped mixtures, have been extensively studied for electrically driven oxygen separation processes [154], while zirconia (ZrO_2)-based oxides operating at temperatures above 700 °C have been used as solid electrolytes in SOFCs. However, fluorite compounds like zirconia display phase transitions at elevated temperatures, hence the need for dopants to chemically stabilize their structure [155, 156]. Thus, the addition of stabilizing dopants in fluorite oxides like gadolinium-doped ceria, samarium-doped ceria, and yttria-stabilized zirconia has ensured their successful operation in SOFCs [157]. Other fluorite compounds like lanthanum (La) and tungsten (W)-based oxides such as $La_{6-x}WO_{12-\delta}$ (LWO) have shown a high degree of chemical stability and high hydrogen permeability during gas separation processes [158, 159].

4.4.3.2 Perovskite's Membrane Technologies

Perovskite compounds have attracted considerable interest due to their interesting properties, such as catalytic activity, high ionic and electronic conductivity, and charge-carrying ability [144, 160]. In particular, the ability of perovskite compounds to conduct both oxygen ions and electrons spontaneously, unlike fluorites, means they do not need external electrical circuits [161]. These compounds are often referred to as mixed ionic-electronic conductors (MIEC) and have been extensively studied in recent years for applications like removing hydrogen or oxygen from gas streams, fuel cells, and membrane reactor applications [162]. The most attractive feature of perovskite compounds is their cubic lattice structure that allows for the incorporation of doping elements. Perovskite compounds are described by the general lattice formula $ABO_{3-\delta}$. The A and B sites are cations, and the O site is possible anions. The A sites are occupied by rare earth elements of the lanthanide series

(La, PR, Nd, etc.) or alkaline elements (Ba, Sr), while B sites are occupied by a transition metal (Mn, Cr, Ti, Co, Fe, Ni) [163]. The δ term represents the nonstoichiometric parameter which is enhanced by the inclusion of metal cations with low valence values. The perovskite cubic structure consists of a face-centered cubic lattice with the larger A cations located at the corners, while the smaller B cations are in the body-centered position and the O anions are in the face-centered positions [164]. Importantly, oxygen vacancies are common in perovskites, and at temperatures greater than 700 °C they readily permit oxygen ions to diffuse through their crystalline structure *via* vacant oxygen sites or lattice defects [165]. Studies have shown that doping both A and B sites with different cations of varying sizes and valances, *via* compositional tailoring, can allow the selection of the level of oxygen non-stoichiometry [166, 167]. Thus, being able to control the level of oxygen non-stoichiometry also means being able to tailor a specific ionic conductivity, electronic conductivity, or catalytic activity for a particular application.

In practice, a bulk pressure gradient across the MIEC membrane is needed to create two different oxygen-level atmospheres. Oxygen permeation *via* ionic and electronic migration through an MIEC membrane consists of five steps: 1) molecular diffusion of oxygen from the high-pressure gas stream onto the membrane surface; 2) adsorption of molecular oxygen at oxygen-surface vacancies, the reaction and reduction of oxygen molecules to O^{2-} ions on membrane surface; 3) bulk diffusion of O^{2-} ions down the pressure gradient while e^- travel through the membrane in the opposite direction; 4) recombination and oxidation of O^{2-} on the membrane surface, then followed by desorption; and 5) diffusion of molecular oxygen from the membrane surface to the oxygen-rich low-pressure gas stream [168, 169]. This process is schematically illustrated in Figure 4.3.

As mentioned above, perovskite materials are inherently catalytic with high ionic and electronic conductivity, which enables them to break down molecular oxygen into oxygen ions at high temperatures. In each of steps 2, 3, and 4 mentioned above, there is a resistance to oxygen permeation through the membrane. In steps 1 and 5, a much smaller resistance occurs and is generally considered

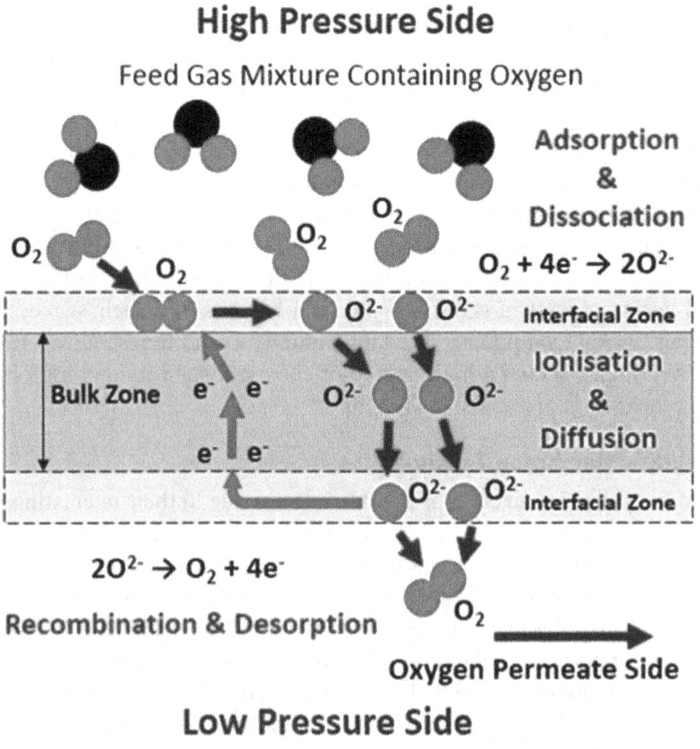

FIGURE 4.3 Pressure-driven oxygen permeation through a MIEC membrane.

negligible [170]. Oxygen transport in steps 2 and 4 is controlled by surface exchange reactions (dissociation/recombination) at the respective interfacial zones. Oxygen-ion transport in step 3 is controlled by bulk diffusion associated with the membrane material and its thickness. Consequently, oxygen permeation through the membrane is limited by either surface exchange reactions or bulk diffusion. Therefore, perovskite composition and geometry become critical factors in determining membrane performance. From a geometric point of view, the thickness can have a direct bearing on oxygen permeation. For relatively thick membranes, bulk diffusion is generally the rate-limiting step. However, thinner membranes have been shown to deliver higher oxygen permeation rates per area of membrane [171]. Despite this, in ion transport membranes, there is an optimal length where further decreases in membrane thickness no longer deliver increases in oxygen permeation. Thus, as the thickness is reduced, surface exchange reactions become more prevalent [172]. When surface exchange reactions become more prevalent, surface modifications can be used to improve permeability. Studies have shown surface modifications like increasing surface roughness or coating with catalysts can significantly improve permeation rates [173, 174].

To date, a wide variety of perovskite materials have been either proposed or developed. However, no material is currently available that meets all the operational requirements needed for optimum membrane performance parameters like durability, chemical stability, electrochemical activity, and cost competitiveness. To address these operational challenges, a strategy of using dopants is one of the most promising methods for improving material properties and increasing oxygen permeation through membranes. In addition to improving electrochemical activity, doping also leads to morphological and microstructural changes that can also enhance membrane performance. For example, oxygen permeation of $La_{1-x}Sr_xCo_{1-y}Fe_yO_{3-\delta}$ can be improved by small increases in Sr or Co doping levels [175]. Similarly, increasing the doping level of Zr in $BaCo_{0.4}Fe_{0.6-x}Zr_xO_{3-\delta}$ has also been found to improve oxygen permeability [176]. However, some material compositions like $Ba_{0.5}Sr_{0.5}Co_{0.8}Fe_{0.2}O_{3-\delta}$ can produce good oxygen permeation levels at around 900 °C, but at temperatures below 825 °C there is a phase change in the crystal structure that leads to membrane failure [177]. Importantly, while studies have revealed perovskite compounds with larger oxygen non-stoichiometry have higher oxygen permeability, they are also prone to poor structure stability [178]. Structures containing alkaline earth metals are prone to structural instability when exposed to acidic gases like CO_2 [179]. Thus, despite significant advances in improving material properties and increasing oxygen permeation, there are several challenges that remain. For instance, it is difficult to find a perovskite-based membrane that displays the long-term chemical stability, electrochemical activity, and mechanical stability needed for industrial applications that involve high pressures, temperatures, and hostile gas streams.

4.5 HIGH-TEMPERATURE SOLID OXIDE CELLS

The problematic nature of CO_2 emissions and their contribution to global warming have highlighted the importance of both reducing the effects and converting them into value-added materials. In addition, conversion technologies are also needed to produce fuels with smaller carbon emissions when combusted from sustainable energy sources [180]. Solid oxide cells offer unique and appealing methods of converting CO_2 into less harmful products. However, these cells can also be used to produce hydrogen; see Figure 4.4. Solid oxide cells can be operated in two modes, as either a solid oxide electrolysis cell (SOEC) or as a solid oxide fuel cell (SOFC).

4.5.1 Solid Oxide Electrolysis Cell (SOEC)

4.5.1.1 Background

Advantageously, SOECs have the potential to convert electrical energy from renewable sources into hydrogen (H_2) from water. Importantly, hydrogen can be used directly as a fuel, or it can be used to store energy derived from renewable sources [181]. Hydrogen has the highest energy density

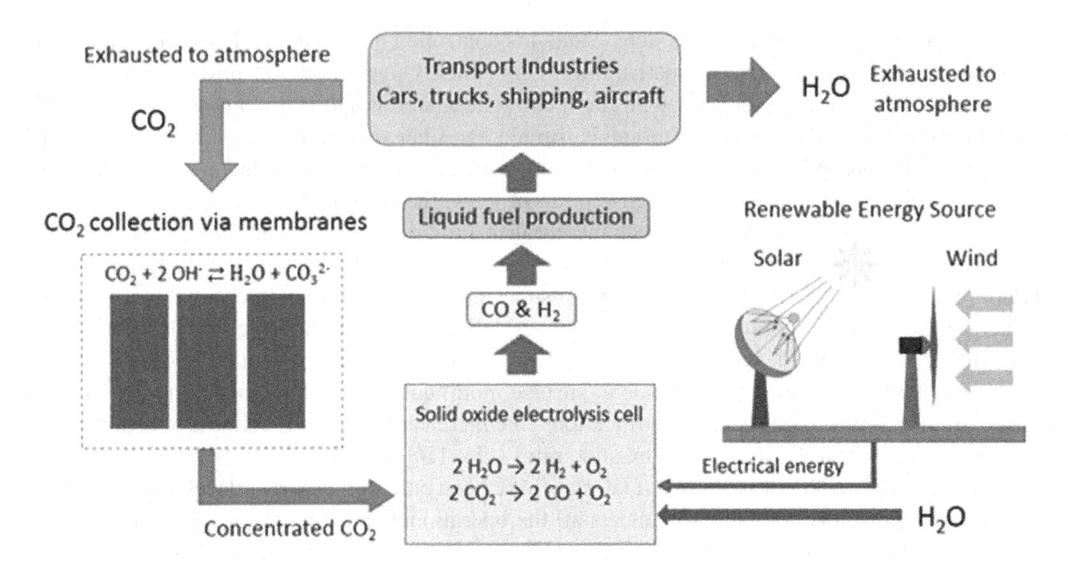

FIGURE 4.4 CO_2 capture from the atmosphere and SOEC recycling to manufacture synthetic transport fuels.

(143 kJ/kg) when compared to other chemical-based fuels. It is also a clean fuel since it only produces heat energy and water (by-product) when combusted with oxygen [182]. In addition, hydrogen produced by SOECs can also be used to generate fuels like methane (CH_4) and synthetic natural gas (SNG) when mixed with CO. Historically, conventional electrolysis carried at relatively low temperatures (less than 100 °C) effectively produces both hydrogen and oxygen. However, large amounts of electrical energy are needed to generate the gases, and expensive noble metal catalysts like platinum (Pt) make the process inefficient and expensive [183]. In contrast, SOECs operating at high temperatures (above 700 °C) are thermodynamically and kinetically more capable of splitting water compared to conventional electrolysis [184]. Importantly, the thermal energy needed to run the SOEC at its operational temperature can be supplied from external sources like industrial waste heat, while its electrical energy can be supplied from renewable energy sources. Operationally, steam (H_2O) enters the SOEC and reduces at the porous fuel (hydrogen) electrode (cathode) to form hydrogen and oxygen ions. The surface accumulates oxygen ions, and the ions then migrate through the dense oxide ion-conducting electrolyte to the porous oxygen electrode (anode). At the oxygen electrode, arriving oxygen ions oxidize to form oxygen gas by releasing electrons. The electrons then follow through the external electrical supply to the hydrogen electrode, thereby completing the electrical circuit. While oxygen is being evolved at the anode, hydrogen gas is being generated at the cathode [185]. In recent years, there has also been an interest in developing high-temperature CO_2/H_2O co-electrolysis to produce syngas ($CO + H_2$) using SOECs. A schematic representation of a co-electrolysis SOEC cell showing supply gases and product gases is presented in Figure 4.5. Co-electrolysis has the potential to capture CO_2 from emissions generated by fossil fuel-powered power stations or industries, thus reducing global greenhouse gas levels. Furthermore, the syngas (CO & H_2) generated by co-electrolysis can either be used as an effective method for large-scale energy storage or be used to produce liquid fuels or chemicals [186, 187]. Importantly, the co-electrolysis process can utilize both waste heat from industrial processes and electricity generated from renewable energy sources [188, 189].

4.5.1.2 Operation of a SOEC

The high-temperature co-electrolysis reduction of both CO_2 and H_2O are endothermic and occur with an external electric voltage across the two electrodes. For the H_2O to H_2 reaction, the porous

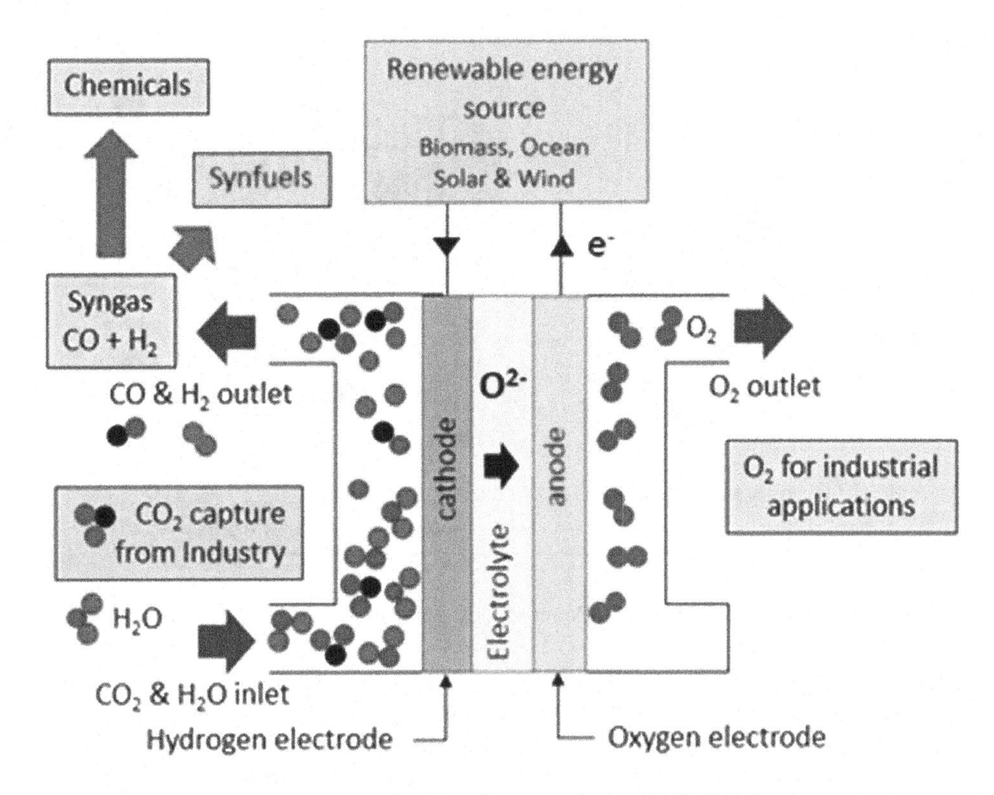

FIGURE 4.5 Schematic representation of solid-oxide electrolysis cell (SOEC) for the production of carbon monoxide & hydrogen (syngas) and oxygen.

fuel (hydrogen) electrode is the cathode (negative), and the reduction reaction is expressed by Equation 4.1:

$$H_2O + 2e^- \rightleftarrows H_2 + O^{2-} \tag{4.1}$$

Since solid oxide cells can be operated in two modes, as either a SOEC or as a SOFC, the right-hand side of the equation is hydrogen evolution (SOEC mode), and the left-hand side of the equation is hydrogen oxidation (SOFC mode). The porous oxygen electrode is the anode (positive), and the oxidation reaction is expressed by Equation 4.2:

$$O^{2-} \rightleftarrows {}^1/_2\, O_2 + 2e^- \tag{4.2}$$

Again, the right-hand side of the equation is SOEC mode (oxygen evolution), and the left-hand side of the equation is SOFC mode (oxygen reduction) [190]. Similarly, the fuel electrode reaction for the CO_2 to CO reaction is expressed by Equation 4.3:

$$CO_2 + 2e^{2-} \rightleftarrows CO + O^{2-} \tag{4.3}$$

The total reactions for the co-electrolysis reductions of CO_2 and H_2O are expressed by Equations 4.4 and 4.5:

$$H_2O \rightleftarrows H_2 + {}^1/_2\, O_2 \tag{4.4}$$

$$CO_2 \rightleftarrows CO + {}^1/_2\, O_2 \tag{4.5}$$

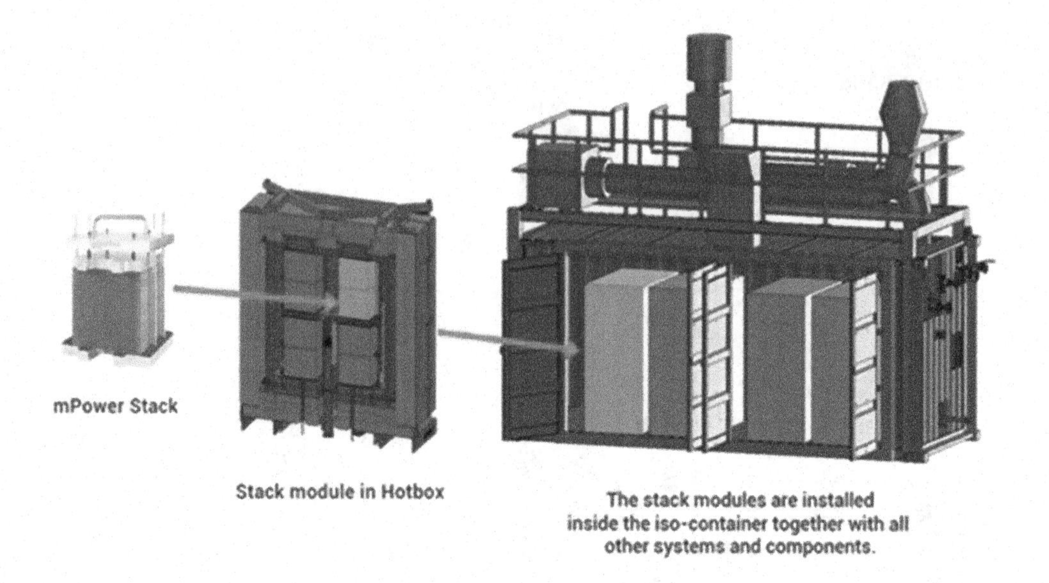

FIGURE 4.6 Schematic representation of mPower's SOEC unit-based technology.

Importantly, the properties of both electrodes and the dense oxide electrolyte determine the overall performance and efficiency of the SOEC during operation. Thus, all materials in the SOEC should be chemically stable in both reducing and oxidizing atmospheres. In addition, all components in the SOEC should have compatible thermal expansion coefficients to prevent material failure at operating temperatures. The pore size and porosity of electrode materials should not only provide optimal transport routes for electrons, ions, products, and reactants but also provide optimal active sites at the triple-phase boundary (interface of the electrode, electrolyte, and gas) for electrochemical reactions to take place [191]. Electrode materials may also be doped with materials that act as catalysts to improve either ionic conduction (IC) or electronic conduction (EC). The dense oxide electrolyte should be as thin as possible to lower the SOEC's ohmic resistance. It should also be a physical barrier that is structurally gas-tight to prevent crossflows within the SOEC and be chemically stable in either reducing or oxidizing atmospheres [192]. Moreover, the dense oxide electrolyte should also promote the ionic conduction of oxide ions from the fuel electrode to the oxygen electrode. Thus, during SOEC operation the properties of the materials making up the electrodes and electrolyte are extremely important in determining pathways for the transport of electrons, ions, products, and reactants. Optimizing the performance and efficiency of SOECs by improving materials that make up the fuel (hydrogen) electrode, dense oxide electrolyte, and oxygen electrode is the focus of current research. The performance of component materials in a SOEC is technically challenging since operational parameters like temperature, chemical stability, and electrochemical activity are demanding [193]. Figure 4.6 presents a schematic of mPower's SOEC unit-based power unit showing the modular design, which has been developed for optimal performance.

4.5.1.3 Materials Needed to Make a SOEC

The technical challenges associated with SOECs have led to significant research into developing new materials. On the inlet side of the SOEC is the porous fuel electrode which provides reaction sites for the decomposition of CO_2 and H_2O during co-electrolysis. At the same time, the fuel electrode provides a pathway for electrons traveling to the electrolyte via the reaction sites. This allows oxygen ions to be transported through the electrode to the triple-phase boundary (TPB) at the electrolyte, while hydrogen and carbon monoxide are liberated from the reaction sites [194]. Nickel (Ni)-based yttria (Y_2O_3) composites stabilized with zirconia (ZrO_2), designated Ni–YSZ, are widely used for porous fuel electrodes due to their lower cost when compared to platinum-based alloys.

Importantly, they also have good oxygen-ion conduction properties, chemical stability, and electrochemical activity [195]. They also have thermal expansion coefficients that are similar to other SOEC components. Similarly, yttrium (Y)-doped barium cerate zirconate (BCZY) has also been used for fuel electrodes due to its proton conduction properties, good chemical stability, and useful electrochemical activity. In both cases, the composite material forms a structural arrangement that permits the transport of electrons, protons/oxygen ions, and gas [196]. In both these structural arrangements, nickel acts as the electronic conductor. In the Ni–YSZ composite, YSZ acts as the ionic conductor, while in the Ni–BCZY, the BCZY acts as the proton conductor. Studies have also shown that decreasing grain size and tuning the pore distribution, controlling metal nanoparticle impregnation and dispersion, and electrode thickness can influence both the size of the TPB and oxidation kinetics [197, 198]. However, studies have shown that both Ni–YSZ and Ni–BCZY are prone to agglomeration which tends to degrade electrode performance over time. There is always the risk that nickel present in both materials will oxidize over time due to exposure to water steam and in the process reduce the conductivity of electrons [197, 199]. Because of the risk of nickel oxidation, other materials such as perovskite-based MIECs have also been extensively studied for potential use as electrodes in solid oxide cells [200–202]. Importantly, studies have revealed that MIECs are sulfur tolerant, and have better catalytic activity and superior conductivity compared to nickel-based electrode materials [203, 204]. For instance, electrodes composed of lanthanum strontium vanadium oxide ($La_{0.6}Sr_{0.4}VO_{3-d}$) with YSZ as the electrolyte have shown good catalytic activity, high sulfur tolerance, and good long-term chemical stability [205]. However, hydrogen electrodes based on lanthanum strontium chromium manganese ($(La_{0.75}Sr_{0.25})_{0.95}$ $Mn_{0.5}Cr_{0.5}O_3$ (LSCM) have displayed superior performance when compared to nickel-based electrodes exposed to low levels of hydrogen [206]. In addition, when LSCM is combined with samarium (Sm)-doped ceria (SDC), there are significant improvements in electronic conductivity, thus making this composite an attractive alternative electrode material [207].

The second electrode in a SOEC is the oxygen (air) electrode. The properties an oxygen electrode must have include 1) promoting efficient oxygen reduction; 2) optimizing oxygen-ion transport; 3) improving oxygen exchange kinetics and evolution; 4) having good catalytic activity; and 5) being chemically stable and compatible with the electrolyte. Oxygen electrodes composed of perovskite materials have shown high electronic and ionic conductivity, while also displaying good catalytic properties and oxygen exchange rates at temperatures above 700 °C [208]. Double-perovskite oxides, phase oxides, and Ruddlesden–Popper structures have shown improved ionic conductivity, electronic conductivity, and large oxygen exchange rates [208, 209]. In addition, lanthanum-based perovskite materials have been extensively investigated for use as oxygen electrodes. Lanthanum-based materials doped with rare earth elements like cobalt (Co), cerium (Ce), and strontium (Sr) have been found to make good oxygen electrodes [210, 211]. For instance, lanthanum-based materials like $(La_{1-x}Sr_x)_{1-y}$ MnO_3 and $(La_{0.6}Sr_{0.4})_{0.98}$ CoO_3 both show high ionic and electronic conductivity, along with high oxygen exchange rates [212, 213]. Further improvements in electronic and/or ionic conductivity can be achieved by replacing A-site ions with alkaline earth metal ions like Ca^{2+}, Sr^{2+}, and Ba^{2+}, while B-site ions can be occupied by transition metal ions like Co^{3+}, Fe^{3+}, and Ni^{3+} [214]. Therefore, it is possible to control the properties of oxygen electrodes by incorporating appropriate doping elements into the material structure. Because of this feature, current research has focused on designing new materials that incorporate a variety of dopants. Research into incorporating nanomaterials into lanthanum-based perovskite materials during synthesis to produce high-performance electrodes is currently being undertaken [215, 216].

The central component of a SOEC, sandwiched between the fuel electrode and the oxygen electrode is the electrolyte. There are two types of electrolytes used in SOECs, namely ion-conducting and proton-conducting. In both cases, the role of electrolyte is essentially the same in that it needs to be very dense, chemically stable during electrochemical reactions, and extremely thin to reduce internal cell resistance [217]. The ideal oxygen-ion-conducting electrolyte should readily allow ions to travel through the electrolyte from the fuel electrode to the oxygen electrode, as seen in Figure

4.5. Therefore, ionic conductivity is a very important material property for an electrolyte and arises from point defects in the electrolyte lattice structure. These defects can result from oxygen lattice vacancies or interstitial sites. Thus, oxygen-ion flow through the electrolyte is *via* a series of hops between individual point defects [218]. Increasing the number of point defects also increases oxygen-ion conductivity. Many types of electrolyte materials have been investigated in recent years, but the two most studied and reported are ceria and zirconia-based materials [219, 220]. Studies have shown that increasing the ionic conductivity of ceria and zirconia-based materials can be achieved by doping. Typical dopants investigated include yttrium (Y^{3+}), gadolinium (Gd^{3+}), dysprosium (Dy^{3+}), europium (III) (Eu^{3+}), scandium (Sc^{3+}), calcium (Ca^{2+}), and magnesium (Mg^{2+}) cations. The most studied electrolyte materials and their operational temperature ranges are yttria-stabilized zirconia (YSZ), operating between 800 and 1000 °C, and gadolinium-doped ceria (GDC), operating between 500 and 800°C [221, 222]. Other electrolytes have been investigated, but at present, they are either less mechanically robust or thermodynamically unstable [223]. However, some electrolyte materials like scandia (Sc_2O_3) stabilized zirconia (ScSZ) have higher oxygen-ion conductivities, but they are much more expensive [224]. In recent years proton-conducting oxides have also attracted considerable research interest [225]. However, problems associated with their chemical stability and compatibility with other SOEC components under normal operating conditions need to be resolved. Perovskite materials like barium cerate ($BaCeO_3$) and barium zirconate ($BaZrO_3$) have also been studied for potential application as electrolytes in SOECs. However, studies have found that barium cerate-based materials are prone to chemical instability when exposed to acidic gases like CO_2 and steam [226]. Barium zirconate-based materials are difficult to fabricate and suffer from weakly conducting grain boundaries which reduces proton conductivity. In addition, because of their refractory nature, they need extremely high sintering temperatures (> 1600 °C) to achieve the necessary sintering properties [227, 228].

4.5.2 Solid Oxide Fuel Cells (SOFC)

4.5.2.1 Increasing Global Energy Demand

Globally, electricity production, transmission, and distribution will increase from USD 4.09 trillion in 2021 to USD 5.93 trillion by 2026. The increasing demand is the result of increasing population growth and industrial expansion. This demand has resulted in a global CO_2 level increase of 1.7% to an all-time high of 33.1 tons in 2018. In spite of a 5.8% drop in global emissions of CO_2 in 2020 due to the COVID-19 pandemic, CO_2 emissions are predicted to return during 2022. In addition, there are indications that global CO_2 levels will increase by 4.8% annually in the near future due to the increased demand for fossil fuels [229, 230]. Crucially, air pollution is a significant danger to both human health and the environment. Currently, the global annual death rate due to air pollution is estimated to be around 4.2 million people [231]. Today fossil fuels are the most predominant source of energy for the production of electrical energy. Importantly, the combustion of fossil fuels to produce electricity accounts for more than 40% of global CO_2 emissions [232]. In spite of the harm they cause, they are expected to be used well into the future. However, the reserves of fossil fuels are limited, as seen in Figure 4.7. Importantly, due to global warming, fossil fuels will be replaced by sustainable renewable energy sources in the near future [233, 234]. One option for generating electrical power in the future is by using solid oxide fuel cells [235, 236].

4.5.2.2 Overview of Solid Oxide Fuel Cells

The history of FC begins in 1800 with the first study describing the electrolysis of water by Nicholson and Carlisle [236]. The investigation and development of the wet-cell gas battery by Grove in 1838 created the world's first FC [237, 238]. Unlike electrolysis reactions that relied on electricity to split water into its component parts (hydrogen and oxygen), this early fuel cell relied on the opposite reactions to combine hydrogen and oxygen to produce electricity [239]. During the latter half of the

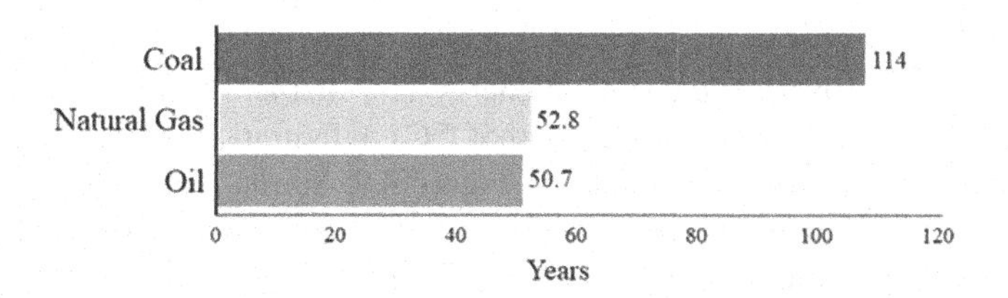

FIGURE 4.7 Life expectancy of fossil fuel reserves.

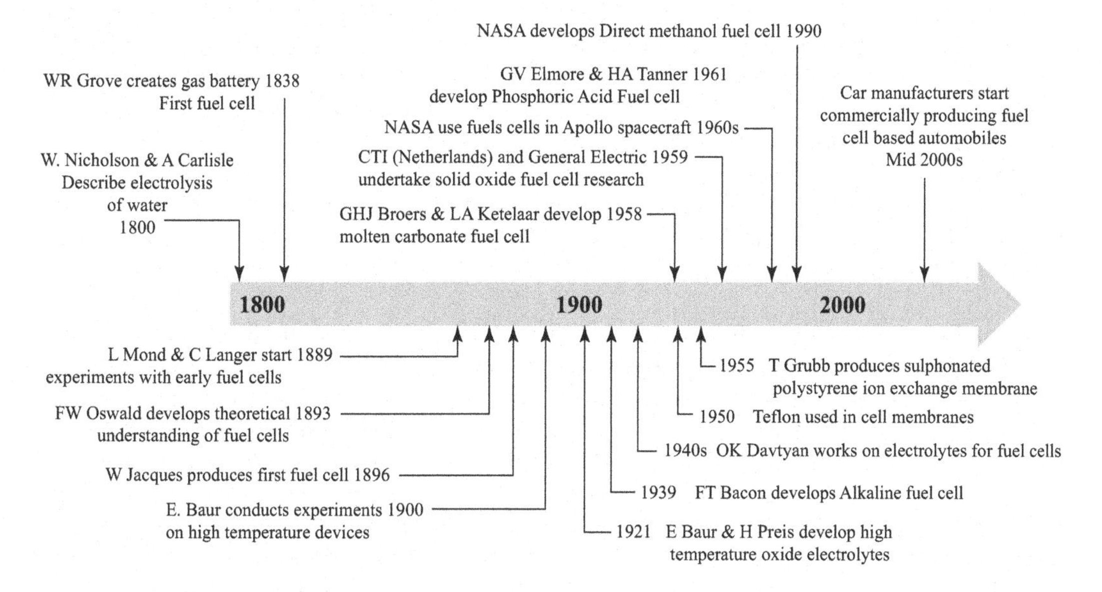

FIGURE 4.8 Historical development of fuel cells since 1800.

nineteenth century, hydrogen-based FCs were the focus of research, as seen in Figure 4.8. Much of the experimental work focused on investigating the properties of FC components like electrodes (anode and cathode), electrolytes, oxidizing, and reducing agents [239]. During the early part of the twentieth century, high-temperature solid oxide electrodes and high-pressure FCs were first investigated [234]. Since these early studies, FCs have been shown to have several advantages over many conventional electrically based generation systems due to their high electrical efficiency and fuel flexibility [240]. The supply of gases like H_2 and O_2 (usually air), and the subsequent electrochemical oxidation of H_2 are highly efficient. The electrochemical reactions are carried out within an all-solid-state structure with ceramic materials being used for both the electrolyte and electrodes [241]. Notably, during the 1970s, the United States spaceflight program investigated the possible generation of O_2 from the predominantly CO_2-based Martian atmosphere using solid oxide cells in electrolysis mode [242, 243]. Further studies investigated using CO and O_2 produced from a SOEC during daylight hours as fuel when the solid oxide cell was switched over to SOFC mode during the night-time period [244–246]. Thus, highlighting the interchangeable nature of the solid oxide cell which can be effectively run in either mode. Despite their advantages, the high cost and poor long-term operational stability of SOFCs have prevented their commercialization and widespread application. However, research carried out during the last decade has significantly improved their performance and cost-effectiveness. Production methods like thin-film formation techniques, the use of alternative nanomaterials and catalysts, and improved doping techniques have significantly improved the performance of both electrolytes and electrodes [247–250], thus making SOFCs an essential and promising energy source for the twenty-first century.

4.5.2.3 Operation of a SOFC

Fundamentally, a SOFC converts the chemical energy of a fuel and oxidant into electrical energy [251, 252]. A schematic representation of a typical SOFC configuration showing major components, supply gases, and product gases is presented in Figure 4.9. Similar to the SOEC mentioned earlier, the SOFC consists of a dense electrolyte sandwiched between a porous anode and a cathode. However, unlike the SOEC, air arriving at the cathode electrode in the SOFC is split and the liberated oxygen is reduced to oxygen ions. The resulting oxygen ions then migrate through the dense oxide (ion-conducting) electrolyte toward the anode, as seen in Figure 4.9. Arriving at the anode, the oxygen ions are consumed during the oxidation of hydrogen which also liberates electrons (e^-) and heat. The liberated electrons then travel back to the cathode via the external load and, in the process, generate electric power [253, 254]. When hydrogen and oxygen are used in a typical SOFC configuration, the electrochemical reactions for the various electrode reactions can be represented by Equations 6 to 8 below [255].

For the H_2 to H_2O reaction at the anode: $$2H_2 + 2O^{2-} \rightarrow 2H_2O + 4e^- \quad (6)$$
For the O_2 to $2O^{2-}$ reaction at the cathode: $$O_2 + 4e^- \rightarrow 2O^{2-} \quad (7)$$
The overall SOFC reaction: $$2H_2 + O_2 \rightarrow 2H_2O \quad (8)$$

Importantly, optimizing the performance and efficiency of SOFCs has been the focus of the scientific community for the past decade and is also currently an active area of research. Significantly, factors such as material properties, oxygen-surface interactions, ion diffusivity, electronic conduction, electro-catalytic activity, and thermal stability influence the overall performance of SOFCs [256, 257]. Because of the importance of these factors, research in recent years has focused on developing new and enhancing the properties of existing materials for use in the manufacture of high-performance anodes, electrolytes, and cathodes. Moreover, the electrolyte and the SOFC structure must be dense and gas-tight to prevent gas mixing. At the same time, both anode and cathode structures must also be well organized and porous to promote gas transport to the reaction sites [258, 259].

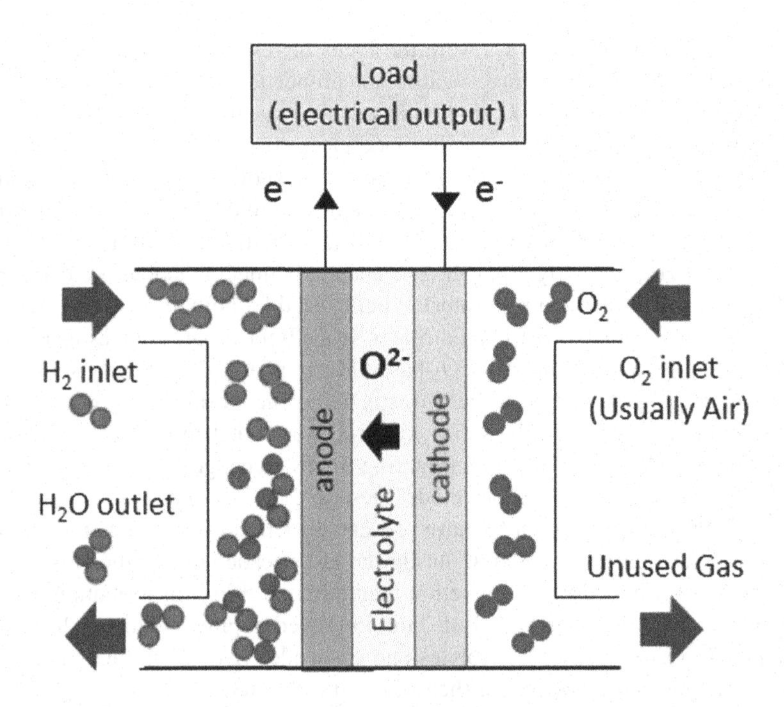

FIGURE 4.9. Schematic representation of solid-oxide fuel cell (SOFC) for the generation of electrical power

4.5.2.4 Electrolyte Materials

The role of the electrolyte in SOFC is like that of an electrolyte in SOEC, in that it must be very dense and have a very small thickness that minimizes the internal resistance of the cell during operation. The electrolyte also needs to be both chemically and mechanically stable and compatible with other cell components. Importantly, it must have high ionic conductivity, which promotes the flow of oxide ions through the electrolyte from the cathode side to the anode (fuel) side of the cell as seen in Figure 4.9. As mentioned above, one method of increasing ionic conductivity is to increase the oxygen vacancy mobility within the electrolyte [260]. One method of increasing ion conductivity is by doping with low-valent cationic oxides to raise the number of oxygen vacancies present in the electrolyte. Historically, a wide range of materials like zirconia, ceria, lanthanum, yttrium, barium, scandium, strontium, and gadolinium, and various combinations and concentrations of these materials and a variety of dopants have been evaluated for use in electrolytes. However, during the last decade, research has focused on three main types of materials for electrolytes. These include 1) yttria-stabilized zirconia (YSZ); 2) gadolinium-doped ceria (GDC); and 3) perovskite compounds with the general lattice formula of $ABO_{3-\delta}$ [261–263]. The most studied electrolyte is YSZ, with 8YSZ having the highest ion conductivity for a chemically and mechanically stable electrolyte. Studies have also shown other doping materials like nickel and scandia can also improve ion conductivity but can be more expensive or less electrochemically stable [223, 264].

However, despite being extensively studied and used, YSZ-based electrolytes still face many operational challenges. These challenges include 1) improving power density; 2) reducing operational temperatures; and 3) improving long-term cell stability [265]. In particular, the presence of water in moist atmospheres can influence the chemical stability of YSZ-based electrolytes. For instance, in moist atmospheres and temperatures ranging between 100 and 200 °C, the zirconia phase will undergo detrimental phase transformations [266]. Thus, during thermal cycling YSZ-based electrolytes must not be exposed to moist air. In addition, several studies have found the presence of water can act as a corroding agent and promote crack growth in the zirconia phase [267, 268]. Because of these shortcomings, research has focused on finding alternative electrolyte materials that avoid these degradation issues. Doped cerium oxide (CeO_2)-based electrolytes are superior to YSZ electrolytes due to their high ionic conductivity and lower operational temperatures. However, a major problem with these electrolytes is the partial reduction of Ce^{4+} to Ce^{3+} during cell operation. This reduction reaction produces two detrimental effects in the electrolyte. Firstly, it causes an expansion of the lattice structure and results in a noticeable decrease in the mechanical properties of the electrolyte. Secondly, it produces an increase in electron conductivity that causes a noticeable decrease in the open circuit voltage of the cell. To alleviate this problem, researchers for several years have investigated various doping methods to incorporate materials like calcium (Ca), gadolinium (Gd), niobium (Nb), and samarium (Sm) into cerium oxides [269, 270]. For instance, increasing the Sm doping concentration can influence the association energy and increase ionic conductivity [271]. However, developing cerium oxide-based electrolytes that overcome problems associated with ion conduction and long-term chemical and mechanical stability is still being studied today.

Another electrolyte material currently being investigated is the perovskite compound with the general lattice formula of $ABO_{3-\delta}$. Many of the perovskite-structured materials have excellent chemical stability in both oxidizing and reducing atmospheres. Some perovskite materials, such as lanthanum gallate doped with strontium and magnesium (LSGM) ($La_{1-x}Sr_xGa_{1-y}Mg_yO_3$), have been found to have good oxygen-ion conductivity [154]. Further modifications to the ion conductivity of LSGM have been achieved with additional dopants like barium, but the long-term stability still needs to be fully investigated [272]. Similarly, dopants like strontium and samarium have also shown promise in modifying the structure and ionic conductivity of other lanthanum-based nanocomposites [273]. Likewise, proton-conducting oxides have also attracted significant research interest due to their lower operating temperatures. Studies of proton-conducting materials like barium cerate ($BaCeO_3$) and barium zirconate ($BaZrO_3$) have been widely studied for use in SOECs

[274, 275]. However, both materials have poor chemical stability in environments containing acidic gases like CO_2 and steam [228]. Alternative materials like barium zirconate-based composites have been found to be chemically stable in the presence of acidic gases and steam under normal SOFC operating conditions [276]. In contrast, other researchers have investigated methods for improving thin-film fabrication techniques to enhance the performance of proton-conducting electrolytes. Replacing Pt with Ni substrates, due to manufacturing costs, was the focus of several studies along with increasing proton conductivity, power density, and lowering operational temperatures [277, 278]. In addition, recent research has focused on electrolytes composed of two or more materials, with each material having a different charge conduction behavior [279]. These composite-based multi-phase electrolyte materials display very high charge-carrying capabilities compared to conventional single-phase electrolyte materials. Ceria-carbonate-based nanocomposites with simultaneous proton and oxygen-ion conductivity have also been investigated in recent years [280]. Studies with composite materials based on gadolinia-doped ceria and alkali carbonates (Li_2CO_3–K_2CO_3 or Li_2CO_3–Na_2CO_3) have shown reasonable stability, higher conductivities (between 7 x 10^{-2} S cm^{-1} to 9 x 10^{-2} S cm^{-1}) and lower operational temperatures (between 400 and 600 °C) [281]. Similarly, a study investigating a doped barium cerate-carbonate, a perovskite-structured composite material, displayed peak power densities of 0.701 W/cm^2 at 550 °C and 0.957 W/cm^2 at 600 °C [282]. However, several challenges still face composite-based multi-phase electrolyte materials. These challenges include 1) fully investigating the charge conduction mechanism; 2) elucidating charge conduction pathways; and 3) improving the long-term stability of these materials at lower operational temperatures [283–285].

4.5.2.5 Anode Materials

The catalytic reaction that oxidizes fuel in the SOFC takes place at the anode as shown in Figure 4.9. The performance of the anode in promoting the fuel oxidation reaction is directly responsible for the efficient operation of the SOFC. Therefore, the selection of appropriate anode materials is critical for the overall performance of SOFCs [240, 286]. The anode material must promote high levels of ion conductivity, be chemically compatible with the fuel environment, and be thermally stable at the operating temperature [287]. In addition, the anode should have an organized and highly porous structure composed of small particle sizes that creates large numbers of active sites [288]. Therefore, to achieve these goals, a wide variety of materials and composites based on nickel oxide (NiO), copper oxide (CuO_2) and lanthanum have been investigated. This is currently an active field of research since many of the earlier and more traditional anode materials have displayed poor performance [193]. For instance, during the electrochemical oxidation of hydrocarbons, traditional Ni–YSZ anode materials are not only influenced by the concentration of hydrocarbons but are also subjected to coking, which has a detrimental effect on SOFC performance [289, 290]. Thus, major research in recent years has focused on improving the design and synthesis of nanostructured anode materials to improve SOFC performance [215, 291]. Novel anode materials have been evaluated for use in alcohol-fueled low-operating temperature SOFCs [292, 293]. Research has also looked at coking-resistant anode materials based on NiSn for use in methane SOFCs [294]. Other anode materials composed of multi-element-doped ceria and nickel-samarium (Sm)-doped ceria (SDC) have been fabricated and investigated [295, 296]. Nanostructured Ni–YSZ doped with ruthenium (Ru) nano-flowers have also been investigated for ethanol-fueled SOFCs in recent years [297]. Materials based on Cu-CeO_2, $CaMoO_{3-d}$, Y, and La that have been substituted into $SrTiO_3$ and Cu-CeO_2-Ni–YSZ composites have also been investigated as potential anode materials in recent years [298–301].

4.5.2.6 Cathode Materials

Apart from the anode and the electrolyte in the SOFC, there is also the air electrode (cathode), which is exposed to the air/oxygen environment as seen in Figure 4.9. The cathode is responsible for oxygen present in air being reduced to form oxygen ions, while simultaneously it receives electrons from the external circuit and directs them to the oxygen reduction sites. Finally, the cathode transports the oxygen ions to the electrolyte interface [302]. To successfully perform this function, the cathode must have several desirable properties. These properties include 1) a high electron conductivity; 2) its chemical structure should be stable and compatible with the air/oxygen environment; 3) its microstructure should be highly porous for gas transfer and oxidation reactions at cathode/electrolyte interface to take place; 4) the porous microstructure must support high levels of catalytic activity during oxygen reduction; 5) it must be airtight and thermally stable, with its coefficient of thermal expansion matching other SOFC components; 6) it must be easily produced and cost-effective during manufacture; and 7) it must have long-term mechanical durability [303–305]. Studies have examined a wide range of materials for potential use as cathodes. Lanthanum-based composites have been extensively studied because of their good thermal stability and electrochemical performance. Lanthanum manganite-based composites doped with rare earth elements like Co, Ce, or Sr have shown good electrochemical properties and oxygen-ion transfer rates [306–308]. Similar performances have been reported for lanthanum-based composites doped with bismuth, ferrites, and Co-containing perovskite oxides [309–311]. Other materials investigated include Fe–Co composites and samarium-barium-cobalt-copper oxide composites [312, 313]. Other researchers have examined combinations of other rare earth materials like bismuth (Bi), ruthenium (Ru), and praseodymium (Pr) [314–316]. For instance, Muhoza and Gross utilized a carbon-samarium template doped with ceria NPs (nSDC) in their SOFC, which increased its maximum power density by around 38% [317]. In each of these studies, synthesis techniques were optimized to promote greater control of oxygen non-stoichiometry and defect formation. Optimizing both these properties can significantly improve the catalytic behavior and ion conductivity of the respective cathode material. At present, research into new cathode materials is a very active field in developing high-performance SOFCs.

4.5.2.7 Recent Advances Made Using Nanomaterials

In recent years that has been considerable interest in developing and fabricating micro-scale SOFCs or μSOFCs. These micro-electromechanical systems (MEMS) are designed with an extremely thin nanometer scale electrolyte film and electrodes. At the nanometer scale, the mechanical strength and stability of the fragile ceramic membrane materials are critical factors for their operational performance [318]. Because of these factors, thin-film manufacturing techniques, generally related to silica wafers and semiconductors, are used to fabricate μSOFCs [319, 320]. Typically, electrode sputtering, lithography, and etching methods are frequently used to fabricate μSOFCs. Using these techniques, dense ion-conducting electrolytes with smooth surfaces and porous electrodes have been produced. For instance, Huang et al. manufactured an ultrathin μSOFC for low-temperature operation with an electrolyte membrane that was between 50 and 150 nm in thickness and a Pt electrode. [319]. Kang et al. used magnetron sputtering to deposit a nano-porous nickel substrate seeded with Pt and combined it with a commercially available porous anodic aluminum oxide (AAO) filter that formed a functionalized anode layer. The advantage of the AAO membrane is that it has a highly ordered porous structure. However, there is only a small active area between the integrated membrane/electrode, and as a result, the power output was too low for widespread usage [321]. Similarly, Kerman et al. used a co-sputtering technique to produce compositionally graded GDC/YSZ nanometer scale electrolytes for μSOFC [322]. Table 4.2 presents a selection of nanometer scale materials used as anodes, cathodes, and electrolytes in μSOFCs working in their operational temperature range. However, these studies have also highlighted that increasing the active area without causing mechanical failure is challenging.

Another challenge facing μSOFC is the thermal instability of electrodes. Therefore, to avoid thermal instability issues, most current μSOFC utilize Pt electrodes [318]. However, thermal coarsening of the Pt electrode can result in de-wetting, metal agglomeration, and structural alterations. The consequence is a significant loss in electrode performance [235, 339]. One solution to this problem is to use heat-resistant oxide electrodes. However, oxide electrodes are less active than Pt electrodes. Therefore, to improve the performance of oxide electrodes, some researchers have investigated incorporating oxides into nano-porous Pt electrodes [340–342]. For instance, Liu et

TABLE 4.2
A selection of nanometer scale materials used as anodes, cathodes, and electrolytes in µSOFCs working in their operational temperature range

Anodes	Cathode	Electrolyte	Operating temperature (°C)	Ref.
Ni	-	GDC	450–550	323
Ni	LSM-YSZ	ScSZ	700	324
Ni	LSCF-GDC	GDC	650–850	325
Pt	LSCF	YSZ	450–550	326
Ni-SDC	SSC	ScSZ	600–700	327
Ru	Pt	CGO-YSZ	470–520	322
Pt	Pt	YSZ	350–500	328
Ni	Pt	YSZ	600	329
Pt-ZrO_2	LSM	YSZ	650–800	330
LSCF	LSCF	CGO	700	331
PSM	PSM	YSZ	500–800	332
SSC-NiO-YSZ	SSC-LSF-GDC	YSZ	700–800	333
LSM-YSZ	LSM-YSZ	YSZ	600–800	334
LSM-YSZ	LSM-YSZ	YSZ	650–850	335
NiO-YSZ	LSF-YSZ	YSZ	700–800	336
NiO-YSZ	LSM-GD	YSZ	750	337
LSM-YSZ	LSM-YSZ	YSZ	650–800	338

al. reported that during a 120-hour test, the degradation rate of a nano-porous Pt electrode coated with a 1.6 nm thick layer of ZrO_2 could be reduced to 30% compared to nano-porous Pt over the same period [341]. Other researchers have investigated using highly porous and vertically oriented columnar structured ceria films instead of metal electrodes without sacrificing activity and performance [322, 343].

In recent years, there has also been considerable interest in using nanomaterials to lower the operational temperature range of SOFCs. Nanomaterials can be employed as pores or contact points that promote a three-phase boundary structure within the electrode and enhance SOFC performance. In addition, nanomaterials can be used to improve sulfur resistance and reduce or avoid carbon deposition on the anode when hydrocarbon-based fuel sources are used. Thus, selecting the appropriate material with suitable compatibility is one of the most difficult aspects of designing and developing SOFCs [344]. For instance, nickel cermets, which are often used as anodes for high-temperature SOFC, are coated with CeO_2 nanoparticles to improve their sulfur resistance. However, the addition of Pd nanoparticles to the anode of a Ni/$Gd_{0.1}C_{0.9}O_{1.95}$-based SOFC has been found to increase its sulfur resistance [345]. Furthermore, Zhan et al. used a nanostructured Ni anode with a thin Sr and Mg doped lanthanum gallate (LSGM) electrolyte, operating at a temperature of 650 °C to produce a power density of around 1.20 W/cm^2 [346]. Rh nanoparticles produced by block polymer lithography have been used for high-temperature electro-catalysis [347]. Pd and CeO_2 core-shell catalyst film deposited on an yttrium-doped zirconium (YSZ) anode can significantly improve its performance compared to the untreated anode of a SOFC [348]. Moreover, other researchers have investigated using nickel-free anodes in SOFCs. For instance, Raza et al. found a $Cu_{0.2}Zn_{0.8}$ nanostructured anode, with particle sizes ranging from 5 to 20 nm, with samarium-doped ceria (Ce) electrolyte producing a superior electrochemical output when operating at 550 °C [349]. In addition, short-time reverse current treatments have produced porous YSZ anodes with finely dispersed Ni nanoparticles embedded in their matrix. The resulting polarization resistance of the anodes was found to be 40% lower than untreated anodes [350].

4.6 CHALLENGES AND FUTURE PERSPECTIVES

Membranes for gas separation are an alternative and efficient method for delivering large quantities of high-purity gases. The advantage of membrane-based technologies is their modular design which makes it relatively simple to add or subtract membrane modules to meet gas separation demands. Inorganic membranes discussed above have superior permeability, selectivity, and stability at higher temperatures compared to polymeric membranes that traditionally operate at lower temperatures. Due to the highly desirable features of inorganic membranes, there has been considerable research and industrial interest in recent years. This interest has translated into many patents for inorganic membrane systems designed for specific gas separation applications [351, 352]. Some typical applications and membrane types include 1) CO_2 separation using silica or zeolite membranes between 20 and 200 °C; 2) hydrocarbon separation using silica or zeolite membranes between 20 and 300 °C; 3) hydrogen separation using dense metallic membranes between 400 and 600 °C; 4) oxygen separation using carbon membranes (molecular sieve) between 20 and 100 °C; and 5) oxygen separation using a dense ceramic membrane (i.e., MIEC membranes) between 600 and 1000 °C [105]. However, inorganic membranes still face several challenges, in particular, the long-term chemical stability of the membrane material under the operating conditions of the separation process. For instance, palladium membranes have high hydrogen selectivity but suffer from chemical instability resulting from hydrogen embrittlement at temperatures below 300 °C. In addition, the membranes are also prone to failure when exposed to reactive gases [125]. Unfortunately, alloying with other expensive noble metals to improve survivability significantly increases production costs. This factor has severely limited the use of dense metallic membranes in large-scale industrial applications. Currently, dense ceramic membranes are being actively researched globally, and MIEC membranes are being studied for potential use in air separation and hydrogen-based technologies [353]. Microporous carbon membranes have low hydrogen fluxes, are unstable in oxygen-containing atmospheres above 200 °C, and are prone to plugging from organic impurities present in supply gases [83]. Similarly, zeolite membranes have low hydrogen fluxes at low temperatures and degrade when exposed to gas streams containing steam at elevated temperatures. Microporous silica membranes have higher hydrogen fluxes that increase with temperature but become unstable when exposed to gas streams containing steam at elevated temperatures. However, despite this limitation, silica membranes are commercially available and used in a variety of applications. One common problem faced by the abovementioned membranes is their susceptibility to steam in gas streams at high temperatures. Other operational challenges facing inorganic membranes operating at high temperatures include 1) producing very thin membranes with large surface-volume ratios; 2) sealing the membranes *in situ* to prevent leakages; 3) matching the thermal expansion coefficient of membrane, seals, and the module itself; 4) developing efficient manufacturing technologies to produce high-quality reproducible membranes on an industrial scale; and 5) reducing the current high cost of inorganic membranes to make them a more economically viable for gas separation processes [27].

The two most important objectives for developing effective solid oxide cells are durability and long-term performance for their specific application. The long-term performance of solid oxide cells is hindered by the effects of high operational temperatures (800 to 1000 °C) that subsequently deteriorate components and change materials properties with time [354]. Importantly, material degradation problems have been identified as the most serious facing solid oxide cells, with degradation rates being twice as large for SOECs compared to SOFCs [280]. In addition, a contributing degradation factor common to SOFCs is the poisoning of the anode by impurities present in the fuel [355]. Importantly, after long operational periods, degradation will be due to electrochemical, structural, and thermal changes within the various cell components. Typical examples of degradation mechanisms in SOECs include 1) structural changes along grain boundaries can form voids that subsequently increase cell resistance and allows mass transport of gases across the electrolyte; 2) water present on the cathode side of the cell reacts with nickel present in the electrode to form nickel (II) hydroxide ($Ni(OH)_2$), which then travels to the electrolyte surface creating polarization losses; and 3) the formation of a secondary phase in the air electrode which produces delamination that results in increased electrode resistance [356–358]. Likewise, SOFCs suffer similar degradation issues after long periods of operation. Thus, the major challenges and limitations facing both SOECs and SOFCs are selecting materials with properties capable of resisting the effects of degradation. Briefly, the challenges that must be overcome in both SOECs and SOFCs to promote long operational periods include 1) reducing chemical instability and incompatibility in the oxidizing and reducing environments; 2) reducing electronic resistance of electrode materials; 3) the design and incorporation of materials with similar coefficients of thermal expansion to prevent thermal stresses that cause fracturing between electrodes and electrolyte interfaces, which results in inter-diffusion between cell components; 4) improvements are needed in sealants and cell interconnects; 5) developing new catalysts and improve the geometry of the active surface; and 6) reducing the ionic resistance of electrolytes [359–361]. In recent years, to address these challenges, research has focused on incorporating nanostructured materials into perovskite-based electrodes in both SOECs and SOFCs [362]. Typically, these nanoscale catalyst materials were incorporated into electrode materials to enhance their performance and reduce catalyst loading [363]. The advantage of incorporating nanostructured materials in electrodes is that the resulting pore structures can act as contact points for triple-phase boundary locations [364]. Thus, current research is focused on designing and developing new nanostructured materials with the necessary architecture to promote the highest possible electrochemical activity and the highest possible volumetric power density [365]. At the same time, the nanostructured material must be electrochemically and mechanically stable and promote enhance ion mobility [365]. It is believed that future research into nanostructured materials holds the technical breakthroughs needed to translate both SOECs and SOFCs into viable large-scale industrial technologies capable of delivering clean, sustainable, and renewable energy sources.

4.7 CONCLUSION

Major advances in inorganic membrane technologies have taken place in the last two decades. The advances have involved developing new materials with enhanced gas selectivity and improved gas fluxes. However, despite these advances, several challenges remain and need to be resolved. Future research needs to focus on improving the chemical, thermal, and hydrothermal stability of membrane materials. This is of particular importance when membranes are exposed to severe temperatures and gas atmospheres containing problematic contaminants like water, CO_2, CH_4, acidic vapors, and sulfur-containing species. In addition, in the case of dense metallic membranes, higher gas fluxes are needed at lower operating temperatures. From the design perspective, there is a need to improve sealing technologies and capitalize on various membrane geometries for optimal modular design of filtration units. Moreover, data is needed to quantify optimal operational parameters, efficiencies, and energy demands of filtration units to produce gas tonnages at the industrial scale. This would make inorganic membrane-based technologies economically viable and capable of sustainably producing fuels and chemical products.

Similar to inorganic membrane-based technologies, solid oxide cells use various structures such as membranes. SOECs offer a unique and appealing environmentally benign method for converting harmful gases like CO_2 into more useful products and offer a unique method for generating electrical energy by eco-friendly means. Although significant progress has been made in recent years, several challenges remain before the industrial-scale use of these technologies is feasible. The challenges facing both SOECs and SOFCs include 1) reduction in operational temperatures; 2) improving the ionic conductivity of electrolytes; 3) improving catalytic and electrochemical activity of anodes and cathodes; 4) improving the chemical stability of anodes and cathodes in their respective gas environments, thus reducing degradation, improving durability, and promoting longer operational life spans; 5) reducing thermal stress/mismatch between cell components during operational conditions; and 6) developing new and optimize existing advanced fabrication techniques to produce large numbers of components at a competitive cost. Crucially, further research is needed to develop new materials and fully investigate methods for improving the performance of existing materials from the nanoscale up to the macro-scale. Recent studies indicate that the incorporation of nanoscale materials into existing cell components has the potential to deliver higher-performing solid oxide cells at lower operational temperatures for longer operational periods. This is of particular importance since SOECs can convert harmful CO_2 emissions into useful synthetic fuels and chemicals. SOFCs are promising devices for generating electrical energy that can contribute to the global demand for electrical energy.

ACKNOWLEDGEMENT

Part of this work was carried out in the Fuel & Energy Technology Institute at Curtin University. The authors thank Prof Chun-Zhu Li, Prof Gordon Parkinson, Dr Dehua Dong and Dr Xin Shao for their encouragement.

REFERENCES

1. Gerland P, Raftery AE, Sevcikova H, Li N, Gu D, Spoorenberg T, Alkema L, Fosdick BK, Chunn J, Lalic N, Bay G. World population stabilization unlikely this century. *Science*. 2014; 346: 234–237.
2. BP statistical review of world energy. 2020. www.bp.com/en/global/corporate/energy. Last accessed 28 September 2020.
3. Panwar NL, Kaushik SC, Kothari S. Role of renewable energy sources in environmental protection: A review. *Renewable and Sustainable Energy Reviews*. 2011; 15: 1513–1524.
4. Ellabban O, Abu-Rub H, Blaabjerg F. Renewable energy resources: Current status, future prospects and their enabling technology. *Renewable and Sustainable Energy Reviews*. 2014; 39: 748–764.
5. Stiegel GJ, Maxwell RC. Gasification technologies: The path to clean, affordable energy in the 21st century. *Fuel Processing Technology*. 2001; 71: 79–97.
6. Rosen L, Degenstein N, Shah M, Wilson J, Kelly S, Peck J, Christie M. Development of oxygen transport membranes for coal-based power generation. In: *10th International Conference on Greenhouse Gas Control Technologies*. 2011; 4: 750–755.
7. Deibert W, Ivanova ME, Baumann S, Guillon O, Meulenberg WA. Ion-conducting ceramic membrane reactors for high-temperature applications. *Journal of Membrane Science*. 2017; 543: 79–97.
8. Ahmad FN, Sazali N, Shalbi S, Ngadiman NHA, Othman MHD. Oxygen separation process using ceramic-based membrane: A review. *Journal of Advanced Research in Fluid Mechanics and Thermal Sciences*. 2019; 62(1): 1–9.
9. Sazali N, Salleh WNW, Ismail AF, Nordin NAHM, Ismail NH, Mohamed MA, Aziz F, Yusof N, Jaafar J. Incorporation of thermally labile additives in carbon membrane development for superior gas permeation performance. *Journal of Natural Gas Science and Engineering*. 2018; 49: 376–384.
10. Liang CZ, Yong WF, Chung TS. High-performance composite hollow fibre membrane for flue gas and air separations. *Journal of Membrane Science*. 2017; 541: 367–377.
11. Castle WF. Air separation and liquefaction: Recent developments and prospects for the beginning of the new millennium. *International Journal of Refrigeration*. 2002; 25(1): 158–172.
12. Shourkaei MA, Rashidi A, Javad Karimi-Sabet J. Life cycle assessment of oxygen-18 production using cryogenic oxygen distillation. *Chinese Journal of Chemical Engineering*. 2018; 26(9): 1960–1966.
13. Smith A, Klosek J. A review of air separation technologies and their integration with energy conversion processes. *Fuel Processing Technology*. 2001; 70(2): 115–134.
14. Baker RW. Future directions of membrane gas separation technology. *Industrial & Engineering Chemistry Research*. 2002; 41: 1393–1411.
15. Bhide BD, Stern SA. A new evaluation of membrane processes for the oxygen- enrichment of air. II. Effects of economic parameters and membrane properties. *Journal of Membrane Science*. 1991; 62(1): 37–58.
16. Pabby AK, Rizvi SS, Requena AS. *Handbook of Membrane Separations: Chemical, Pharmaceutical, Food, and Biotechnological Applications*. Boca Raton, FL: CRC Press, 2008.
17. Mulder M. (Ed.). *Basic Principles of Membrane Technology*. Netherlands: Kluwer Academic Publishers, 1996.

18. Meinema HA, Dirrix RWJ, Brinkman HW, Terpstra RA, Jekerle J, Kosters PH. Ceramic membranes for gas separation – Recent developments and state of the art. *Interceram.* 2005; 54(2): 86–91.
19. Koros WJ. Gas separation membranes: Needs for combined materials science and processing approaches. *Macromolecular Symposia.* 2002; 188: 13–22.
20. Bernado P, Clarizia G. 30 years of membrane technology for gas separation. *Chemical Engineering Transactions.* 2013; 32: 1999–2004.
21. Puig-Arnavat M, Soprani S, Sogaard M, Engelbrecht K, Ahrenfeldt J, Henriksen UB, Hendriksen PV. Integration of mixed conducting membranes in an oxygen-steam biomass gasification process. *RSC Advances.* 2013; 3(43): 20843–20854.
22. Pandey P, Chauhan RS. Membranes for gas separation. *Progress in Polymer Science.* 2002; 26: 853–893.
23. Shimekit B, Mukhtar H, Ahmad F, Maitra S. Ceramic membranes for the separation of carbon dioxide—A review. *Transactions of the Indian Ceramic Society.* 2009; 68(3): 115–138.
24. Freemantle M. Membranes for gas separation. *Chemical & Engineering News.* 2005; 43: 49–57.
25. Chan KK, Brownstein AM. Ceramic membranes - Growth prospects and opportunities. *American Ceramic Society Bulletin.* 1991; 70: 703–707.
26. Chen XY. Membrane gas separation technologies for biogas upgrading. *RSC Advances.* 2015; 5(31): 24399–24448.
27. Guerra K, Pellegrino J. Development of a techno-economic model to compare ceramic and polymeric membranes. *Separation Science and Technology.* 2013; 48: 51–65.
28. Keizer K, Verweij H. Progress in inorganic membranes. *Chemtech.* 1996; 26(1): 37–41.
29. Abedini R, Nezhadmoghadam A. Application of membrane in gas separation processes: Its suitability and mechanisms. *Petroleum & Coal.* 2010; 52: 69–80.
30. Lin YS. Inorganic membranes for gas separation and purification. *Membrane.* 2006; 3: 170–173.
31. Tsai CY, Tam SY, Lu Y, Brinker CJ. Dual-layer asymmetric microporous silica membranes. *Journal of Membrane Science.* 2000; 169: 255–268.
32. Zdravkov BD, Cermak JJ, Sefara CM, Janku J. Pore classification in the characterization of porous materials. *Central European Journal of Chemistry.* 2007; 5(2): 385–395.
33. Fane AG, Beatson P, Li H. Membrane fouling and its control in environmental applications. *Water Science and Technology.* 2000; 41: 303–308.
34. Guo J, Wang L, Zhu J, Zhang J, Sheng D, Zhang X. Highly integrated hybrid process with ceramic ultrafiltration-membrane for advanced treatment of drinking water: A pilot study. *Journal of Environmental Science and Health Part A - Toxic/Hazardous Substances & Environmental Engineering.* 2013; 48: 1413–1419.
35. Moure A, Gullon P, Domínguez H, Parajo JC. Advances in the manufacture, purification and applications of xylo-oligosaccharides as food additives and nutraceuticals. *Process Biochemistry.* 2006; 41: 1913–1923.
36. De Souza MP, Petrus JCC, Gonçalves LAG, Viotto LA. Degumming of corn oil/hexane miscella using a ceramic membrane. *Journal of Food Engineering.* 2008; 86: 557–564.
37. Iwamoto Y, Kawamoto H. Trends in research and development of nanoporous ceramic separation membranes – Saving energy by applying the technology to the chemical synthesis process. *Science and Technology Trends, Quarterly Review.* 2009; 32: 43–57.
38. Simon A, Price WE, Nghiem LD, Changes in surface properties and separation efficiency of a nanofiltration membrane after repeated fouling and chemical cleaning cycles. *Separation and Purification Technology.* 2013; 113: 42–50.
39. Caro J, Noack M, Kolsch P. Zeolite membranes: From the laboratory scale to technical applications. *Adsorption.* 2005; 11: 215–227.
40. Bird RB, Stewart WE, Lightfoot EN. *Transport Phenomena*, 2nd ed. New York: Wiley International, 2002.
41. Kast W, Hohenthanner CR. Mass transfer within the gas-phase of porous media. *International Journal of Heat and Mass Transfer.* 2000; 43(5): 807–823.
42. Nagy E. *Basic Equations of the Mass Transport Through a Membrane Layer*, 1st ed. London: Elsevier, 2012.
43. Van den Berg GB. Diffusional phenomena in membrane separation processes. *Journal of Membrane Science.* 1992; 73: 103–118.
44. Meinema HA, Dirrix RWJ, Brinkman HW, Terpstra RA, Jekerle J, Kosters PH. Ceramic membranes for gas separation - Recent developments and state of the art. *Interceram.* 2005; 54: 86–91.
45. Hashim S, Mohamed A, Bhatia S. Oxygen separation from air using ceramic-based membrane technology for sustainable fuel production and power generation. *Renewable and Sustainable Energy Reviews.* 2011; 15(2): 1284–1293.

46. Alimov V, Busnyuk A, Notkin M, Livshits A. Pd–V–Pd composite membranes: Hydrogen transport in a wide pressure range and mechanical stability. *Journal of Membrane Science.* 2014; 457: 103–112.
47. Al-Mufachi N, Rees N, Steinberger-Wilkens R. Hydrogen selective membranes: A review of palladium-based dense metal membranes. *Renewable and Sustainable Energy Reviews.* 2015; 47: 540–551.
48. Li H, Haas-Santo K, Schygulla U, Dittmeyyer R. Inorganic microporous membranes for H_2 and CO_2 separation – Review of experimental and modelling programs. *Chemical Engineering Science.* 2015; 127: 401–417.
49. Geffroy P, Fouletier J, Richet N, Chartier T. Rational selection of MIEC materials in energy production processes. *Chemical Engineering Science.* 2013; 87: 408–433.
50. Luque S, Gomez D, Alvarez JR. Industrial applications of porous ceramic membranes (pressure driven processes). *Membrane Science and Technology.* 2008; 13: 177–216.
51. Benes N, Nijmeijer A, Verweij H. Microporous silica membranes. In: Kanellopoulos NK (Ed.), *Recent Advanced in Gas Separation by Microporous Ceramic Membranes.* Amsterdam: Elsevier, 2000, pp. 335–372.
52. Paola B, Drioli E, Golemme G. Membrane gas separation: A review: State of the art. *Industrial & Engineering Chemistry Research.* 2009; 48(10): 4638–4663.
53. Daufin G, Escudier JP, Carrere H, Berot S, Fillaudeau L, Decloux M. Recent and emerging applications of membrane processes in the food and dairy industry. *Food and Bioproducts Processing.* 2001; 79: 89–102.
54. Lee S, Lee CH. Microfiltration and ultrafiltration as a pre-treatment for nanofiltration of surface water. *Separation Science and Technology.* 2006; 41: 1–23.
55. Lin YS, Kumakiri I, Nair BN, Alsyouri H. Microporous inorganic membranes. *Separation and Purification Methods.* 2002; 31, 229–379.
56. Mohammad AW, Teowa YH, Anga WL, Chunga YT, Hilal N. Nanofiltration membranes review: Recent advances and future prospects. *Desalination.* 2015; 356: 226–254.
57. De Meis D, Richetta M, Serra E. Microporous inorganic membranes for gas separation and purification. *Interceram.* 2018; 4: 16–21.
58. Iwamoto Y, Kawamoto H. Trends in research and development of nanoporous ceramic separation membranes – Saving energy by applying the technology to the chemical synthesis process. *Science and Technology Trends, Quarterly Review.* 2009; 32: 43–57.
59. Meulenberg WA, Schulze-Kuppers F, Deibert W, Van Gestel T, Baumann S. Ceramic membranes: Materials-components-potential applications. *ChemBioEng Review.* 2019; 6(6): 198–208.
60. Bein T. Synthesis and applications of molecular sieve layers and membranes. *Chemical Materials.* 1996; 8: 1636–1653.
61. Ayral A, Julbe A, Roussac V, Roualdes S, Durand J. Microporous silica membranes: Basic principles and recent advances. *Membrane Science and Technology.* 2008; 13: 33–79.
62. Khatib SJ. Silica membranes for hydrogen separation prepared by chemical vapour deposition (CVD). *Separation & Purification Technology.* 2013; 111: 20–42.
63. Li P, Wang Z, Qiao Z, Liu Y, Cao Y, Wen L, Wang J, Wang S. Recent developments in membranes for efficient hydrogen purification. *Journal of Membrane Science.* 2015; 495: 130–168.
64. Meixner DL, Dyer PN. Characterization of the transport properties of microporous inorganic membranes. *Journal of Membrane Science.* 1998; 140: 81–95.
65. Battersby S, Tasaki T, Smart S, Ladewig B, Liu S, Duke MC, Rudolph V, Diniz da Costa JC. Performance of cobalt silica membranes in gas mixture separation. *Journal of Membrane Science.* 2009; 329: 91–98.
66. Xomeritakis G, Tsai CY, Jiang YB, Brinker CJ. Tubular ceramic-supported sol-gel silica-based membranes for flue gas carbon dioxide capture and sequestration. *Journal of Membrane Science.* 2009; 341: 30–36.
67. Tsai CY, Tam SY, Lu Y, Brinker CJ. Dual-layer asymmetric microporous silica membranes. *Journal of Membrane Science.* 2000; 169: 255–268.
68. Kahlib NAZ, Daud FDM, Mel M, Azhar AZA, Hassan NA. Synthesis and characterization of silica membranes via sol-dip coating. *Journal of Advanced Research in Materials Science.* 2017; 39: 1–7.
69. De Lange RSA, Hekkink JHA, Keizer K, Burggraaf AJ. Permeation and separation studies on microporous sol-gel modified ceramic membranes. *Microporous Materials.* 1995; 4: 169–186.
70. Campaniello J, Engelen CWR, Haije WG, Pex P, Vente JF. Long-term pervaporation performance of microporous methylated silica membranes. *Chemical Communications.* 2004; 7: 834–835.
71. De Vos RM, Maier WF, Verweij H. Hydrophobic silica membranes for gas separation. *Journal of Membrane Science.* 1999; 158: 277–288.
72. Camus O, Perera S, Crittenden B, Van Delft YC, Meyer DF, Pex P, Kumakiri I, Miachon S, Dalmon JA, Tennison S, Chanaud P. Ceramic membranes for ammonia recovery. *Aiche Journal.* 2006; 52: 2055–2065.

73. Ahn SJ, Takagaki A, Sugawara T, Kikuchi R, Oyama ST. Permeation properties of silica-zirconia composite membranes supported on porous alumina substrates. *Journal of Membrane Science*. 2017; 526: 409–416.
74. Morooka S, Kusakabe K. Microporous inorganic membranes for gas separation. *MRS Bulletin*. 1999; 24: 25–29.
75. Wang H, Lin Y. Effects of synthesis conditions on MFI zeolite membrane quality and catalytic cracking deposition modification results. *Microporous and Mesoporous Materials*. 2011; 142: 481–488.
76. Elyassi B, Sahimi M, Tsotsis TT. Silicon carbide membranes for gas separation applications. *Journal of Membrane Science*. 2007; 288: 290–297.
77. Smart S, Vente J, Diniz da Costa J. High temperature H_2/CO_2 separation using cobalt oxide silica membranes. *International Journal of Hydrogen Energy*. 2012; 37: 12700–12707.
78. Battersby S, Tasaki T, Smart S, Ladewig B, Liu S, Duke MC, Rudolph V, Diniz da Costa JC. Performance of cobalt silica membranes in gas mixture separation. *Journal of Membrane Science*. 2009; 329: 91–98.
79. Kanezashi M, Asaeda M. Hydrogen permeation characteristics and stability of Ni- doped silica membranes in steam at high temperature. *Journal of Membrane Science*. 2006; 271: 86–93.
80. Khatib SJ. Silica membranes for hydrogen separation prepared by chemical vapour deposition (CVD). *Separation & Purification Technology*. 2013; 111: 20–42.
81. Darmawan A, Motuzas J, Smart S, Julbe A, Diniz da Costa J. Temperature dependent transition point of purity versus flux for gas separation in Fe/Co-silica membranes. *Separation and Purification Technology*. 2015; 151: 284–291.
82. Yang Y, Le TH, Kang F, Inagaki M. Polymer blend techniques for designing carbon materials. *Carbon*. 2017; 111: 546–568.
83. Salleh WNW, Ismail AF. Carbon membranes for gas separation processes: Recent progress and future perspective. *Journal of Membrane Science and Research*. 2015; 1(1): 2–15.
84. Koresh JE, Soffer A. Mechanism of permeation through molecular-sieve carbon membrane. Part 1. The effect of adsorption and the dependence on pressure. *Journal of the Chemical Society. Faraday Transactions I*. 1986; 82: 2057–2063.
85. Hamm JBS, Ambrosi A, Griebeler JG, Marcilio NR, Tessaro IC, Pollo LD. Recent advances in the development of supported carbon membranes for gas separation. *International Journal of Hydrogen Energy*. 2017; 42(39): 24830–24845.
86. Ismail AF, David LIB. A review on the latest development of carbon membranes for gas separation. *Journal of Membrane Science*. 2001; 193: 1–18.
87. Hagg MB, Lie JA, Lindbrathen A. Carbon molecular sieve membranes: A promising alternative for selected industrial applications. *Annals of the New York Academy of Sciences*. 2003; 984: 329–345.
88. Zeng LC, Yong WF, Chung TS. High-performance composite hollow fibre membrane for flue gas and air separations. *Journal of Membrane Science*. 2017; 541: 367–377.
89. Nair RR, Wu HA, Jayaram PN, Grigorieva IV, Geim AK. Unimpeded permeation of water through helium leak-tight graphene-based membranes. *Science*. 2012; 335: 442–444.
90. Van Gestel T, Barthel J. New types of graphene-based membranes with molecular sieve properties for He, H_2 and H_2O. *Journal of Membrane Science*. 2018; 554: 378–384.
91. Isobe T, Shimizu M, Matsushita S, Nakajima A. Preparation, and gas permeability of the surface modified porous Al_2O_3 ceramic filter for CO_2 gas separation. *Journal of Asian Ceramic Society*. 2013; 1(1): 65–70.
92. Pingelley CN. *Membranes for Gas Separation*. PhD Thesis, University of Bath, November 2016.
93. Aoki K, Kusakabe K, Morooka S. Separation of gases with an A-type zeolite membrane. *Industrial & Engineering Chemistry Research*. 2000; 39(7): 2245–2251.
94. Korelskiy D, Ye P, Fouladvand S, Karimi S, Sjoberg E, Hedlund J. Efficient ceramic zeolite membranes for CO_2/H_2 separation. *Journal of Materials Chemistry A*. 2015; 3: 12500–12506.
95. Skoulidas AI, Bowen TC, Doelling CM, Falconer JL, Noble RD, Sholl DS. Comparing atomistic simulations and experimental measurements for CH_4/CF_4 mixture permeation through silicalite membranes. *Journal of Membrane Science*. 2003; 227: 123–136.
96. Kosinov N. Recent developments in zeolite membranes for gas separation. *Journal of Membrane Science*. 2016; 499: 65–79.
97. Sorenson S, Payzant E, Noble R, Falconer J. Influence of crystal expansion/contraction on zeolite membrane permeation. *Journal of Membrane Science*. 2010; 357: 98–104.
98. Kondo M, Kita H. Permeation mechanism through zeolite NaA and T-type membranes for practical dehydration of organic solvents. *Journal of Membrane Science*. 2010; 361: 223–231.

99. Kanezashi M, O'Brien J, Lin YS. Template-free synthesis of MFI-type zeolite membranes: Permeation characteristics and thermal stability improvement of membrane structure. *Journal of Membrane Science*. 2006; 286: 213–222.
100. Choi J, Jeong H, Snyder M, Stoeger J, Masel R, Tsapatsis M. Grain boundary defect elimination in a zeolite membrane by rapid thermal processing. *Science*. 2009; 325: 590–593.
101. Hong Z, Zhang C, Gu X, Jin W, Xu N. A simple method for healing nonzeolitic pores of MFI membranes by hydrolysis of silanes. *Journal of Membrane Science*. 2011; 366: 427–435.
102. Kanezashi M, O'Brien J, Lin YS. Template-free synthesis of MFI-type zeolite membranes: Permeation characteristics and thermal stability improvement of membrane structure. *Journal of Membrane Science*. 2007; 194: 213–222.
103. Zhang Y, Avila A, Tokay B, Funke H, Falconer J, Noble R. Blocking defects in SAPO-34 membranes with cyclodextrin. *Journal of Membrane Science*. 2010; 358: 7–12.
104. Lin Y, Duke M. Recent progress in polycrystalline zeolite membrane research. *Current Opinion in Chemical Engineering*. 2013; 2: 209–216.
105. Bernardo P, Drioli E, Golemme G. Membrane gas separation: A review/state of the art. *Industrial & Engineering Chemistry Research*. 2009; 48: 4638–4663.
106. Drioli E, Giorno L. *Comprehensive Membrane Science and Engineering*. Amsterdam, The Netherlands: Elsevier, 2010.
107. Koros WJ, Mahajan R. Pushing the limits on possibilities for large scale gas separation: Which strategies? *Journal of Membrane Science*. 2000; 175: 181–196.
108. Adhikari S, Fernando S. Hydrogen membrane separation techniques. *Industrial & Engineering Chemistry Research*. 2006; 45: 875–881.
109. Coronas J, Falconer JL, Noble RD. Characterization and permeation properties of ZSM-5 tubular membranes. *AlChE Journal*. 1997; 43: 1797–1812.
110. Kamazawa K, Aoki M, Noritake T, Miwa K, Sugiyama J, Towata S, Ishikiriyama M, Callear S, Jones M, David W. In-operando neutron diffraction studies of transition metal hydrogen storage materials. *Advanced Energy Materials*. 2013; 3: 39–42.
111. Reshak A. MgH_2 and LiH metal hydrides crystals as novel hydrogen storage material: Electronic structure and optical properties. *International Journal of Hydrogen Energy*. 2013; 38: 11946–11954.
112. Mazzucco A, Voskuilen T, Waters E, Pourpoint T, Rokuni M. Heat exchanger selection and design analyses for metal hydride heat pump systems. *International Journal of Hydrogen Energy*. 2016; 41: 4198–4213.
113. Nayebossadri S, Speight J, Book D. Effects of low Ag additions on the hydrogen permeability of Pd-Cu-Ag hydrogen separation membranes. *Journal of Membrane Science*. 2014; 451: 216–225.
114. Yun S, Oyama ST. Correlations in Palladium membranes for hydrogen separation: A review. *Journal of Membrane Science*. 2011; 375: 28–45.
115. Conde JJ, Marono M, Sanchez-Hervas JM. Pd-based membranes for hydrogen separation: Review of alloying elements and their influence on membrane properties. *Separation & Purification Reviews*. 2017; 46: 152–177.
116. Li H, Caravella A, Xu HY. Recent progress in Pd-based composite membranes. *Journal of Materials Chemistry A*. 2016; 4: 14069–14094.
117. Hatlevik O, Gade S, Keeling M, Thoen P, Davidson A, Way J. Palladium and palladium alloy membranes for hydrogen separation and production: History, fabrication strategies, and current performance. *Separation and Purification Technology*. 2010; 73: 59–64.
118. Pinnau I, He Z, Pure and mixed gas permeation properties of polydimethylsiloxane for hydrocarbon/methane and hydrocarbon/hydrogen separation. *Journal of Membrane Science*. 2004; 244: 227–233.
119. Phair JW, Donelson R. Developments and design of novel (non-palladium based) metal membranes for hydrogen separation. *Industrial & Engineering Chemistry Research*. 2006; 45: 5657–5674.
120. Ockwig NW, Nenoff TM. Membranes for hydrogen separation. *Chemical Reviews*. 2007; 107: 4078–4110.
121. Uemiya S, Sato N, Ando H, Kude Y, Matsuda T, Kikuchi E. Separation of hydrogen through palladium thin film supported on a porous glass tube. *Journal of Membrane Science*. 1991; 56: 303–313.
122. Yin H, Yip ACK. A review on the production and purification of biomass-derived hydrogen using emerging membrane technologies. *Catalysts*. 2017; 7(297): 1–31.
123. Andrew PL, Haasz AA. Models for hydrogen permeation in metals. *Journal of Applied Physics*. 1992; 72: 2749–2757.
124. Paglieri SN, Way JD. Innovations in Palladium membrane research. *Separation and Purification Methods*. 2002; 31: 1–169.

125. Nam SE, Seong YK, Lee JW, Lee KH. Preparation of highly stable palladium alloy composite membranes for hydrogen separation. *Desalination*. 2009; 236: 51–55.
126. Lewis F. *The Palladium Hydrogen System*. London: Academic Press, 1967.
127. Lu GQ, Diniz da Costa JC, Duke M, Giessler S, Socolow R, Williams RH, Kreutz TG. 2007. Inorganic membranes for hydrogen production and purification: A critical review and perspective. *Journal of Colloid and Interface Science*. 2007; 314: 589–603.
128. Lynch S. Hydrogen embrittlement phenomena and mechanisms. *Corrosion Reviews*. 2012; 30: 105–123.
129. Rebeiz K, Dahlmeyer J, Garrison TR, Garrison TY, Darkey S, Paciulli D, Talukder M, Kubic J, Wald K, Massicotte F, Nesbit S, Craft A. Tensile properties of a series of palladium-silver alloys exposed to hydrogen. *Journal of Energy Engineering*. 2015; 141: 1–7.
130. Perrot P, Moelans N, Lebrun N. *Noble Metal Ternary Systems: Phase Diagrams, Crystallographic and Thermodynamic Data*. Edited by G. Effenberg and S. Ilyenko. Berlin, Germany: Springer, 2006.
131. Timofeev N, Berseneva F, Makarov V. New Palladium-based membrane alloys for separation of gas mixtures to generate ultrapure hydrogen. *International Journal of Hydrogen Energy*. 1994; 19: 895–898.
132. Hashi K, Ishikawa K, Matsuda T, Aoki K. Microstructures and hydrogen permeability of Nb-Ti-Ni alloys with high resistance to hydrogen embrittlement. *Materials Transactions*. 2005; 46: 1026–1031.
133. Adhikari S, Fernando S. Hydrogen membrane separation techniques. *Industrial & Engineering Chemistry Research*. 2006; 45: 875–881.
134. Cheng XY, Wu QY, Sun YK. Hydrogen permeation behaviour in a Fe3Al-based alloy at high temperature. *Journal of Alloys and Compounds*. 2005; 389: 198–203.
135. Jimenez G, Dillon E, Dahlmeyer J, Garrison T, Garrison T, Darkey S, Wald K, Kubik J, Paciulli D, Talukder M, Nott J, Ferrer M, Prinke J, Villaneuva P, Massicotte F, Rebeiz K, Nesbit S, Andrew Craft A. A comparative assessment of hydrogen embrittlement: Palladium and Palladium-Silver (25 weight% silver) subjected to hydrogen absorption/desorption cycling. *Advances in Chemical Engineering and Science*. 2016; 6: 246–261.
136. Melendez J, Fernandez E, Gallucci F, Annaland MVS, Arias PL, Tanaka DAP. Preparation and characterization of ceramic supported ultra-thin (~1 μm) Pd-Ag membranes. *Journal of Membrane Science*. 2017; 528: 12–23.
137. Tosti S, Bettinali L, Violante V. Rolled thin Pd and Pd-Ag membranes for hydrogen separation and production. *International Journal of Hydrogen Energy*. 2000; 25: 319–325.
138. Shen Y, Emerson SC, Magdefrau NJ, Opalka SM, Thibaud-Erkey C, Vanderspurt TH. Hydrogen permeability of sulphur tolerant Pd-Cu alloy membranes. *Journal of Membrane Science*. 2014; 452: 203–211.
139. Zhang K, Way JD. Palladium-copper membranes for hydrogen separation. *Separation and Purification Technology*. 2017; 186: 39–44.
140. Roa F, Block MJ, Way JD. The influence of alloy composition on the H_2 flux of composite Pd Cu membranes. *Desalination*. 2002; 147: 411–416.
141. Flanagan TB, Wang D. Hydrogen permeation through FCC Pd-Au alloy membranes. *The Journal of Physical Chemistry C*. 2011; 115: 11618–11623.
142. Zhang K, Shao Z, Li C, Liu S. Novel CO_2-tolerant ion-transporting ceramic membranes with an external short circuit for oxygen separation at intermediate temperatures. *Energy Environmental Science*. 2012; 5(1): 5257–5264.
143. Shimekit B, Mukhtar H, Ahmad F, Maitra S. Ceramic membranes for the separation of carbon dioxide—A review. *Transactions of the Indian Ceramic Society*. 2009; 68(3): 115–138.
144. Bhalla SA, Guo R, Roy R. The Perovskites structure – A review of its role in ceramic science and technology. *Materials Research Innovations*. 2000; 4: 3–26.
145. Magraso A, Haugsrud R. Effects of the La/W ratio and doping on the structure, defect structure, stability and functional properties of proton-conducting lanthanum tungstate $La_{28-x}W_{4+x}O_{54+\delta}$: A review. *Journal of Materials Chemistry A*. 2014; 2: 12630–12641.
146. Zhang K, Sunarso J, Shao Z, Zhou W, Sun C, Wang S, Liu S. Research progress and materials selection guidelines on mixed conducting perovskite-type ceramic membranes for oxygen production. *RSC Advances*. 2011; 1: 1661–1676.
147. Vollestad E, Vigen CK, Magraso A, Haugsrud R. Hydrogen permeation characteristics of $La_{27}Mo_{1.5}W_{3.5}O_{55.5}$. *Journal of Membrane Science*. 2014; 461: 81–88.
148. Hashim SS, Mohamed AR, Bhatia S. Oxygen separation from air using ceramic-based membrane technology for sustainable fuel production and power generation. *Journal of Renewable Sustainable Energy Reviews*. 2011; 15: 1284–1293.
149. Dyer P, Richards R, Russek S, Taylor D. Ion transport membrane technology for oxygen separation and syngas production. *Solid State Ionics*. 2000; 134(1–2): 21–33.

150. Skinner S, Kilner J. Oxygen ion conductors. *Materials Today*. 2003; 6(3): 30–37.
151. Sammes NM, Tompsett GA, Nafe H, Aldinger F. Bismuth based oxide electrolytes – Structure and ionic conductivity. *Journal of the European Ceramic Society*. 1999; 19: 1801–1826.
152. Eguchi K, Setoguchi T, Inoue T, Arai H. Electrical properties of ceria-based oxides and their application to solid oxide fuel cells. *Solid State Ion*. 1992; 52: 165–172.
153. Megaw HD. *Crystal Structures: A Working Approach*. Philadelphia: WB Saunders Company, 1973.
154. Brett DJL, Atkinson A, Brandon NP, Skinner S. Intermediate temperature solid oxide fuel cells. *Chemical Society Reviews*. 2008; 37(8): 1568–1578.
155. Inaba H, Tagawa H. Ceria-based solid electrolytes. *Solid State Ion*. 1996; 83: 1–16.
156. Aldebert P, Traverse JP. Structure and ionic mobility of zirconia at high temperature. *Journal of the American Ceramic Society*. 1985; 68: 34–40.
157. Hutchings KN, Bai J, Cutler RA, Wilson MA, Taylor DM. Electrochemical oxygen separation and compression using planar, co-sintered ceramics. *Solid State Ionics*. 2008; 179(11–12): 442–450.
158. Seeger J, Ivanova ME, Meulenberg WA, Sebold D, Stover D, Scherb T, Schumacher G, Escolastico S, Solis C, Serra JM. Synthesis and characterization of non-substituted and substituted proton-conducting $La_{6-x}WO_{12-y}$. *Inorganic Chemistry*. 2013; 52(18): 10375–10386.
159. Ivanova ME, Seeger J, Serra Alfaro JM, Solis Diaz C, Meulenberg WA, Fischer W, Roitsch W, Buchkremer S, Peter H. Influence of $La_6W_2O_{15}$ phase on the properties and integrity of $La_{6-x}WO_{12-d}$-based membranes. *Chemistry of Materials*. 2012; 2(1): 56–81.
160. Iwahara H, Esaka T, Uchida H, Maeda N. Proton conduction in sintered oxides and its application to steam electrolysis for hydrogen production. *Solid State Ionics*. 1981; 3–4: 359–363.
161. Sunarso J, Baumann S, Serra JM, Meulenberg WA, Liu S, Lin YS, Da Costa JD. Mixed ionic-electronic conducting (MIEC) ceramic-based membranes for oxygen separation. *Journal of Membrane Science*. 2008; 320: 13–41.
162. Sunarso J, Hashim SS, Zhu N, Zhou W. Perovskite oxides applications in high temperature oxygen separation, solid oxide fuel cell and membrane reactor: A review. *Progress in Energy and Combustion Science*. 2017; 61: 57–77.
163. Sunarso J, Baumann S, Serra JM, Meulenberg WA, Liu S, Lin YS, Diniz da Costa JC. Mixed ionic–electronic conducting (MIEC) ceramic-based membranes for oxygen separation. *Journal of Membrane Science*. 2008; 320: 13–41.
164. Anderson MT, Greenwood KB, Taylor GA, Poeppelmeier KR. B-cation arrangements in double perovskites. *Progress in Solid State Chemistry*. 1993; 22: 197–233.
165. Rao C, Gopalakrishnan J, Vidyasagar K. Superstructures, ordered defects and non- stoichiometry in metal oxides of perovskite and related structures. *Indian Journal of Chemistry A*. 1984; 23: 265–84.
166. Teraoka Y, Nobunaga T, Yamazoe N. Effect of cation substitution on the oxygen semi- permeability of perovskites-type oxides. *Chemistry Letters*. 1988; 17: 503–506.
167. Yamazaki Y, Hernandez-Sanchez R, Haile SM. Cation non-stoichiometry in yttrium doped barium zirconate: Phase behavior, microstructure, and proton conductivity. *Journal of Materials Chemistry*. 2010; 20(37): 8158–8166.
168. Leo A, Liu S, Costa J. Development of mixed conducting membranes for clean coal energy delivery. *International Journal of Greenhouse Gas Control*. 2009; 3(4): 357–367.
169. Hashim S, Mohamed A, Bhatia S. Current status of ceramic-based membranes for oxygen separation from air. *Advances in Colloid and Interface Science*. 2010; 160(1–2): 88–100.
170. Xu SJ, Thomson WJ. Oxygen permeation rates through ion-conducting perovskite membranes. *Chemical Engineering Science*. 1999; 54: 3839–3850.
171. Hong WK, Choi GM. Oxygen permeation of BSCF membrane with varying thickness and surface coating. *Journal of Membrane Science*. 2010; 346: 353–360.
172. Bouwmeester HJM, Kruidhof H, Burggraaf AJ. Importance of the surface exchange kinetics as rate limiting step in oxygen permeation through mixed-conducting oxides. *Solid State Ionics*. 1994; 72: 185–194.
173. Shen Z, Lu P, Hu J, Hu X. Performance of $Ba_{0.5}Sr_{0.5}Co_{0.8}Fe_{0.2}O_{3-d}$ membrane after laser ablation for methane conversion. *Catalysis Communications*. 2010; 11: 892–895.
174. Markov AA, Patrakeev MV, Leonidov IA, Kozhevnikov VL. Reaction control and long-term stability of partial methane oxidation over an oxygen membrane. *Journal of Solid State Electrochemistry*. 2011; 15: 253–257.
175. Teraoka Y, Zhang H, Furukawa S, Yamazoe N. Oxygen permeation through perovskite type oxides. *Chemical Letters*. 1985; 14: 1743–1746.

176. Tong J, Yang W, Cai R, Zhu B, Lin L. Novel and ideal zirconium-based dense membrane reactors for partial oxidation of methane to syngas. *Catalysis Letters*. 2002; 78: 129–137.
177. Schiestel T, Kilgus M, Peter S, Caspary KJ, Wang H, Caro J. Hollow fibre perovskite membranes for oxygen separation. *Journal of Membrane Science*. 2005; 258: 1–4.
178. Svarcova S, Wiik K, Tolchard J, Bouwmeester HJM, Grande T. Structural instability of cubic perovskite $Ba_xSr_{1-x}Co_{1-y}Fe_yO_{3-\delta}$. *Solid State Ion*. 2008; 178: 1787–1791.
179. Klande T, Ravkina O, Feldhoff A. Effect of A-site lanthanum doping on the CO_2 tolerance of $SrCo_{0.8}Fe_{0.2}O_{3-\delta}$ oxygen transporting membrane. *Journal of Membrane Science*. 2013; 437: 122–130.
180. Christopher K, Dimitrios RA. Review on energy comparison of hydrogen production methods from renewable energy sources. *Energy & Environmental Science*. 2012; 5(5): 6640–6651.
181. Holladay JD, Hu J, King DL, Wang Y. An overview of hydrogen production technologies. *Catalysis Today*. 2009; 139(4): 244–260.
182. Liu W, Cui Y, Du X, Zhang Z, Chao Z, Deng Y. High efficiency hydrogen evolution from native biomass electrolysis. *Energy & Environmental Science*. 2016; 9(2): 467–472.
183. Hauch A, Ebbesen SD, Jensen SH, Mogensen M. Highly efficient high temperature electrolysis. *Journal of Materials Chemistry*. 2008; 18(20): 2331–2340.
184. Bi L, Boulfrad S, Traversa E. Steam electrolysis by solid oxide electrolysis cells (SOECs) with proton-conducting oxides. *Chemical Society Reviews*. 2014; 43(24): 8255–8270.
185. Mogensen MB, Chen M, Frandsen HL, Graves C, Hansen JB, Hansen KV, Hauch A, Jacobsen T, Jensen SH, Skafte TL, Sun X. Reversible solid-oxide cells for clean and sustainable energy. *Clean Energy*. 2019; 3(3): 175–201.
186. Ebbesen SD, Jensen SH, Hauch A, Mogensen MB. High temperature electrolysis in alkaline cells, solid proton conducting cells, and solid oxide cells. *Chemical Reviews*. 2014; 114(21): 10697–10734.
187. Ge X, Zhang L, Fang Y, Zeng J, Chan SH. Robust solid oxide cells for alternate power generation and carbon conversion. *RSC Advances*. 2011; 1(4): 715–724.
188. Hansen JB. Solid oxide electrolysis – A key enabling technology for sustainable energy scenarios. *Faraday Discuss*. 2015; 182: 9–48.
189. Graves CR, Ebbesen SD, Mogensen MB, Lackner KS. Sustainable hydrocarbon fuels by recycling CO_2 and H_2O with renewable or nuclear energy. *Journal of Renewable Sustainable Energy Review*. 2011; 15: 1–23.
190. Tesfi A, Irvine JTS. Solid oxides fuel cells: Theory and material. *Journal of Renewable Sustainable Energy Review*. 2012; 4: 241–256.
191. Yu AS, Vohs JM, Gorte RJ. Interfacial reactions in ceramic membrane reactors for syngas production. *Energy & Environmental Science*. 2014; 7(3): 944–953.
192. Nakajo A, Mueller F, Brouwer J, Van Herle J, Favrat D. Mechanical reliability and durability of SOFC stacks. Part II: Modelling of mechanical failures during ageing and cycling. *International Journal of Hydrogen Energy*. 2012; 37: 9269–9286.
193. Shaikh SPS, Muchtar A, Somalu MR. A review on the selection of anode materials for solid-oxide fuel cells. *Journal of Renewable Sustainable Energy Review*. 2015; 51: 1–8.
194. Orera A, Slater PR. New chemical systems for solid oxide fuel cells. *Chemical Materials*. 2010; 22: 675–690.
195. Boukamp B. Fuel cells: Anodes sliced with ions. *Nature Materials*. 2006; 5(7): 517–518.
196. Jiang SP. Challenges in the development of reversible solid oxide cell technologies: A mini review. *Asia-Pacific Journal of Chemical Engineering*. 2016; 11(3): 386–391.
197. Lay-Grindler E, Laurencin J, Villanova J, Cloetens P, Bleuet P, Mansuy A, Mougin J, Delette G. Degradation study by 3D reconstruction of a nickel–yttria stabilized zirconia cathode after high temperature steam electrolysis operation. *Journal of Power Sources*. 2014; 269: 927–936.
198. Usseglio-Viretta F, Laurencin J, Delette G, Villanova J, Cloetens P, Leguillon D. Quantitative microstructure characterization of a Ni–YSZ bi-layer coupled with simulated electrode polarisation. *Journal of Power Sources*. 2014; 256: 394–403.
199. Tietz F, Sebold D, Brisse A, Schefold J. Degradation phenomena in a solid oxide electrolysis cell after 9000 h of operation. *Journal of Power Sources*. 2013; 223: 129–135.
200. Suntivich J, Gasteiger H, Yabuuchi N, Nakanishi H, Goodenough JB, Shao-Horn Y. Design principles for oxygen-reduction activity on perovskite oxide catalysts for fuel cells and metal-air batteries. *Nature Chemistry*. 2011; 3: 546–550.
201. Jiang SP, Chan SH. A review of anode materials development in solid oxide fuel cells. *Journal of Material Science*. 2004; 39: 4405–4439.

202. Azad AK, Kim JH, Irvine JTS. Structural, electrochemical, and magnetic characterization of the layered-type $PrBa_{0.5}Sr_{0.5}Co_2O_{5+\delta}$ perovskite. *Journal of Solid State Chemistry*. 2014; 213: 268–274.
203. Rossmeisl J, Bessler WG. Trends in catalytic activity for SOFC anode materials. *Solid State Ion*. 2008; 178: 1694–1700.
204. Zhou Y, Guan X, Zhou H, Ramados K, Adam S, Liu H, Lee S, Shi J, Tsuchiya M, Fong DD, Ramanathan S. Strongly correlated perovskite fuel cells. *Nature*. 2016; 534: 231–234.
205. Ge XM, Chan SH, Liu QL, Sun Q. Solid oxide fuel cell anode materials for direct hydrocarbon utilization. *Advances in Energy Materials*. 2012; 2: 1156–1181.
206. Yang X, Irvine JT. $(La_{0.75}Sr_{0.25})_{0.95}$ $Mn_{0.5}Cr_{0.5}O_3$ as the cathode of solid oxide electrolysis cells for high temperature hydrogen production from steam. *Journal of Materials Chemistry*. 2008; 18(20): 2349–2354.
207. Chen S, Xie K, Dong D, Li HX, Qin QQ, Zhang Y, Wu YC. A composite cathode based on scandium-doped chromate for direct high-temperature steam electrolysis in a symmetric solid oxide electrolyzer. *Journal of Power Sources*. 2015; 274: 718–729.
208. Kim G, Wang S, Jacobson AJ, Reimus L, Brodersen P, Mims CA. Rapid oxygen ion diffusion and surface exchange kinetics in $PrBaCo_2O_{5+x}$ with a perovskite related structure and ordered A cations. *Journal of Materials Chemistry*. 2007; 17(24): 2500–2505.
209. Xu Z, Li Y, Wan Y, Zhang S, Xia C. Nickel enriched Ruddlesden-Popper type lanthanum strontium manganite as electrode for symmetrical solid oxide fuel cell. *Journal of Power Sources*. 2019; 425:153–61.
210. Meixner DL, Cutler RA. Sintering and mechanical characteristics of lanthanum strontium manganite. *Solid State Ion*. 2002; 146: 273–284.
211. McCarthy BP, Pederson LR, Chou YS, Zhou XD, Surdoval WA, Wilson LC. Low temperature sintering of lanthanum strontium manganite-based contact pastes for SOFCs. *Journal of Power Sources*. 2008; 180: 294–300.
212. Lacorre P, Goutenoire F, Bohnke O, Retoux R, Laligant Y. Designing fast oxide-ion conductors based on $La_2Mo_2O_9$. *Nature*. 2000; 404(6780): 856–858.
213. Jiang SP. Development of lanthanum strontium cobalt ferrite perovskite electrodes of solid oxide fuel cells—A review. *International Journal of Hydrogen Energy*. 2019; 44: 7448–7493.
214. Jeong C, Lee J-H, Park M, Hong J, Kim H, Son J-W. Design and processing parameters of $La_2NiO_{4+\delta}$-based cathode for anode-supported planar solid oxide fuel cells (SOFCs). *Journal of Power Sources*. 2015; 297: 370–378.
215. Masaru T, Bo-Kuai L, Shriram R. Scalable nanostructured membranes for solid oxide fuel cells. *Nature Nanotechnology*. 2011; 6: 282–286.
216. Ishihara T. Nanomaterials for advanced electrode of low temperature solid oxide fuel cells (SOFCs). *Journal of Korean Ceramics Society*. 2016; 53(5): 469–477.
217. Menzler NH, Tietz F, Uhlenbruck S, Buchkremer HP, Stover D. Materials and manufacturing technologies for solid oxide fuel cells. *Journal of Materials Science*. 2010; 45: 3109–3135.
218. Arunkumar P, Meena M, Babu KS. A review on cerium oxide-based electrolytes for ITSOFC. *Nanomaterials and Energy*. 2012; 1(5): 288–305.
219. Arabaci A, Oksuzome MF. Preparation and characterization of 10 mol% Gd doped CeO_2 (GDC) electrolyte for SOFC applications. *Ceramics International*. 2012; 38: 6509–6515.
220. Kilner JA. Optimisation of oxygen ion transport in materials for ceramic membrane devices. *Faraday Discussions*. 2007; 134: 9–15.
221. Mahato N, Gupta A, Balani K. Doped zirconia and ceria-based electrolytes for solid oxide fuel cells: A review. *Nanomaterials and Energy*. 2012; 1(1): 27–45.
222. Majumdar P, Penmetsa SK. Solid oxide fuel cell: Design, materials, and transport phenomena. *Nanomaterials and Energy*. 2012; 1(5): 247–264.
223. Kilner JA, Druce J, Ishihara T. Chapter 4: Electrolytes. In: *High-Temperature Solid Oxide Fuel Cells for the 21st Century: Fundamentals, Design and Applications*, 2nd ed. London, UK: Elsevier, 2016.
224. Laguna-Bercero MA, Skinner SJ, Kilner JA. Performance of solid oxide electrolysis cells based on scandia stabilised zirconia. *Journal of Power Sources*. 2009; 192(1): 126–131.
225. Chiodelli G, Malavasi L, Tealdi C, Barison S. Role of synthetic route on the transport properties of $BaCe_{1-x}Y_xO_3$ proton conductor. *Journal of Alloys and Compounds*. 2009; 470(1): 477–485.
226. Ni M. An electrochemical model for syngas production by co-electrolysis of H_2O and CO_2. *Journal of Power Sources*. 2012; 202(8): 209–216.
227. D'Epifanio A, Fabbri E, Di Bartolomeo E, Licoccia S, Traversa E. Design of $BaZr_{0.8}Y_{0.2}O_{3-d}$ protonic conductor to improve the electrochemical performance in intermediate temperature solid oxide fuel cells (IT-SOFCs). *Fuel Cells*. 2008; 8(1): 69–76.

228. Zakowsky N, Williamson S, Irvine JT. Elaboration of CO_2 tolerance limits of $BaCe_{0.9}Y_{0.1}O_{3-d}$ electrolytes for fuel cells and other applications. *Solid State Ionics.* 2005; 176(39–40): 3019–3026.
229. IEA. *Global Energy & CO_2 Status Report 2019.* Paris: IEA, 2019. https://www.iea.org/reports/global-energy-co2-status-report-2019.
230. IEA. *Electricity Market Report.* Paris: IEA, 2021. https://www.iea.org/reports/electricity-market-report-july-2021; https://www.who.int/health-topics/air-pollution#tab=tab_2.
231. IEA. *World Energy Outlook 2020.* Paris: IEA. https://www.iea.org/reports/world-energy-outlook-2020.
232. EG &G Services Inc. *Fuel Cell Handbook*, 5th ed. Morgantown, West Virginia: U.S. Department of Energy Office of Fossil Energy National Energy Technology Laboratory Inc., 2000. http://www.cientificosaficionados.com/libros/pilas%20de%20com.
233. Grove WR. On voltaic series and the combination of gases by platinum. *The London, Edinburgh, and Dublin Philosophical Magazine and Journal of Science.* 1839; 14(86–87): 127–130.
234. Kerman K, Lai BK, Ramanathan S. Nanoscale compositionally graded thin-film electrolyte membranes for low-temperature solid oxide fuel cells. *Advanced Energy Materials.* 2012; 2(6): 656–661.
235. Andjar JM, Segura F. Fuel cells: History and updating. A walk along two centuries. *Journal of Renewable Sustainable Energy Review.* 2009; 13: 2309–2322.
236. Gross JH. Fuel cell technology. *Joint Legislative Air Water Pollution Communication.* 2002; 2: 1–7.
237. Cook B. Introduction to fuel cells and hydrogen technology. *Engineering Science and Education Journal.* 2002; 11(6): 205–216.
238. Hikosaka N. *Fuel Cells: Current Technology Challenges and Future Research Needs.* Amsterdam: Elsevier B.V., 2013.
239. Steele BCH, Heinzel A. Materials for fuel-cell technologies. *Nature.* 2001; 414(6861): 345–352.
240. Haile SM. Fuel cell materials and components. *Acta Materials.* 2003; 51: 5981–6000.
241. Ash RL, Dowler WL, Varsi G. Feasibility of rocket propellant production on Mars. *Acta Astronautica.* 1978; 5: 705–724.
242. Stancati ML, Niehoff JC, Wells WC, Ash RL. Remote automated propellant production: A new potential for round trip spacecraft. *AIAA.* 1979; 79–0906: 262–270.
243. Richter R. Basic investigation into the production of oxygen in a solid electrolyte. *AIAA.* 1981; 81–1175: 1–14.
244. Sridhar KR, Vaniman BT. Oxygen production on mars using solid oxide electrolysis. *Solid State Ionics.* 1997; 93(3–4): 321–328.
245. Sridhar KR, Iacomini CS, Finne JE. Combined H_2O/CO_2 solid oxide electrolysis for Mars *in situ* resource utilization. *Journal of Propulsion and Power.* 2004; 20(5): 892–901.
246. Pandey A. Progress in solid oxide fuel cell (SOFC) research. *JOM.* 2018; 71(6396): 88–89.
247. Hossain S, Abdalla AM, Binti Jamain SN, Zaini JJH, Azad AK. A review on proton conducting electrolytes for clean energy and intermediate temperature-solid oxide fuel cells. *Journal of Renewable Sustainable Energy Review.* 2017; 79: 750–764.
248. Lee YH, Chang I, Cho GY, Park J, Yu W, Tanveer WH, Cha SW. Thin film solid oxide fuel cells operating below 600°C: A review. *International Journal of Precision Engineering and Manufacturing.* 2018; 5(3): 441–453.
249. Lyu Y, Wang F, Wang D, Jin Z. Alternative preparation methods of thin films for solid oxide fuel cells: Review. *Material Technology.* 2019; 37(24): 19371–19379.
250. Goodenough JB. Electrochemical energy storage in a sustainable modern society. *Journal of Energy Environmental Science.* 2014; 7: 14–18.
251. Yokokawa H, Tu H, Iwanschitz B, Mai A. Fundamental mechanisms limiting solid oxide fuel cell durability. *Journal of Power Sources.* 2008; 182: 400–412.
252. Kan WH, Thangadurai V. Challenges and prospects of anodes for solid oxide fuel cells (SOFCs). *Ionics.* 2015; 21(2): 301–318.
253. Stambouli AB, Traversa E. Solid oxide fuel cells (SOFCs): A review of an environmentally clean and efficient source of energy. *Journal of Renewable Sustainable Energy Review.* 2002; 6: 433–455.
254. Tesfi A, Irvine JTS. Solid oxides fuel cells: Theory and material. *Comprehensive Renewable Energy.* 2012; 4: 241–256.
255. Hajimolana SA, Hussain MA, Daud WMAW, Soroush M, Shamiri A. Mathematical modeling of solid oxide fuel cells: A review. *Journal of Renewable Sustainable Energy Review.* 2011; 15(4): 1893–1917.
256. Park JS, An J, Lee MH, Prinz FB, Lee W. Effects of surface chemistry and microstructure of electrolyte on oxygen reduction kinetics of solid oxide fuel cells. *Journal of Power Sources.* 2015; 295: 74–78.
257. Huang J, Mao Z, Liu Z, Wang C. Development of novel low-temperature SOFCs with co-ionic conducting SDC-carbonate composite electrolytes. *Electrochemical Communications.* 2007; 9(10): 2601–2605.

258. Huang X, Ni C, Zhao G, Irvine JTS. Oxygen storage capacity and thermal stability of the $CuMnO_2$–CeO_2 composite system. *Journal of Material Chemistry A*. 2015; 3: 12958–12964.
259. Huang J, Gao Z, Mao Z. Effects of salt composition on the electrical properties of samaria-doped ceria/carbonate composite electrolytes for low-temperature SOFCs. *International Journal of Hydrogen Energy*. 2010; 35(9): 4270–4275.
260. Tikkanen H, Suciu C, Wærnhus I, Hoffmann AC. Dip-coating of 8ysz nanopowder for sofc applications. *Ceramics International*. 2011; 37(7): 2869–2877.
261. Lee D-S, Kim W, Choi S-H, Lee H, Lee JH. Characterization of ZrO_2 Co-doped with Sc_2O_3 and CeO_2 electrolyte for the application of intermediate temperature SOFCs. *Solid State Ionics*. 2005; 176: 33–39.
262. Singh RK, Singh P. (2014) Electrical conductivity of barium substituted LSGM electrolyte materials for IT-SOFC. *Solid State Ionics*. 2014; 262: 428–432.
263. Oskouyi OE, Maghsoudipour A, Shahmiri M, Hasheminiasari M. Preparation of YSZ electrolyte coating on conducting porous Ni–YSZ cermet by DC and pulsed constant voltage electrophoretic deposition process for SOFCs applications. *Journal of Alloys and Compounds*. 2019; 795: 361–369.
264. Lyu Y, Xie J, Wang D, Jiarao Wang J. Review of cell performance in solid oxide fuel cells. *Journal of Material Science*. 2020; 55: 7184–7207.
265. Chevalier J, Gremillard L, Virkar AV, Clarke DR. The tetragonalmonoclinic transformation in Zirconia: Lessons learned and future trends. *Journal of the American Ceramic Society*. 2009; 92: 1901–1920.
266. Fleischhauer F, Bermejo R, Danzer R, Mai A, Graule T, Kuebler J. High temperature mechanical properties of zirconia tapes used for electrolyte supported solid oxide fuel cells. *Journal of Power Sources*. 2015; 273: 237–243.
267. Ni DW, Charlas B, Kwok K, Molla TT, Hendriksen PV, Frandsen HL. Influence of temperature and atmosphere on the strength and elastic modulus of solid oxide fuel cell anode supports. *Journal of Power Sources*. 2016; 311: 1–12.
268. Toor S, Croiset E. Reducing sintering temperature while maintaining high conductivity for SOFC electrolyte: Copper as sintering aid for Samarium Doped Ceria. *Ceramics International*. 2019; 46: 1148–1157.
269. Graves C, Martinez L, Sudireddy BR. High performance nanoceria electrodes for solid oxide cells. *ECS Transactions*. 2016; 72: 183–192.
270. Fu Z, Sun Q, Ma D, Zhang N, An Y, Yang Z. Effects of Sm doping content on the ionic conduction of CeO_2 in SOFCs from first principles. *Applied Physical Letters*. 2017; 111: 023903-1-023903-5.
271. Singh RK, Singh P. Electrical conductivity of barium substituted LSGM electrolyte materials for IT-SOFC. *Solid State Ionics*. 2014; 262: 428–432.
272. Aarthi U, Arunkumar P, Sribalaji M, Keshri AK, Babu KS. Strontium mediated modification of structure and ionic conductivity in samarium doped ceria/sodium carbonate nanocomposites as electrolytes for LTSOFC. *RSC Advances*. 2016; 6(88): 84860–84870.
273. Shima D, Haile SM. The influence of cation nonstoichiometry on the properties of undoped and gadolinia-doped barium cerate. *Solid State Ionics*. 1997; 97(1–4): 443–455.
274. Katahira K, Kohchi Y, Shimura T, Iwahara H. Protonic conduction in Zr-substituted $BaCeO_3$. *Solid State Ionics*. 2000; 138(1–2): 91–98.
275. Fabbri E, D'Epifanio A, Di Bartolomeo E, Licoccia S, Traversa E. Tailoring the chemical stability of $Ba(Ce_{0.8-x}Zr_x)Y_{0.2}O_{3-d}$ protonic conductors for intermediate temperature solid oxide fuel cells (IT-SOFCs). *Solid State Ionics*. 2008; 179(15–16): 558–564.
276. Bi L, Shangquan Z, Fang S, Zhang L, Xie K, Xia C, Liu W. Preparation of an extremely dense BaCe 0.8Sm 0.2O3 - d thin membrane based on an in-situ reaction. *Electrochemical Communications*. 2008; 10: 1005–1007.
277. Ito N, Iijima M, Kimura K, Iguchi S. New intermediate temperature fuel cell with ultra- thin proton conductor electrolyte. *Journal of Power Sources*. 2005; 152: 200–203.
278. Zhong HT, Ai DS, Lin XP. LSGM-carbonate composite electrolytes for intermediate-temperate SOFCs. *Key Engineering Materials*. 2014; 602–603: 862–865.
279. Wang X, Ma Y, Li S, Kashyout A, Zhu B, Muhammed M. Ceria-based nanocomposite with simultaneous proton and oxygen ion conductivity for low-temperature solid oxide fuel cells. *Journal of Power Sources*. 2011; 196:2754–2758.
280. Benamira M, Ringuede A, Hildebrandt L, Lagergren C, Vannier RN, Cassir M. Gadolinia-doped ceria mixed with alkali carbonates for SOFC applications: II—An electrochemical insight. *International Journal of Hydrogen Energy*. 2012; 37(24): 19371–19379
281. Jiang X, Wu F, Wang H. Yb-doped $BaCeO_3$ and its composite electrolyte for intermediate-temperature solid oxide fuel cells. *Materials (Basel)*. 2019; 12(5): 739.

282. Wang W, Qu J, Juliao PSB, Shao Z. Recent advances in the development of anode materials for solid oxide fuel cells utilizing liquid oxygenated hydrocarbon fuels: A mini review. *Energy Technology*. 2019; 7: 33–44.
283. Tan W, Fan L, Raza R, Khan A, Zhu B. Studies of modified lithiated NiO cathode for low temperature solid oxide fuel cell with ceria-carbonate composite electrolyte. *International Journal of Hydrogen Energy*. 2013; 38: 370–376.
284. Xia C, Wang B, Cai Y, Zhang W, Afzal M, Zhu B. Electrochemical properties of LaCePr-oxide/K_2WO_4 composite electrolyte for low-temperature SOFCs. *Electrochemical Communications*. 2016; 77: 44–48.
285. Bharadwaj SR, Varma S, Wani BN. Electroceramics for fuel cells, batteries and sensors. *Functional Materials*. 2012; 639–674.
286. Mocoteguy P, Brisse A. A review and comprehensive analysis of degradation mechanisms of solid oxide electrolysis cells. *International Journal of Hydrogen Energy*. 2013; 38(36): 15887–15902.
287. Pelegrini L, Joao NR, Hotza D. Process and materials improvements on Ni/Cu-YSZ composites towards nanostructured SOFC anodes: A review. *Reviews on Advanced Materials Science*. 2016; 46: 6–21.
288. You HX, Gao HJ, Chen G, Abudula A, Ding XW. Effects of dry methane concentration on the methane reactions at Ni-YSZ anode in solid oxide fuel cell. *Journal of Fuel Chemical Technology*. 2013; 41: 374–379.
289. Koh JH, Yoo Y-S, Park J, Lim H. Carbon deposition and cell performance of Ni-YSZ anode support SOFC with methane fuel. *Solid State Ionics*. 2002; 149: 157–166.
290. Menzler NH, Schafbauer W, Han F, Buchler O, Mucke R, Buchkremer HP, Stover D. Development of high-power density solid oxide fuel cells (SOFCs) for long-term operation. *Materials Science Forum*. 2010; 654: 2875–2878.
291. Akikusa J, Adachi K, Hoshino K, Ishihara T, Takita Y. Development of a low temperature operation solid oxide fuel cell. *Journal of Electrochemical Society*. 2001; 148: A1275.
292. Yang B, Koo J, Shin J, Go D, Shim J, An J. Direct alcohol-fueled low-temperature solid oxide fuel cells: A review. *Energy Technology*. 2017; 7: 5–19.
293. Bogolowski N, Iwanschitz B, Drillet JF. Development of a coking resistant NiSn anode for the direct methane SOFC. *Fuel Cells*. 2015; 15(5): 711–717.
294. Li Y, Yu C, Fang M, Zhuan X, Zhongyang L, Kefa C. Fabrication and characterization of SOFC anode, Ni-SDC cermet. *Journal of Chinese Rare Earth Society*. 2006; 24(1): 32–36.
295. Handal HT, Thangadurai V. Electrochemical characterization of multi-element-doped ceria as potential anodes for SOFCs. *Solid State Ionics*. 2014; 262: 359–364.
296. Sun LL, Hu ZM, Luo LH, Yefan W, Jijun S, Liang C, Xu X. Application of Ru nano flowers doped ni-ysz anode in ethanol-fueled SOFC. *Rare Metal Materials and Engineering*. 2017; 46(8): 2322–2326.
297. Lei Z, Zhu Q-S, Han MF. Fabrication and performance of direct methane SOFC with a Cu-CeO_2-based anode. *Acta Physico-Chimica Sinica*. 2010; 26: 583–588.
298. Im HN, Jeon SY, Choi MB, Kim HS, Song SJ. Chemical stability and electrochemical properties of $CaMoO_{3-d}$ for SOFC anode. *Ceramic International*. 2012; 38: 153.
299. Ma Q, Tietz F. Comparison of Y and La-substituted $SrTiO_3$ as the anode materials for SOFCs. *Solid State Ionics*. 2012; 225: 108–112.
300. Xiu-Xia M, Gong X, Tan XY, Ma ZF. Preparation and properties of direct-methane solid oxide fuel cell based on a graded Cu-CeO_2-Ni–YSZ composite anode. *Acta Physico-Chimica Sinica*. 2013; 29: 1719–1726.
301. Sun C, Hui R, Roller J. Cathode materials for solid oxide fuel cells: A review. *Journal of Solid State Electrochemistry*. 2010; 14: 1125–1144.
302. Zuo N, Zhang M, Mao Z, Xie F. Fabrication and characterization of composite electrolyte for intermediate temperature SOFC. *Journal of European Ceramic Society*. 2011; 31: 3103–3107.
303. Raharj J, Muchtar AT, Dawood WRW, Muhamad N, Majlanlie EH. Fabrication of porous LSCF-SDC carbonates composite cathode for solid oxide fuel cell (SOFC). *Trans Tech Publications*. 2011; 471: 179–185.
304. Wang G, Wu X, Cai Y, Ji Y, Yaqub A, Zhu B. Design, fabrication and characterization of a double layer solid oxide fuel cell (DLFC). *Journal of Power Sources*. 2016; 332: 8–15.
305. Fushao L, Shubiao X, Yuxing Y, Feixiang C. Thermal stability and electrochemical performance of LaSr-$CoO_{4\pm d}$ as the cathode material for solid-oxide fuel cells. *Journal of Kunming University Science & Technology (Natural Science)*. 2018; 43(5): 14–21.
306. Huang Y, Vohs J, Gorte R. An examination of LSMLSCo mixtures for use in SOFC cathodes. *Journal of Electrochemical Society*. 2006; 153: A951–A955.

307. Songbo L, Yanru Y, Yingjie Z, An S. Preparation and properties of cathode material $La_{0.7}Sr_{0.3}Cu_xCo_yMn_{(1-xy)}O_3$ for SOFC. *Rare Metal Material Engineering.* 2013; 42: 487–490.
308. Wang W, Mogensen M. High-performance lanthanum-ferrites-base cathode for SOFC. *Solid State Ionics.* 2005; 176: 457–462.
309. Fan B, Ren X, Cong Y, Liang W. The influence of A-site composition changes of LSCF cathode materials on the transmittability of oxygen-ion and electrical performance of SOFC. *Jouranl of Functional Materials.* 2015; 46(A2): 125–128.
310. Xiao H, Sun LP, Zhao H, Huo L-H, Bassat J, Rougier A, Fourcade S, Grenier J. Preparation and electrochemical properties of LaBiMn2O6 cathode for IT-SOFCs. *Chinese Journal of Inorganic Chemistry.* 2015; 31: 1139–1144.
311. Hansen KK, Hansen KV, Mogensen M. High-performance Fe–Co-based SOFC cathodes. *Journal of Solid State Electrochemistry.* 2010; 14: 2107–2112.
312. He Z. Preparation and properties of SmBaCo-CuO_{5+d} material as cathode for IT-SOFC. *China Ceramic Industry.* 2015; 6: 15–19.
313. Jaiswal A, Wachsman E. Bismuth-ruthenate-based cathodes for IT-SOFCs. *Journal of Electrochemical Society.* 2005; 152: A787–A790.
314. Meng F, Xia T, Wang J, Shi Z, Zhao H. Praseodymium-deficiency $Pr_{0.94}BaCo_2O_{6-\delta}$ double perovskite: A promising high performance cathode material for intermediate- temperature solid oxide fuel cells. *Journal of Power Sources.* 2015; 293: 741–750.
315. Chen S, Tu H, Li S, Yu Q. Fabrication and performance of $Pr_{2-x}La_xNiO_{4+d}$ based cathode for intermediate temperature solid oxide fuel cells. *Chinese Jouranl of Power Sources.* 2018; 42(2): 219–222.
316. Muhoza SP, Gross MD. Creating and preserving nanoparticles during co-sintering of solid oxide electrodes and its impact on electrocatalytic activity. *Catalysts.* 2021; 11(9): 1073.
317. Hawkes A, Staffell I, Brett D, Brandon N. Fuel cells for micro-combined heat and power generation. *Energy & Environmental Science.* 2009; 2(7): 729–744.
318. Huang H, Nakamura M, Su P, Fasching R, Saito Y, Prinz FB. High-performance ultrathin solid oxide fuel cells for low-temperature operation. *Journal of the Electrochemical Society.* 2006; 154(1): B20.
319. Tsuchiya M, Lai BK, Ramanathan S. Scalable nanostructured membranes for solid-oxide fuel cells. *Nature Nanotechnology.* 2011; 6(5): 282–286.
320. Kang S, Su PC, Park YI, Saito Y, Prinz FB. Thin-film solid oxide fuel cells on porous nickel substrates with multistage nanohole array. *Journal of the Electrochemical Society.* 2006; 153(3): A554.
321. Jung W, Gu KL, Choi Y, Haile SM. Robust nanostructures with exceptionally high electrochemical reaction activity for high temperature fuel cell electrodes. *Energy & Environmental Science.* 2014; 7(5): 1685–1692.
322. Chockalingam R, Basu S. Impedance spectroscopy studies of Gd-CeO_2–(LiNa) CO_3 nano composite electrolytes for low temperature SOFC applications. *International Journal of Hydrogen Energy.* 2011; 36(22): 14977–14983.
323. Myung JH, Shin T, Kim SD, Park HG, Moon J, Hyun SH. Optimization of Ni–zirconia based anode support for robust and high-performance 5x5 cm^2 sized SOFC via tape-casting/co-firing technique and nano-structured anode. *International Journal of Hydrogen Energy.* 2015; 40(6): 2792–2799.
324. Shah M, Voorhees PW, Barnett SA. Time-dependent performance changes in LSCF-infiltrated SOFC cathodes: The role of nano-particle coarsening. *Solid State Ionics.* 2011; 187(1): 64–67.
325. Tsuchiya M, Lai BK, Ramanathan S. Scalable nanostructured membranes for solid-oxide fuel cells. *Nature Nanotechnology.* 2011; 6(5): 282–286.
326. Zhang H, Zhao F, Chen F, Xia C. Nano-structured Sm0. 5Sr0. 5CoO3−δ electrodes for intermediate-temperature SOFCs with zirconia electrolytes. *Solid State Ionics.* 2011; 192(1): 591–594.
327. Wang X, Huang H, Holme T, Tian X, Prinz FB. Thermal stabilities of nanoporous metallic electrodes at elevated temperatures. *Journal of Power Sources.* 2008; 175(1): 75–81.
328. Cho GY, Lee YH, Cha SW. Multi-component nano-composite electrode for SOFCS via thin film technique. *Renewable Energy.* 2014; 65: 130–136.
329. Chen Y, Gerdes K, Song X. Nanoionics and nanocatalysts: Conformal mesoporous surface scaffold for cathode of solid oxide fuel cells. *Scientific Reports.* 2016; 6(1): 1–8.
330. Chrzan A, Karczewski J, Szymczewska D, Jasinski P. Nanocrystalline cathode functional layer for SOFC. *Electrochimica Acta.* 2017; 225: 168–174.
331. Matheswaran P, Rajasekhar M, Subramania A. Assisted combustion synthesis and characterization of Pr0. 6Sr0. 4MnO3±δ nano crystalline powder as cathode material for IT-SOFC. *Ceramics International.* 2017; 43(1): 988–991.

332. Yoon KJ, Biswas M, Kim HJ, Park M, Hong J, Kim H, Son JW, Lee JH, Kim BK, Lee HW. Nano-tailoring of infiltrated catalysts for high-temperature solid oxide regenerative fuel cells. *Nano Energy*. 2017; 36: 9–20.
333. Liu Z, Zhao Z, Shang L, Ou D, Cui D, Tu B, Cheng M. LSM-YSZ nano-composite cathode with YSZ interlayer for solid oxide fuel cells. *Journal of Energy Chemistry*. 2017; 26(3): 510–514.
334. Li J, Zhang N, Sun K, Wu Z. High thermal stability of three-dimensionally ordered nano-composite cathodes for solid oxide fuel cells. *Electrochimica Acta*. 2016; 187: 179–185.
335. Fan H, Zhang Y, Han M. Infiltration of La0• 6Sr0• 4FeO3-δ nanoparticles into YSZ scaffold for solid oxide fuel cell and solid oxide electrolysis cell. *Journal of Alloys and Compounds*. 2017; 723: 620–626.
336. Ozmen O, Zondlo JW, Lee S, Gerdes K, Sabolsky EM. Bio-inspired surfactant assisted nano-catalyst impregnation of solid-oxide fuel cell (SOFC) electrodes. *Materials Letters*. 2016; 164: 524–527.
337. Li J, Zhang N, He Z, Sun K, Wu Z. Preparation and characterization of one-dimensional nano-structured composite cathodes for solid oxide fuel cells. *Journal of Alloys and Compounds*. 2016; 663: 664–671.
338. Takagi Y, Lai BK, Kerman K, Ramanathan S. Low temperature thin film solid oxide fuel cells with nanoporous ruthenium anodes for direct methane operation. *Energy & Environmental Science*. 2011; 4(9): 3473–3478.
339. Chang I, Ji S, Park J, Lee MH, Cha SW. Ultrathin YSZ coating on Pt cathode for high thermal stability and enhanced oxygen reduction reaction activity. *Advanced Energy Materials*. 2015; 5(10): 1402251.
340. Liu KY, Fan L, Yu CC, Su PC. Thermal stability and performance enhancement of nano-porous platinum cathode in solid oxide fuel cells by nanoscale ZrO_2 capping. *Electrochemistry Communications*. 2015; 56: 65–69.
341. Choi HJ, Kim M, Neoh KC, Jang DY, Kim HJ, Shin JM, Kim GT, Shim JH. High-performance silver cathode surface treated with scandia-stabilized zirconia nanoparticles for intermediate temperature solid oxide fuel cells. *Advanced Energy Materials*. 2017; 7(4): 1601956.
342. Jung W, Dereux JO, Chueh WC, Hao Y, Haile SM. High electrode activity of nanostructured, columnar ceria films for solid oxide fuel cells. *Energy & Environmental Science*. 2012; 5: 8682.
343. Mahato N, Banerjee A, Gupta A, Omar S, Balani K. Progress in material selection for solid oxide fuel cell technology: A review. *Progress in Materials Science*. 2015; 72: 141–337.
344. Zheng LL, Wang X., Zhang L, Wang JY. Effect of Pd-impregnation on performance, sulfur poisoning and tolerance of Ni/GDC anode of solid oxide fuel cells. *International Journal of Hydrogen Energy*. 2012; 37(13): 10299–10310.
345. Zhan Z, Bierschenk DM, Cronin JS, Barnett SA. A reduced temperature solid oxide fuel cell with nano-structured anodes. *Energy & Environmental Science*. 2011; 4(10): 3951–3954.
346. Boyd DA, Hao Y, Li C, Goodwin DG, Haile SM. Block copolymer lithography of rhodium nanoparticles for high temperature electrocatalysis. *ACS Nano*. 2013; 7(6): 4919–4923.
347. Adijanto L, Sampath A, Yu AS, Cargnello M, Fornasiero P, Gorte RJ, Vohs JM. Synthesis and stability of Pd@ CeO_2 core–shell catalyst films in solid oxide fuel cell anodes. *ACS Catalysis*. 2013; 3(8): 1801–1809.
348. Raza R, Wang X, Ma Y, Zhu B. A nanostructure anode (Cu0. 2Zn0. 8) for low- temperature solid oxide fuel cell at 400–600 C. *Journal of Power Sources*. 2010; 195(24): 8067–8070.
349. Klotz D., Butz B, Leonide A, Hayd J, Gerthsen D, Ivers-Tiffée E. Performance enhancement of SOFC anode through electrochemically induced Ni/YSZ nanostructures. *Journal of the Electrochemical Society*. 2011; 158(6): B587.
350. Tan X, Wang Z, Meng B, Meng X, Li K., Pilot-scale production of oxygen from air using perovskite hollow fibre membranes, *Journal of Membrane Science*. 2010; 352: 89–196.
351. Sunarti AR, Ahmad AL. Peformances of automated control system for membrane gas absorption: Optimization study. *Jounal of Advanced Research in Applied Mechanics*. 2015; 6: 1–20.
352. Vente JF, Haije WG, IJpelaan R, Rusting FT. On the full-scale module design of an air separation unit using mixed ionic electronic conducting membranes. *Journal of Membrane Science*. 2006; 278(1–2): 66–71.
353. Wang Y, Leung DYC, Xuan J, Wang H. A review on unitized regenerative fuel cell technologies, part B: Unitized regenerative alkaline fuel cell, solid oxide fuel cell, and microfluidic fuel cell. *Renewable Sustainable Energy Reviews*. 2017; 75: 775–795.
354. Wincewicz KC, Cooper JS. Taxonomies of SOFC material and manufacturing alternatives. *Journal of Power Sources*. 2005; 140: 280–296.
355. Jiao Z, Takagi N, Shikazono N, Kasagi N. Study on local morphological changes of nickel in solid oxide fuel cell anode using porous Ni pellet electrode. *Journal of Power Sources*. 2011; 196(3): 1019–1029.

356. Chen K, Jiang SP. Failure mechanism of (La, Sr) MnO_3 oxygen electrodes of solid oxide electrolysis cells. *International Journal of Hydrogen Energy.* 2011; 36(17): 10541–10549.
357. Virkar AV. Mechanism of oxygen electrode delamination in solid oxide electrolyser cells. *International Journal of Hydrogen Energy.* 2010; 35(18): 9527–9543.
358. Kan WH, Thangadurai V. Challenges and prospects of anodes for solid oxide fuel cells (SOFCs). *Journal of Ion.* 2014; 21: 301–318.
359. Schefold J, Brisse A, Tietz F. Nine thousand hours of operation of a solid oxide cell in steam electrolysis mode. *Journal of the Electrochemical Society.* 2011; 159(2): A137–A144.
360. Mahato N, Banerjee A, Gupta A, Omar S, Balani K. Progress in material selection for solid oxide fuel cell technology: A review. *Progress in Materials Science.* 2015; 72: 141–337.
361. Li S, Zhu B. Electrochemical performances of nanocomposite solid oxide fuel cells using nano-size material $LaNi_{0.2}Fe_{0.65}Cu_{0.15}O_3$ as cathode. *J Nanoscience Nanotechnology.* 2009; 9: 3824–3827.
362. Chen Y, Gerdes K, Song X. Nanoionics and nanocatalysts: Conformal mesoporous surface scaffold for cathode of solid oxide fuel cells. *Science Reports.* 2016; 6: 32997.
363. Jiang SP. Nanoscale and nano-structured electrodes of solid oxide fuel cells by infiltration advances and challenges. *International Journal of Hydrogen Energy.* 2012; 37: 449–470.
364. Chrzan A, Karczewski J, Szymczewska D, Jasinski P. Nanocrystalline cathode functional layer for SOFC. *Electrochimica Acta.* 2017; 225: 168–174.
365. Adams T, Nease J, Tucker D, Barton PI. Energy conversion with solid oxide fuel cell systems: A review of concepts and outlooks for the short- and long-term. *Industrial & Engineering Chemistry Research.* 2012; 52: 3089–3111.

5 Graphene Oxide and Reduced Graphene Oxide Additives to Improve Basin Water Evaporation Rates for Solar Still Desalination

Wisut Chamsa-ard, Derek Fawcett, Chun Che Fung, and Gérrard Eddy Jai Poinern

CONTENTS

5.1 INTRODUCTION

Finding new sources of high-quality drinkable water is an international objective. Globally, the demand for water is rising by 2% each year and is expected to reach 6,900 billion m^3 by 2030 [1]. However, Earth's natural water cycle only produces 4,200 billion m^3 annually which will result in a shortfall of 2700 billion m^3 [2]. Presently, Earth's natural water cycle is unable to meet this global demand, and the deficit is made up by using high energy-consuming desalination processes. Current estimates also indicate that if no new sources of high-quality water are found or desalination levels are not increased, then around two-thirds of the global population will have insufficient drinkable water by 2025 [3]. Furthermore, the problem will be exacerbated by global population growth which is expected to be around 12.3 billion in 2100 [4, 5]. Therefore, developing new sustainable and cost-effective technologies for generating high-quality water is urgently needed to meet both current and future demands. Solar irradiation is the largest source of sustainable and renewable energy, and harvesting it for water distillation offers a practical technology for producing high-quality water [6]. Solar thermal stills have low operational costs and do not produce harmful greenhouse gas emissions [7]. Typically, solar energy passes through a sloping glass cover before heating saline water contained in a basin. The resulting vaporized water

DOI: 10.1201/9781003181422-6

condenses on the cooler inner surface of the sloping glass cover. The condensed water droplets then move downwards under the influence of gravity and are collected by a trough fitted along the entire length of the cover's lower edge [8]. The collected desalinated water is free of impurities and drinkable [9]. However, solar thermal stills tend to have lower water production rates (2 to 5 L/m^2/day) compared to more traditional high-energy-based desalination processes, which reduces their economic competitiveness [10].

In recent years, there has been a concerted effort to improve solar thermal still water productivity by improving evaporation rates during the solar heating cycle [11]. During the heating cycle, only water molecules located close to the air-water interface are evaporated to form the vapor phase [12]. However, heat is also transferred to the water volume below the air-water interface where it causes the water to boil. During boiling, bubbles form and migrate to the air-water interface, which further dissipates heat. Importantly, studies have shown the addition of small amounts of particular materials in basin water can significantly improve the absorption of solar energy. Thus, increasing energy absorption translates to higher water temperatures and higher evaporation rates. Studies have shown that the addition of dyes to basin water not only darkens the water but also increases the absorption of solar energy and improves solar thermal still performance [13, 14]. For instance, Nijmeh et al. found the addition of violet dye into basin water increased its evaporation rate by 29% [14]. Studies have also found the addition of floating materials like charcoal can also significantly improve productivity [15, 16]. For example, porous floating absorbers made of blackened jute cloth were found to increase water productivity by 68% on clear days and 35% on cloudy days when added to the basin water of a single slope solar thermal still [17]. Other researchers have focused on incorporating nanomaterials into basin water as a method for improving overall thermal efficiency and water productivity [18, 19]. In particular, carbon-based nanomaterials like graphite, graphene, and carbon nanotubes have all been found to display superior broadband solar absorbance. The enhanced solar absorbance of these carbon-based nanomaterials is due to their sp^{2-} hybridized carbon atomic structure and π-band optical transitions [20]. However, high levels of nanomaterial concentrations in basin water can reduce performance. High concentrations restrict solar irradiation to the upper layer of basin water and also lead to greater nanofluid instability. The instability results from nanoparticle agglomeration and sedimentation [21].

The present study has developed a method for producing well-dispersed graphene oxide (GO) and reduced graphene oxide (RGO) nanoparticles in water-based solutions at room temperature. These two carbon-based additives were designed to be mixed into the basin water of a solar thermal still in order to improve its thermal efficiency and water evaporation rate. Commercially available graphene oxide powder was reduced using tetra ethylene ammonium hydroxide (TEAH), which also acted as the stabilizing agent. TEAH reduction was confirmed by X-ray diffraction (XRD), UV-visible spectroscopy, and Fourier-transform infrared spectroscopy (FTIR). Electron microscopy was used to determine the size and structure of the generated carbon nanomaterials. The photothermal response and evaporation rates of water with and without carbon-based nanomaterials were also experimentally investigated and discussed.

5.2 MATERIALS AND METHODS

5.2.1 MATERIALS

Carbon-based materials were derived from graphene oxide (GO) powder (product number: GNOS0010) supplied by ACS Material, LLC (Pasadena, California, United States of America). The reduction agent used was tetra ethyl ammonium hydroxide [TEAH, $C_8H_{21}NO$, (35% in water)] supplied by Tokyo Chemical Industry Co., Ltd. (Kita-Ku, Tokyo, Japan). All other chemicals and solvents used in the preparation procedures were analytical grade and used without further purification. All aqueous-based solutions used in the preparation procedures were made from Milli-Q® water (10 MΩ cm^{-1}) produced from a Milli-Q® Reagent water generation system supplied by the Millipore Corporation.

5.2.2 Preparation of Carbon-Based Nanomaterials and Stock Solutions

The carbon-based materials were derived from GO powder. The first stock solution prepared was the 0.1% GO-based aqueous solution. The procedure started with 0.1g of GO powder being placed into a mortar. Then a 2 ml solution of Milli-Q® water was added. The mixture was then ground with a pestle for five minutes to generate a smooth paste. The paste was then added to a 100 ml solution of Milli-Q® water and then sonicated for ten minutes using an ultrasonic processor (Hielscher UP400S) set at a power level of 400 W. After processing, the orange-tinted brown solution (as seen in Figure 5.1) was stored at room temperature before being used to prepare the treated water samples. The second stock solution prepared was the 0.1% RGO-based aqueous solution. The procedure commenced with preparing a mixture consisting of TEAH (20% by volume) and Milli-Q® water being poured into a Schott bottle to make a 100 ml stock solution. This was followed by adding 0.1g of GO powder into a mortar. Then 2 ml of stock solution was added and the mixture was ground with a pestle for five minutes to produce a smooth paste. After mixing, the paste was added to a glass beaker containing stock solution (10 ml) and Milli-Q® water (90 ml). The mixture was then sonicated for ten minutes at a power level of 400 W. After treatment, the resulting black solution (as seen in Figure 5.1) was stored at room temperature before being used to prepare the treated water samples.

5.2.3 Preparation of Water Solutions Containing Carbon-Based Additives

The GO and RGO stock solutions prepared above were then used to produce aqueous-based test solutions. The range of test solutions made up was equivalent to 10%, 20%, and 30% carbon-based

FIGURE 5.1 (a) A representative SEM image of supplied GO oxide powders, (b) TEM image of a well dispersed and transparent flake after grinding, and (c) a typical UV-visible absorption spectrum for GO and RGO samples, with the insert showing respective resultant stock solution colors.

additive stock solution. Thus, 15, 30, and 45 ml of respective GO and RGO stock solutions were added to 135, 120, and 105 ml of Milli-Q® water. In each case, after the addition of stock solution to Milli-Q® water, the glass vial was hand shaken for one minute. No further dispersion techniques like sonication were needed since the GO- and RGO-based solutions were found to be well dispersed and very stable for long periods of time (~3 months).

5.2.4 GO and RGO Characterization Studies

Characterization studies were carried out to confirm the reduction of GO by TEAH to produce RGO and to evaluate the size and shape of the powders. Images produced by a JEOL JCM-6000, NeoScope™ scanning electron microscope were used to determine the particle size and shape of supplied powders. A Tecnai G2, FEI (Electron Optics, USA) transmission electron microscope operating at 100 kV was used to evaluate the size and shape of individual GO flakes produced after processing. An Agilent/HP 8453 UV-visible spectrophotometer over a spectral range between 190 and 1100 nm, with a 1 nm range resolution, was used to assess the chromophores present in the GO and RGO stock solution-based nanofluid samples. X-ray diffraction (XRD) spectroscopy was used to investigate the crystalline structure of both GO and RGO powder samples. Diffraction patterns were taken over a 2θ range starting at 5° and finishing at 75°. An incremental step size of 0.02° was used, with an acquisition speed of 2° min^{-1} using a GBC® eMMA X-ray powder diffractometer (Cu $K\alpha$ = 1.54056 Å radiation source) operating at 35 kV and 28 mA. The recorded Bragg peak locations were identified in the patterns and were compared with those reported in the International Centre for Diffraction Data (ICDD) databases. Then Miller indices were assigned to the respective peaks. A PerkinElmer FTIR/NIR Spectrometer Frontier with Universal signal bounce Diamond ATR attachment was used to identify functional groups and their respective vibration modes present in the samples. FTIR spectra were collected over a scanning range from 400 to 4000 cm^{-1}, with a resolution step of 1 cm^{-1} over the range.

5.2.5 Photothermal Response and Evaporation Rate Measurements

Photothermal and evaporation measurements were carried out in an in-house solar simulator. The light source used was a Philips 13096 ELH (120V, 300W-G5D) light bulb. The simulator was adjusted and calibrated using an LI-200 R Pyranometer to produce an irradiance of 985 Wm^{-2} at a testing distance of 25 cm from the illumination source. The respective solution samples (25 g) were placed in an uncovered Petri dish located on an insulated shelf at the testing distance. Measurements were taken at 5, 10, 15, 20, and 60 minutes. Temperature measurements of the respective samples were taken using a hand-held thermal camera (Fluke Ti 25). Weight loss measurements of the respective samples (to determine evaporation rates) were carried out using a laboratory balance (Ohrus pioneer PA214C analytical balance) at a room temperature of 27°C.

5.3 RESULTS AND DISCUSSIONS

SEM analysis of supplied GO oxide powders revealed the powders were highly agglomerated and had large numbers of folding micrometer scale sheets. These sheets also displayed a high degree of surface wrinkling as seen in Figure 5.1 (a). After the grinding process, the resulting powders consisted of well-dispersed and transparent flakes as seen in the TEM image presented in Figure 5.1 (b). Image analysis revealed the flakes ranged in size from 20 to 80 μm in size, with thicknesses ranging from 5 to 20 nm. The reduction of GO was verified by UV-visible absorption, XRD spectroscopy, and FTIR spectroscopy. A typical UV-visible absorption spectrum for GO and RGO samples is presented in Figure 1 (c). Inspection of the absorption spectrum reveals major absorption peaks and smaller shoulder peaks. The major peaks are located at 242 nm (GO) and 236 nm (RGO), which corresponds to the $\pi \rightarrow \pi^*$ transition of sp^2 poly-aromatic (C–C bonds) carbon structures. The smaller shoulder peaks are located at 302 nm (GO) and 297 nm (RGO), and correspond to the n → π* electron transitions for carbonyl and carboxyl (C=O bonds) functional groups [22]. The absorption peaks are typical of GO and RGO suspensions reported in the literature, and reveal that TEAH has modified the chemical structure of the absorbing species present on GO sheets [23, 24]. Also shown in Figure 1 (c) is the color of the resulting stock solutions after processing, with GO being brown and RGO being black.

XRD spectroscopy was used to confirm the reduction of GO by TEAH to produce by RGO. Representative XRD patterns for GO and RGO samples are presented in Figure 5.2 (a). The GO pattern displays the characteristic 2θ diffraction peak located at 10.68°, which was attributed to the (001) crystalline plane of GO and is consistent with values reported in the literature [25, 26]. The RGO pattern revealed a strong 2θ diffraction peak located at 24.52° and was identified as the (002) peak for graphite [27]. The graphitic-like character seen in the RGO pattern confirms the conversion of GO to RGO. Likewise, this result was confirmed by FTIR spectroscopy analysis which identified several vibrational transitions that confirmed the conversion of GO to RGO. Representative spectra for GO and RGO samples are presented in Figure 5.2 (b). Vibrational transitions identified in the GO spectrum included a broad band located at 3169 cm^{-1}, which was identified as O-H stretching vibrations of adsorbed water molecules and OH groups. The next band located at 1711 cm^{-1} was identified as C=O stretching vibrations of carbonyl groups. This was followed by the 1619 cm^{-1} band that identified C=O stretching vibrations of carboxylic and/or carbonyl functional groups. This was followed by the 1396 cm^{-1} band which was assigned to C-OH bending vibrations or O-H deformation. Finally, the 1048 cm^{-1} band was identified as C-O stretching vibrations [28, 29]. Also included in Figure 2 (b) is a typical RGO spectrum that contains a number of bands not seen in the GO

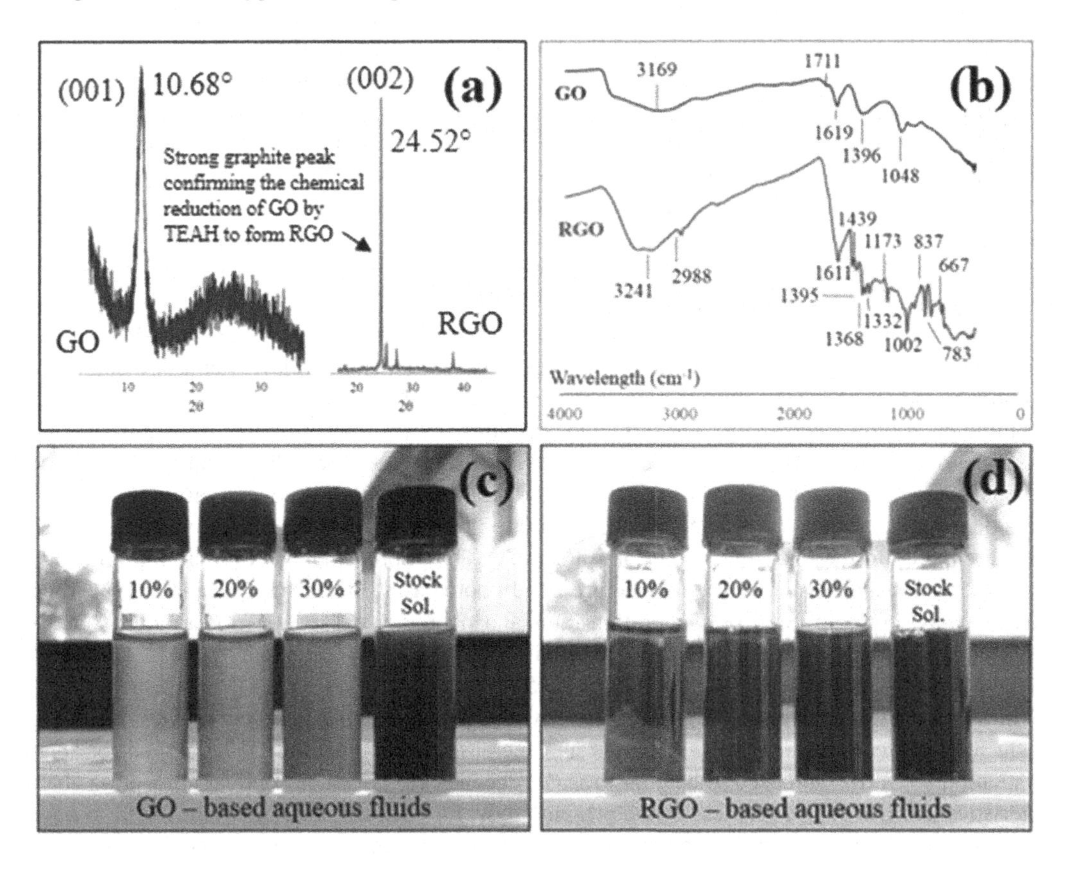

FIGURE 5.2 (a) XRD diffraction patterns of GO and RGO, (b) FTIR spectroscopy analysis of GO and RGO stock solution samples, and (c) and (d) water-based solutions containing GO and RGO additives.

spectrum. For instance, there was a band located at 2988 cm^{-1} that was identified as a C-H stretching vibration. This was followed by bands located at 1611 cm^{-1} and 1439 cm^{-1} that were identified as C=C stretching and C-H bending vibrations respectively. Also seen were aliphatics ($-NO_2$), which were located at bands 1395, 1368, and 1332 cm^{-1}. This was followed by a C-N stretching vibration located at 1173 cm^{-1} and a C-O stretching vibration located at 1002 cm^{-1}. The remaining three bands (837, 783, and 667 cm^{-1}) were identified as C-H bending [29, 30].

The range of test solutions was made up of varying percentages (10, 20, and 30%) of GO and RGO stock solutions. Figure 2 (c) presents the color range of the three GO-based solutions and a sample of stock solution. Starting from the left, the 10% solution was light gray, the 20% solution was pale yellow, and the 30% solution was light brown, while the original GO stock solution was brown in color. The RGO-based solutions are presented in Figure 2 (d). The 10% solution was light brown, the 20% solution was medium brown, and the 30% solution was dark brown. The original RGO stock solution was black in color. Each of these different colored solutions produced varying degrees of photothermal response and evaporation rates. The results of photothermal testing for all solutions irradiated at 985 Wm^{-2} for a period of 60 minutes are presented in Figure 5.3. During the first ten minutes, all solutions displayed a rapid increase in temperature at an average rate of 1 °C min^{-1}. The next ten minutes saw a much lower temperature increase rate of around 0.2 °C min^{-1}. Beyond 20 minutes, the temperature rise was small and started to level off. The control (Milli-Q® water) increased from room temperature (27 °C) up to a maximum temperature of 39 °C by the end of the test period.

All of the GO-based and RGO-based solutions had higher temperatures than the control at the end of the test period. The best-performing solution was the 30% RGO-based aqueous solution which achieved a temperature of 41 °C as seen in Figure 5.3 (d, e, and f). This equated to a temperature enhancement of 5.2% when compared to the control solution. Similarly, the 30% GO-based aqueous solution reached a temperature of 40.5 °C and equated to a temperature enhancement of 3.8% when compared to Milli-Q® water.

Additionally investigated were water mass losses and evaporation rates of the respective solutions during the test period. The mass losses for all solutions over the test period displayed a linear relationship against time. Both GO-based and RGO-based solutions lost larger amounts of water mass over the same period when compared to the Milli-Q® water control, as seen in Figures 5.4 (a) and (c). Milli-Q® water lost 7.46 g in mass after 60 minutes, whereas the 30% GO-based solution lost 9.17 g, and the 30% RGO-based solution lost 9.74 g.

Interestingly, the evaporation rate of all solutions rapidly increased in the first 20 minutes. However, beyond the initial 20 minutes, the evaporation rate appeared to level out. For instance, at 20 minutes, Milli-Q® water reached an evaporation rate of 1.85 mg min^{-1} cm^{-2}, but after a further 40 minutes, it only reached 1.94 mg min^{-1} cm^{-2}. A similar trend was seen for the GO-based and RGO-based solutions. The maximum evaporation rates achieved were 2.39 mg min^{-1} cm^{-2} for the 30% GO-based solution and 2.54 mg min^{-1} cm^{-2} for the 30% RGO-based solution after 60 minutes. This demonstrates a significant increase in evaporation rates for solutions containing carbon additives compared to the Milli-Q® water control sample. The improvement in evaporation rates and photothermal responses is presented in Figure 5.4. In both cases, increasing the amount of carbon additive was shown to improve evaporation rates and photothermal responses. In particular, the addition of RGO in the 30% RGO-based solution improved its evaporation rate by 30.5% compared to the Milli-Q® water control, as seen in Figure 5.5 (b).

In a typical solar thermal still system, fitted with a single sloping glass cover over the basin, basin water absorbs incident solar radiation and increases the temperature. The amount of solar irradiation converted and used to heat basin water depends on the water's optical absorption characteristics and thermal properties. In the present study, RGO-based solutions were found to have superior photothermal responses and evaporation rates compared to GO-based solutions. In addition, both had superior photothermal responses and evaporation rates compared to Milli-Q® water (control), as seen in Table 5.1. Several studies have evaluated various methods for harvesting solar

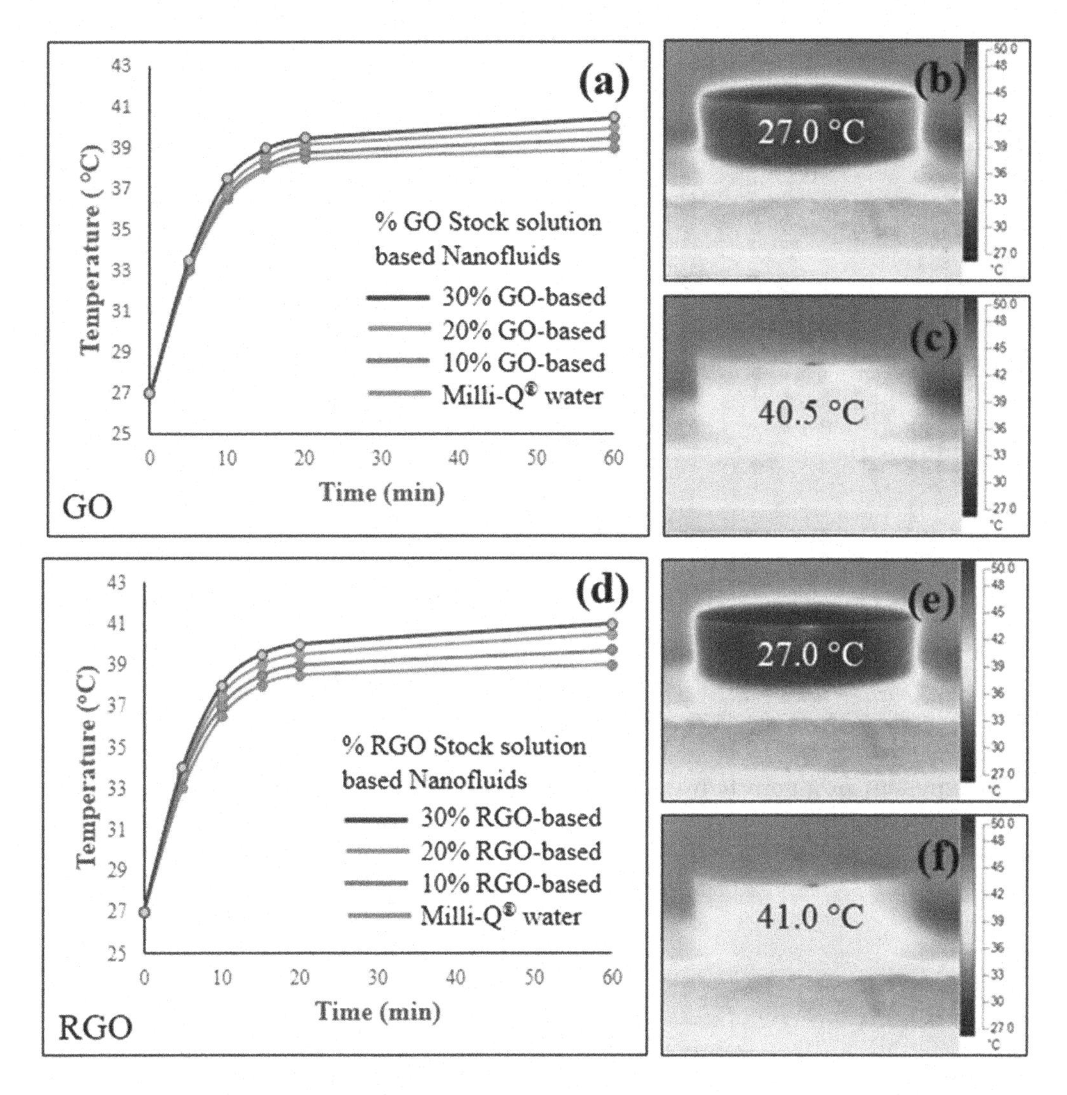

FIGURE 5.3 Photothermal response of test solutions: (a) GO-based solutions, (b) initial and (c) final temperature, d) RGO-based solutions, (e) initial and (f) final temperature.

thermal energy through nanofluid-based volumetric absorption systems using a variety of carbon materials like graphene oxide. However, these studies have generally focused on using solar energy levels greater than one sun [31, 32]. Furthermore, other studies have evaluated the use of floating materials to improve evaporation rates, but these configurations also promote higher levels of bio-fouling if not properly maintained [15, 33]. The present study has focused on using solar radiation equivalent to one sun and has avoided bio-fouling issues normally associated with floating materials. In addition, other studies have reported color changes occurring in GO when exposed to UV light [34, 35]. However, the present study detected no color changes occurring in any of the solutions after illumination. In addition, during and after irradiation the GO-based and RGO-based solutions displayed good dispersion stability over the entire temperature range. Furthermore, no sedimentation was found in any of the vials containing the GO-based or RGO-based solutions for over two months, thus confirming the long-term dispersion stability of the respective additive solutions. In addition, there were no signs of biological activity or bio-fouling on the vial surfaces over this period. Importantly, the evaporation rates achieved were at relatively low solution temperatures that

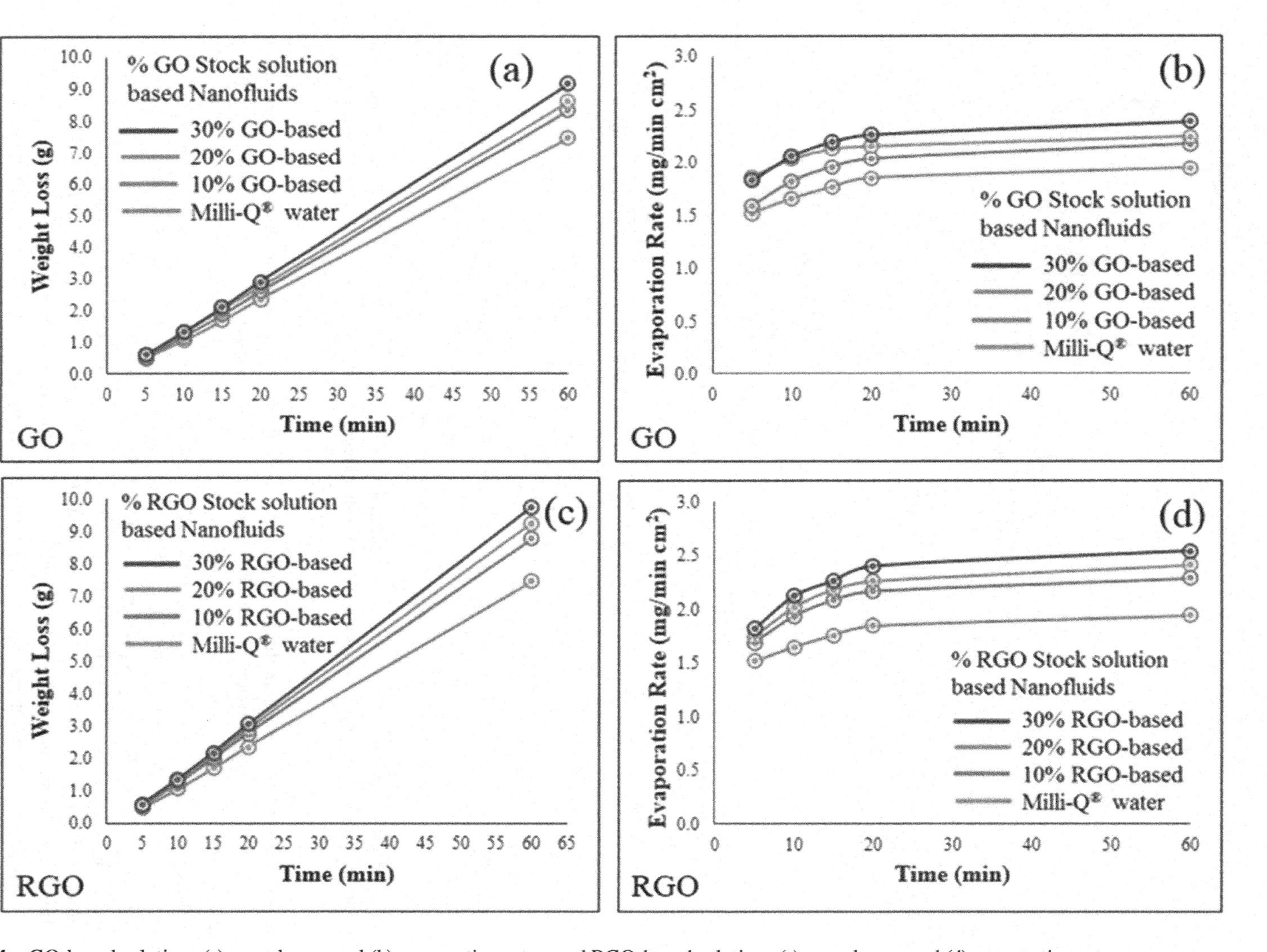

FIGURE 5.4 GO-based solutions (a) mass losses and (b) evaporation rates, and RGO-based solutions (c) mass losses and (d) evaporation rates.

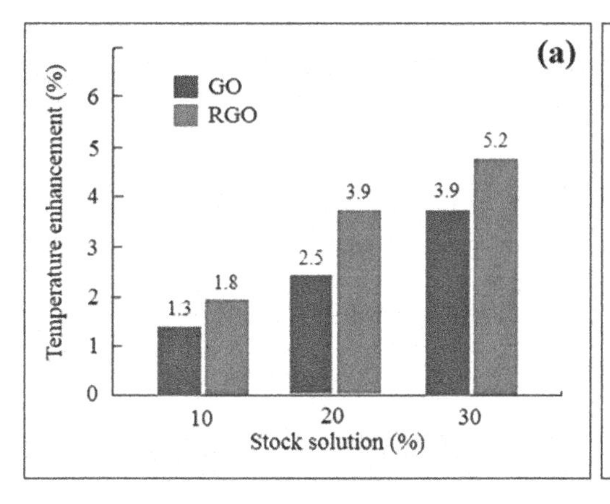

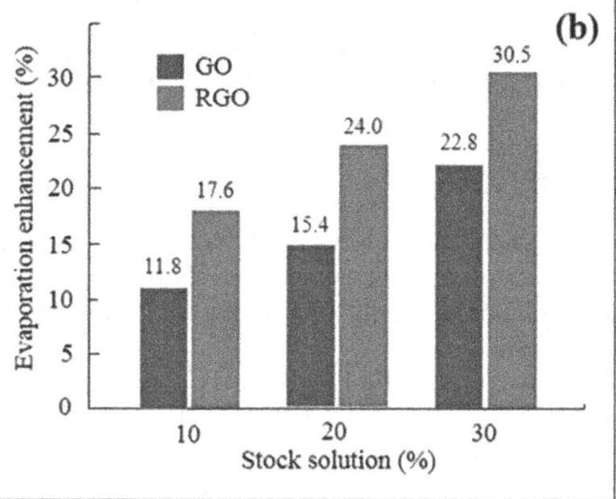

FIGURE 5.5 (a) Enhancement in photothermal response and (b) improvement in evaporation rate with increasing amounts of either GO or RGO.

TABLE 5.1
Maximum temperature, mass losses, and evaporation rates achieved after 60 minutes of exposure to an irradiance of 985 Wm^{-2}

Material	%	Max. temperature (°C)	Max. mass loss (g)	Max. evap. rate (mg min^{-1} cm^{-2})
Milli-Q® water	-	39	7.46	1.94
GO-based	10	39.5	8.35	2.17
	20	40.0	8.62	2.24
	30	40.5	9.17	2.39
RGO-based	10	39.7	8.78	2.29
	20	40.5	9.26	2.41
	30	41.0	9.74	2.54

were typically between 39 and 41 °C, thus highlighting the advantage of adding either GO or RGO to basin water to improve evaporation rates for lower operational temperatures as well as at higher temperatures. Future work will investigate the long-term stability and operational performance of these GO-based and RGO-based additives in a variety of solar thermal still configurations.

5.4 CONCLUSION

The present study has shown that RGO-based solutions have higher photothermal responses and superior evaporation rates compared to both GO-based solutions and the Milli-Q® water control. In particular, the 30% RGO-based solution achieved a temperature enhancement of 5.2% and an improved evaporation rate of 30.5% compared to the Milli-Q® water control. All GO- and RGO-based solutions had excellent dispersion stability and showed no signs of degradation or sedimentation. Therefore, the excellent dispersion stability, superior photothermal properties, and excellent evaporation rates of both GO-based and RGO-based solutions compared to pure water make them ideal fluids for solar evaporation applications. This work demonstrate that atomically thin layers of graphene oxide and reduced graphene oxide can indeed be harnessed to increase significantly the performance of solar still and have a direct potential to enhance the capture solar energy and improve solar stills desalination technology.

REFERENCES

1. Ng KC, Thu K, Kim Y, Chakraborty A, Amy G. Adsorption desalination: An emerging low-cost thermal desalination method. *Desalination*. 2013; 308: 161–179.
2. Wallace JS, Gregory PJ. Water resources and their use in food production systems. *Aquat. Sci.* 2002; 64: 363–375.
3. Elimelech M, Phillip WA. The future of seawater desalination: Energy, technology, and the environment. *Science*. 2011; 333: 712–717.
4. Gerland P, Raftery AE, Sevcikova H, Li N, Gu D, Spoorenberg T, Alkema L, Fosdick BK, Chunn J, Lalic N, Bay G. World population stabilization unlikely this century. *Science*. 2014; 346: 234–237.
5. Watkins K. *Human Development Report 2006-Beyond Scarcity: Power, Poverty and the Global Water Crisis*, Vol. 28, 2006.
6. Qiblawey HM, Banat F. Solar thermal desalination technologies. *Desalination*. 2008; 220: 633–644.
7. Vishwanath Kumar P, Kumar A, Prakash O, Kaviti AK. Solar stills system design: A review. *Renew. Sustain. Energy Rev.* 2015; 51: 153–181.
8. Sharon H, Reddy K. A review of solar energy driven desalination technologies. *Renew. Sustain. Energy Rev.* 2015; 41: 1080–1118.
9. Badran OO. Experimental study of the enhancement parameters on a single slope solar still productivity. *Desalination*. 2007; 209(1–3): 136–143.
10. Velmurugana V, Srithar K. Performance analysis of solar stills based on various factors affecting the productivity: A review. *Renew. Sustain. Energy Rev.* 2011; 15: 1294–1304.
11. Prakash P, Velmurugan V. Parameters influencing the productivity of solar stills: A review. *Renew. Sustain. Energy Rev.* 2015; 49: 585–609.
12. Ghasemi H, Ni G, Marconnet AM, Loomis J, Yerci S, Miljkovic N, Chen G. Solar steam generation by heat localization. *Nat. Commun.* 2014; 5: 4449.
13. Tamini A. Performance of a solar still with reflectors and black dye. *Sol. Wind Technol.* 1987; 4(4): 443–446.
14. Nijmeh S, Odeh S, Akash B. Experimental and theoretical study of a single-basin solar sill in Jordan. *Int. Commun. Heat Mass Transf.* 2005; 32: 565–572.
15. El-Haggar S, Awn A. Optimum conditions for a solar still and its use for a greenhouse using the nutrient film technique. *Desalination*. 1993; 94(1): 55–68.
16. Sharshir S, Yang N, Peng G, Kabeel A. Factors affecting solar stills productivity and improvement techniques: A detailed review. *Appl. Therm. Eng.* 2016; 100: 267–284.
17. Srivastava Pankaj K, Agrawal SK. Experimental and theoretical analysis of single sloped basin type solar still consisting of multiple low thermal inertia floating porous absorbers. *Desalination*. 2013; 311: 198–205.
18. Kabeel AE, Omara ZM, Essa FA. Enhancement of modified solar still integrated with external condenser using nanofluids: An experimental approach. *Energy Convers. Manag.* 2014; 78: 493–498.
19. Wang J, Li Y, Deng L, Wei N, Weng Y, Dong S, Qi D, Qiu J, Chen X, Wu T. High performance photothermal conversion of narrow-band-gap Ti_2O_3 nanoparticles. *Adv. Mater.* 2017; 29(3): 1603730.
20. Theye ML, Paret V. Spatial organization of the sp2-hybridized carbon atoms and electronic density of states of hydrogenated amorphous carbon films. *Carbon*. 2002; 40: 1153–1166.
21. Colombara D, Dale PJ, Kissling GP, Peter LM, Tombolato S. Photo-electrochemical screening of solar cell absorber layers: electron transfer kinetics and surface stabilization. *J. Phys. Chem. C*. 2016; 120: 15956–15965.
22. Thakur S, Karak N. Green reduction of graphene oxide by aqueous phytoextracts. *Carbon*. 2012; 50: 5331–5339.
23. Marcano DC, Kosynkin DV, Berlin JM, Sinitskii A, Sun Z, Slesarev A, Alemany LB, Lu W, Tour JM. Improved synthesis of graphene oxide. *ACS Nano*. 2010; 4: 4806–4814.
24. Chen J-L, Yan X-P. A dehydration and stabilizer-free approach to production of stable water dispersions of graphene nanosheets. *J. Mater. Chem.* 2010; 20: 4328–4332.
25. Shao G, Lu Y, Wu F, Yang C, Zeng F, Wu Q. Graphene oxide: The mechanisms of oxidation and exfoliation. *J. Mater. Sci.* 2012; 47: 4400–4409.

26. Lee JW, Ko JM, Kim JD. Hydrothermal preparation of nitrogen-doped graphene sheets via hexamethylenetetramine for application as supercapacitor electrodes. *Electrochim. Acta.* 2012; 85: 459–466.
27. Sharma R, Chadha N, Saini P. Determination of defect density, crystalline size and numbers of graphene layers in graphene analogues using X-ray diffraction and Raman spectroscopy. *Ind. J. Pure Appl. Phys.* 2017; 55: 625–629.
28. Wang DW, Du A, Taran E, Lu GQ, Gentle IR. A water-dielectric capacitor using hydrated graphene oxide film. *J. Mater. Chem.* 2012; 22: 21085–21091.
29. Wojtoniszak M, Mijowska E. Controlled oxidation of graphite to graphene oxide with novel oxidants in a bulk scale. *J. Nanopart. Res.* 2012; 14: 1248.
30. El Achaby M, Arrakhiz FZ, Vaudreuil S, Essassi EM, Qaiss A. Piezoelectric β-polymorph formation and properties enhancement in graphene oxide – PVDF nanocomposite films. *Appl. Surf. Sci.* 2012; 258: 7668–7677.
31. Khullar V, Tyagi H, Hordy N, Otanicar TP, Hewakuruppu Y, Modi P, Taylor RA. Harvesting solar thermal energy through nanofluid-based volumetric absorption systems. *Int. J. Heat Mass Transf.* 2014; 77: 377–384.
32. Lee SW, Kim KM, Bang IC. Study on flow boiling critical heat flux enhancement of graphene oxide/water nanofluid. *Int. J. Heat Mass Transf.* 2013; 65: 348–356.
33. Srivastava Pankaj K, Agrawal SK. Experimental and theoretical analysis of single sloped basin type solar still consisting of multiple low thermal inertia floating porous absorbers. *Desalination.* 2013; 311: 198–205.
34. Ding Y, Zhang P, Zhuo Q, Ren H, Yang Z, Jiang Y. A green approach to the synthesis of reduced graphene oxide nanosheets under UV irradiation. *Nanotechnology.* 2011; 22: 215601.
35. Chen L, Xu C, Liu J, Fang X, Zhang Z. Optical absorption property and photothermal conversion performance of graphene oxide/water nanofluids with excellent dispersion stability. *Solar Energy.* 2017; 148: 17–24.

6 Electrophoretic Deposition and Inkjet Printing as Promising Fabrication Routes to Make Flexible Rechargeable Cells and Supercapacitors

Debasish Das, Subhasish Basu Majumder, Sarmistha Basu, Kh M Asif Raihan, Rajavel Krishnamoorthy, and Suprem Ranjan Das

CONTENTS

DOI: 10.1201/9781003181422-7

6.1 BACKGROUND

Some fascinating flexible electronic devices could pave the way for a new era of personalized Internet of Things (IoT), where devices including but not limited to patchable/implantable sensors, e-textiles, rollable displays, robotic suits, and cellular phones will be closely connected to the human body [1, 2]. The conventional power source devices with their mechanically rigid structure cannot satisfactorily meet the requirements of flexible electronic devices. The growing demand for the aforementioned wearable devices has urged the key requirement for developing flexible electrochemical energy storage devices (EES) with superior capacity that are monolithically integrated into electronic peripherals. The required capacity of the flexible devices depends on the type of application. For IoT, medical implants, smart cars, radio frequency identification (RFID), etc., the required capacity lies in the range of 1 to 10 mAh, while the capacity requirements of wearable textiles and foldable smartphones are much higher (100 to 1000 mAh) [1]. Among the various energy storage systems, rechargeable Li-ion batteries are considered the most promising power sources for such flexible devices due to their higher nominal voltage, higher energy and power density, and long life span. Companies such as Panasonic, LG Chem, Jenax, Samsung, and Prologium have already developed different types of flexible battery prototypes and commercially adaptive products. Apart from lithium-ion batteries (Li-ion; LIBs), flexible supercapacitors (featured with intrinsic high-power density, faster charge/discharge capability, and extended cycle life) are also extensively studied [3] and are considered to be a potential technology to power up various flexible devices often in conjunction with flexible Li-ion rechargeable batteries.

6.2 CONFIGURATIONS OF FLEXIBLE ELECTROCHEMICAL ENERGY STORAGE

Flexible electrochemical energy storage can mainly be of three different types: bendable, stretchable, and compressible. Not much research accomplishment has so far been reported on compressible flexible storage. In this chapter, we will limit our review only to bendable and stretchable energy storage devices. These batteries work within the elastic deformation range of the constituents. It means that their shapes are completely recovered when the external force is removed. By definition, the electrochemical properties of flexible batteries should remain the same before, during, and after the mechanical deformation. The salient features of each type of flexible storage are outlined below.

6.2.1 Bendable Storage Device

Consider a composite electrode film (consisting of active material, binder, and conducting agent) that is deposited on a metallic current collector. If the assembly is bent with a radius of curvature (R), then the strain on top of the composite film (ε_{top}) is given by

$$\varepsilon_{top} = \frac{(d_f + d_s)}{2R} \frac{(1 + 2\eta + \chi\eta^2)}{(1+\eta)(1+\chi\eta)} \tag{1}$$

where d_f and d_s are the thickness of the film and substrate; η is the ratio of film thickness: substrate thickness (d_f / d_s); and χ is the ratio of Young's modulus of film: substrate (Y_f/Y_s). If the film has the same elastic modulus as the underlying current collector, then:

$$\varepsilon_{top} = \frac{(d_f + d_s)}{2R} \tag{2}$$

Chang et al. [4] tabulated the elastic limits (ε) of various constituents used in LIB and single-cell (SC) batteries. Active materials (electrochemically active materials used in batteries and supercapacitors, viz. $LiCoO_2$, $Li_4Ti_5O_{12}$, silicon, activated carbon, NiO, etc.) have $\varepsilon < 0.1\%$. Carbonaceous materials such as graphite ($\varepsilon \sim 0.1 - 0.2\%$), graphene film ($\varepsilon \sim 0.4 - 0.8\%$), graphene oxide film ($\varepsilon \sim 0.6\%$), and CNT ($\varepsilon \sim 1.4 - 3\%$) yield relatively better elastic limits compared to other active materials used in LIB and SC. Current collectors like Al ($\varepsilon < 1\%$) and Cu ($\varepsilon < 2\%$) have reasonably higher elastic limits as separators (viz. PP $\varepsilon < 1\%$ and cellulose film ($\varepsilon \sim 3\%$). Polymer electrolytes (such as polyvinylidene fluoride (PVDF), chitin, and $LiClO_4$) also have a higher elastic limit (($\varepsilon < 5\%$), similar to that of packaging materials (cloth, cellulose paper, and polydimethylsiloxane (PDMS), ($\varepsilon \sim 3 - 50\%$)).

Equations 1 and 2 form the basis of designing a bendable flexible battery. From Equation 1, for 0.1 μm of free-standing Si film (($\varepsilon < 0.1\%$), one can have R < 50 μm for the film to crack during bending. For active material deposited on a current collector, the critical bending radius scales linearly with the total thickness ($d_f + d_s$). To take the advantage of thickness reduction, vanadium disulfide (VS_2) flakes are exfoliated to fabricate a flexible supercapacitor [5] in which the electrode comprises a stack of less than five S – V – S single layers. Thinner electrodes would lower the capacity for rechargeable flexible cells. To circumvent the problem, another fruitful strategy is to deposit relatively thicker active material onto a compliant substrate (viz. cellulose paper, carbon nanotubes (CNTs), or graphene film, where $Y_f >> Y_s$). Using Equation 1, for a compliant substrate (Y_f/Y_s = 100), Suo et al. [6] estimated the normalized top surface strain [ε_{top} [$2R/(d_f + d_s)$]] in the active electrode for different d_f/d_s values. Interestingly, on a compliant substrate, for given R and $d_f + d_s$ values, a significant reduction in the top surface strain of the active electrode is calculated when d_f/d_s are varied from 10^{-4} to 10^{-1}. Strain in active layer film can further be reduced when the flexible battery (having a configuration of, for example, current collector/$LiCoO_2$/LIPON/Li) is sandwiched between suitable encapsulation layers (viz. PDMS) [7]. The neutral plane can be moved to the rigid active material by maintaining the relation $Y_s d_s^2 = Y_e d_e^2$, where Y_s and Y_e are Young's moduli of the substrate and encapsulation materials, and d_s and d_e are their respective thicknesses. Summarizing the above description, three strategies are fruitful for bendable batteries: first to use free-standing thinner electrodes of active material; second, use of thinner electrode material deposited on a relatively thicker compliant current collector; and finally, positioning the neutral plane (region neither in tension or compression) into the rigid (brittle) active material by manipulating the thickness and Young's modulus of the encapsulation layer and substrate material.

6.2.2 Stretchable Storage Device

Stretchable batteries are another important class of flexible batteries. Stretchable batteries demand intrinsically elastic active materials, and none of the conventional electrode materials used in LIB or SC qualify for this. To circumvent this challenge, various specialized structural layouts are explored to buffer the large deformations during stretching. A wavy structured electrode is one of the popular designs to realize stretchability. In this configuration, a thin electrode film is deposited on a prestrained compliant substrate, and upon releasing the strain, a wavy structured electrode is formed which can withstand a large amount of strain (50–400%) [8].

An interconnected island-bridge structure is another attractive approach for developing stretchable electrodes. In this architecture, the active materials are strongly bonded to a compliant substrate known as the island. These islands are connected through highly flexible and conductive interconnects (bridges) that help to accommodate the large elastic strain. The bridges can be straight or serpentine, and coplanar or non-coplanar [2, 9]. Systematic tuning of interconnect parameters like ribbon width, arc radius, arc angle, and arm length can remarkably improve the stretchability of the electrode. However, the fabrication of such specialized architecture is complex and costly.

Another interesting stretchable shape is fiber-shaped flexible batteries. These can be weaved into textiles. Both co-axial and twisted 1D flexible batteries are reported in the literature. Weng et al. [10] used CNT current collector deposited with $LiFePO_4$ to make a cathode strand. An Si-coated CNT sheet was sandwiched between two CNT sheets and scrolled to form anode composite yarn. First, the cathode strand was wound on a cotton strand followed by gel polymer electrode coating. Then the anode composite yarn was coated, followed by suitable packaging to make a fiber-shaped battery with a typical diameter of 2 mm. At 1C, the fiber battery yielded ~ 110 mAh/g capacity with decent capacity retention up to 100 cycles. Fiber-shaped full LIB was woven into a textile. In a twisted flexible battery, individual electrode strands are bundled together to get high mass loading per unit length [11]. Fiber-shaped hybrid energy storage devices have also been fabricated by twisting CNT/ordered mesoporous carbon, CNT/LTO, and CNT/$LiMn_2O_4$ together. These twisted hybrid batteries deliver reasonably good energy (90 Wh/kg) and power densities (5.97kW/kg) [12].

6.3 COMPONENTS OF FLEXIBLE STORAGE DEVICES

Materials are at the core of the fabrication and performance of any functional device. As mentioned earlier, flexible storage is composed of various classes of materials: organic or inorganic; in the form of solids, liquids, or gels; and they can be insulators, metals, or ionically conducting. This section focuses on the important materials used in flexible batteries.

6.3.1 Flexible Electrode Materials

Conventional active materials for LIB typically include $LiCoO_2$, $LiMn_2O_4$, and $LiFePO_4$, as positive and graphite/Si, $Li_4Ti_5O_{12}$, etc. as negative electrodes. Except for graphite (elastic strain, ~ 0.1 – 0.2%), all other electrode materials have a low elastic limit (< 0.1%) and are intrinsically brittle. These brittle active materials are the major obstacle for developing flexible devices. For LIBs and SCs, two main types of flexible electrodes are made using non-conducting (paper, polymer, textile) and conducting matrices (carbonaceous materials, CNT, carbon nano-fiber, graphene, etc.). CNT, noble metals, graphene, etc., as conductive additives are used with non-conducting support to improve their conductivity. Active materials for LIB, as well as SC, are deposited by various techniques including spray painting, coating, and electrodeposition [13]. Owing to their remarkable mechanical stiffness (Young's modulus of SWCNT and MWCNT are 1TPa and 0.9 TPa, respectively), excellent mechanical resiliency, high electrical conductivity, and good electrochemical properties, CNTs are extensively used for constructing flexible electrodes. Numerous strategies including but not limited to chemical vapor deposition, vacuum filtration, and hydrothermal/solvothermal methods are adopted to fabricate CNT-based composite electrodes. For example, a modified CVD technique was utilized to grow the CNTs into a graphitic carbon layer to make a self-standing flexible anode. At 0.2C, such a flexible anode delivers a specific capacity, ~ 570 mAh/g, after 100 charge-discharge cycles [14]. For Li-ion cells, excellent electrochemical performance was reported in a self-standing flexible Si nanoparticles-MWCNT composite anode [15]. The 3D interconnected network of the MWCNT not only helps to accommodate the volume fluctuation during cycling but also ensures higher electrical conductivity of the flexible electrode. Free-standing Sb_2S_3 decorated MWCNT, prepared by solvothermal synthesis, was also reported as yielding good electrochemical performance as an anode for flexible LIB [16]. The electrodeposition technique has also been used to make high-capacity paper like NiS/CNT composite anode, which can be reversibly bent up to 180° without compromising its electrical performance [17].

Excellent flexibility is achieved in freeze-dried Si nanoparticle-incorporated graphene foam, calcined at 350°C [18]. Highly conductive 3D graphene foam facilitates electron transport. In addition, it acts as a buffer to accommodate the large volumetric changes during the alloying-de-alloying reaction with lithium. Free-standing flexible $Li_4Ti_5O_{12}$/rGO, prepared using a hydrothermal route, yielded excellent cycling stability and rate performance [19]. Excellent capacity retention was reported in a free-standing $LiFePO_4$/graphene composite cathode even when measured at a bending angle of 45° [20]. It remains challenging to incorporate high contents of the active materials into flexible non-conducting as well as conducting current collectors. High active material loading is required to increase the energy density of the flexible cells. Another major challenge is to do efficient tab welding to the flexible electrode. During repeated deformation, tabs often get delaminated.

Active material is often delaminated from the flexible base which leads to rapid capacity fading. If a liquid electrolyte is used, a porous flexible electrode demands more electrolytes, which raises concerns for cell safety [1].

6.3.2 Flexible Current Collectors

For rigid Li-ion batteries, Al and Cu current collectors yield necessary electronic conductivities and exhibit excellent electrochemical stability in the working potential range. As flexible current collectors, they should offer superior mechanical stability during deformation, and also the coated composite electrode should not get delaminated upon repeated flexing. Both Al and Cu offer relatively lower yield strain (0.9% and 1.2% respectively), which affect the electrical performance upon repeated deformation. Lowering the current collector thickness could be one of the strategies to increase the yield strain as high as 5%; however, ultrathin metal foils are not conducive for electrode coating using the traditional tape casting route. A careful trade-off between the thickness and yield strain is of utmost importance. The delamination of coated composite electrodes from the current collector (both during cycling and flexing) remains another burning issue. Delamination deteriorates the electrochemical performance of these cells. Functional layer coating in between the current collector and composite electrodes or the use of engineered current collectors are reported to be fruitful to improve electrode adhesion with underlying substrates [1, 21].

Carbonaceous materials act as an effective alternative to metal current collectors. The use of carbonaceous material-based current collectors offers excellent flexibility, good electrochemical stability, reasonably high conductivities, and a lightweight and higher specific area for the development of flexible batteries. A covalently bonded sp^2 carbon nano-tube can easily be fabricated in the form of a free-standing film, sponges, and also in the form of fiber. In these forms, CNT can act as flexible current collectors [1, 22–24]. Graphene or reduced graphene oxide (rGO) can also be fabricated as film, foam, or fiber [1, 25–27]. Occasionally, a combination of 1D CNT with 2D graphene-based electrodes exhibits fruitful mechanical and electrical characteristics far superior to their metallic counterparts [1]. In one of the interesting works, Huang et al. [28] developed a flexible Li-sulfur battery by confining sulfur in a CNT graphene hybrid nano-cage which not only provided a highly conductive network but also impeded poly-sulfide dissolution.

In our opinion, carbon fiber woven carbon cloth could also act as a suitable alternative to 1D CNT or 2D graphene-based flexible current collectors. Carbon cloth is economic, offers acceptable electronic conductivity, and has high mechanical strength with repeated flexing. Active materials have been grown in situ on carbon cloth, which eliminates the use of conducting additives and binders. For example, Wang et al. [29] grew cuboid arrays of lithium titanate on flexible carbon fiber cloth to fabricate an additive-free flexible Li-ion battery anode. In a different study, Pan et al. [30] hydrothermally synthesized a $LiFePO_4$ cathode on the surface of carbon cloth, which showed excellent electrochemical performance. Electro-spun polymers are carbonized to make free-standing low-cost flexible current collectors for Li-ion cells. For example, Miao et al. [31] developed a solvothermal-grown MOS_2 electrode on soft and free-standing porous carbon nano-fiber current collectors. Cost-effective upscaling of CNT and graphene-based flexible electrodes remain challenging. We believe that carbon cloth-based current collectors would be an effective alternative for flexible electrochemical energy storage.

6.3.3 Flexible Electrolytes

Owing to its ionically conducting and electronically insulating nature, the electrolyte facilitates ion transport between the electrodes and also minimizes self-discharge. Due to their decent ionic conductivity, facile percolation to the electrode, adequate wettability, and stable solid electrolyte-interface formation, liquid electrolyte is still the primary choice for flexible LIBs. However, for such liquid electrolytes, Li dendrite formation during repeated deformation and electrolyte leakage can lead to serious safety concerns. Solid-state electrolytes featuring negligible electrolyte leakage probability to mitigate the safety concerns and high Young's modulus to impede dendrite formation could be beneficial for developing flexible batteries.

Various types of flexible electrolytes are under investigation [1]. Solid polymer electrolytes (SPEs) are prepared by the solvation of a lithium salt (with low lattice energy) into a co-ordination polymer scaffold. The commonly used coordinating polymers are poly(ethylene oxide) (PEO), poly(vinylidene fluoride) (PVDF), and polyacrylonitrile (PAN). Li-ions, generated from the used Li salt, coordinate with the electron-donor group of the polymer scaffold. Under an electric field, Li-ions hop to the neighboring electron-donor sites of the adjacent polymer scaffold. Usually, SPE yields relatively lower ionic conductivity (~10^{-6} S/cm at room temperature), which is not suitable for any practical cell. Room temperature ionic conductivity can significantly be increased by structural modification of the polymer scaffold, selecting suitable Li salt, and use of plasticizers and/or fillers. For example, Jia et al. [32] reported that graphene oxide as a nanofiller into $LiClO_4$-PAN-based SPE significantly improves its ionic conductivity and mechanical strength.

Organic/inorganic type solid-state electrolytes have also been investigated. Lithium lanthanum zirconium tantalum oxide (LLZTO)-based ionic conductor is mixed with Li salt-free PEO to make a membrane with typical ionic conductivity of 10^{-4} S/cm at room temperature. LLZTO favors the local amorphous region in the polymer and also provides an extra ionic transport pathway in the polymer matrix [33]. Such a composite electrode also immobilizes Li salt anion in a LiTFSI–LLZTO–PEO ternary flexible electrolyte membrane. Simultaneous retardation of polymer crystallization and pinning of anions of LiTFSI salt further improve the ionic conductivities [34].

Gel polymer electrolytes (GPEs) are normally prepared by trapping a Li salt-containing liquid plasticizer into a polymer matrix. The liquid plasticizer acts as the main media for Li-ion transport and is thereby responsible for high ionic conductivity, while the polymer scaffold provides the necessary mechanical strength of the electrolyte. Electrochemical stability, mechanical properties, and cost are the key parameters to be considered while selecting a polymer matrix for GPEs. Up to now, numerous polymers have been explored, including PEO, PAN, PVDF, and poly(vinylidene-co-hexafluoropropylene (PVDF-HFP). Organic solvents and ionic liquids are mostly used as plasticizers in GPEs. Due to their higher volatility, solvent loss over time cannot be avoided for organic solvents. Consequently, a large amount of solvent is required to maintain a stable ionic conductivity of GPEs. Unfortunately, the excessive use of plasticizers can compromise safety, deteriorate mechanical properties, and increase the risk of flammability. On the other hand, ionic liquid-based plasticizers are nonvolatile and have relatively low flammability. The use of IL-based plasticizers is hindered due to complexities in synthesis and high cost. Recently Xu et al. [35] used a $LiPF_6$-carbonate-based electrolyte and polyvinyl acetate (PVAc)-incorporated PVDF-HFP/SiO_2 matrix to develop a flexible GPE. An NMC523/MCMB type pouch cell fabricated with PVAc/PVDF-HFP/SiO_2 GPE demonstrates good electrochemical properties and a high tolerance for mechanical abuse.

6.3.4 Flexible Separators

The main purpose of the separator is to prevent electron flow but allow an easier passage to Li^+ ions. Intrinsically flexible and mechanically robust polyolefin (viz. polyethylene, polypropylene, and their blends) separators soaked in liquid electrolytes can be directly used in flexible batteries. These separators must be chemically stable within the working potential range and must also be chemically inert toward the electrolyte and other battery components. Apart from chemical inertness, commercial separators have porosity in the range of 40–60%, the size of pores < 1 μm in diameter, and typically 20–25 μm thick [36]. The use of liquid electrolytes has disadvantages, therefore, routine separators, used in commercial rigid Li-ion batteries are replaced by bi-functional solid electrolytes. The bi-functional electrolytes serve dual purposes: to avoid short-circuiting between anode and cathode and to allow Li-ion transportation. Relatively thicker soft GPE is required to avoid short-circuiting. A thinner high-performance flexible separator has also been reported. In such a separator, the functional membrane is made out of a polyimide host. It has vertically aligned nanochannels filed with PEO–LiTFSI electrolytes.

6.3.5 Flexible Packaging

Stainless-steel and polymer-coated aluminum film are traditionally used as packaging material for commercial LIBs. These packaging materials lead to the hermetic sealing of the liquid electrolyte inside. Flexible batteries are mostly reported in the pouch cell type configuration in which laminated aluminum films are used as the packaging material. The aluminum-polymer film consists of three layers made of nylon, metallic Al foil, and polypropylene. This film has lower flexibility depending on the thickness of the Al foil and polymer layers. Polydimethylsiloxane (PDMS) offers excellent mechanical strength and better flexibility and, therefore, has been used in most of the recent reports. Koo et al. [37] demonstrated a bendable thin film Li-ion battery using PDMS as packaging material. PDMS can also withstand more than 100% stretch due to its good tensile properties. Xu et al. [38] fabricated stretchable batteries with serpentine interconnects where PDMS was used to attain the hermetic sealing. These batteries can be stretched reversibly up to 300% while maintaining a capacity density of ~1.1 mAh/cm^2. Following a different approach, Rao et al. [39] developed a spring-like all-solid-state battery consisting of SnO_2 quantum dot@rGO anode, $LiCoO_2$ cathode, and PVDF-HFP soaked in $LiClO_4$ dissolved in EDC/EC liquid electrolyte. A self-healing carboxylate polyurethane was used to package the spring-like flexible battery, which exhibited a specific capacity of ~82 mAh/g under a series of deformations.

6.4 NOVEL FABRICATION ROUTES FOR BENDABLE BATTERIES

Out of the different types of flexible batteries, in this chapter, we focus on the bendable battery. Facile electrophoretic deposition (EPD), due to its reproducibility and versatility, has been accepted as a successful surface coating technique in paint industries. EPD could be a potential alternative to fabricate a uniform coating of the electrode on a flexible 3D current collector [40]. By tuning the suspension composition, applied voltage, and deposition time, one can easily control the thickness of the electrode and active mass loading. By carefully designing the counter electrode, one can deposit active materials in any complex-shaped substrates. Alcohol-based suspensions used in EPD are much more eco-friendly compared to conventional tape casting which uses toxic NMP solvent for binder (PVDF) dissolution. In this chapter, we will describe the utility of EPD in developing graphite electrode on a flexible carbon cloth substrate. Printable solid-state batteries are essential for developing flexible and wearable mobile electronic devices with shape diversity. As compared to lithographic methodologies like photolithography, e-beam, and ion-beam lithography, relatively inexpensive printing techniques will significantly lower the fabrication cost of large-scale flexible batteries and supercapacitors with desired electrochemical performances. Further, we have described the electrochemical performance of a flexible carbonaceous supercapacitor printed using an inkjet printer.

6.5 CARBON CLOTH AS COMPLIANT CURRENT COLLECTOR

Among various compliant substrates, carbon cloths (featured with abundance, ease of surface functionalization, large surface area, 3-D architecture, good flexibility, and facile structural tunability through mature knitting, weaving, and embroidery technologies) are considered one of the most promising candidates. Electrode materials are normally coated or in-situ grown onto the conductive and porous textile network to form an interconnected 3-D electrode structure [41]. The highly porous architecture of textile substrates can also dissipate the top strain because their porous structure can significantly reduce the elastic modulus as compared to their solid counterpart. The normalized elastic modulus (Y_p/Y_s) can be expressed by the following equation [41, 42].

$$\frac{Y_p}{Y_s} = \left[\frac{2\left(1+\sqrt{2}\right)^2}{15} \frac{5\pi + 3(1-\rho)\left(1+\sqrt{2}\right)^3}{25\pi + 7(1-\rho)\left(1+\sqrt{2}\right)^3} (1-\rho) \right] \tag{3}$$

where Y_p and Y_s are the elastic moduli of porous and solid architecture, and ρ is the porosity of the textile. For a textile possessing 50% porosity, i.e. ρ =0.5, the elastic modulus will be around 10% of its solid counterpart. Consequently, the top strain will be much less for such a porous structure. Under similar experimental conditions, the carbon cloth-based electrodes outperform the conventional foil-based electrodes in terms of rate performance and cycling stability. Such improved electrochemical performance can be attributed to the higher contact areas between the active material and current collector, minimal presence (or total absence) of insulating polymeric binder in the electrode, and easier ionic diffusion due to the highly porous architecture (as shown in Figure 6.1). Moreover, for a similar active material loading on the identical projective area of the current collector at the same charge-discharge current, the effective current density realized by the active material in a carbon cloth base electrode is much lower than the foil-based electrode. Consequently, remarkable improvement in rate performance and cycling stability is observed. For instance, the V_2O_5 cathode when fabricated on a 3-D textile-based current collector delivers a discharge capacity of ~222 mAh/g after 500 cycles at a specific current of 200 mA/g. However, the same V_2O_5 material when cast on Al foil exhibits a specific capacity of ~102 mAh/g after 100 cycles at a specific current of 200 mAh/g [43]. Again, the highly porous structure can also confine the active materials in its pores if the particle size of the active material matches well with the pore dimension of the textile. Such confinement of active material can be beneficial in preventing the delamination of the active material from the current collector during repeated bending tests.

The fabrication approaches mostly involve the coating of active materials through chemical vapor deposition [44], dip coating [45–46], a screen-printing technique [47], in-situ growth of active materials by electrodeposition [48], or hydrothermally [49–50]. Carbon-coated Si nanowire grown onto carbon cloth (CC) by atmospheric CVD process has been proposed as a flexible anode for LIBs, which shows high specific capacity (3362 mAh/g after 100 cycles @ 100 mA/g), good cycling stability (~2000 mAh/g after 500 cycles @ 1A/g), and superior rate performance (1500 mAh/g @ 5A/g) [44]. A high-performance 3D macroporous MoS_2@C/CC electrode (obtained by dipping the CC in precursor solution followed by calcination in an argon atmosphere) was able to retain its structural integrity and initial morphology even after 300 bending cycles. Hu et al. [45] developed a flexible lithium-ion textile battery where they prepare the $Li_4Ti_5O_{12}$ and $LiFePO_4$ electrodes by dipping the CNT-coated 3-D textile in the respective slurries. In another report, SnO_2 nanowires hydrothermally grown on a CC surface did not show any observable changes even after 200 bending cycles. The 3D porous architecture of the SnO_2/CC electrode is beneficial for effective electrolyte percolation and also helps with the large stress generated during cycling [51]. Balogun et al. demonstrated a TiO_2@TiN/CC anode-based pouch cell that can power a LED even after 100

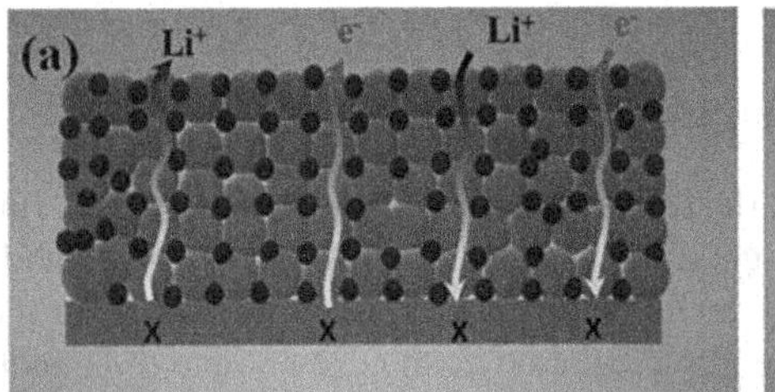

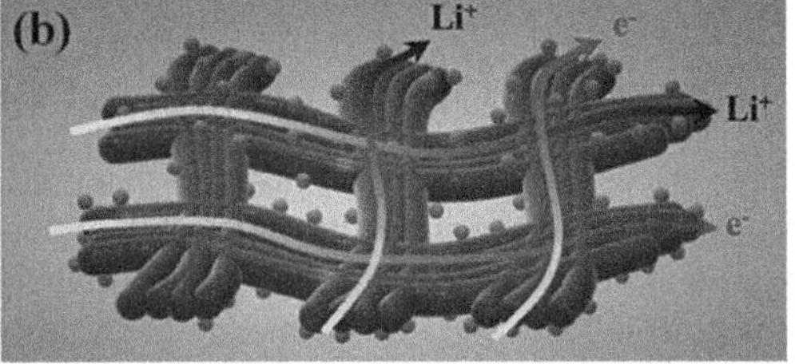

FIGURE 6.1 (a) Poor charge transfer in thick electrode cast on Al foil and (b) good charge transfer in a textile-based electrode due to its highly porous architecture and the large contact area between the active material and current collector.

deformation cycles [52]. Although dip coating or screen printing provides relatively higher mass loading, a uniform coating of each strand of the textile is difficult to achieve. Again, the binding of the fabricated electrode with the textile substrate is also poor. On the other hand, the in-situ growth of active materials offers uniform coating onto each strand of the textile and promises good adhesion between the active material and substrate. Moreover, this approach also circumvents the use of inactive additives like binders or conductive agents and thus increases the active material percentage in the electrode. However, the major drawbacks of this method are its complex processing steps, time-intensive nature, and, particularly, low active mass loading. These challenges open the door for new 3-D electrode manufacturing techniques which can offer a uniform, well-adhered coating of high active material loading.

6.6 CASE STUDY – I: EPD OF GRAPHITE ANODE ON 3D CARBON CLOTH CURRENT COLLECTORS

6.6.1 Salient Features of EPD

Suspension-based electrophoretic deposition (EPD) is considered to be a promising technique for developing flexible electrodes. EPD is a colloidal process where the depositing particles are dispersed in an electrostatically stabilized suspension. Upon application of an external electric field, the charged particulates in the suspension are forced to move toward the oppositely charged electrode and get deposited onto it to form a well-adhered, uniform coating. The active material loading (which remains a big concern for in-situ grown electrodes) can be easily controlled in EPD by merely tuning the experimental parameters such as applied potential, deposition time, and the suspension composition. In addition, the fast deposition rate of this technique enables the growth of several micron-thick coated electrodes within a short period of time (a few minutes). The capability of uniformly coating any complex-shaped substrate makes the EPD process more lucrative for developing 3D flexible electrodes. More importantly, this process can be used to coat various types of electrode materials including but not limited to metals alloys, conversation type oxides, carbonaceous materials, and their composites. Note that in-situ growth of most of the conventional cathode materials on carbon cloth, or any other carbonaceous material, is very challenging, as high-temperature annealing is required for the cathode phase formation. However, EPD can easily be employed to develop a uniform, porous, and well-adhered cathode as well as anode films on CC or any 3D carbonaceous current collectors. In conventional tape casting, carcinogenic N-methyl-2-pyrrolidone (NMP) solvent is used to prepare the slurry. In contrast, eco-friendly dispersing mediums such as ethanol and isopropanol are normally used in EPD. Apart from the aforementioned advantages, it is found that merely changing the electrode fabrication technique from tape cast to EPD can grossly improve the electrochemical performance of conversion-type and alloying-type anodes, which usually suffer large volumetric fluctuations during cycling [53–57].

Most of the carbon cloth-based flexible electrodes involve the in-situ growth of conversion-type anode materials. These electrodes offer significantly higher specific capacity compared to the conventional graphite anode (theoretical capacity ~372 mAh/g). However, conversion-type electrodes suffer large volumetric fluctuations during cycling, poor initial coulombic efficiency, and significant voltage hysteresis. As a result, the graphite anode still holds the major market share (~98%) in commercial LIBs [58]. As an anode of flexible Li-ion batteries, therefore, we have fabricated EPD-grown graphite/carbon black composite anode on a carbon cloth current collector.

6.6.2 Preparation of Suspension and Electrode Fabrication

In order to develop a uniform coating through EPD, the preparation of an electrostatically stabilized suspension of the depositing materials is one of the major prerequisites. Such stable suspension of commercial graphite and carbon black powder was prepared by dispersing these powders in an

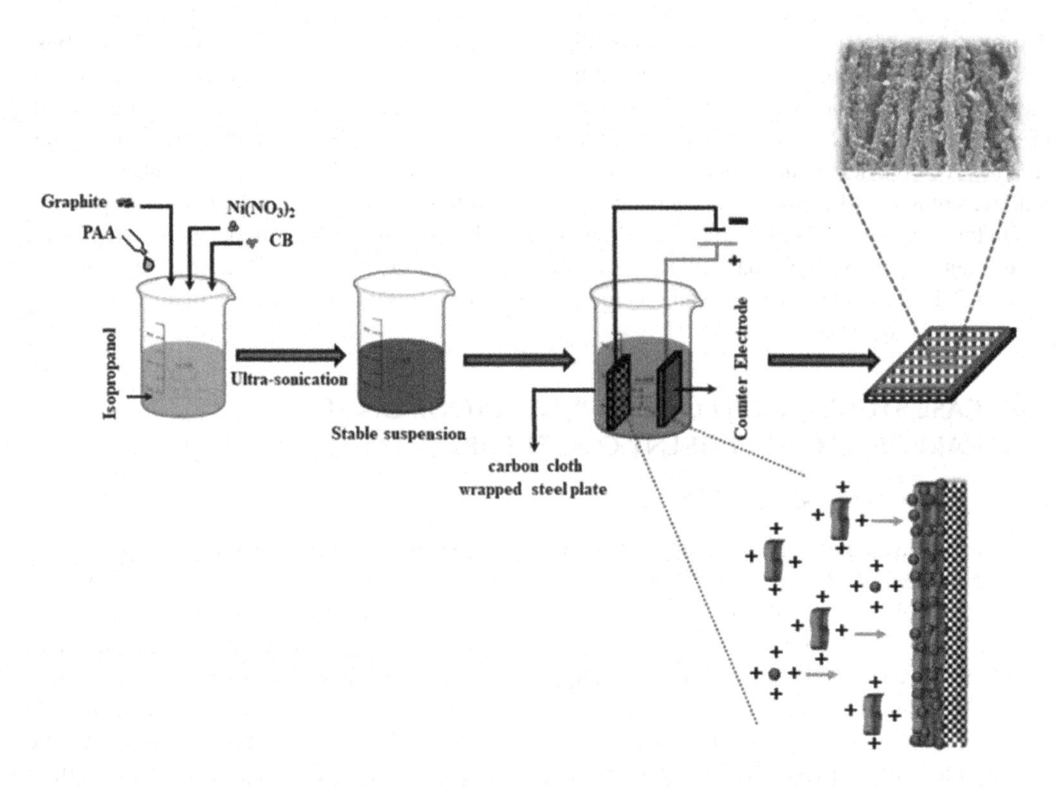

FIGURE 6.2 Schematic of flexible graphite electrode fabrication process on 3D carbon cloth current collector by electrophoretic deposition.

isopropyl medium where polyacrylic acid (PAA) in combination with nickel nitrate was added as a dispersant. The resultant mixture was then subjected to ultrasonic agitation for one hour to prepare an agglomeration-free suspension. The electrophoretic deposition procedure is schematically shown in Figure 6.2. The zeta potential values of the dispersed particulates in different suspension compositions were measured to check the applicability of the prepared suspension in the EPD process.

The suspension with optimal rheology was taken in a 100 ml beaker where a stainless-steel plate wrapped with carbon cloth (depositing electrode) and another steel plate (counter electrode) with a slightly larger exposed surface area were placed parallel to each other. The distance between these two electrodes was fixed at 2 cm. An external DC voltage was then applied to deposit the graphite-carbon black composite electrode on the carbon cloth current collector. To remove the solvent, subsequently, the deposited film was dried in a vacuum oven at 100°C for 12 hours. The developed electrode film was bent at different bending radii to check the flexibility. Moreover, a Scotch-tape test was performed to investigate the adhesion property of the deposited electrode.

6.6.3 Electrochemical Characteristics

From the dried film, an electrode of 1 x 1 cm area was cut to evaluate the electrochemical performance of the EPD-grown electrode. A CR2032-type coin cell was fabricated in an argon-filled glove box where the moisture and oxygen levels were maintained at < 0.5 ppm. Li foil was used as both a counter and reference electrode. The polypropylene separator (Cellgard 2400) was soaked in 1 M $LiPF_6$ dissolved in 3:7 (wt.%) EC-DMC electrolyte. Galvanostatic charge/discharge experiments were conducted at different current densities using an automated battery cycler.

The quality of the deposited electrode is largely governed by the zeta potential of the dispersed particulates in the prepared suspension. Again, for successful co-deposition of the dispersed graphite and carbon black particles, they must possess a similar type of surface charge (both should have either positive or negative zeta potential). It was found that when the commercial graphite or carbon black powder is dispersed in isopropanol, they readily settle down which can be attributed to their low zeta potentials. However, the addition of PAA along with nickel nitrate salt in the suspension significantly improves the zeta potential of the dispersed particulates (> +50 mV), which leads to the formation of a sufficiently stable suspension. Such electrostatically stabilized suspension ensures the co-deposition of graphite and carbon black onto the CC current collector. Here, PPA acts as dispersing agent as well as a binder. It has been reported that due to having –COOH groups, PAA can offer excellent adhesive properties toward the active materials as well as the current collector through hydrogen and ester-like bond formation [59]. A deposition kinetics study was carried out to find out the optimal deposition parameters. It was found that the deposited mass per unit area was linearly increasing with applied voltage and deposition period. However, the deposited electrodes exhibit good uniformity and adherence properties up to a deposition voltage of 100 V and for deposition periods of 5 min. Beyond these parameters, an irregular and loosely bound coating is formed. The deposited mass at 100V for 5 min was around 4 mg/cm^2. In order to estimate the amount of active material in the EPD-grown electrode, the film was deposited onto a steel plate under similar experimental conditions. After properly drying in a vacuum oven, the deposited film was scratched out to perform thermogravimetric analysis in an air atmosphere. From the TG study, it was found that the electrode consists of ~ 3 wt.% absorbed solvent molecules and ~7 wt.% of PAA binder. The remaining 90 wt.% material is graphite and carbon black. As observed in the TG plot, both graphite and carbon black decompose simultaneously in a temperature range of 400–650 °C. It is difficult to determine the individual weight percentage of these materials in the electrode. Therefore, the total mass of graphite and carbon black is considered during specific capacity calculations.

As seen in Figures 6.3a and 6.3b, both the carbon cloth current collector and the graphite/CC electrode demonstrate excellent flexibility. The EPD-grown electrode does not show any noticeable delamination of the graphite electrode even after repeated bending. A Scotch-tape test is also performed to investigate the binding property of the deposited electrode where the electrode is stuck and flayed in between two adhesive tapes (Figure 6.3c). As depicted in Figure 6.3d, a marginal amount of electrode material is found to be stuck on the tape which suggests the excellent adhesive property of the EPD-grown electrode.

The SEM images of the bare carbon cloth at different magnifications are shown in Figures 6.4a and 6.4b. It can be observed that the CC is composed of numerous well-woven strands of ~10 μm diameter. The microstructures of the graphite-carbon black composite film deposited at 100 V for 5 min are displayed in Figure 6.4c. It is apparent from the SEM images that the EPD technique can uniformly coat each strand of the carbon cloth. The average diameter of carbon cloth strands after EPD is around ~18 μm, which means that the thickness of the electrode layer on each strand is around 8 μm.

Figure 6.5a demonstrates the cyclic voltammogram over the initial three cycles obtained at a scan rate of 0.2 mV/s in the potential range of 0.01 to 3.0 V. In the first reduction process, the broad peak observed at 1.0 V can be attributed to the decomposition of electrolyte and SEI layer formation which disappears in the following cycles. The reversible cathodic and anodic peaks that appeared in the subsequent cycles between 0.4 to 0.01 V can be ascribed to the Li-ion intercalation-deintercalation phenomenon. The insertion of Li-ions in the graphite layers occurs through different stage compound formations such as the diluted stage I ($Li_{0.083}C_6$) to stage IV ($Li_{0.166}C_6$), stage III ($Li_{0.222}C_6$), stage II ($Li_{0.5}C_6$), and stage I (LiC_6) during the cathodic process and follows the reverse reactions sequence in the anodic scan [60]. However, the cyclic voltammetry should be performed at a very low scan rate (few μV/s) in order to visualize each of these individual processes. As in this present study, the CV was carried out at a relatively higher scan rate (0.2 mV/s), and only a few peaks appeared in the graph. The peak at around 0.2 V is ascribed to the diluted stage I to stage IV phase formation whereas, the peak at 0.03-0.04 V is attributed to the complete lithiated LiC_6 phase (stage I) formation.

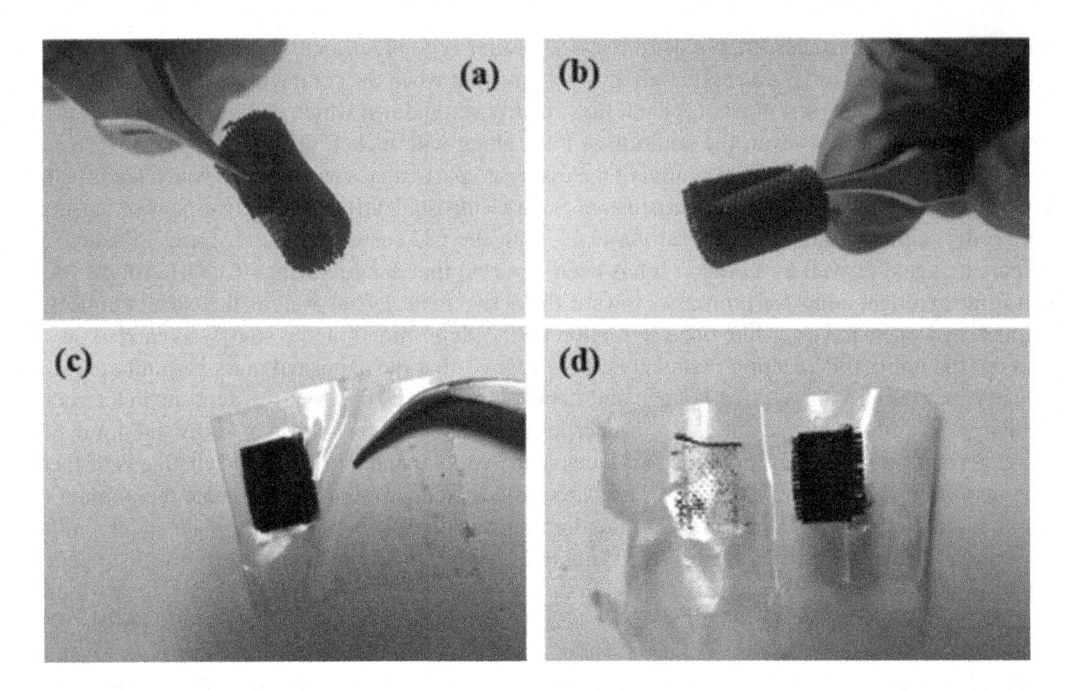

FIGURE 6.3 Images of carbon cloth (a) before and (b) after deposition of graphite electrode layer on carbon cloth current collector demonstrate excellent flexibility; (c) and (d) depict the images of EPD-grown electrode prior and after performing the Scotch-tape test respectively. The images confirm the excellent binding property of the deposited electrode with the carbon cloth as the minimum amount of active material is found to be stuck on the adhesive tape after the test.

FIGURE 6.4 (a) and (b) display the SEM images of the bare carbon cloth. The microstructure of the carbon cloth after deposition of the graphite-carbon black composite electrode is shown in (c).

The initial galvanostatic charge-discharge profiles at the C/10 rate are shown in Figure 6.5b. In the first discharge, the cell is able to deliver a specific capacity of ~ 463 mAh/g with a coulombic efficiency of ~85%. However, the coulombic efficiency rises quickly to >99% in the subsequent cycles. The irreversible capacity loss in the first cycle can be attributed to the electrolyte decomposition and SEI layer formation. When cycled at a 1C rate (375 mA/g), the EPD-grown cell is able to deliver a reversible capacity of ~ 384 mAh/g after 100 cycles (Figure 6.5c). More importantly, no significant capacity fading is observed even after 100 cycles (98.5% capacity retention after 100 cycles). The graphite electrode deposited on flexible carbon cloth also demonstrates excellent rate capability and is able to deliver a reversible capacity of ~ 330 mAh/g at the 2C rate. Upon reducing the specific current to C/10, it is able to regain its initial capacity (384 mAh/g), which suggests that the EPD-grown electrode can endure a large range of current densities without suffering any permanent structural setbacks.

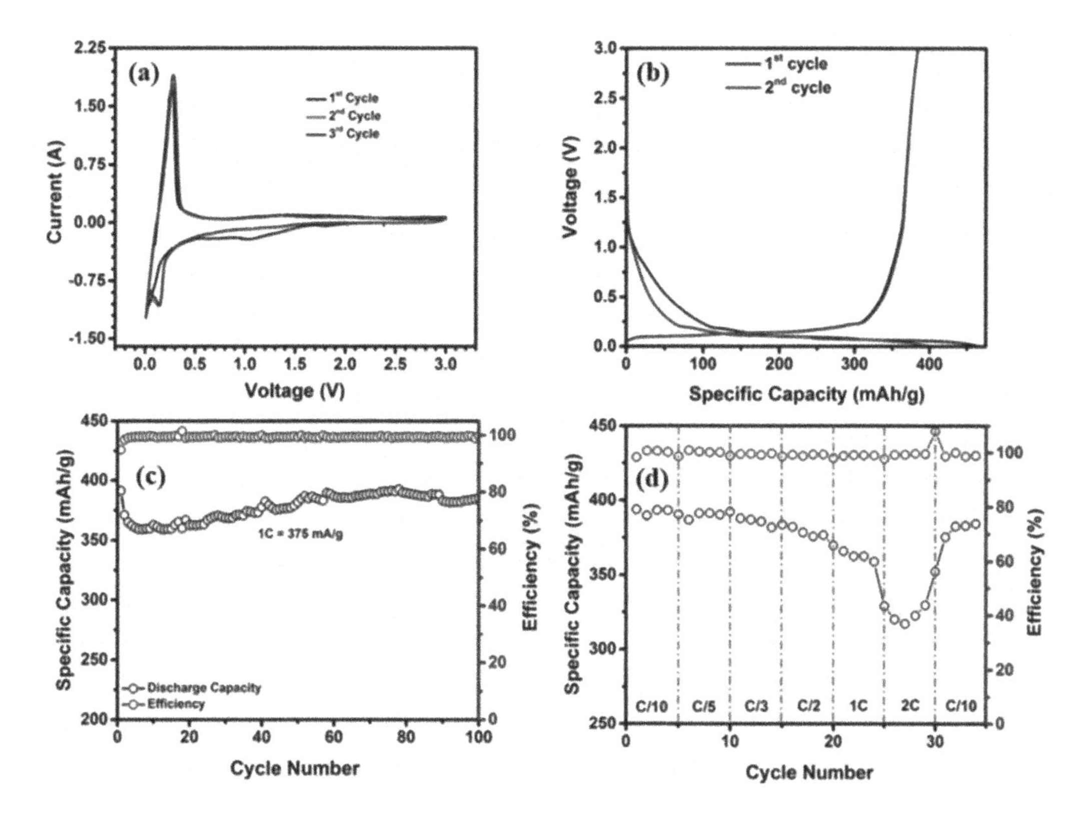

FIGURE 6.5 Cyclic voltammetry plots of EPD-grown graphite-carbon black electrode performed between 3.0 and 0.01 V at a scan rate of 0.2 mV/s. (b) Galvanostatic charge-discharge profiles of the graphite electrode at C/10 rate. (c) Long-term cyclability at 1C rate and (d) rate capability test performed in the range of C/10 – 2C current.

6.6.4 Future Prospective for EPD

Although we have demonstrated the fabrication of graphite electrodes on 3D carbon cloth by the EPD technique, this process can easily be extended for the fabrication of various flexible battery electrodes. Figure 6.6 shows the schematic of the bendable battery we propose making in our laboratory. Electrode materials are coated on a compliant carbon cloth current collector. The protruded portion of the current collector will itself be the terminal for positive and negative electrodes. In between two flexible electrodes, we will use a polypropylene sheet (Celgard 24) soaked with conventional liquid organic electrolyte (1M $LiPF_6$ dissolved in EC:DMC (3:7) with FEC additives). The separator is also flexible in nature. The flexible electrodes will be stacked sequentially as shown in the schematic (Figure 6.6). Finally, we will use polyethylene terephthalate (PET) film/Al laminated polymer substrate or poly-(dimethyl siloxane) sheets (PDMS) as packaging materials. All these fabrications will be done in a moisture-controlled glove box (O_2 and H_2O contents maintained less than 0.5 ppm). Finally, the fabricated cell will be laminated using a commercial laminator kept inside the glove box.

6.7 PRINTED FLEXIBLE ENERGY STORAGE DEVICES

The emerging trend of flexible solid-state energy storage devices primarily originated from the idea of conformally embedding them onto bendable, twistable, stretchable, and compressible states on a wide variety of substrate materials, making them fundamentally different from the conventional rigid form but challenging to achieve reliable devices. Furthermore, the safety considerations of

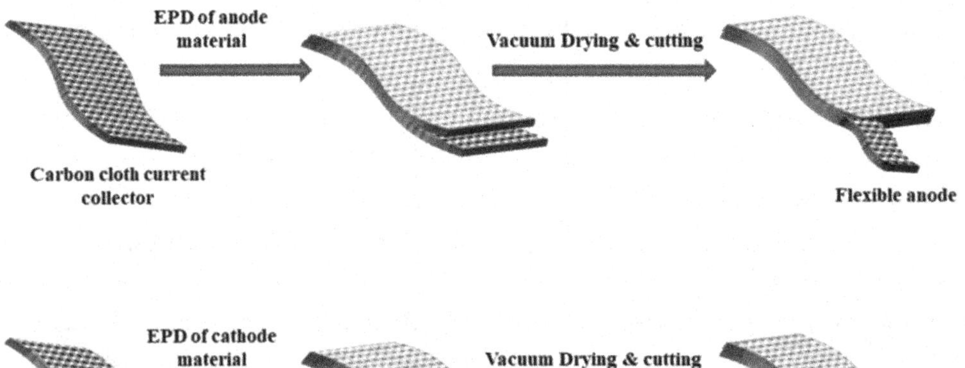

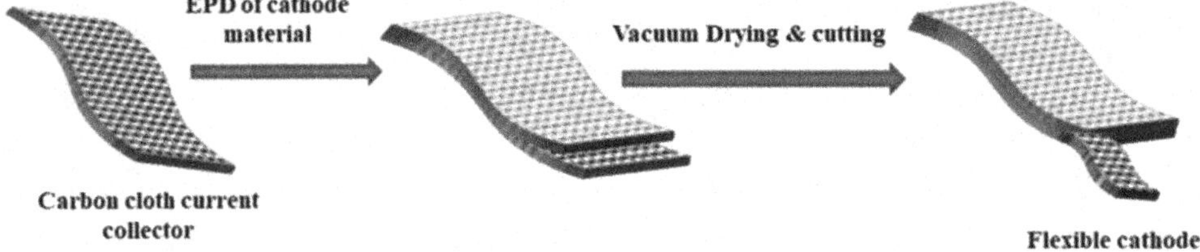

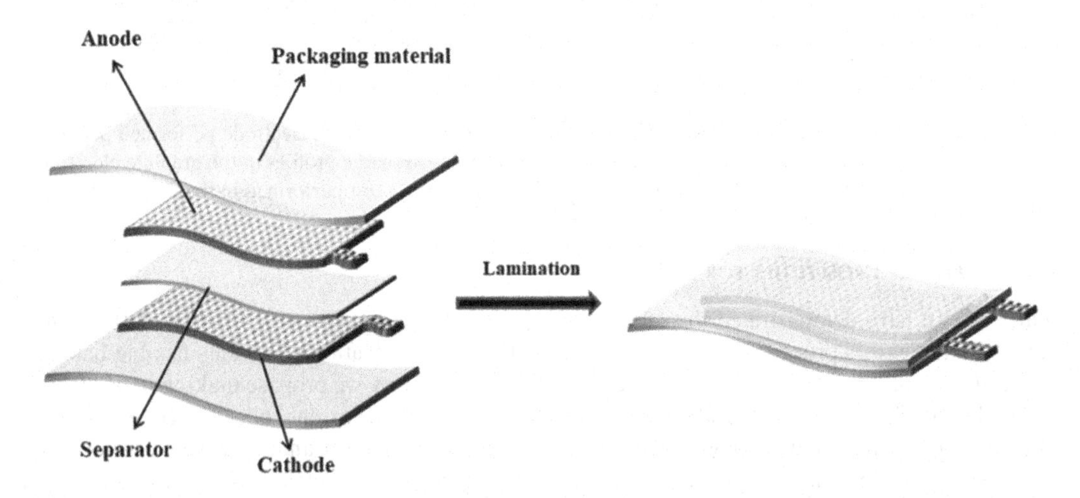

FIGURE 6.6 Schematic of flexible cell fabrication using EPD technique.

involved materials to human health (such as skin) bring additional constraints. Both subtractive (i.e., lithographic) [61] and additive (i.e., printing) [62] manufacturing approaches have been adopted to design different flexible electronic and energy storage devices. Lithographic methods require a cleanroom facility, extensive laboratory conditions, high production costs, and labor-intensive processes, making the devices even costlier. On the other hand, printing techniques have significant advantages over the lithographic approaches for large-scale production with desired precision and accuracy, although presently it is difficult to challenge lithographically produced devices. Specifically, printing flexible power sources in a desired way is a fascinating technique of modern manufacturing due to its exceptional form factors, performance compatibility, monolith integration with flexible devices, scalability, versatility, and processible at low cost [63]. Printed energy storage devices such as batteries and supercapacitors can be bent around body parts (such as the human wrist or an animal leg) while maintaining stellar energy storage capacity retention, which is necessary to power up the IoT sensor devices [64]. Recently, printing technology has been broadly classified into subcategories based on printing approaches that utilize template and non-template methodologies. The template printing method includes screen printing, gravure printing, flexography, and imprinting, aerosol jet printing, inkjet (IJP) printing, and 3D printing comprise non-template printing techniques [65–67]. Non-template printing approaches have added advantages compared to the template method (which involves direct contact printing) because such methods require no physical contact with the substrate platform. The idea of using different printing approaches applied to make flexible electrochemical devices and systems is illustrated in the following section.

6.7.1 Screen Printing

Screen printing is one of the more popular printing methods used for the fast and scalable fabrication of printed EES. Ink with higher viscosity is squeezed through a mesh screen and undergoes shear thinning to facilitate penetration when forced through the screen image/mask and printed on the desired substrate (Figure 6.7). The mesh screen contains the specific pattern of the targeted structure and determines the resolution and feature size. Various material combinations have been reported to make battery materials such as current collectors, electrodes (cathodes and anodes), and separators using screen printing, which show significant performance in terms of areal capacities, cyclic performance, and high energy densities compared with solid-state batteries [68–69]. A typical thickness for screen-printed patterns is on the scale of one hundred to a few hundred microns. Although screen printing is a simple printing approach, it requires more inks with predefined properties to make uniform patterns. In addition, the printed features obtained from the screen-printing method have rougher surface finishing which could possibly affect the overall electrochemical energy storage device performance, such as short-circuiting by the physical piercing of

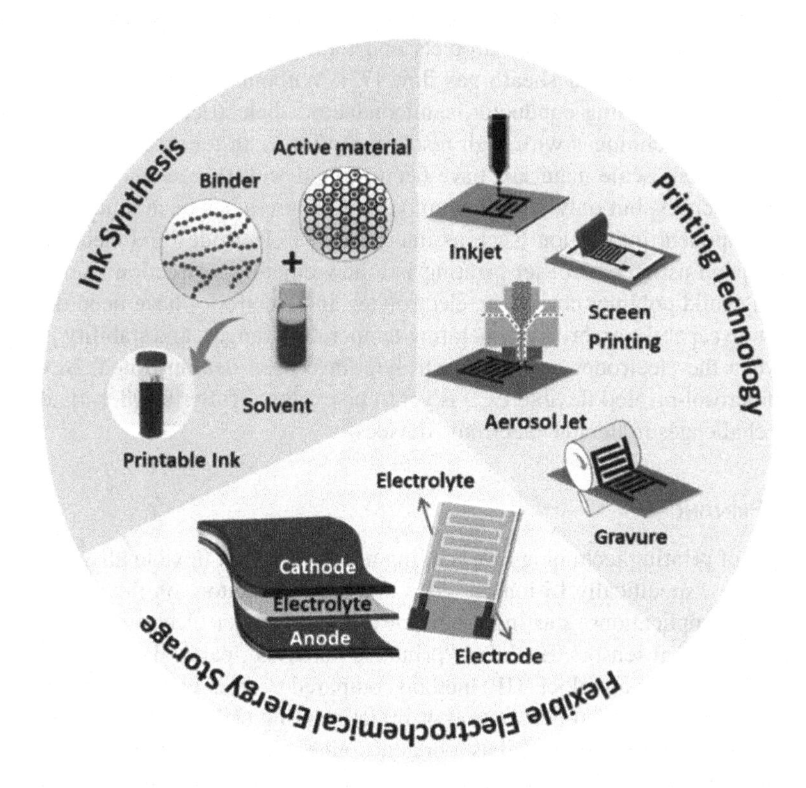

FIGURE 6.7 Flexible printed electrochemical energy storage devices.

other components in a cell, charge accumulation, withstanding larger strain (in a flexible electronics) etc. Therefore, better quality screen-printing machine development is critical for the successful demonstration of printed EES systems [64].

6.7.2 Gravure Printing

Gravure printing is another promising roll-to-roll printing method having the capability of high-resolution printing at high speed (10 ms^{-1}) that can be used for industrial-scale fabrication of printed EES [70, 71]. The required electrode pattern is engraved on the surface of a gravure cylinder which gets filled with ink, and a doctor blade wipes the excess ink from the cylinder. The cylinder rolls over the substrates, and ink is transferred to the substrate in a specified pattern. Recently, battery components, such as anode materials ($Zn_{0.9}Fe_{0.1}O$-C), separators (PVDF-HFP/Al_2O_3/PE/PVDF-HFP Composite), and cathodes ($LiFePO_4$), have been printed with the gravure printing method, and this improved battery performance in terms of long-life cyclic stability, flexibility, and integration into industrial level production has been reported [70–72]. Moreover, printed batteries show excellent high cycling stability compared with the conventional battery fabrication approach, which is demonstrated by tuning the ink synthesis and controlling printing techniques. While this technique has superior performance in terms of resolution and printing speed, it suffers from limited customizability of the pattern. As far as discussing the above template-based contact printing techniques is concerned, critical parameters such as mask fabrication, customizable pattern printing, porosities, and thickness control need to be established with a number of pre-/post-processing steps in order to achieve the desired properties in flexible EES.

6.7.3 Aerosol Jet Printing

Aerosol jet printing is a relatively new printing technique where the electrode material gets aerosolized into small (~ 1-5 μm diameter) droplets and then gets transferred to the substrate by a carrier gas focused by an annular sheath gas flow [73]. Without use of any mask, a wide range of range of materials (including conductor, semiconductor, dielectric, packaging materials etc.,) can be printed by this techniques with high resolution printing in a contactless manner. Specific to EES applications, nanoscale materials have been printed with this technology for printed batteries and supercapacitors, but only a limited amount of progress has been made in this direction. Making multi-component integration (such as integration of electrodes and electrolyte materials) with spatial precision using aerosol jet printing is a new emerging direction. Aerosol jet-printed Li-ion conducting solid polymer composite electrolytes and electrodes have been demonstrated to produce higher rate capabilities, broader operating temperature ranges, and stability [64, 74]. Strong interfaces between the electrodes and the electrolyte have been demonstrated. Nevertheless, the development of aerosol-printed flexible EES is yet to be explored fully in order to address process manufacturing challenges in flexible electronic devices.

6.7.4 Inkjet Printing

The effective use of printing technology for making next-generation flexible all-solid state and all-printed EES devices, specifically Li-ion batteries and supercapacitors on flexible substrates, will allow a plethora of applications, starting from powering consumer electronics to wearable electronics and environmental sensors. Enabling "printable batteries", particularly miniaturized power sources, is highly feasible in an inkjet (IJP) method compared to other printing techniques [63]. The following section gives a detailed description of inkjet printing (IJP) processes along with a case study. IJP technology has been identified as a breakthrough manufacturing technique for attaining high-performance electronic and energy devices. IJP affords several advantages compared to other printing techniques, namely contactless printing, higher resolution of printed patterns (as small as a few tens of microns), printing compatibility, low cost, low material wastage, versatile inks, and ease of integration with different substrates [75–77]. IJP requires three important steps for high-performance device manufacturing: suitable ink, printing process optimization, and post-treatment process and/or curing. Apart from the type of printers used, all of the three parameters above are highly dependent on the type of material chosen. The technique requires inks with low viscosity and surface tension, making a low concentration material ink to result in a number of layer assembly formations, yet making the final thickness of the stack as low as tens of nanometer to micron scale [78–79]. Suitable and stable printable inks for IJP can be prepared by mixing the active material of choice with solvents and an additive and/or binder. The specific physical and chemical characteristics of inks such as viscosity, thixotropy, shear thinning and thickening nature, boiling point, surface tension, solubility, contact angle, density, and specific gravity need to be thoroughly invested to obtain high precision and reliable printed patterns with uniform film morphology. Basic processes involved in IJP printing such as droplet formation, stable ink jetting, and satellite droplet formation are expressed by the dimensional physical numbers such as the Reynolds number (R_e), the Weber number (W_e), and the Ohnesorge (Oh) number, as shown in Equations (4), (5), and (6) [80, 81].

$$W_e = \frac{\upsilon^2 \rho \alpha}{\gamma} \tag{4}$$

$$R_e = \frac{\upsilon \rho \alpha}{\eta} \tag{5}$$

$$Oh = \frac{\sqrt{W_e}}{R_e} = \frac{\eta}{\sqrt{(\gamma \rho \alpha)}} \tag{6}$$

where υ, γ, ρ, α, and η are velocity (m/s), ink surface tension (mN/m), ink density (g/cm^3), the diameter of inkjetting nozzle (μm), and ink viscosity (mPa s), respectively. The reciprocal of the Oh gives the dimensionless quantity Z, which is used to evaluate the figure of merit about printability with the values of $1 < Z < 10$ [82]. For smaller Z values, the ink must be too viscous to prevent droplet splitting, whereas Z above 10 indicates the formation of unwanted satellite drops. Furthermore, W_e should be above 4 for the droplets to possess appropriate threshold energy to overcome the ejection energy barrier [83]. Typically, a piezoelectric actuator is used to jet the ink through the printing nozzle continuously supplied from an ink reservoir (Figure 6.7) [65, 84].

With its layer-by-layer printing methodology, the multi-material composition of the same electrode and multiple electrode configurations can be achieved to facilitate the manufacturing of engineered battery electrodes. As the printed flexible battery/supercapacitor technologies require multilayer printing with current collectors, anodes, electrolytes, separators, and cathode materials, the use of IJP for miniaturized power sources constitutes the most advanced additive manufacturing. Therefore, developing inks for each component of the printed battery is required. Different IJP-printed thin films have been used as working electrodes/cathodes ($LiMnO_2$, $LiCoO_2$, and $LiFePO_4$) and anodes to exhibit higher rate capability, high discharge capacity, and capacity retention compared with the tape casting method [85–88]. In addition, IJP can also be used to provide printed electrolytes and a packaging solution to make the cell/battery pack a complete solution. The conformal IJP printing of thin film electrolytes not only contributes to total lower internal resistance ($R=\delta_i/\sigma_i$, where, δ_i and σ_i are electrolyte thickness and ionic conductivity, respectively) but also creates seamless effective interfacial union [88]. As a result, IJP has become an emerging promising technology to fabricate flexible all-printed solid-state energy storage devices. A wide variety of material spectrum, including polymeric materials, along with vast design considerations, will thus make printed batteries and supercapacitors an innovative and emerging technology for a wide variety of applications.

6.7.5 Case Study – II: Inkjet-Printed Supercapacitor

This section provides a case study of an all-printed solid-state graphene-based supercapacitor device that was fabricated using IJP to demonstrate its competitive advantages and high reliability in performance. Interdigitated supercapacitor electrodes having micron-level resolution (width) were printed using an IJP (Microplotter II from Sonoplot Inc.). The printer was equipped with a 2 mm x 2 mm piezoelectric actuator on which a glass nozzle having a 40 μm tip head was installed. Lab-grown graphene aerosol gel ink with optimized viscosity (~ 25 cP) and particle concentration (70 mg/mL) was loaded into the glass tip from an ink reservoir by capillary force through the tip nozzle. The in-built "Sonodraw" drawing software was used to create the interdigitated pattern to define the device dimensions. The supercapacitor electrodes were printed onto a cleaned 25 μm thick polyimide film without the coffee ring effect. The printed electrodes were subsequently annealed to remove the nonconductive binder used in the ink. Aqueous-based PVA-H_3PO_4 polymer electrolyte was printed on the top of the interdigitated electrodes to complete the fabrication of all-printed solid-state flexible supercapacitors [89]. The fabricated supercapacitor device was tested for a 1,000 charge-discharge cycle, which demonstrated an extraordinary 99.6% capacity retention of the initial specific capacitance (376 mF/cm^3 at 0.25 μA). This performance is far superior compared to screen printing, gravure, and aerosol jet supercapacitors, which show < 90%, 85–90%, and < 95% capacity retention respectively [90–93]. The device was also tested under a 15 mm to 4 mm bending radius and showed 80% retention of the initial capacitance at extreme bending conditions, which signifies superior flexibility and functionality under any type of bending stress. The results are shown in Figure 6.8.

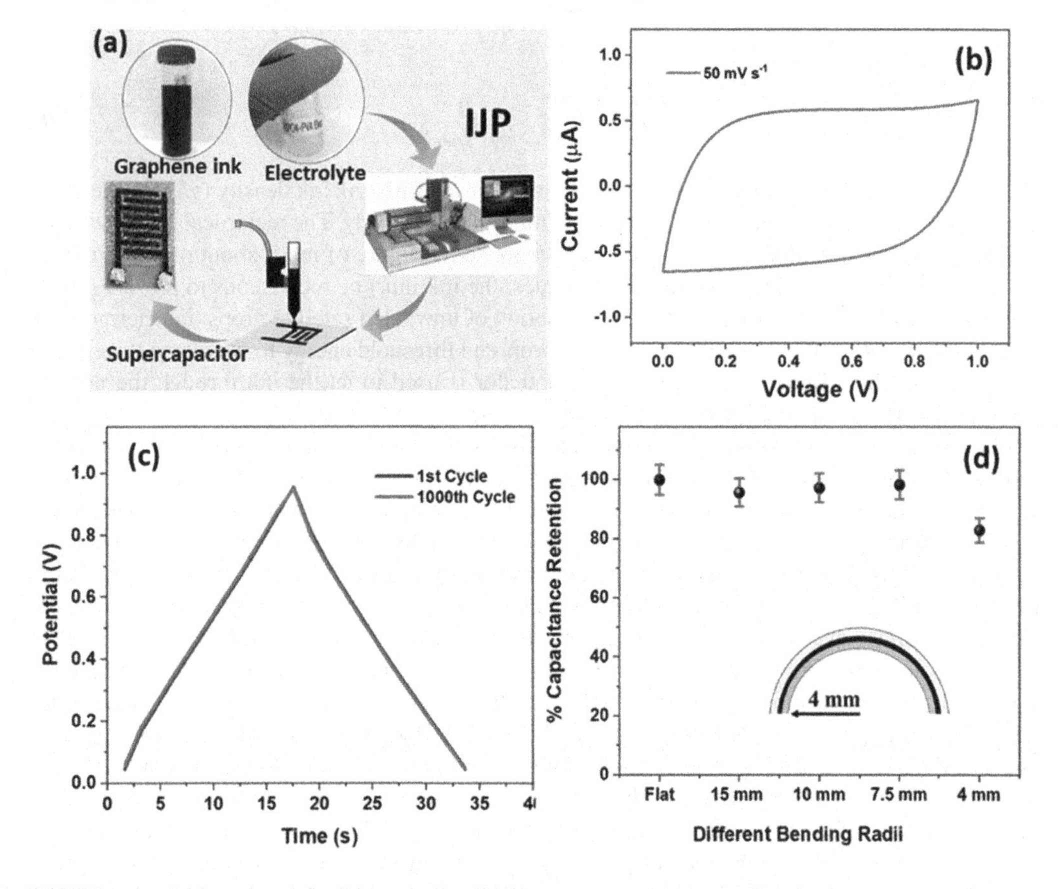

FIGURE 6.8 Inkjet-printed flexible and all-solid-state supercapacitors. (a) Fabrication process, (b) cyclic voltammetry characteristics, (c) cyclic stability over 1,000 cycles of performance, and (d) capacitance retention over different bending cycles.

6.7.6 3D/Extrusion Printing

Apart from electronics applications of previously mentioned printing methods, additive manufacturing in three-dimensional space (alternatively, 3D printing or extrusion printing) is currently trending to fabricate all-printed solid-state energy storage devices where higher energy density requirement is needed (mostly applications involving higher dimensional objects). 3D-printed energy devices have the potential to achieve higher charge storage capacity and long-term cycling stability due to the increased surface area offered by macropores. The primary advantages over traditional battery manufacturing are the simple (additive) fabrication approach, and the customizable and intelligent architecture, as few steps are required to assemble complete device structures. The added advantages of 3D printing over other methods are (1) direct writing of electrode materials on the desired plate with engineered design and controllable interspacing porosity; (2) integrated printing steps via the printing of electrodes and electrolytes on a single platform; and (c) optimized size and thickness of the cell. These are more important parameters to improve the areal capacitance of energy devices [94–95]. Despite early success, the challenging tasks for successful fabrication of 3D printed energy devices rely on inter-related parameters such as making and controlling ink types, ink rheological properties, nozzle diameter control, nozzle-to-substrate height, printing speed, and controlling ink extrusion force that varies from printer to printer. In addition, exploration of making printable and aggregation-free ink, surface finishing of rough printed structure (when needed), size shrinkage after curing, structural formation to achieve mechanical flexibility, thickness control, etc. are yet to be achieved in 3D printed energy devices. The applicability of introducing different combinations of additive-based printing materials and manufacturing processes is expected to address the above issues without compromising the performances of energy storage devices in the future [65].

6.8 CONCLUSION

Flexible Li-ion batteries and supercapacitors are emerging and rapidly growing platform technologies due to the rise of flexible electronics. Although these flexible energy storage systems are still in their infancy, an enormous amount of research effort has been directed toward the development of high-performance and durable flexible batteries. This chapter consists of a brief introduction to the bendable and stretchable energy storage systems and state-of-the-art flexible LIB components, followed by a research strategy adopted by our group to develop flexible Li-ion batteries. Although current battery materials are mostly brittle in nature, several effective cell architectures, such as thin film, wavy structure interconnected island-bridge structure, and coiled fiber structures, have been found to be beneficial to develop flexible LIBs. However, we have designed a simple electrophoretic deposition-assisted methodology to fabricate a highly flexible anode. The EPD-grown graphite anode on a soft 3D carbon cloth exhibits excellent flexibility as well as superior Li storage properties. Therefore, EPD can be an attractive technique to develop flexible 3D electrodes with superior performance and can also be extended to fabricate other flexible electrodes for future applications. The popular research area of flexible printed electrochemical energy storage devices relies on suitable ink formulation and choice of printing approaches depending on commercial requirements. The printed graphene-based supercapacitors presented in this case study exhibit excellent cyclic stability (99.6% capacity retention after 1,000 cycles at 0.25 μA) and flexibility (80% retention of the initial capacity at extreme bending conditions). This indicates the remarkable opportunity for the development of miniaturized high-resolution printable energy storage devices. However, more exploration needs to be done on designing standard protocols to achieve super-performance printed electrochemical energy storage devices.

ACKNOWLEDGMENTS

Sarmistha Basu gratefully acknowledges financial support for the present work by DST under WOS – A scheme-wide grant, award no. SR/WOS–A/PM–54/2019 (G) dated 12 November 2020.

Subhasish Basu Majumder acknowledges the partial financial support of the Fulbright–Kalam Climate Fellowship for Academic and Professional Excellence grant, award no. 2760/F-K Climate APE/2022 dated 27 May 2022. Debasish Das would like to acknowledge the FESEM facility sponsored by DST–FIST at the Materials Science Center, IIT Kgp, for scanning electron microscopy experiments.

REFERENCES

1. L. Kong, C. Tang, H. J. Peng, J. Q. Huang, Q. Zhang, Advanced energy materials for flexible batteries in energy storage: A review, *SmartMat*, 2020; 1: e1007.
2. L. Wen, F. Li, H.-M. Cheng, Carbon nanotubes and graphene for flexible electrochemical energy storage: From materials to devices, *Adv. Mater.*, 2016; 28: 4306–4337.
3. Y. Ko, M. Kwon, W. K. Bae, B. Lee, S. W. Lee, J. Cho, Flexible supercapacitor electrodes based on real metal-like cellulose papers, *Nat. Commun.*, 2017; 8: 536.
4. J. Chang, Q. Huang, Y. Gao, Z. Zheng, Pathways of developing high-energy-density flexible lithium batteries, *Adv. Mater.*, 2021; 33: 2004419.
5. J. Feng, X. Sun, C. Wu, L. Peng, C. Lin, S. Hu, J. Yang, Y. Xie, Metallic few-layered VS_2 ultrathin nanosheets: High two-dimensional conductivity for in-plane supercapacitors, *J. Am. Chem. Soc.*, 2011; 133: 17832–17838.
6. Z. Suo, Mechanics of rollable and foldable film-on-foil electronics, *Appl. Phys. Lett.*, 1999; 74: 1177.
7. M. Koo, K. I. Park, S. H. Lee, M. Suh, D. Y. Jeon, J. W. Choi, K. Kang, K. J. Lee, Bendable inorganic thin-film battery for fully flexible electronic systems, *Nano Lett.*, 2012; 12: 4810–4816.
8. J. Song, H. Jiang, Z. J. Liu, D. Y. Khang, Y. Huang, J. A. Rogers, C. Lu, C. G. Koh, Buckling of a stiff thin film on a compliant substrate in large deformation, *Int. J. Solids. Struct.*, 2008; 45: 3107.
9. T. Widlund, S. Yang, Y. Y. Hsu, N. Lu, Stretchability and compliance of freestanding serpentine-shaped ribbons, *Int. J. Solids Struct.*, 2014; 51: 4026–4037.
10. W. Weng, Q. Sun, Y. Zhang, H. Lin, J. Ren, X. Lu, M. Wang, H. Peng, Winding aligned carbon nanotube composite yarns into coaxial fiber full batteries with high performances, *Nano Lett.*, 2014; 14: 3432–3438.
11. H. Lin, W. Weng, J. Ren, L. Qiu, Z. Zhang, P. Chen, X. Chen, J. Deng, Y. Wang, H. Peng, Twisted aligned carbon nanotube/silicon composite fiber anode for flexible wire-shaped lithium-ion battery, *Adv. Mater.*, 2014; 26: 1217–1222.
12. Y. Zhang, Y. Zhao, X. Cheng, W. Weng, J. Ren, X. Fang, Y. Jiang, P. Chen, Z. Zhang, Y. Wang, H. Peng, Realizing both high energy and high power densities by twisting three carbon-nanotube-based hybrid fibers, *Angew. Chem. Int. Ed.*, 2015; 54: 11177–11182.
13. L. Wen, F. Li H. Z. Luo, H. M. Cheng, *Nanocarbons for Advanced Energy Storage*. Wiley - VCH, Weinheim, Germany, 2015.
14. J. Chen, A. I. Minett, Y. Liu, C. Lynam, P. Sherrell, C. Wang, G.G. Wallace, Direct growth of flexible carbon nanotube electrodes, *Adv. Mater.*, 2008; 20: 566–570.
15. K. Yao, J. P. Zheng, Z. Liang, Binder-free freestanding flexible Si nanoparticle-multi-walled carbon nanotube composite paper anodes for high energy Li-ion batteries, *J. Mater. Res.*, 2018; 33: 482–494.
16. I. Elizabeth, B. P. Singh, S. Gopukumar, Electrochemical performance of Sb_2S_3/CNT free-standing flexible anode for Li-ion batteries, *J. Mater. Sci.*, 2019; 54: 7110–7118.
17. P. Fan, H. Liu, L. Liao, J. Fu, Z. Wang, G. Lv, L. Mei, H. Hao, J. Xing, J. Dong, Flexible and high capacity lithium-ion battery anode based on a carbon nanotube/electrodeposited nickel sulfide paper like composite, *RSC Adv.*, 2017; 7: 49739–49744.
18. Y. Ma, R. Younesi, R. Pan, C. Liu, J. Zhu, B. Wei, Constraining Si particles within graphene foam monolith: Interfacial modification for high-performance Li^+ storage and flexible integrated configuration, *Adv. Funct. Mater.*, 2016; 26: 6797–6806.
19. K. Zhu, H. Gao, G. Hu, A flexible mesoporous $Li_4Ti_5O_{12}$-rGO nanocomposite film as free-standing anode for high rate lithium ion batteries, *J. Power Sources*, 2018; 375: 59–67.
20. Y. H. Ding, H. M. Ren, Y. Y. Huang, F. H. Chang, P. Zhang, Three-dimensional graphene/$LiFePO_4$ nanostructures as cathode materials for flexible lithium ion batteries, *Mater. Res. Bull.*, 2013; 48: 8–11.
21. J. Q. Huang, P. Y. Zhai, H. J. Peng, W. C. Zhu, Q. Zhang, Metal/nanocarbon layer current collectors enhanced energy efficiency in lithium-sulfur batteries, *Sci. Bull.*, 2017; 62: 1267–1274.
22. X. Ding, Z. Pan, N. Liu, L. Li, X. Wang, G. Xu, J. Yang, J. Yang, N. Yu, M. Liu, W. Li, Freestanding carbon nanotube film for flexible strap like lithium/sulfur batteries, *Chem. Eur. J.*, 2019; 25(3775): 3780.

23. Y. Wang, Z. Ma, Y. Chen, M. Zou, M. Yousaf, Y. Yang, L. Yang, A. Cao, R. P. S. Han, Controlled synthesis of core-shell carbon@MoS_2 nanotube sponges as high performance battery electrodes, *Adv. Mater.*, 2016; 28: 10175–10181.
24. J. Ren, L. Li, C. Chen, X. Chen, Z. Cai, L. Qiu, Y. Wang, X. Zhu, H. Peng, Twisting carbon nanotube fibers for both wire-shaped micro-supercapacitor and micro-battery, *Adv. Mater.*, 2013; 25: 1155–1159.
25. G. Hu, C. Xu, Z. Sun, S. Wang, H.-M. Cheng, F. Li, W. Ren, 3D graphene-foam-reduced-graphene-oxide hybrid nested hierarchical networks for high-performance Li-S batteries, *Adv. Mater.*, 2016; 28: 1603–1609.
26. A. L. M. Reddy, A. Srivastava, S. R. Gowda, H. Gullapalli, M. Dubey, P. M. Ajayan, Synthesis of nitrogen-doped graphene films for lithium battery application, *ACS Nano*, 2010; 4: 6337–6342.
27. W. Gie Chong, F. Xiao, S. Yao, J. Cui, Z. Sadighi, J. Wu, M. Ihsan-Ul-Haq, M. Shao, J.-K. Kim, Nitrogen-doped graphene fiber webs for multi-battery energy storage, *Nanoscale*, 2019; 11: 6334–6342.
28. J.-Q. Huang, H.-J. Peng, X. Y. Liu, J. Q. Nie, X.-B Cheng, Q. Zhang, F. We, Flexible all-carbon interlinked nanoarchitectures as cathode scaffolds for high-rate lithium–sulfur batteries. *J. Mater. Chem. A.*, 2014; 2: 10869–10875.
29. C. Wang, X. Wang, C. Lin, X. S. Zhao, Lithium titanate cuboid arrays grown on carbon fiber cloth for high-rate flexible lithium-ion batteries, *Small*, 2019; 15: 1902183.
30. G. X. Pana, F. Cao, Y. J. Zhang, X. H. Xia, Integrated carbon cloth supported $LiFePO_4$/N-C films as high-performance cathode for lithium ion batteries, *Mater. Res. Bull.*, 2018; 98: 70–76.
31. Y. Miao, Y. Huang, L. Zhang, W. Fan, F. Lai, T. Liu, Electrospun porous carbon nanofiber@MoS_2 core/sheath fiber membranes as highly flexible and binder-free anodes for lithium-ion batteries, *Nanoscale*, 2015; 7: 11093–11101.
32. W. Jia, Z. Li, Z. Wu, L. Wang, B. Wu, Y. Wang, Y. Cao, J. Li, Graphene oxide as a filler to improve the performance of PAN-$LiClO_4$ flexible solid polymer electrolyte, *Solid State Ion.*, 2018; 315: 7–13.
33. J. Zhang, N. Zhao, M. Zhang, Y. Li, P. K. Chu, X. Guo, Z. Di, X. Wang, H. Li, Flexible and ion-conducting membrane electrolytes for solid-state lithium batteries: Dispersion of garnet nanoparticles in insulating polyethylene oxide, *Nano Energy*, 2016; 28: 447–454.
34. C.-Z. Zhao, X.-Q. Zhang, X.-B. Cheng, R. Zhang, R. Xu, P.-Y. Chen, H.-J. Peng, J.-Q. Huang, Q. Zhang, An anion-immobilized composite electrolyte for dendrite-free lithium metal anodes, *Proc Natl. Acad. Sci. USA*, 2017; 114: 11069–11074.
35. G. Xu, M. Zhao, B. Xie, X. Wang, M. Jiang, P. Guan, P. Han, G. Cui, A rigid-flexible coupling gel polymer electrolyte towards high safety flexible Li-Ion battery, *J. Power Sources*, 2021; 499: 229944.
36. E. Foreman, W. Zakri, M. H. Sanatimoghaddam, A. Modjtahedi, S. Pathak, A. G. Kashkooli, N. G. Garafolo, S. Farhad, A review of inactive materials and components of flexible lithium-ion batteries, *Adv. Sustain. Syst.*, 2017; 1: 1700061.
37. M . Koo, K. I. Park, S. H. Lee, M. Suh, D. Y. Jeon, J. W. Choi, K. Kang, K. J. Lee, Bendable inorganic thin-film battery for fully flexible electronic systems, *Nano Lett.*, 2012; 12: 4810.
38. S. Xu, Y. H. Zhang, J. Cho, J. Lee, X. Huang, L. Jia, J. A. Fan, Y. W. Su, J. Su, H. G. Zhang, H. Y. Cheng, B. W. Lu, C. J. Yu, C. Chuang, T. I. Kim, T. Song, K. Shigeta, S. Kang, C. Dagdeviren, I. Petrov, P. V. Braun, Y. G. Huang, U. Paik, J. A. Rogers, Stretchable batteries with self-similar serpentine interconnects and integrated wireless recharging systems, *Nat. Commun.*, 2013; 4: 1543.
39. J. Y. Rao, N. S. Liu, Z. Zhang, J. Su, L. Y. Li, L. Xiong, Y. H. Gao, All-fiber-based quasi-solid-state lithium-ion battery towards wearable electronic devices with outstanding flexibility and self-healing ability, *Nano Energy*, 2018; 51: 425.
40. H. Mazor, D. Golodnitsky, L. Burstein, A. Gladkich, E. Peled, Electrophoretic deposition of lithium iron phosphate cathode for thin-film 3D-microbatteries, *J. Power Sources*, 2012; 198: 264–272.
41. Y. Gao, C. Xie, Z. Zheng, Textile composite electrodes for flexible batteries and supercapacitors: Opportunities and challenges, *Adv. Energy Mater.*, 2020; 11: 2002838.
42. R. Hedayati, M. Sadighi, M. Mohammadi-Aghdam, A. A. Zadpoor, Mechanical properties of regular porous biomaterials made from truncated cube repeating unit cells: Analytical solutions and computational models, *Mater. Sci. Eng. C*, 2016; 60: 163–183.
43. Y. Zhu, M. Yang, Q. Huang, D. Wang, R. Yu, J. Wang, Z. Zheng, D. Wang, V_2O_5 textile cathodes with high capacity and stability for flexible lithium-ion batteries, *Adv. Mater.*, 2020; 32: 1906205.
44. X. Wang, G. Li, M. H. Seo, G. Lui, F. M. Hassan, K. Feng, X. Xiao, Z. Chen, Carbon-coated silicon nanowires on carbon fabric as self-supported electrodes for flexible lithium-ion batteries, *ACS Appl. Mater. Interfaces*, 2017; 9: 9551–9558.
45. L. Hu, F. L. Mantia, H. Wu, X. Xie, J. McDonough, M. Pasta, Y. Cui, Lithium-ion textile batteries with large areal mass loading, *Adv. Energy Mater.*, 2011; 1: 1012.

46. L. Hu, M. Pasta, F. La Mantia, L. Cui, S. Jeong, H. D. Deshazer, J. W. Choi, S. M. Han, Y. Cui, Stretchable, porous, and conductive energy textiles, *Nano Lett.*, 2010; 10: 708.
47. K. Jost, D. Stenger, C. R. Perez, J. K. Mcdonough, K. Lian, Y. Gogotsi, G. Dion, Knitted and screen printed carbon-fiber supercapacitors for applications in wearable electronics. *Energy Environ. Sci.*, 2013; 6: 2698.
48. L. Liu, Y. Yu, C. Yan, K. Li, Z. Zheng, Wearable energy-dense and power-dense supercapacitor yarns enabled by scalable graphene–metallic textile composite electrodes, *Nat. Commun.*, 2015; 6: 7260.
49. J. Xia, R. Tian, Y. Guo, Q. Du, W. Dong, R. Guo, X. Fu, L. Guan, H. Liu, Zn_2SnO_4-carbon cloth free-standing flexible anodes for high-performance lithium-ion batteries, *Mater. Des.*, 2018; 156: 272–277.
50. J. Luo, J. Peng, P. Zeng, Z. Wu, J. Y. Li, W. Li, Y. Huang, B. Chang, X. Wang, $TiNb_2O_7$ nano-particle decorated carbon cloth as flexible self-support anode material in lithium-ion batteries, *Electrochim. Acta*, 2020: 332135469.
51. X. Min, B. Sun, S. Chen, M. Fang, X. Wu, Y. Liu, A. Abdelkader, Z. Huang, T. Liu, K. Xi, R. Vasant Kumar, A textile-based SnO_2 ultra-flexible electrode for lithium-ion batteries, *Energy Storage Mater.*, 2019; 16: 597–606.
52. M.-S. Balogun, C. Li, Y. Zeng, M. Yu, Q. Wu, M. Wu, X. Lu, Y. Tong, Titanium dioxide@titanium nitride nanowires on carbon cloth with remarkable rate capability for flexible lithium-ion batteries, *J. Power Sources*, 2014; 272: 946–953.
53. D. Das, A. Mitra, S. Jena, S. B. Majumder, R. N. Basu, Electrophoretically deposited $ZnFe_2O_4$-carbon black porous film as a superior negative electrode for lithium-ion battery, *ACS Sustain. Chem. Eng.*, 2018; 6: 17000.
54. S. Das, D. Das, S. Jena, A. Mitra, A. Bhattacharya, S. B. Majumder, Electrophoretic deposition of nickel ferrite anode for lithium-ion half-cell with superior rate performance, *Surf. Coat. Technol.*, 2021; 421: 127365.
55. T. Majumder, D. Das, S. Jena, A. Mitra, S. Das, S. B. Majumder, Electrophoretic deposition of metal-organic framework derived porous copper oxide anode for lithium and sodium ion rechargeable cells, *J. Alloys Compd.*, 2021; 879: 160462.
56. T. Majumder, D. Das, S. B. Majumder, Investigations on the electrochemical characteristics of electrophoretically deposited $NiTiO_3$ negative electrode for lithium-ion rechargeable cells, *J. Phys. Chem. Solids*, 2021; 158: 110239.
57. L. Dashairya, D. Das, P. Saha, Electrophoretic deposition of antimony/reduced graphite oxide hybrid nanostructure: A stable anode for lithium-ion batteries, *Mater. Today Commun.*, 2020; 24: 101189.
58. H. Zhang, Y. Yang, D. Ren, L. Wang, X. He, Graphite as anode materials: Fundamental mechanism, recent progress and advances, *Energy Storage Mater.*, 2021; 36: 147.
59. D. Das, S. B. Majumder, A. Dhar, S. Basu, Electrophoretic deposition: An attractive approach to fabricate graphite anode for flexible Li-ion rechargeable cells, *J. Mater. Sci. Mater. Electron.*, 2022; 33: 13110–13123.
60. D. Aurbach, B. Markovsky, I. Weissman, E. Levi, Y. Ein-Eli, On the correlation between surface chemistry and performance of graphite negative electrodes for Li ion batteries, *Electrochim. Acta*, 1999; 45: 67.
61. M. Berggren, D. Nilsson, N. D. Robinson, Organic materials for printed electronics, *Nat. Mater.*, 2007; 6: 3–5.
62. M. A. Shah, D.-G. Lee, B.-Y. Lee, S. Hur, Classifications and applications of inkjet printing technology: A review, *IEEE Access*, 2021; 9: 140079–140102.
63. K.-H. Choi, D. B. Ahn, S.-Y. Lee, Current status and challenges in printed batteries: Toward form factor-free, monolithic integrated power sources, *ACS Energy Lett.*, 2017; 3: 220–236.
64. B. Clement, M. Lyu, E. S. Kulkarni, T. Lin, Y. Hu, V. Lockett, C. Greig, L. Wang, Recent advances in printed thin-film batteries, *Engineering*, 2022; 13: 238–261.
65. Y. Khan, A. Thielens, S. Muin, J. Ting, C. Baumbauer, A. C. Arias, A new frontier of printed electronics: Flexible hybrid electronics, *Adv. Mater.*, 2020; 32: 1905279.
66. V. Vandeginste, A review of fabrication technologies for carbon electrode-based micro-supercapacitors, *Appl. Sci.*, 2022; 12: 862.
67. Y. Pang, Y. Cao, Y. Chu, M. Liu, K. Snyder, D. MacKenzie, C. Cao, Additive manufacturing of batteries, *Adv. Funct. Mater.*, 2020; 30: 1906244.
68. W. Johannisson, D. Carlstedt, A. Nasiri, C. Buggisch, P. Linde, D. Zenkert, L. E. Asp, G. Lindbergh, B. Fiedler, A screen-printing method for manufacturing of current collectors for structural batteries, *Multifunct. Mater.*, 2021; 4: 035002.

69. H. Emani, X. Zhang, G. Wang, D. Maddipatla, T. Saeed, Q. Wu, W. Lu, Atashbar, M. In development of a screen-printed flexible porous graphite electrode for Li-Ion battery. *IEEE International Conference on Flexible and Printable Sensors and Systems (FLEPS)*. IEEE, 2021, pp. 1–4.
70. M. Montanino, G. Sico, A. De Girolamo Del Mauro, J. Asenbauer, J. R. Binder, D. Bresser, S. Passerini, Gravure-printed conversion/alloying anodes for lithium ion batteries, *Energy Technol.*, 2021; 9: 2100315.
71. Z. Wang, P. Pang, Z. Ma, H. Chen, J. Nan, A four-layers Hamburger-structure PVDF-HFP/Al_2O_3/PE/PVDF-HFP composite separator for pouch lithium-ion batteries with enhanced safety and reliability, *J. Electrochem. Soc.*, 2020; 167: 090507.
72. M. Montanino, G. Sico, A. De Girolamo Del Mauro, M. Moreno, LFP-based gravure printed cathodes for lithium-ion printed batteries, *Membranes*, 2019; 9: 71.
73. Y. Zuo, Y. Yu, J. Feng, C. Zuo, Ultrathin Al–air batteries by reducing the thickness of solid electrolyte using aerosol jet printing, *Sci. Rep.*, 2022; 12: 1–10.
74. L. J. Deiner, T. Jenkins, T. Howell, M. Rottmayer, Aerosol jet printed polymer composite electrolytes for solid state li ion batteries, *Adv. Eng. Mater.*, 2019; 21: 1900952.
75. G. Cummins, M. P. Desmulliez, Inkjet printing of conductive materials: A review, *Circuit World*, 2012.
76. B. J. De Gans, P. C. Duineveld, U. S. Schubert, Inkjet printing of polymers: State of the art and future developments, *Adv. Mater.*, 2004; 16: 203–213.
77. F. Bonaccorso, A. Bartolotta, J. N. Coleman, C. Backes, 2D crystal based functional inks, *Adv. Mater.*, 2016; 28: 6136–6166.
78. G. Hu, J. Kang, L. W. Ng, X. Zhu, R. C. Howe, C. G. Jones, M. C. Hersam, T. Hasan, Functional inks and printing of two-dimensional materials, *Chem. Soc. Rev.*, 2018; 47: 3265–3300.
79. K. Hassan, M. J. Nine, T. T. Tung, N. Stanley, P. L. Yap, H. Rastin, L. Yu, D. Losic, Functional inks and extrusion-based 3D printing of 2D materials: A review of current research and applications, *Nanoscale*, 2020; 12: 19007–19042.
80. J. Castrejón-Pita, G. Martin, S. Hoath, I. Hutchings, A simple large-scale droplet generator for studies of inkjet printing, *Rev. Sci. Instrum.*, 2008; 79: 075108.
81. D. Jang, D. Kim, J. Moon, Influence of fluid physical properties on ink-jet printability, *Langmuir*, 2009; 25: 2629–2635.
82. B. Derby, Inkjet printing of functional and structural materials: Fluid property requirements, feature stability, and resolution, *Annu. Rev. of Mater. Res.*, 2010; 40: 395–414.
83. H. Hu, R. G. Larson, Marangoni effect reverses coffee-ring depositions, *J. Phys. Chem. B*, 2006; 110: 7090–7094.
84. K. Yan, J. Li, L. Pan, Y. Shi, Inkjet printing for flexible and wearable electronics, *APL Mater.*, 2020; 8: 120705.
85. Y. Zhao, Q. Zhou, L. Liu, J. Xu, M. Yan, Z. Jiang, A novel and facile route of ink-jet printing to thin film SnO_2 anode for rechargeable lithium ion batteries, *Electrochim. Acta*, 2006; 51: 2639–2645.
86. F. Xu, T. Wang, W. Li, Z. Jiang, Preparing ultra-thin nano-MnO_2 electrodes using computer jet-printing method, *Chem. Phys. Let.*, 2003; 375: 247–251.
87. L. J. Deiner, T. L. Reitz, Inkjet and aerosol jet printing of electrochemical devices for energy conversion and storage, *Adv. Eng. Mater.*, 2017; 19: 1600878.
88. C. Julien, Z. Stoynov, *Materials for Lithium-Ion Batteries*, Springer Science & Business Media, 2012, p. 85.
89. A. P. S. Gaur, W. Xiang, A. Nepal, J. P. Wright, P. Chen, T. Nagaraja, S. Sigdel, B. LaCroix, C. M. Sorensen, S. R. Das, Graphene aerosol gel ink for printing micro-supercapacitors, *ACS Appl. Energy Mater.*, 2021; 4: 7632–7641.
90. L. Liu, Y. Feng, W. Wu, Recent progress in printed flexible solid-state supercapacitors for portable and wearable energy storage, *J. Power Sources*, 2019; 410–411: 69–77.
91. A. Alam, G. Saeed, S. Lim, Screen-printed activated carbon/silver nanocomposite electrode material for a high performance supercapacitor, *Mater. Lett.*, 2020; 273: 127933.
92. F. Li, Y. Li, J. Qu, J. Wang, V. K. Bandari, F. Zhu, O. G. Schmidt, Recent developments of stamped planar micro-supercapacitors: Materials, fabrication and perspectives, *Nano Mater. Sci.*, 2021; 3: 154–169.
93. A. Azhari, E. Marzbanrad, D. Yilman, E. Toyserkani, M. A. Pope, Binder-jet powder-bed additive manufacturing (3D printing) of thick graphene-based electrodes, *Carbon*, 2017; 119: 257–266.
94. S. Zhang, Y. Liu, J. Hao, G. G. Wallace, S. Beirne, J. Chen, 3D printed wearable electrochemical energy devices, *Adv. Funct. Mater.*, 2022; 32: 2103092.
95. Y. Wang, C. Chen, H. Xie, T. Gao, Y. Yao, G. Pastel, X. Han, Y. Li, J. Zhao, K. Fu, 3D printed all-fiber Li-ion battery toward wearable energy storage, *Adv. Funct. Mater.*, 2017; 27: 1703140.

Theme 2

Developing Advanced Materials for Medicine

7 Sustainable Nanotechnology for Targeted Therapies using Cell-Encapsulated Hydrogels

Itisha Chummun Phul and Archana Bhaw-Luximon

CONTENTS

7.1 INTRODUCTION

Nanomedicine, tissue regeneration, and materials engineering have allowed the emergence of transformational nanotechnologies for human sustainability and targeted therapies. Nanotechnologies have evolved exponentially during the past decades to address new worldwide challenges such as vaccine development. In the new concept of Society 5.0, focusing on human-centered socio-economic development, nanotechnology can interconnect global regions for reasons of well-being and fair access to health (Pokrajac et al., 2021). Nanofibers, nanoparticles, and hydrogels dominate drug delivery and tissue regeneration, and more recently vaccines. Their first generation was expected to

DOI: 10.1201/9781003181422-9

provide biocompatibility traits with the human body, and the latest generation mimics the human body in all aspects by bridging traditional, complementary, and modern medicine. More and more natural molecules and materials are being integrated into nanostructures in data-supported strategies. Nanotechnology has thus been able to embrace targeted disease mechanism approaches taking advantage of the versatile physiochemical characteristics of nanostructures. Cancer nanomedicine, tissue regeneration of wounds, and viral vaccines are already realities, and other diseases and conditions such as atherosclerosis are also being positively impacted by nanotechnologies (Ou et al., 2021).

In the health sector, nano-based systems can have very diverse applications, from nano-filtration systems to nano-scaled biosensors to nanomedicine. Similarly, hydrogels with nanostructures can mimic the properties of living systems responding to environmental changes such as pH, temperature, light, pressure, electric fields, or a combination of different stimuli (Montoro et al., 2014). These hydrogels find applications as scaffolds for tissue engineering, drug delivery systems, biosensors, pre-clinical cancer models, and cancer cell capture. Hydrogels offer the advantage of being injectable and compressible, allowing liquid retention and cell encapsulation.

7.2 SUSTAINABLE NANOTECHNOLOGY

Sustainability should be at the heart of every human activity. The 2030 Agenda for Sustainable Development, adopted in 2015 by all United Nations Member States, provides a blueprint for peace and prosperity for people and the planet, now and into the future. The 17 Sustainable Development Goals (SDGs) are an urgent call for action by all countries – developed and developing – in a global partnership (United Nations, n.d.). Sustainability translates to technology management processes to capture evolving technology and to understand its benefits and risks (Subramanian et al., 2014). Nanotechnology can offer new possibilities to move the SDG agenda in the health, energy, agriculture, and industrial sectors (Figure 7.1. (a)), united under the umbrella of research, innovation, and development. All these are dependent on the sustainability of materials and processes. As nanotechnology advances, it is important to continuously monitor the sustainability of this technology, as its primary objective is designed to improve the quality of life, without causing any unintentional harm to human health and the environment (George et al., 2015). Equitable distribution of nanotechnology-based knowledge is even more challenging in the health sector as has been recently witnessed during the COVID-19 pandemic with the advent of nano-based mRNA vaccines. To date, nano-based vaccines are already in clinical trials for other diseases such as cancer, regenerative therapeutics, and pulmonary therapeutics.

Hydrogels can offer a sustainable approach to nanotechnology as they are developed from abundant and/or inexpensive bio-renewable resources such as plant fibers and marine biopolymers (Figure 1.1 (b)) (Ciolacu, 2021).

7.3 HYDROGELS FOR CELL ENCAPSULATION

Hydrogels have shown potential as three-dimensional (3D) scaffolds for cell growth and new tissue development. Owing to their high-water content, biodegradability, and ability to crosslink in situ, they are attractive biomaterials for cell encapsulation. There exist two main strategies in tissue engineering whereby cells are either seeded on preformed scaffolds or they are encapsulated during scaffold formation. In cell encapsulation strategies, the fabrication process including hydrogel precursor solutions and crosslinking mechanisms should be cytocompatible. Thus, natural polymers such as polysaccharides and proteins biosynthesized by living organisms are the most suited biomaterials for cell-encapsulated hydrogels. A wide range of polysaccharide and protein/peptide-based hydrogels have been synthesized to encapsulate cells using various technologies (Figure 7.2).

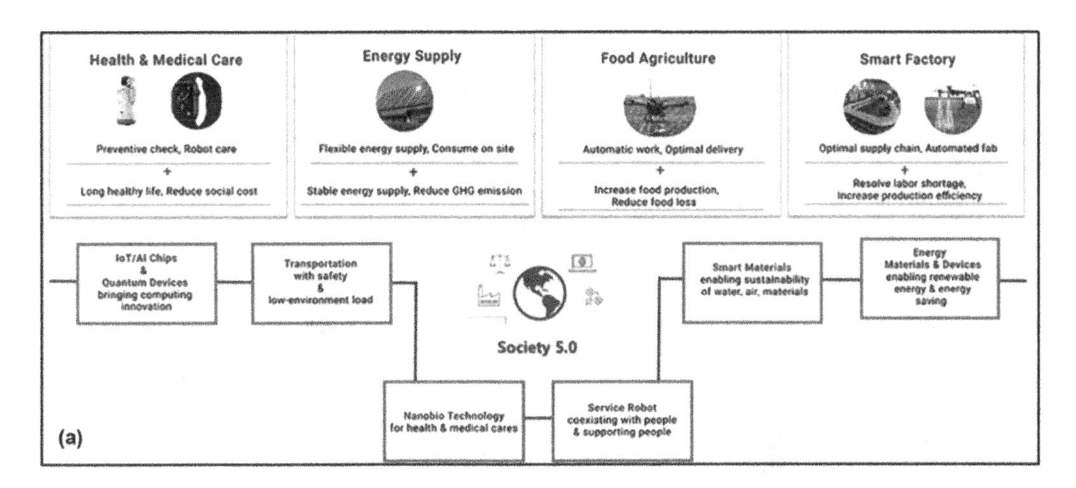

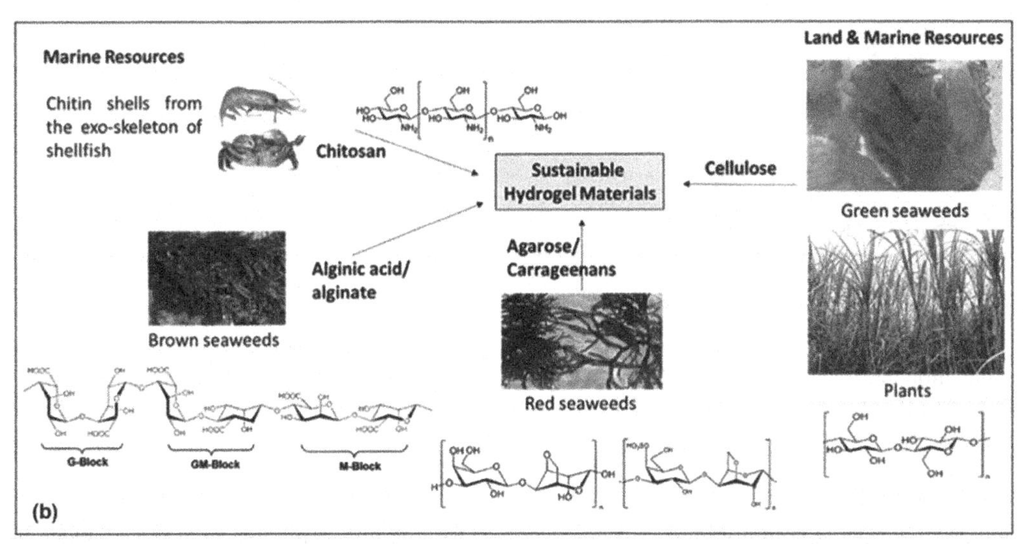

FIGURE 7.1 (a) New values created in various fields and grand challenges for nanotechnology and materials to support Society 5.0. IoT, Internet-of-Things; AI, artificial intelligence (Reproduced with permission from Pokrajac et al., 2021) (b) Hydrogel bio-renewable materials.

7.3.1 Polysaccharide-Based Hydrogels

Polysaccharides are natural polymers composed of long carbohydrate molecules of monosaccharides unit joined together by glycosidic bonds. They are widely used as scaffolds in tissue engineering (TE) owing to their inherent structures and functions analogous to the extracellular matrix made up of proteoglycans, glycosaminoglycans, glycoproteins, and glycolipids. They have numerous intrinsic functional groups on their repeating units enabling crosslinking during hydrogel synthesis. These crosslinks also serve as a means of encapsulating cells within a porous aqueous environment allowing for nutrient and oxygen diffusion.

7.3.1.1 Agarose

Since 1982, agarose has been used to encapsulate cells, namely chondrocytes, with 80% survival (Benya & Shaffer, 1982). Expression of type II collagen and cartilage-specific proteoglycan which are characteristics of differentiated chondrocyte phenotype was shown. The latter is usually lost

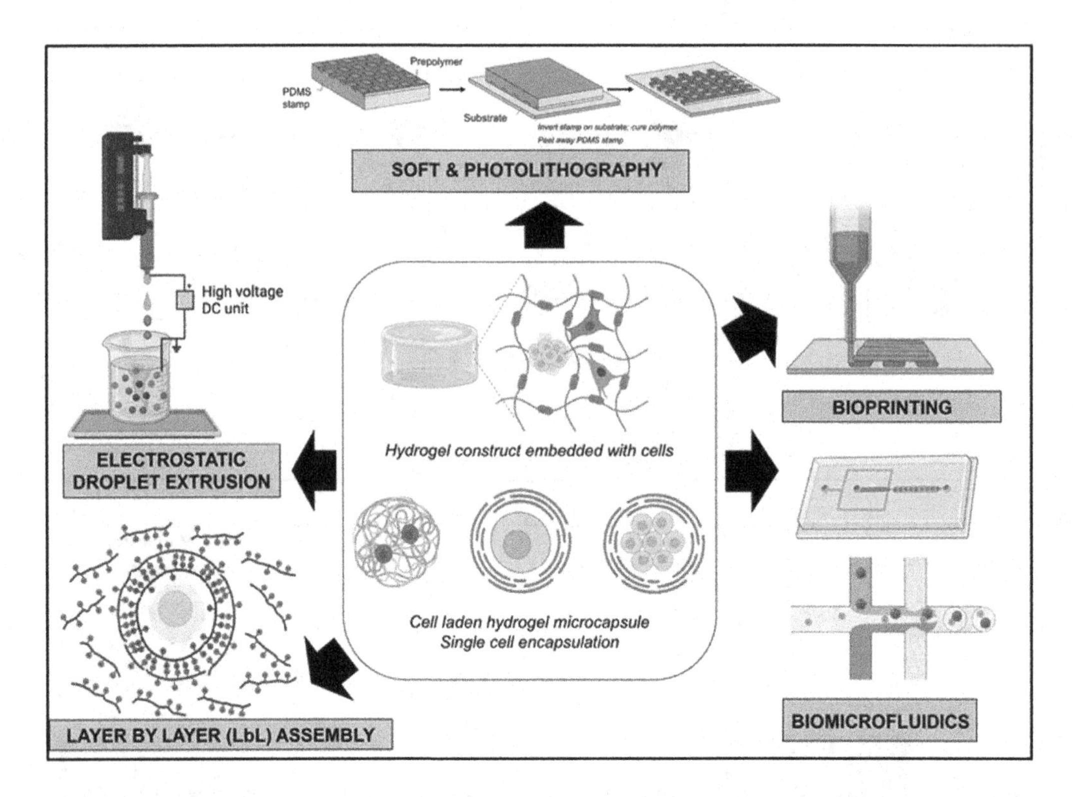

FIGURE 7.2 Cell encapsulation technologies.

during monolayer culture. Alginate and agarose hydrogels showed the highest collagen type II synthesis in embedded marrow stromal cells, while agarose promoted aggrecan gene expression (Diduch et al., 2000).

7.3.1.2 Alginate

Alginate is the most commonly used polysaccharide for cell encapsulation owing to its biocompatibility, biodegradability, and ease of crosslinking. It is a gold-standard encapsulating matrix, approved for human use by the US Food and Drug Administration (FDA) (Pandolfi et al., 2017). Alginate is a naturally occurring polysaccharide in algae that undergoes ionic crosslinking in the presence of bivalent and trivalent cations such as Ca^{2+}, Cu^{2+}, Zn^{2+}, Ba^{2+}, and Sr^{2+}, which bond to blocks of alginate polymers in a planar 2D manner referred to as the "egg-box" model. Although the alginate core provides a matrix for the immobilization of anchorage-independent cells, it may not provide an adequate environment for anchorage-dependent cells which require a substrate for attachment and spreading. In the last decade, a large number of studies reported that the conjugation of alginate with peptides such as arginine-glycine-aspartate (RGD) or tyrosine-isoleucine-glyci ne-serine-arginine (YIGSR) sequence to improve integrin-mediated cell adhesion during encapsulation. Peptide-modified alginate has been shown to improve the viability and function of cells embedded within the matrix (Sarkar et al., 2019; Trachsel et al., 2019).

Alginate has also been combined with proteins such as collagen, gelatin, and silk fibroin to form hydrogels for cell encapsulation (Li et al., 2020; Dogan et al., 2021; Yang et al., 2018; Kim et al., 2019; Kim et al., 2021; Lòpez-Marcial et al., 2018) In addition to improving cell adhesion, silk fibroin formed double crosslinked network with alginate to improve its structural stability. Li et al. (2020) formulated an alginate-silk fibroin (Alg/SF) bioink in which alginate was ionically crosslinked using calcium chloride ($CaCl_2$) while silk fibroin formed sheet structures in the presence of F127 (PEO-PPO-PEO tri-block copolymers). Fast proliferation of liver cancer cells C3A with a survival rate of 99.5% was seen on Alg/SF hydrogel. Alginate/tyramine (Alg-Tyr) and silk fibroin methacrylol (SF-MA), crosslinked via ionic interactions, photocrosslinking, and beta-sheet formation, was used to encapsulate NIH3T3 fibroblasts, which showed high cell viability and proliferative activity (Kim et al., 2021).

The complexation of alginate with cationic polysaccharide such as chitosan has been considered a promising strategy for cell encapsulation. K-562 cells have been successfully encapsulated within layers of chitosan and alginate. The negatively charged cell membrane is known to bind to the positively charged chitosan surface which in turn interacts with negatively charged alginate layer. The desired capsule thickness, surface configuration, and permeability could be controlled by layering the capsule with subsequent oppositely charged polymer (Chander et al., 2018). Another encapsulation procedure involved cells embedded within ionically crosslinked alginate capsules coated with a chitosan layer. Alginate-chitosan hydrogel systems have also been used to encapsulate various cell lines such as red blood cells, pancreatic islets, and hepatocytes (Baruch & Machluf, 2006).

7.3.1.3 Chitosan

Chitosan is a European and American pharmacopoeia approved excipient widely used in various cell-encapsulating hydrogel systems intended for various applications such as spinal cord regeneration, pancreatic cell support and insulin production, neurodegenerative diseases, and cartilage and bone regeneration (Pirela et al., 2021). It is a prominent polysaccharide obtained by deacetylation of chitin, used for cell encapsulation via ionotropic crosslinking, cell assemblies within layers of chitosan or covalent crosslinking of a gelling mixture containing cells. The most common covalent crosslinking reaction of chitosan occurs via Schiff base reactions between their amine groups and dialdehydes. Peptide-modified methacrylated glycol chitosan (MGC) hydrogels crosslinked via thermally induced free radical polymerization using ammonium persulfate (APS) and tetramethylethylenediamine (TEMED) have been successfully used to encapsulate adipose-derived stem/stromal cells (ASCs). Though the ASCs encapsulated hydrogels failed to upregulate angiogenesis-associated genes *in vitro*, a pilot *in vivo* study in NOD/SCID mice indicated $CD31^+$ cells recruitment through paracrine factor secretion supported angiogenesis (Dhillon et al., 2019). Chitosan amino groups are protonated in dilute acidic solution and can form ionic complexes with a wide variety of natural or synthetic anionic species. For instance, a negatively charged glycosaminoglycans solution containing mesenchymal stem cells (MSCs) was electrosprayed in a chitosan solution to form cell-laden GAG-chitosan microcapsules (Vossoughi & Matthew, 2018).

7.3.1.4 Cellulose

Carboxymethylcellulose (CMC), hydroxylethylcellulose (HEC), carboxyethylcellulose (CEC), and hydroxypropyl methyl cellulose (HPMC) are water-soluble cellulose derivatives generating hydrogels suited for in situ crosslinking of cell-containing cellulose hydrogel precursors. These cellulose derivatives have either been combined with other polysaccharides to improve structural stability of encapsulating systems or functionalized to form photocrosslinked hydrogels. Ni et al. (2020) combined photocrosslinkable methacrylated HPMC (HPMC-MA) to silk fibroin to form a double-network hydrogel with improved ductility for bioprinting of bone mesenchymal stem cells for cartilage tissue repair. Norbornene-CMC (cCMC) is another photoresponsive functionalized cellulose used as bioink to encapsulate human mesenchymal stem cells (hMSCs), NIH 3T3 fibroblasts, and human umbilical vein endothelial cells (HUVECs) (Ji et al., 2020). CEC grafted with adipic acid dihydrazide (CEC-ADH) have been synthesized to form self-healable and injectable hydrogel via reversible ketoester-type acylhydrazone linkages with ketoester-grafted poly (vinyl alcohol) (PVA-Ket) successfully laden with viable L929 fibroblasts (Jiang et al., 2021).

Cellulose nanofiber (CNF) and cellulose nanocrystals (CNC) have been incorporated in small amounts in hydrogel systems to improve their rheological properties and processability essential for different encapsulation technologies (Boonlai et al., 2022; Fan et al., 2020; Wu et al., 2021; Maturavongsadit et al., 2021; García-Lizarribar et al., 2018). Better shear-thinning and thixotropic properties of kappa carrageenan and methylcellulose hydrogels in the presence of 2% w/w and 4% w/w CNC for extrusion-based bioprinting has been demonstrated (Boonlai et al., 2022). TEMPO-mediated oxidation of CNF resulted in a stimuli-responsive gel which can transform into liquid under shear stress and transform back to a fully structured hydrogel in less than 60 s without the use of any crosslinker. Human breast cancer cells (MCF-7) and mouse embryonic stem cells homogenized within the hydrogel system showed high cell viability and long-term survival upon injection (Sanandiya et al., 2019).

7.3.2 Protein/Peptide-Based Hydrogels

Peptide-based hydrogels are highly conducive novel matrix for cell encapsulation due to their non-toxic stimulus-responsive molecular self-assembly and ligand-receptor recognition. They are formed through non-covalent interactions, including π–π interactions, charge-based interactions, hydrophobic interactions, hydrogen bond interactions, and van der Waals forces, and self-assemble into secondary structural motifs such as alpha-helices, beta-sheets, beta-hairpins, and coils in response to the physiological environment such as temperature, pH, ions, and enzymes. A panoply of stimuli-responsive self-assembled peptides is possible (Figure 7.3).

H9e peptide-based hydrogel have been used to encapsulate human epithelial cancer cells MCF-7 grown into tumor-like clusters for drug screening (Huang et al., 2013). H9e is an amphiphilic peptide which self-assembles into nanofibers that can form a nanoweb-like structure. The H9e peptide-based hydrogel was formed within 15 mins at room temperature by hand-shaking H9e solution (in sodium bicarbonate) to MEM culture medium and fetal bovine serum (FBS) for 10 s. Other cell lines such as hepatocarcinoma (HepG2) and colon adenocarcinoma (SW480) were embedded within the H9e hydrogel system to evaluate anti-cancer properties of chemo preventive phenolic acid (Xu et al., 2019).

A group of researchers showed the potential of a 20 amino acid peptide (H2N-VKVKVKV KVDPPTKVKVKVKV-CONH2), MAX1 for cell encapsulation. The peptide is composed of alternating lysine and valine residues on two b-strands which fold to form amphiphilic b-hairpins self-assembled into a supramolecular structure composed of fibrils that are monodisperse in diameter in the presence of cell-containing cell culture medium, as seen in Figure 7.4. The kinetics of hydrogenation were optimized for homogenous encapsulation of C3H10t1/2 mesenchymal stem cells. This was achieved by substituting a point amino acid on the lysine side chain by a negatively charged side chain of glutamic acid (MAX8). The peptide hydrogel system exhibited shear-thinning properties allowing for syringe-based injection of mesenchymal stems which maintained their viability and homogenous distribution (Kretsinger et al., 2005; Haines-Butterick et al., 2007).

The anionic peptide sequence AcVES3-RGDV was synthesized to overcome shortcomings, such as their cationic nature and the need for large changes in pH impacting on cell viability and phenotype, of the MAX1 and MAX8 peptide family for cell encapsulation. It contains 28 residues designed to form unsymmetrical amphiphilic b-hairpins due to the inclusion of a linker and integrin binder motif, RGDV, and incorporates one of the strongest turn-propensity motifs, namely VDPPT. These peptides can assemble under physiological conditions at 37 °C and encapsulate human dermal fibroblasts (HDFs) homogeneously within the structure, showing higher proliferation rate compared to collagen gels. *In vivo* studies indicated higher survival rate of HDFs encapsulated within the AcVES3-RGDV peptide hydrogels injected within the dorsal flank of athymic mice compared to control injection of cells suspended in media (Yamada et al., 2019).

PeptiGel® is a commercially available positively or neutrally charged peptide-based gel available in different formulation and properties, adapted to meet any cell requirements. The gels are produced by Manchester BIOGEL, UK. Recently, PeptiGel®Alpha1, a neutrally charged peptide gel, has been used for 3D growth of breast cancer cell lines MCF-7 and MDA-MB-231. The gel self-assembled to form uniformly sized nanofibers of diameter 5.7 ± 1.4 nm. The cell-laden peptide hydrogel was found to support formation of MCF-7 cellular spheroids, closely mimicking the

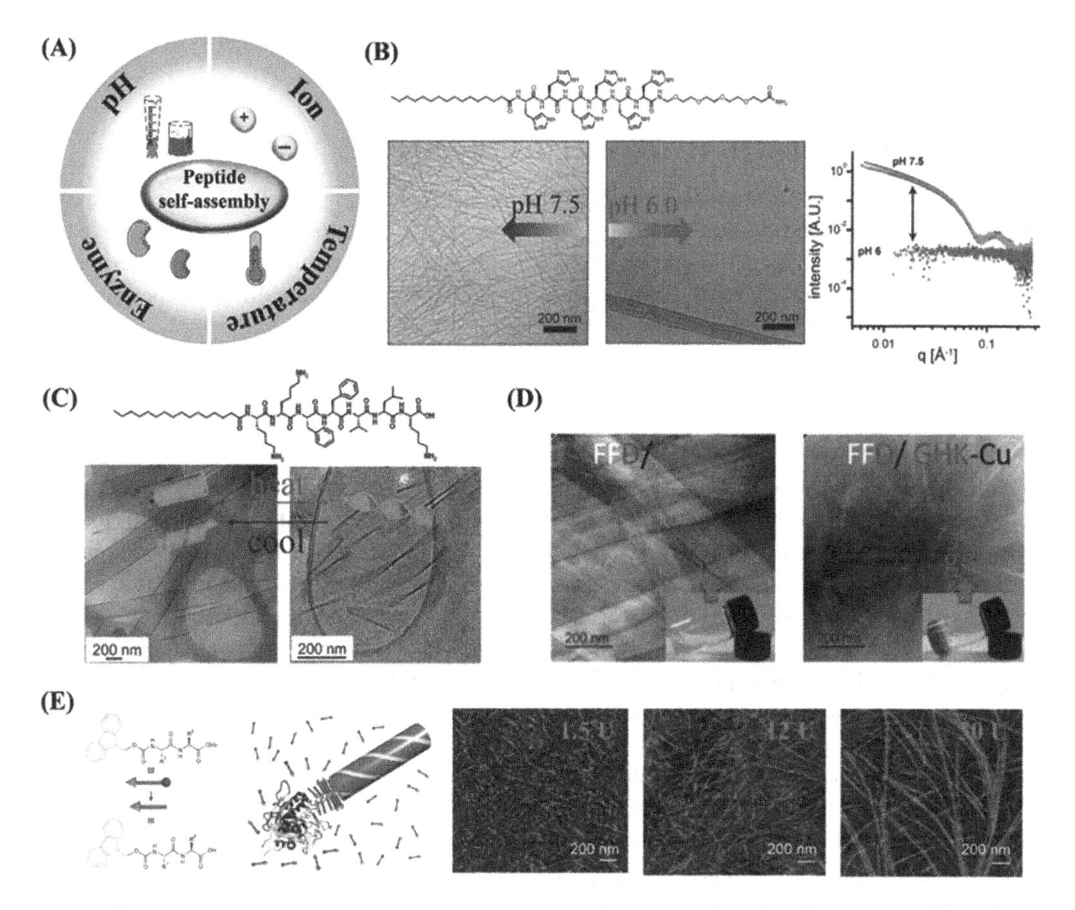

FIGURE 7.3 Stimuli-responsive self-assembly of peptides. (A) Parameters responsible for induced self-assembly, (B) cryo-TEM images and small angle X-ray scattering pattern showing pH triggered reversible assembly and disassembly of $C_{16}H_{16}$-OEG nanofibers, (C) cryo-TEM images of temperature driven self-assembled (C16-KKFFVLK) peptide to form twisted tapes upon heating and nanotubes/ribbons upon cooling, (D) TEM images of copper ions driven (FFD/GHK) peptide self-assembly into long fiber hydrogel, (E) enzymatic hydrolysis promoting self-assembly, nucleation, and structural growth of Fmoc-dipeptide methyl esters; higher enzyme concentration increase π–π interactions and leads to a more ordered fiber supramolecular structure. (Reproduced with permission from Guan et al., 2022.)

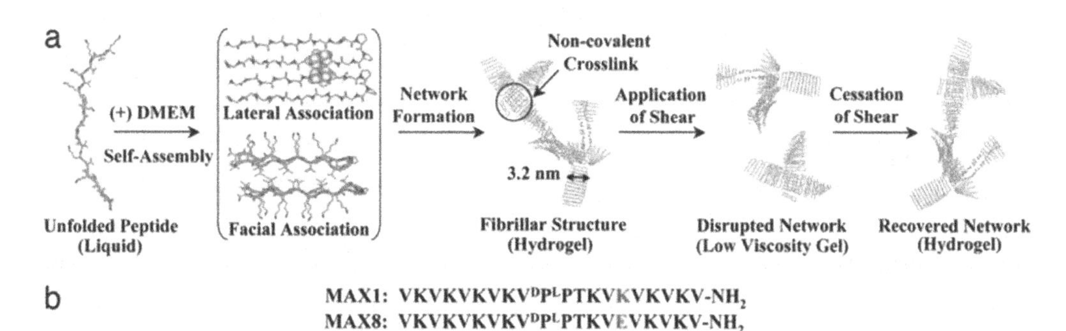

FIGURE 7.4 (a) Self-assembly of MAX1 and MAX8 peptides upon addition of DMEM (pH 7.4, 37°C) into β-hairpin structure that undergoes lateral and facial self-assembly to form rigid hydrogel with a fibrillar supramolecular structure; shear-thinning and self-healing mechanism of the non-covalent crosslinked network, (b) peptide sequence of MAX1 and MAX8. (Reproduced with permission from Haines-Butterick et al., 2007.)

stiffness of breast tumor tissue compared to collagen I and Matrigel® hydrogel systems (Clough et al., 2021).

7.4 CELL-ENCAPSULATED HYDROGEL TECHNOLOGIES

Several microfabrication techniques including lithography, bioprinting, electrostatic droplet extrusion, electrospraying, microfluidics, and layer-by-layer (LbL) assembly have been employed to produce 3D cell-laden hydrogel architecture, such as cells containing microcapsules, microfibers, microtubes, or tissue building blocks. This section further elaborates on the different encapsulation techniques used for various applications.

7.4.1 LITHOGRAPHY

7.4.1.1 Soft Lithography

Soft lithography is a group of patterning techniques that uses elastomeric polymeric molds made of chemically crosslinked hydrophobic materials such as poly (dimethyl siloxane) (PDMS) to form micro- or nano-scaled structures. PDMS composed of a liquid base and hardening agent can be poured or spin-coated on a solid master and cured at room temperature for few hours. It is generally used as a master for micromolding of hydrogel-based structures formed by either applying a hydrogel precursor solution to the PDMS master, or the PDMS master can be pressed onto the solution to fill empty spaces of the master prior to crosslinking (Selimović et al., 2012).

NIH 3T3 fibroblasts have been photoencapsulated within methacrylated hyaluronic acid hydrogels using micromolding (Khademhosseini et al., 2006). The cell containing the gel precursor was applied to the PDMS stamp, and excess liquid was squeezed out using a methacrylated glass slide followed by UV crosslinking in the presence of a 0.5%w/v Irgacure crosslinker. It was also shown that microfluidic channels could be generated within cell-laden agarose gels using micromolding. Molten agarose solution containing AML-12 hepatocytes cast on a silicon substrate was de-molded and bonded to a thin agarose surface. Embedded cells in the vicinity of the channels through which cell culture medium was pumped, remained viable after three days, indicating the importance of a perfused network of microchannels for nutrient and oxygen delivery (Ling et al., 2007).

Thermoresponsive poly (*N*-isopropylacrylamide) micromolds have been synthesized to form a multicellular compartment hydrogel exhibiting different patterns such as microgrooves, and circular and square microwells. Agarose gel was used to encapsulate 3T3 fibroblasts, HepG2 cells, and HUVEC, whereby gel containing either 3T3 fibroblasts or HepG2 cells was first laid on the PNIPAAM micromold at 24 °C, molded using a PDMS slab and crosslinked at 4 °C. Upon a rise in temperature to 37 °C, PNIPAAM hydrogels shrank, leaving spaces between the crosslinked hydrogel and surrounding walls of the PNIPAAM microwells, which was then filled with agarose gel embedded with HUVEC and crosslinked by incubating at 37 °C (Tekin et al., 2011)

7.4.1.2 Photo Lithography

The soft lithography technique has been scarcely explored for 3D culture due to distortion or breaking of hydrogel structures with high aspect ratios. This limitation is overcome by photolithography whereby a photomask is used to selectively crosslink the photosensitive hydrogel precursor adherent to a substrate based on the pattern of the mask. Polyethylene diacrylate (PEG-DA) have been widely used as hydrogels for photolithography. Liu et al. (2002) designed an apparatus for photopatterning of PEG-DA hydrogels by injecting the precursor solution into a chamber made of a Teflon base and a pre-treated 2" borosilicate glass wafer onto which the photomask was pressed flat by a glass slide. Multilayer hydrogel structures were formed by increasing the thickness of the spacer and different cells entrapped within each layer of the hydrogel, showing potential for fabrication of 3D living tissues.

In another study by Koh et al. (2002), PEG-DA was spin-coated on glass and silicon surfaces tethered with methacrylate groups, generating photo-crosslinked cell-containing hydrogel microstructures with aspect ratios ranging from 0.02 to 1.4 by using masks with different features and sizes and by controlling the spin-coating rate. Encapsulated 3T3 NIH fibroblasts showed 80% viability 24 h after crosslinking, and the viability increased with increasing molecular weight of PEG-DA, whereby cells embedded in MW575 PEG-DA hydrogels lost viability after three days compared to MW 4000 PEG-DA hydrogels in which they retained viability for seven days. Photoinitiators were found to be cytotoxic in a dose-dependent manner, while UV exposure did not trigger immediate cell death. Thus, cells maintained their viability after photocrosslinking due to exposure to initiators within a short time frame and PEG-DA acting as a site for free radical termination.

Photolithography has also been combined with microfluidics to form cell-laden hydrogel microstructures. For instance, a stop-flow lithography process has been developed which involved stopping the flow of PEG-DA precursor solution mixed with NIH 3T3 fibroblasts, polymerized by a flash of UV light through a photomask and collected in the outlet reservoirs filled with culture media (Panda et al., 2008). The microfluidic device was mounted on an inverted microscope, and cell-encapsulated microgels were visualized using a camera. Similarly, Liu et al. (2012) used microscope projection lithography to encapsulate HepG2 single cell within PEG-DA hydrogels. PEG-DA precursor solution containing cells were infused into microchannels (8 mm long, 800 μm wide, and 75 μm deep) by capillary action. Single cells were selectively photo encapsulated by manually shifting the fluorescence microscope stage using Xenon lamp as a UV source (wavelength 340–380 nm).

7.4.2 Bioprinting

Over the last few years, highly innovative cell encapsulation technologies, including bioprinting, have revolutionized the landscape of tissue regeneration and regenerative medicine. There has been a gradual rise in the number of publications on 3D bioprinting concerning biopolymers and cell types, growth factors denoted as the "bioink" from 2000 to 2020. The bioink is layered to form 3D living constructs with high precision through the formulation of a predefined architecture in computer-aided design software.

Different techniques are utilized for bioprinting and categorized into three types: droplet-based, extrusion-based, and photo-curing-based. The droplet-based bioprinting uses droplets of controlled volumes of bioink deposited at predefined positions. The droplets are driven by means of pressure generated by thermal, piezoelectric, or electrostatic actuators (inkjet bioprinting), an electric field (electrohydrodynamic bioprinting), an acoustic field (acoustic bioprinting), an electromechanical valve (micro-valve bioprinting), and laser pulses (laser-assisted bioprinting). Extrusion-based bioprinting is widely utilized for cell encapsulation. It uses continuous filaments generated by either pneumatic force using compressed air (pneumatic-driven extrusion) or by mechanical force using a piston or a screw (mechanical micro-extrusion). Photocuring-based bioprinting involves stereolithography in which photosensitive bioink is crosslinked via controlled UV rays as per a specific design and digital light processing (DLP), during which the UV rays are projected uniformly across the surface of the layer.

Water-soluble polymers crosslinked into hydrogels are one of the predominant biomaterial components of 3D bioprinters. A wide variety of polysaccharides and proteins such as alginate, chitosan, cellulose, kappa carrageenan, agarose, gelatin, silk fibroin, and collagen have been used for bioprinting. Alginate-based bioinks are mostly crosslinked via ionic interactions using calcium ions ($CaCl_2$, $CaCO_3$, $CaSO_4$), whereby the alginate/cell mixture is printed directly into the Ca^{2+} solution and incubated for uniform crosslinking. Sarker et al. (2019) showed the correlation between concentrations of $CaCl_2$ solution (50, 100, 150 mM) and the needle translation speed (6–26 mm/s) during bioprinting using a pneumatic 3D-Bioplotter and the effect on printability, dimensional uniformity of strands, and pore geometry. High concentration of $CaCl_2$ required slower needle translation speed of 6 or 8 mm/s for good printability, otherwise the solution gelled around the needle before reaching the printing plate. Table 7.1 summarizes different polysaccharide-based bioinks and bioprinting methods investigated over the last years.

Polysaccharides/proteins have also been functionalized with UV-sensitive moieties such as methacrylate, norbornene/carbic, tyramine, or thiol groups for photocrosslinking using 2-hydroxy-4′-(2-hydroxyethoxy)-2-methylpropiophenone (Irgacure 2959) and lithium phenyl (2,4,6-trimethylbenzoyl) phosphinate (LAP) as photoinitiators. Gelatin methacrylol (GelMA) has been synthesized and combined with other polymers to form photocurable bioinks with adjustable viscosity to form bioprinted hydrogels with long-lasting mechanical properties. Encapsulated human mesenchymal stem cells (hMSCs) showed an 80% survival rate with high osteogenic and chondrogenic differentiation capacity in these bioprinted hydrogels (Gao et al., 2015). Other methacrylated compounds such as alginate-methacrylate (AlgMA), carboxymethylcellulose-methacrylate (CMCMA), or hyaluronic acid methacrylate (HAMA) combined with GelMA have been used to encapsulate cells such as C2C12 myoblasts and mouse chondrogenic cell line ATDC5 using I2959 and LAP as photoinitiators (García-Lizarribar et al., 2018; Fan et al., 2020). ATDC5 cells maintained high viability (above 95%) after printing into a hybrid construct containing CNC reinforced GelMA/HAMA array printed parallel to GelMA/HAMA in a layer-by-layer manner (Fan et al., 2020).

Blood vessel-like tubular constructs made of GelMA/gelatin (Gel)/hyaluronic acid (HA)/glycerol composite materials were bioprinted using an Integrated Tissue Organ Printing (ITOP) system. The system allowed simultaneous 3D printing of two cell-laden bioinks of different polymer composition loaded in two separate syringes. The inner and outer rim of the tubular construct was made of HUVEC-laden bioink (6% GelMA, 20mg/ml gel, 3mg/ml HA, 100 μl/ml glycerol, 200 μl/ml I2959) and smooth muscle cells (SMCs)-laden bioink (4% GelMA, 40mg/ml gel, 3mg/ml HA, 100 μl/ml glycerol, 200 μl/ml I2959), respectively giving rise to bilayered vessels exhibiting mechanical properties ranging between those of a 4% and a 6% gelatin construct (Xu et al., 2020). An advanced coaxial 3D bioprinter was used to generate small diameter blood vessels made of GelMA/PEGDA/alginate and lyase gel containing two distinct cells layers of vascular endothelial cells (VECs) and vascular smooth muscle cells (VSMCs). The latter embedded within the polymer layer was extruded along with pluronic F-127 solution through a coaxial needle so that the F-127 inner core was leached out to form a hollow tube seeded with VECs solution (Figure 7.5) (Zhou et al., 2020).

Another composite material made of GelMA combined with kappa carreeganan (kCA) and nanosilicate, referred to as nanoengineered ionic covalent entanglement (NICE) bioink, formed covalent bonds between MA groups under UV and ionic bonds within kCA in the presence of potassium ions. The osteoinductive bioink induced endochondral differentiation of hMSCs (Chimene et al., 2020). Such bioinks allow for formation of hydrogels with double network of ionically crosslinked rigid/brittle regions and covalently-crosslinked ductile/elastic regions. Similarly, silk fibroin (SF) combined with hydropropyl methyl cellulose methacrylate (HPMC-MA) formed a double-network hydrogel composed of chemically crosslinked HPMC-MA and β-sheet formation of ultrasonicated SF. While high concentration of SF increased the compressive stress of the hydrogel, HPMC-MA improved reproducibility and anti-fatigue properties of the hydrogel. A hMSC-laden SF/HPMC-MA bioprinted construct resulted in 46±3% cell death due to shear stress and UV irradiation on day one, which decreased over time as live cells proliferated to a great extent on day ten (Ni et al., 2020).

7.4.3 Biomicrofluidics

Droplet-based microfluidics is the most promising technique used to isolate and encapsulate cells into monodisperse droplets of precisely controlled size for various applications (Figure 7.6). Droplet generation in microfluidics is achieved by introduction of a dispersed stream into a continuous stream, whereby the break-up of the dispersed stream into droplets depends on the fluid instability at the interface of two immiscible phases. The microfluidic devices can be designed into three typical geometries, i.e., cross-flow, co-flow, and flow-focusing microchannels, which can form droplets in three hydrodynamic modes: squeezing, dripping, and jetting mode, depending on the force

TABLE 7.1
Cell-laden bioprinted scaffolds: biomaterial, crosslinking mechanism, 3D-Bioplotter, cells and their applications

Polysaccharide	Other compounds	Crosslinking mechanism	Bioprinting method	Cells	Application	References
Alginate	-	Ionic crosslinking 100 mM $CaCl_2$	Extrusion bioprinting using FRESH method (four-layer lattice)	Human neuroblastoma (SK-N-BE-(2))	3D cell culture for cancer research	Lewicki et al., 2019
	RGD conjugation	Ionic crosslinking 50mM $CaCl_2$	Extrusion bioprinting using pneumatic 3D-Bioplotter (Envision TEC, Germany)	Primary Schwann cells	Filler material in nerve graft	Sarker et al., 2018
	Silk fibroin	Ionic crosslinking 5% $CaCl_2$	Extrusion bioprinter with coaxial nozzle (Regenovo, China)	Liver cancer cells (C3A)	-	Li et al., 2020
	Collagen type I	Ionic crosslinking 20 mM $CaCO_3$	Extrusion bioprinting using VIEWEG GmbH brand DC 200 model analog dispenser with time	Human-induced pluripotent stem cell-derived mesodermal progenitor cells (hiMPCs)	Blood vessel formation	Dogan et al., 2021
	Agarose	-	Extrusion bioprinting	Chondrocytes	Cartilage tissue regeneration	Lòpez-Marcial et al., 2018
	Collagen and agarose	Ionic crosslinking 10% $CaCl_2$ and 0.1% fetal bovine serum (FBS)	Extrusion bioprinting (Regenovo, China)	Primary articular chondrocytes	Cartilage tissue regeneration	Yang et al., 2018
	Kappa carrageenan	Ionic crosslinking $CaSO_4$	Extrusion bioprinting	Mesenchymal stem cells	-	Kim et al., 2019
	Poly(2-ethyl-2-oxozoline) -peptide conjugation (PEOXA-LPETG and PEOXA-GGGG) Cellulose nanofiber (CNF)	Enzymatic crosslinking of PEOXA-peptide with Sortase A Ionic crosslinking 100 mM $CaCl_2$	Extrusion bioprinting (Cellink, Sweden)	Human auricular chondrocytes (hACs)	-	Trachsel et al., 2019
Alginate-tyramine	Silk fibroin methacryol	Ionic crosslinking using $CaCO_3$ and Glucono-δ-lactone (GDL) Photocrosslinking using riboflavin (RF) and sodium persulfate (SPS)	Extrusion bioprinting using a laboratory set up in Korea Institute of Machinery and Materials	NIH 3T3 fibroblasts	Tissue engineering	Kim et al., 2021
Gelatin methacryloyl/ gelatin-methacrylate	Polyethylene glycol dimethacrylate	Photocrosslinking using Irgacure 2959	Inkjet bioprinting	Human bone marrow-derived mesenchymal stem cells (hBMSCs)	Bone and cartilage tissue engineering	Gao et al., 2015
	Polyethylene glycol diamine Alginate lyase	Photocrosslinking using Irgacure 2959 Ionic crosslinking using $CaCO_3$	Extrusion bioprinting using coaxial nozzle	Vascular smooth muscle cells (VSMCs) Vascular endothelial cells (VEC)	Small diameter blood vessel replacement	Zhou et al., 2020

(Continued)

TABLE 7.1 CONTINUED
Cell-laden bioprinted scaffolds: biomaterial, crosslinking mechanism, 3D-Bioplotter, cells and their applications

Polysaccharide	Other compounds	Crosslinking mechanism	Bioprinting method	Cells	Application	References
	Gelatin Hyaluronic acid glycerol	Photocrosslinking using Irgacure 2959	Bioprinting using an Integrated Tissue Organ Printing system	Smooth muscle cells (SMCs) Human umbilical vascular endothelial cells (HUVECs)	Small diameter blood vessel replacement	Xu et al., 2020
	Kappa carrageenan	Photocrosslinking using Irgacure 2959	Extrusion bioprinting using Biofactory bioprinter machine, RegenHu	Mouse myoblasts cells C2C12	-	Li et al., 2018
	Kappa carrageenan laponite	Photocrosslinking using Irgacure 2959 Ionic crosslinking using KCl	ANET A8 3D printer kit adapted for extrusion bioprinting	Murine 3T3 preosteoblasts (MC3T3-E1) Primary bone marrow-derived hMSC stem cells	Bone tissue engineering	Chimene et al., 2020
	Carboxymethyl cellulose OR Alginate-methacrylate OR Polyethylene glycol diacrylate	Photocrosslinking using lithium phenyl-2, 4, 6-trimethylbenzoylphosphinate (LAP)	Extrusion bioprinting (3DDiscovery BioSafety, RegenHU, Switzerland)	C2C12 myoblasts	Skeletal muscles constructs	García-Lizarribar et al., 2022
	Hyaluronic acid methacrylate Cellulose nanocrystals (CNC)	Photocrosslinking using LAP	Extrusion-based hybrid bioprinting	Mouse chondrogenic cell line (ATDC5)	Cartilage tissue regeneration	Fan et al., 2020
	Alginate CNC	Ionic crosslinking using $CaCl_2$	Extrusion bioprinting Medprin bioprinter (BMP-C300-T300-IN3)	RSC96 cells	Neural tissue engineering	Wu et al, 2021
Chitosan	-	Thermogelation at 37 °C using β glycerophosphate, potassium phosphate, and sodium bicarbonate as gelling agent	Extrusion bioprinting (laboratory set up) using a multi-nozzle printer head	Human periodontal ligament stem cells (PDLSCs)	-	Ku et al., 2020
	Gamma-poly(glutamic acid)	Electrostatic interaction between amino-carboxylic groups	Extrusion bioprinting using pneumatic 3D Bioplotter equipped with two dual heated print heads (Cellink, Sweden)	Human adult fibroblasts	-	Pisani et al., 2020
	Hydroxyethyl cellulose CNC	Thermogelation at 37 °C using β glycerophosphate	Extrusion bioprinting pneumatic 3D-Bioplotter (Cellink, Sweden)	Pre-osteoblast cells (MC3T3-E1)	Bone regeneration	Maturavongsadit et al., 2021
	Methyl Furan functionalized gelatin	Covalent crosslinking using 4arm-PEG10K-Maleimide (PEG-star-MA		U87 glioblastoma cell		

TABLE 7.1 CONTINUED
Cell-laden bioprinted scaffolds: biomaterial, crosslinking mechanism, 3D-Bioplotter, cells and their applications

Polysaccharide	Other compounds	Crosslinking mechanism	Bioprinting method	Cells	Application	References
Chitosan methacrylol/ Chitosan methacrylate	-	Photocrosslinking using LAP	Digital light processing (DLP) bioprinting	HUVEC	-	Shen et al., 2020
	-	Photocrosslinking using LAP Thermogelation at 37 °C using β glycerophosphate, disodium salt hydrate	Extrusion bioprinting using Rokit Invivo 3D Bioprinter (RokitHealthcare) equipped with a temperature-controlled bio-dispenser	NIH 3T3 fibroblasts	-	Tonda-Turo et al., 2020
Carboxymethyl chitosan	Agarose	Thermogelation	Extrusion bioprinting pneumatic 3D-Bioplotter (Cellink, Sweden)	Neuro2A mouse neuroblastoma cells	Nerve tissue regeneration	Butler et al., 2021
	Oxidized hyaluronic acid (HAox)	Covalent crosslinking via Schiff-based reaction	Extrusion bioprinting (3DDiscovery BioSafety, RegenHU, Switzerland) equipped with two syringes coupled with a static mixer	L929 fibroblast cells	-	Puertas-Bartolomé et al., 2020
Methyl Furan functionalized chitosan	Methyl Furan functionalized gelatin	Covalent crosslinking using 4arm-PEG10K-Maleimide (PEG-star-MA	Bioprinting	U87 glioblastoma cell	-	Magli et al., 2020
Pegylated chitosan (CS-mPEG)	Gelatin	Host–guest interactions between PEG and α-cyclodextrin Secondary crosslinking via immersion into β-glycerophosphate solution	Extrusion bioprinting (Regenovo, China)	Bone marrow-derived mesenchymal stem cells (BMSCs)	-	Hu et al., 2020
Methylcellulose	Kappa carrageenan CNC	Ionic crosslinking 0.1% w/w KCl	Extrusion bioprinting (Cellink, Sweden)	L929 fibroblasts	-	Boonlai et al., 2022
Hydroxyethyl cellulose	Alginate Gelatin	Ionic crosslinking using $CaCl_2$	Extrusion bioprinting Laboratory set up	MCF-7 cells	3D tumor spheroids models	Li et al., 2020
Hydroxypropyl methyl cellulose (HPMC)		Inorganic sol-gel polymerization using NaF and glycine	Extrusion bioprinting (Allevi 1)	Human mesenchymal stem cells (hMSCs)	-	Montheil et al., 2020
Hydroxypropyl methyl cellulose methacrylate	Silk fibroin	Photocrosslinking using LAP β-sheet formation of SF	Bioscaffolder 3.3 (GESIM, Germany) with four channel-led driver (Thorlabs, USA)	Bone marrow mesenchymal stem cells (BMSCs)	Cartilage tissue regeneration	Ni et al., 2020
Norbenene carboxymethylcellulose		Photocrosslinking using LAP as crosslinker	Extrusion bioprinting (Allevi 2)	hMSCs HUVEC NIH 3T3 fibroblast	-	Ji et al., 2020
CELLINK® GelMA A (Gelatin methacryloyl and alginate) CELLINK® GelXA FIBRIN (Xanthan gum and Fibrinogen) CELLINK® FIBRIN (Nanofibrillated cellulose and alginate-fibrinogen)	Photocrosslinking Ionic crosslinking Thrombin 50mM $CaCl_2$	C2C12 myoblasts	Extrusion bioprinting (Cellink, Sweden)	Skeletal muscle tissue regeneration	-	Ronzoni et al., 2022

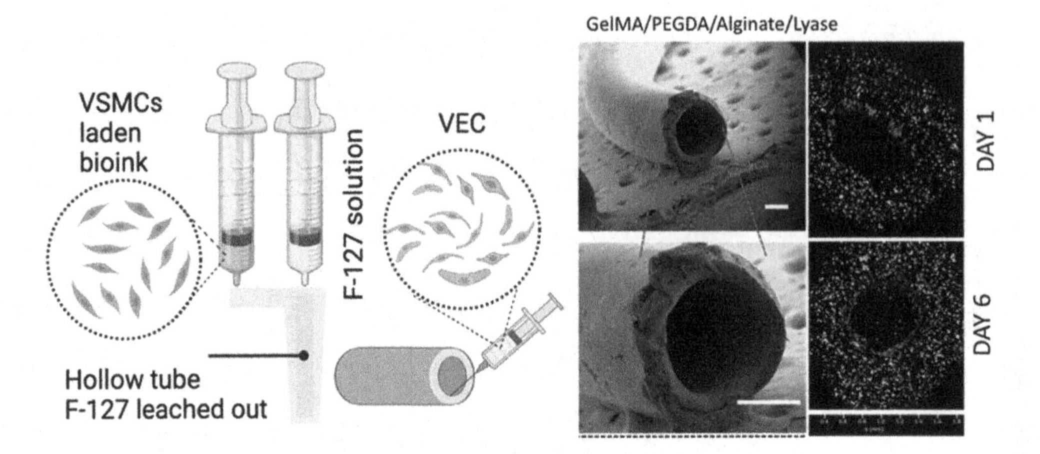

FIGURE 7.5 Bioprinting of cell-laden small blood vessels made of gelatin methacryloyl (GelMA), polyethylene (glycol) diacrylate, alginate, and lyase using extrusion bioprinting equipped with coaxial nozzle. Vascular smooth muscle cells (VSMCs) (pre-stained by Cell Tracker Green CMFDA dye) containing bioink extruded to form the shell of the vessels. Vascular endothelial cells (VEC) (pre-stained by Cell Tracker Orange CMTMR dye) seeded on the surface of the inner rim of shell after leaching out the F-127 core to form hollow tubes. (Adapted with permission from Zhou et al., 2020.)

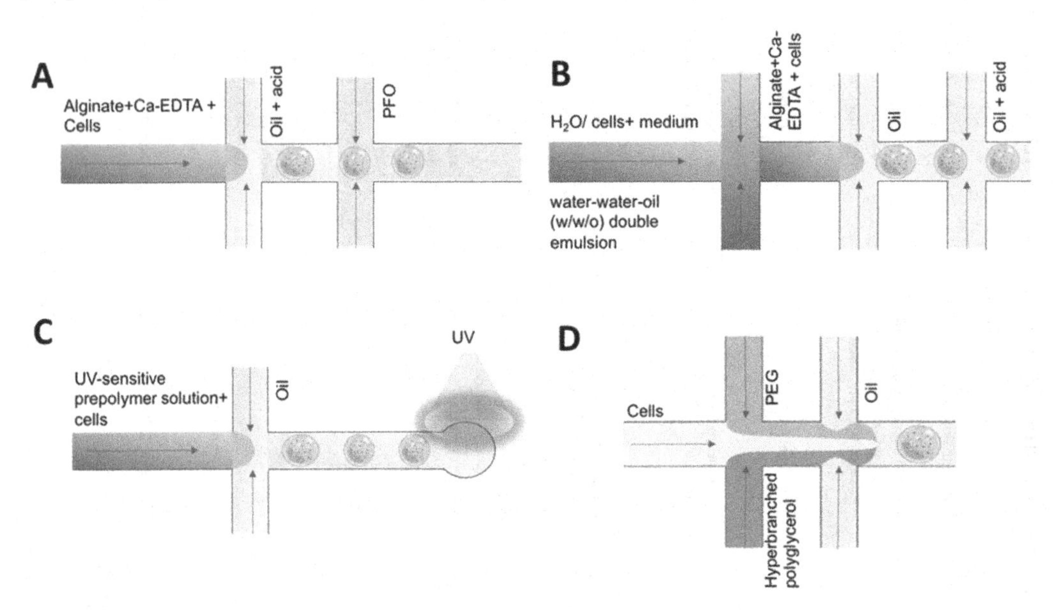

FIGURE 7.6 Cell microencapsulation using microfluidic device: microfluidic channels for cell encapsulation within (A & B) alginate crosslinked using calcium-ethylenediaminetetraacetic acid (EDTA) triggered by acetic acid, (B) water-water-oil (w/w/o) double emulsion allowing different cell encapsulation in inner core and outer shell of the alginate microcapsule, (C) UV-sensitive polymers, and (D) acrylated hyperbranched glycerol crosslinked using dithiolated PEG macro-crosslinkers in which the polymer, crosslinker, and cells flowing through three separate inlets are mixed at the first cross-junction.

balance inside the microchannel which is characterized by dimensionless numbers: Reynolds number, capillary number, Bond number, and Weber number. Microfluidic devices are generally made of glass capillary or polydimethylsiloxane (PDMS). The microfluidic chips can be manufactured using subtractive manufacturing, 3D printing, and photolithography (Ling et al., 2020).

A simplified microfluidic device uses water-in-oil emulsion whereby an aqueous polymer precursor solution containing suspended cells are emulsified in a continuous oil phase at the multichannel junction to form microcapsules. BMSC-embedded GelMA photocrosslinked microspheres of size 160 μm in diameter were synthesized using a capillary microfluidic flow-focusing device. The GelMA solution supplemented with I2959 photoinitiator containing BMSCs was injected in the disperse phase and formed monodisperse droplets in the oil phase collected for polymerization under UV light for 20 s (Zhao et al., 2016). Similarly, PEG norbonene (PEBNB) and PEG-DA microspheres of diameter ranging from 90 to 130 μm were fabricated using w/o emulsion microfluidic device to encapsulate human lung adenocarcinoma epithelial cells (A549s) for long-term survival (Jiang et al., 2017). A multichannel microfluidic system which combined dithiolated PEG, acrylated hyperbranched polyglycerol (hPG), and cell-containing medium at a first junction followed by microsphere formation at a second oil junction, has also been demonstrated (Rossow et al., 2012). The microgels of size 150–200 μm were formed by nucleophilic Michael addition reactions between PEG and hPG without the need for cytotoxic initiators. Mesenchymal stem cells (MSCs) embedded alginate microgels were also fabricated using an on-chip gelation technique whereby acid triggered release of Ca^{2+} from calcium complexes and induced gelation of aqueous alginate containing MSCs in the oil/acetic acid phase (Shao et., 2020).

Microgel capsules can also be made into sophisticated structures composed of different cellular compartments using either single or double emulsions in a microfluidic device. A multibarrelled inlet has been used to inject polymer solutions mixed with different cells so that their parallel flows are emulsified into single multicellular microgels by the viscous drag force of the continuous oil phase. Core-shell microgels based on double emulsions is another technique used to form microgels with different cell compartments. For instance, Chen et al. (2016) successfully encapsulated hepatocytes surrounded by fibroblasts using a water-water-oil (w/w/o) double emulsion. The flow-focusing microfluidic device was composed of hepatocytes in a cell culture medium as the inner phase and alginate/Ca-EDTA aqueous solution containing NIH 3T3 fibroblasts as the middle phase, emulsified by the oil phase at the first junction followed by crosslinking of the alginate shell by the oil/acetic acid phase at the second junction.

A microfluidic device has also been designed to produce cell-laden microfibers. Liu et al. (2017) fabricated a simple glass capillary microfluidic device with sodium alginate aqueous solution running through the inner capillary as core flow and $CaCl_2$ aqueous solution introduced into the side channels as the sheath flow in the region between the inner capillary and outer rectangular glass. As the two streams flowing in the same direction are merged, the alginate gelled spontaneously via fast diffusion-controlled ion crosslinking to form microfibers of size 111–537 μm in diameter depending on the size of the glass capillary tube tip. MSCs and growth factors such as VEGF164 and FGF-2 were successfully introduced to the core flow and encapsulated within the microfibers. The microfluidic device can be slightly modified to form an alginate-based multicompartment with hollow microfibers, by introducing a multi-barrel capillary as the injection channel or by introducing hierarchical injection channels, respectively. Six-compartment alginate microfibers were successfully formed using a six-barrel injection capillary. The hollow microfibers were achieved by a three-layered coaxial flow of $CaCl_2$ as the core fluid, sodium alginate as the middle fluid, and $CaCl_2$ as the outer sheath fluid (Chen et al., 2014).

7.4.4 Electrostatic Droplet Extrusion

Electrostatic droplet extrusion is another simple technique used to form hydrogel-based microcapsules of a diameter less than 50 μm. A charged stream of droplets is formed by breaking the liquid jet extruded through a needle tip connected to the high-voltage electrostatic generator. The pressure applied to the needle transforms the liquid into a Taylor cone-like droplet which expands to create a thin strand. Parameters such as electrostatic potential, distance between the collecting device and the syringe pumps, size of the nozzle, flow rate, and concentration and viscosity of the polymer solutions determine the size of the droplets. Small and symmetrical microspheres are ideal for molecular transport, allowing for diffusion of oxygen and nutrients from and to the cells encapsulated within the spheres as well as for improved injection kinetics. Oxygen can diffuse through mammalian tissues over a maximum distance of 100 μm. Therefore, microspheres with a radius of 100 μm are ideal for molecular exchange.

This technique has been mostly used to generate cell-laden alginate microdroplets due to its fast gelation process via ionic crosslinking mechanism, whereby alginate-cell suspension is pumped through a positively charged needle into a ground-hardening solution of calcium chloride (Yao et al., 2009). Moyer et al. (2010) developed adipose-derived stem cell (ADSCs)-encapsulated microbeads of < 200 μm in diameter using an electrostatic bead generator operated at a flow rate of 5 ml/h and electrostatic potential of 7 kV. The cell-laden microbeads resuspended in sodium hyaluronate (NaHA)/PBS solution were evenly injected through a 21G needle *in vivo* at a low pressure, showing higher fidelity. Nests of healthy cells were visible at both center and periphery of the microspheres three months after subcutaneous injection into a nude mouse.

Leslie et al. (2013) increased the degradability of alginate microbeads by incorporating alginate-lyase to alginate at different ratios for the controlled release of ADSCs. The latter maintained high viability and retained their osteogenic phenotype. To improve its cell-matrix interactions, alginate combined with agarose and RGD-motif has been used to encapsulate cells using electrostatic droplet extrusion (Yu et al., 2010; Orive et al., 2003). Yu et al. (2010) generated hMSC-embedded RGD-modified alginate microbeads of size ranging between 75 and 100 μm which were successfully injected using a 25G needle into the myocardium with 99% of the microbeads remaining intact. RGD-modified alginate microbeads alone have shown to induce FGF2 gene expression *in vitro* and angiogenesis *in vivo* compared to unmodified alginate beads and cell-encapsulated RGD-modified alginate beads. Co-extrusion techniques have been used to form microbeads surrounded by a semipermeable membrane which do not require the membrane and core material to be oppositely charged. Lewińska et al. (2012) used an electrostatic droplet generator with 3-coaxial-nozzle head to form a cell-laden alginate microbead core surrounded by a semipermeable polyethersulfone membrane with glycerol used as a core-membrane separating fluid.

7.4.5 Layer-by-Layer Self-Assembly

The layer-by-layer (LbL) self-assembly technique introduced by Iler and Decher in the early 1990s emerged as a cutting-edge cell-encapsulating nanotechnology involving sequential deposition of multilayers of oppositely charged materials to form nanofilms around cells. This is achieved through multiple intermolecular interactions (electrostatic interactions, covalent bonding, hydrogen bonding, van der Waals forces, hydrophobic interactions, charge-transfer interactions, host–guest interactions, etc.) which can occur between a plethora of both synthetic and natural polyelectrolytes. LbL cell encapsulation can be divided into direct and indirect cell encapsulation.

Direct cell encapsulation involves multi-layered nanofilms surrounding a single cell or cell aggregates. Indirect cell encapsulation uses a hydrogel core to embed the cells prior to further encapsulating the cell-containing hydrogel within a self-assembled polymer shell. The hydrogel core can also be sacrificial by liquefaction to obtain a hollow core-shell structure. The multi-layered polymer capsule can be tailored to have a functionalized outer layer for improved characteristics such as responsiveness to physical, chemical, or biological stimuli (Marin et al., 2022).

A variety of both synthetic and natural polyelectrolytes including polycations (gelatin, cationic cellulose, chitosan, polyamidoamine, poly-L-lysine (PLL), and polyanions (alginate, hyaluronic acid (HA) and PSS) have been used to encapsulate mammalian cells such as stem cells and islet cells. Other materials such nanoparticles have been used for single cell encapsulation. It has been linked to poly (allylamine hydrochloride) (PAH) to prevent internalization by cells for improved cell survival (Li et al., 2022). Two major techniques used for cell encapsulation using LbL are immersive and fluidic assembly. Immersive assembly, also known as dip assembly, requires immersion of cells into polymer solution for a given time period followed by centrifugation to remove any unbound material. Fluidic technology allows for shorter assembly time, as cells are loaded into fluidic channels with sequential flow of polymers and washing solutions requiring no incubation time (Liu et al., 2019).

Nanocapsules of poly (L-lysine) (PLL) and hyaluronic acid (thickness 6–9 nm) have been used to encapsulate single mesenchymal stem cells using electrostatic LbL assembly. MSCs of a five million count were first added to the cationic PLL containing the amino groups that surround the negatively charged cell membrane. The coated cells were incubated in the anionic HA, and the process was repeated to form several polymeric bilayers until the desired thickness was reached for optimum cell permeability and viability (Veerabadran et al., 2007). However, highly cationic polymers such PLL and PAH have shown high cytotoxicity for numerous cell types. Gattás-Asfura et al. (2013) grafted a primary layer N_3-PEG-NHS to the surface of pancreatic islets followed by three bilayers of azido-functionalized highly branched alginate (Alg-N_3) and phosphine-functionalized poly (amido amine) (PAMAM) to improve cell viability and LbL uniformity (Gattás-Asfura et al., 2013). No significant decline in cell viability was noted on day five for coated pancreatic islets compared to control. Wilkens et al. (2016) fabricated a complex vessel-like multi-layered hydrogel structure with controlled position of the cellularized layer. The construct consisted of three layers of UV-crosslinked 10% gelatin-methacrylate (GelMA) and 0.2% alginate with three intercalated sets of 25 micro-layers of 10% GelMA. The layers were built on a resorbable alginate mandrel structure by dipping into pre-crosslinked polymer solutions. HUVECs were added at a concentration of 8 x 10^6 cells per ml of GelMA/alginate dipping solution to encapsulate evenly distributed cells in the GelMA/alginate layer beside the luminal area.

7.5 CELL-BASED THERAPIES: MESENCHYMAL STEM CELLS (MSCS)

Mesenchymal stem/stromal cells (MSCs) implantation is the most studied cell-based therapy for various diseases such as graft-versus host disease (GvHD), Crohn's disease, type I diabetes, multiple sclerosis (MS), lupus, cardiovascular disease, liver disorders, respiratory disorders, COVID-19, spinal cord injury, kidney failure, skin diseases, Alzheimer's disease (AD), and Parkinson's disease (PD). MSCs are known for their immunosuppressive and regenerative properties. They express multitude of cytokines, chemokines, and growth factors which aid in tissue regeneration, restoring tissue metabolism and suppressing inflammation. The trophic effectors are either released as soluble molecules or carried by extracellular vesicles (ECVs) produced by the MSCs. MSC-derived ECVs have shown to promote generation of regulatory T cells (Tregs), regulatory B cells (Bregs), and M2 macrophages, which inhibit proliferation of lymphocytes, macrophages, and B and T cells, preventing monocytes differentiation (Wu et al., 2020; Zhou et al., 2021; Mancuso et al., 2019).

However, MSC transplants have shown to employ a "hit and run" mechanism in which the cells are short-lived after production of EVs and secretion of cytokines, chemokines, and growth factors (Wu et al., 2020). Several factors leading to the failure of clinical trials such as heterogeneity, immune incompatibility, stemness instability, and loss of directed migratory capacity of MSCs have shown to result from serial passaging of the donor cells under different culture conditions. Large scale expansion in 2D plates impact stem cell characteristics of MSCs, necessitating 3D cultures allowing large cell expansion with guaranteed cell quality. Encapsulation of MSCs in biomaterials suitable for implantation provide a protective layer which improves the survival and functionality of MSCs for prolonged durations in clinical treatment (Zhou et al., 2021).

7.5.1 3D Tissue Regeneration

MSCs secrete a variety of mediators for tissue repair including anti-apoptotic (B cell lymphoma 2 (BCL-2), survivin, VEGF, HGF, insulin-like growth factor-I (IGF-I), stanniocalcin-1(STC-1), transforming growth factor β (TGF-β), FGF, and granulocyte–macrophage colony-stimulating factor (GM-CSF)), anti-inflammatory, immunomodulatory (prostaglandin E-2 (PGE-2), soluble human leukocyte antigen G5 (sHLA-G5), TGF-β, HGF, IL-10, IL-6, indoleamine 2,3-dioxygenase (IDO), nitric oxide (NO), inducible nitric-oxide synthase (iNOS), hemeoxygenase-1 (HO-1), galectin-1 (Gal-1), Gal-9, and TNFα stimulated gene 6 (TSG-6)), migratory (CCR1, CCR2, CCR4, CCR7, CXCR5, and CCR10), anti-fibrotic and pro-angiogenic (endothelial growth factor (VEGF), fibroblast growth factor (FGF), hepatocyte growth factor (HGF), placental growth factor (PGF), monocyte monocyte chemotactic protein 1 (MCP-1), stromal cell-derived factor 1 (SDF-1), and angiopoietin-1 (Ang-1)) agents (Mancuso et al., 2019; Wu et al., 2020). They have been encapsulated within hydrogel systems for various tissue engineering applications such as bone and cartilage repair, small diameter blood vessel formation, and nerve regeneration (Table 7.2).

Bone marrow-derived stem cells (BMSCs) encapsulated in alginate microcapsules showed higher osteogenic differentiation in 3D culture compared to 2D culture *in vitro*, resulting in a high amount of new bone formation *in vivo* (An et al., 2020). Hydrogels also allow for the incorporation of growth factors boosting cellular proliferation. Bone morphogenic protein-2 (BMP-2) was added to gelatin methacryloyl microcapsules laden with BMSCs which induced the largest new bone formation in rabbit femoral defects. The encapsulated BMSCs exhibited > 60% viability on day seven. VEGF and FGF were also added to BMSCs containing alginate microfibers inducing BMSCs to differentiate into endothelial cells for small diameter blood vessels formation. Collagen type I was also combined with alginate to bioprint human-induced pluripotent stem cell-derived mesodermal progenitor cells (hiMPCs), which assembled in spheroids forming a lumen on day 10, vessel-like structures on day 14, and multi-layered wall structure composed of vascular progenitor cells, and pro-endothelial and endothelial cells *in vitro* (Dogan et al., 2021). BMSCs encapsulated in other complex microcapsules have been used for cartilage repair. The cells showed > 90% cell viability on day seven in gelatin norbonene (GelNB)/polyethylene glycol-dithiol (PEG-(SH)2) microcapsules with higher level of cartilaginous protein gene expressions (aggrecan, COMP, type II collagen) compared to bulk hydrogel or cell pellet (Li et al., 2017).

Optogenitically modified human pluripotent stem cells derived neural progenitor cells (hPSC-NPC denoted as Axol), Axol-ChR2 have shown to differentiate to mature neurons in 3D RGD conjugated alginate microcapsules showing single peak spikes driven by CAMKII promoter (Lee et al., 2019). High viability and proliferative activity of neural stem cells (NSCs) with higher expression of βIII-tubulin have been reported in 3D culture using an injectable and self-healing oxidized sodium alginate (OSA)/N carboxyethyl-chitosan (CEC) hydrogel (Wei et al., 2016). These recent studies show the potential of mesenchymal stem cells therapy to treat neurological disorders.

7.5.2 Type I and II Diabetes

Type I diabetes (TIDM) is an autoimmune disease that leads to destruction of islet cells and insulin deficiency, while type II diabetes (T2DM) is related to insulin resistance causing exhaustion of islet cells. Islet implantation can potentially cure diabetes but is limited by lack of donors or immune rejection. Recently, mesenchymal stem cells (MSCs) have demonstrated their efficacy in the treatment of both T1DM and T2DM. They have shown their capacity to reverse hyperglycaemia in a number of animal and clinical studies (Li et al., 2021) MSC-treated groups showed a decrease in fasting blood glucose level (FPG) and hemoglobin (HbA1c) and an increase in C-peptide. It is evidenced that MSCs promote insulin production by enabling regeneration of endogenous pancreatic islet β-cells due to their repairing potential and antiapoptotic effects through the paracrine and autocrine actions as well as mitochondrial transfer to β-cells. Although MSCs can differentiate into

TABLE 7.2
Mesenchymal stem cell therapy for tissue engineering applications

Mesenchymal stem cell	Biomaterial	Encapsulation method	*in vitro/in vivo* studies	References
Tissue engineering				
Bone regeneration				
Bone marrow-derived mesenchymal stem cells (BMSCs)	Alginate	Microfluidics	Higher ALP activity, ALP and OCN gene expression of BMSCs in osteogenic medium cultured in microgels compared to 2D culture. Largest amount of new bone formation characterized by osteocyte-populated woven bone filling the medullar cavity of rat tibial marrow ablation model two weeks post implantation. Cells in microcapsule > microcapsule only > microcapsule + cells	An et al., 2020
	Gelatin methacryloyl (GelMA)	Microfluidics	Viability of BMSCs in GELMA microspheres > 60% on day seven. High proliferative activity and migration from center to periphery of microspheres. ALP activity detected on day one and peaked on day seven. Largest new bone formation in rabbit femoral defect treated with BMSCs laden GelMA microspheres containing BMP-2 growth factor. GelMA/BMSC/BMP-2 > GelMA/BMSC > GelMA > control	Zhao et al., 2016
Cartilage regeneration				
Adipose-derived stem cells (ADSCs)	Alginate	Electrostatic droplet extrusion	Viable and proliferative ADSCs in alginate microbeads three months following subcutaneous implantation in NICE mice	Moyer et al., 2010
Bone marrow-derived mesenchymal stem cells (hBMSCs)	Gelatin norbonene (GelNB) / Polyethylene Glycol-dithiol (PEG-(SH)2)	Microfluidics	Viability of BMSCs in GelNB/PEG-$(SH)_2$ microspheres >90% on day 7 High proliferative activity and migration from center to periphery of microspheres. High level of type II collagen and 100-fold increase in aggrecan and COMP; 118986-fold increase in COL2A1 in chondro-inductive cultured microgel compared to a less significant increase for bulk hydrogel and cell pellet.	Li et al., 2017
	Silk fibroin (SF)/ hydroxypropyl methyl cellulose methacrylate (HPMC-MA)	3D Bioprinting using (BioScaffolder 3.2)	Decrease in dead cells from day one (46%) to day seven (15%) and day ten (3%). High cell proliferation on day ten. High level of type II collagen, aggrecan, and Sox 9 in week two.	Ni et al., 2020

(Continued)

TABLE 7.2 CONTINUED
Mesenchymal stem cell therapy for tissue engineering applications

Mesenchymal stem cell	Biomaterial	Encapsulation method	*in vitro*/*in vivo* studies	References
Small diameter blood vessels formation				
Bone marrow-derived mesenchymal stem cells (hBMSCs)	Alginate	Microfluidics	High proliferative activity on day seven; reduction on day fourteen. Higher viability in presence of VEGF and FGF growth factors on days one, seven, and fourteen. MSCs/VEGF/FGF > MSCs/VEGF > MSCs. Higher level of CD31, VE-cadherin, vWF in MSCs/VEGF/FGF containing microfibers. No adverse reaction observed over 14 days in Kunming mice implanted with 50 ml/kg microfibers with or without MSCs/VEGF/bFGF in their abdominal cavity.	Liu et al., 2017
Human-induced pluripotent stem cell-derived mesodermal progenitor cells (hiMPCs)	Alginate/collagen type-1	Bioprinting (VIEWEG GmbH brand DC 200 model analog dispenser with timer)	High cell viability (> 75%) using optimal extruded cell density. Spheroid formation with lumen on days seven to ten. Formation of vessel-like structures on day 14. Large vessels with multi-layered wall structure on day 21. CD31+ endothelial cells lined the luminal surface of the vessels, NG2+ and αSMA+ peri-endothelial cells covered the endothelial layer from outside and CD34+ vascular progenitor cells formed the outmost layer of the vessel walls CD31+ endothelial lining, αSMA+ peri-endothelial cells and nucleated chicken blood vessels in the lumen of the printed vessel transplanted into chicken embryo chorioallantois membrane (CAM) and *in vivo* blood perfusion.	Dogan et al., 2021
Nerve regeneration				
Human Embryonic stem cells (hESCs denoted as HUES) Human pluripotent stem cells (hPSCs) derived neural progenitor cells (hPSC-NPC denoted as Axol)	Alginate (RGD conjugation)	Electrostatic droplet extrusion	Increase in cell viability in RGD-modified alginate beads on day 14. Viability of Axol cells (45–50%) > HUES on day 14. Expression of βIII-tubulin by HUES at day 14 and day 21 after neural differentiation. Optogenetically modified neurons (Axol-ChR2) expressed ChR2-eYFP signals in 3D RGD-alginate beads. Light stimulated Axol-ChR2 cells appeared in mixed and burst calcium waves in RGD-alginate hydrogel. High number of single peak calcium spikes driven by CAMKII promoter indicate of more mature neurons.	Lee et al., 2019
Neuron stem cells (NSCs)	Oxidized sodium alginate (OSA)/N-carboxyethyl-chitosan (CEC))	Bulk encapsulation Injectable and self-healing properties	High cell viability (80%) on day five after extrusion and self-healing of gel. Increase of proliferative rate from 28% (day one) to 73 and 89% on days three and five similar to 2D culture. Higher expression of βIII-tubulin in 3D culture compared to 2D culture on day nine. Limited growth of neurites in 3D culture in which morphology of neurons remained round.	Wei et al., 2016

insulin-producing cells and β-cells *in vitro, in vivo* studies indicated limited trans-differentiation of infused MSCs (Gao et al., 2022).

One of the strategies employed to improve the survival and function of MSCs is via delivery with biomaterials. MSCs have been co-encapsulated with islet cells in hydrogel system which have shown to improve their viability and increase their insulin secretion (Chua et al., 2018). Human adipose-derived stem cells (hADSCs) co-encapsulated with rat pancreatic islets in fibrin gel achieved better glucose lowering effect when transplanted subcutaneously in diabetic mice compared to fibrin gel with islet cells alone (Bhang et al., 2013). hADSCs secrete growth factors that improve islet viability under hypoxic conditions and enhance vascularization via overexpression of VEGF. Mesenchymal stromal cells (MSCs), pancreatic islets, and ECM proteins such as collagen type IV and laminin encapsulated in silk hydrogel resulted in a 3.2-fold increase in insulin secretion compared to islet in silk alone (Davis et al., 2012). Alangpulinsa et al. (2019) used stem β cells (SC- β) embedded with CXCL12 chemokine in alginate microcapsules to treat diabetes. The cells remained viable and maintained their differentiation state over 154 days *in vitro*, following which the cell-laden microcapsules were implanted in the peritoneal cavity of C57BL/6 diabetic mice. The diabetic mice became normoglycemic seven days post transplantation.

7.6 CLINICAL APPLICATION OF CELL-ENCAPSULATED HYDROGEL IMPLANTS TO TREAT DIABETES MELLITUS

Biomaterials/scaffold strategies have improved the clinical challenges in cell-based therapy such as islet transplantation to treat diabetes mellitus. The clinical application of encapsulated micro/nanodevices which allow for transfer of large number of islets have been investigated in a few clinical studies (Ghasemi et al., 2021). In 1994, 20,000 allogenic islets encapsulated in alginate microcapsules were implanted into the peritoneum of an immunosuppressed patient who had been insulin-independent for 9 months (Soon-Shiong et al., 1994) Another clinical study implanted alginate-poly-L-ornithine hydrogels containing 400,000 and 600,000 human islets in two diabetic patients who showed reduction in insulin requirements (Calafiore et al., 2006). Similar results were observed with human islets encapsulated in barium alginate hydrogels and implanted in four patients treated with immunosuppressive drugs (Tuch et al., 2009).

Several commercial microencapsulation devices which include bioartificial pancreas air device (Beta-O_2 Technologies Ltd), TheraCyte™ devices (Viacyte), PEC-Encap and PEC-Direct (Viacyte), and BioHub™ have been used to encapsulate islets and subjected to clinical trials (Ghasemi et al., 2021). The bioartificial pancreas Air device uses alginate hydrogel to immobilize islet cells into two chambers separated by an oxygen module composed of a central and two side compartments parted by silicon membranes. The chamber is surrounded by hydrophilic 0.4 μm porous polytetrafluoroethylene (PTFE) membranes and is impregnated with hydrophobically modified (HM) alginate. This device allows the controlled supply of oxygen from the integrated oxygen reservoir coupled with a gaseous oxygen source and maintains viable human islets for more than six months post-subcutaneous implantation in patients with T1DM (Ludwig et al., 2013).

It is becoming increasingly recognized that in clinical settings, retrievable encapsulation devices are essential to ensure safety and success of cell-based therapies. Wang et al. (2021) designed a retrievable nanofiber-integrated cell-encapsulating (NICE) device composed of a highly porous electrospun nanofibrous medical-grade thermoplastic silicon-polycarbonate-urethane (TSPU) membrane and an alginate hydrogel core immobilizing human SC cells. The device regulated blood sugar levels within a week of transplantation and reversed diabetes in both immunodeficient and immunocompetent mice for up to 120 and 60 days respectively. Emerging technologies such as bioprinting have allowed scientists to design artificial pancreatic tissues with vascular structures (Ghasemi et al., 2021). The bioengineers at the Rice University designed a 3D-printed vascularized hydrogel system in which passageways were created for blood flow located no more than 100 μm from the encapsulated beta cells (Boyd J, 2021). These implants are yet to be tested clinically.

7.7 FUTURE PERSPECTIVES AND CHALLENGES

7.7.1 Islet Transplantation – Scalability to Humans

Islet transplantation using hydrogel-based carriers is a promising approach for the treatment of type 1 diabetes. Semipermeable biomaterial-based hydrogel containing islet cells allows nutrients, oxygen, and secreted hormones to diffuse through the membrane while sheltering from immune cells. The latter allows long-term graft survival and prevents the long-term use of immunosuppression (Zhang et al., 2022) (Figure 7.7).

Reversal of diabetes using encapsulated pancreatic islets has been successful in rat models. (Roshni S. Rainbow & Michael J. Lysaght, 2013). In larger animal models, such as canine models and primates, limited success has been achieved using encapsulated islets (Lanza et al., 1995; Wang et al., 2008; Elliott et al., 2005). The reversal and stabilization of glucose levels required additional insulin administration to supplement the amount produced by the encapsulated cells. The large number of cells required made cell harvesting and processing more difficult in large animal models. 20,000 encapsulated islets per kg were needed to achieve insulin independence in diabetic dogs (Soon-Shiong et al., 1992; 1993). In 1994, alginate microcapsules with a high content of guluronic acid were used to encapsulate islet cells and resulted in insulin independence and tight glycemic control in a single type 1 diabetic patient nine months after the islet transplantation (Soon-Shiong et al., 1994).

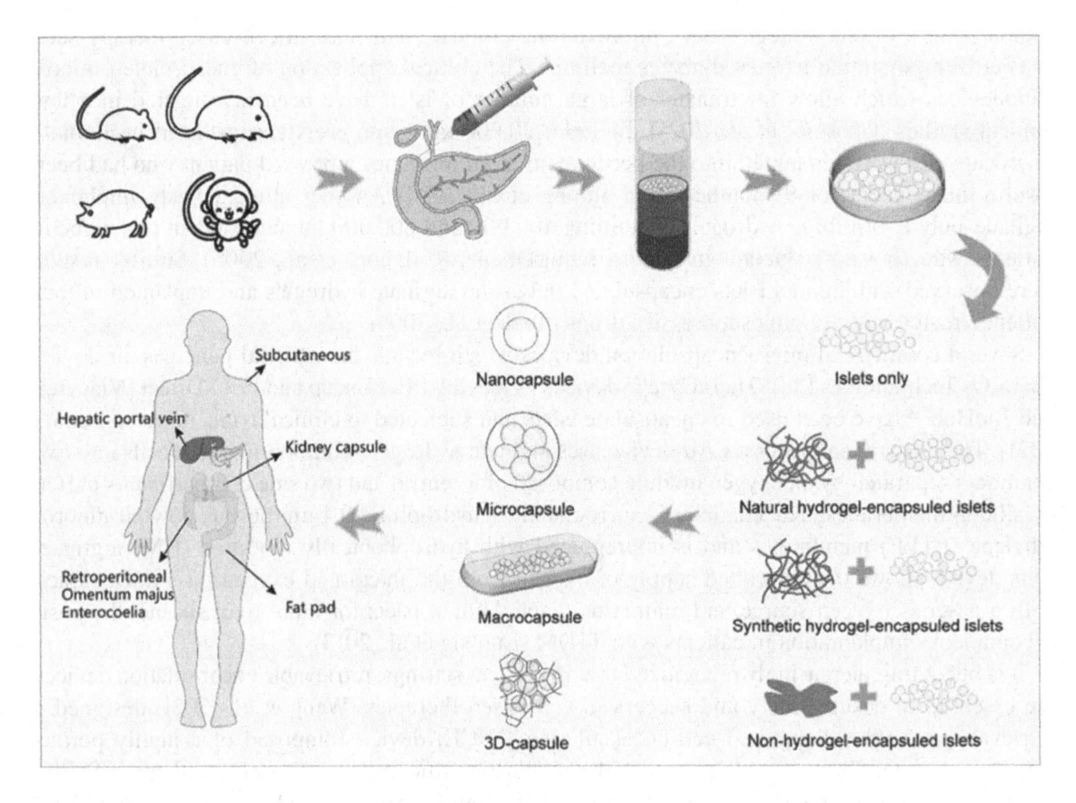

FIGURE 7.7 Islet encapsulation and transplantation. The islets are capsuled with natural hydrogels, synthetic hydrogels, and other types of hydrogels, forming capsules of different sizes (nanocapsule, microcapsule, macrocapsule, and 3D-capsule) and then transplanted. Common sites of transplantation are subcutaneous, kidney capsule, hepatic portal vein, retroperitoneal, omentum majus, enterocoelia, fat pad, etc. Reproduced from Zhang et al., 2022, Copyright © 2022 Zhang, Gonelle-Gispert, Li, Geng, Gerber-Lemaire, Wang, and Buhler. This is an open-access article distributed under the terms of the Creative Commons Attribution License (CC BY).

The surface charge and mechanical stability of hydrogels have been identified as important features which determine the *in vivo* fate of the encapsulated cells (Jansen et al., 2018). Foreign body response (FBR) capsule limits the diffusion of nutrients and oxygen to the implant or prevents insulin release and waste discharge. The stiffness of the hydrogel determines the strength of the foreign body reaction. Thus, the hydrogels should be able to provide adequate cell protection with minimum FBR.

Alginate microcapsules used in two patients showed that both patients reduced their exogenous insulin requirements but did not achieve complete insulin independence (Buder et al., 2013). In another study, laparoscopy showed the barium alginate microcapsules were surrounded by fibrous tissue, and in some cases the islets were necrotic (Tuch et al., 2009), thus showing that fibrous tissue deposition negatively impacts encapsulated islet function.

7.8 CONCLUSION

Hydrogels from renewable resources offer tremendous opportunities for sustainability in the healthcare sector. These renewable resources can be both land and marine-based. Many of these are already in use in other commercially available medical devices such as wound dressings. New therapies for diabetes and cancer can emerge using hydrogel cell encapsulation. The main challenge to overcome in cell encapsulation techniques remain the transition from small animal experiments to humans and the judicious choice of materials. The latter can be enhanced through the use of predictive modeling tools which can facilitate the choice of materials for efficient cell-material interactions.

REFERENCES

Alagpulinsa DA, Cao JJ, Driscoll RK, Sîrbulescu RF, Penson MF, Sremac M, Engquist EN, Brauns TA, Markmann JF, Melton DA, Poznansky MC. Alginate-microencapsulation of human stem cell–derived β cells with CXCL 12 prolongs their survival and function in immunocompetent mice without systemic immunosuppression. *American Journal of Transplantation.* 2019 Jul; 19(7):1930–40.

An C, Liu W, Zhang Y, Pang B, Liu H, Zhang Y, Zhang H, Zhang L, Liao H, Ren C, Wang H. Continuous microfluidic encapsulation of single mesenchymal stem cells using alginate microgels as injectable fillers for bone regeneration. *Acta Biomaterialia.* 2020 Jul 15; 111:181–96.

Baruch L, Machluf M. Alginate–chitosan complex coacervation for cell encapsulation: Effect on mechanical properties and on long-term viability. *Biopolymers: Original Research on Biomolecules.* 2006 Aug 15; 82(6):570–9.

Benya PD, Shaffer JD. Dedifferentiated chondrocytes reexpress the differentiated collagen phenotype when cultured in agarose gels. *Cell.* 1982 Aug 1; 30(1):215–24.

Bhang SH, Jung MJ, Shin JY, La WG, Hwang YH, Kim MJ, Kim BS, Lee DY. Mutual effect of subcutaneously transplanted human adipose-derived stem cells and pancreatic islets within fibrin gel. *Biomaterials.* 2013 Oct 1; 34(30):7247–56.

Boonlai W, Tantishaiyakul V, Hirun N. Characterization of κ-carrageenan/methylcellulose/cellulose nanocrystal hydrogels for 3D bioprinting. *Polymer International.* 2022 Feb; 71(2):181–91.

Boyd J. (2021, 26 July). Rice team creating insulin producing implant for Type 1 diabetes. *Rice News.* Retrieved September 30, 2022, from https://news.rice.edu/news/2021/rice-team-creating-insulin-producing-implant-type-1-diabetes.

Buder B, Alexander M, Krishnan R, Chapman DW, Lakey JR. Encapsulated islet transplantation: Strategies and clinical trials. *Immune Network.* 2013 Dec 1; 13(6):235–9.

Butler HM, Naseri E, MacDonald DS, Tasker RA, Ahmadi A. Investigation of rheology, printability, and biocompatibility of N, O-carboxymethyl chitosan and agarose bioinks for 3D bioprinting of neuron cells. Materialia. 2021 Aug 1; 18:101169.

Calafiore R, Basta G, Luca G, Lemmi A, Montanucci MP, Calabrese G, Racanicchi L, Mancuso F, Brunetti P. Microencapsulated pancreatic islet allografts into nonimmunosuppressed patients with type 1 diabetes: First two cases. *Diabetes Care.* 2006 Jan 1; 29(1):137–8.

Chander V, Singh AK, Gangenahalli G. Cell encapsulation potential of chitosan-alginate electrostatic complex in preventing natural killer and CD8+ cell-mediated cytotoxicity: An in vitro experimental study. *Journal of Microencapsulation.* 2018 Aug 18; 35(6):522–32.

Chen Q, Utech S, Chen D, Prodanovic R, Lin JM, Weitz DA. Controlled assembly of heterotypic cells in a core–shell scaffold: Organ in a droplet. *Lab on a Chip.* 2016; 16(8):1346–9.

Cheng Y, Zheng F, Lu J, Shang L, Xie Z, Zhao Y, Chen Y, Gu Z. Bioinspired multicompartmental microfibers from microfluidics. *Advanced Materials.* 2014 Aug; 26(30):5184–90.

Chimene D, Miller L, Cross LM, Jaiswal MK, Singh I, Gaharwar AK. Nanoengineered osteoinductive bioink for 3D bioprinting bone tissue. *ACS Applied Materials & Interfaces.* 2020 Feb 24; 12(14):15976–88.

Chua ST, Song X, Li J. Hydrogels for stem cell encapsulation: Toward cellular therapy for diabetes. In *Functional Hydrogels as Biomaterials* (pp. 113–127). Springer, Berlin, Heidelberg, 2018.

Ciolacu D. Sustainable hydrogels from renewable resources. In *Sustainability of Biomass Through Bio-Based Chemistry* 2021 Mar 21 (pp. 161–189). CRC Press.

Clough HC, O'Brien M, Zhu X, Miller AF, Saiani A, Tsigkou O. Neutrally charged self-assembling peptide hydrogel recapitulates in vitro mechanisms of breast cancer progression. *Materials Science and Engineering: C.* 2021 Aug 1; 127:112200.

Davis NE, Beenken-Rothkopf LN, Mirsoian A, Kojic N, Kaplan DL, Barron AE, Fontaine MJ. Enhanced function of pancreatic islets co-encapsulated with ECM proteins and mesenchymal stromal cells in a silk hydrogel. *Biomaterials.* 2012 Oct 1; 33(28):6691–7.

Dhillon J, Young SA, Sherman SE, Bell GI, Amsden BG, Hess DA, Flynn LE. Peptide-modified methacrylated glycol chitosan hydrogels as a cell-viability supporting pro-angiogenic cell delivery platform for human adipose-derived stem/stromal cells. *Journal of Biomedical Materials Research Part A.* 2019 Mar; 107(3):571–85.

Diduch DR, Jordan LC, Mierisch CM, Balian G. Marrow stromal cells embedded in alginate for repair of osteochondral defects. *Arthroscopy: The Journal of Arthroscopic & Related Surgery.* 2000 Sep 1; 16(6):571–7.

Dogan L, Scheuring R, Wagner N, Ueda Y, Schmidt S, Wörsdörfer P, Groll J, Ergün S. Human iPSC-derived mesodermal progenitor cells preserve their vasculogenesis potential after extrusion and form hierarchically organized blood vessels. *Biofabrication.* 2021 Sep 27; 13(4):045028.

Elliott RB, Escobar L, Tan PL, Garkavenko O, Calafiore R, Basta P, Vasconcellos AV, Emerich DF, Thanos C, Bambra C. Intraperitoneal alginate-encapsulated neonatal porcine islets in a placebo-controlled study with 16 diabetic cynomolgus primates. In *Transplantation Proceedings* 2005 Oct 1 (Vol. 37, No. 8, pp. 3505–3508). Elsevier.

Fan Y, Yue Z, Lucarelli E, Wallace GG. Hybrid printing using cellulose nanocrystals reinforced GelMA/HAMA hydrogels for improved structural integration. *Advanced Healthcare Materials.* 2020 Dec; 9(24):2001410.

Gao G, Schilling AF, Hubbell K, Yonezawa T, Truong D, Hong Y, Dai G, Cui X. Improved properties of bone and cartilage tissue from 3D inkjet-bioprinted human mesenchymal stem cells by simultaneous deposition and photocrosslinking in PEG-GelMA. *Biotechnology Letters.* 2015 Nov; 37(11):2349–55.

Gao S, Zhang Y, Liang K, Bi R, Du Y. Mesenchymal stem cells (MSCs): A novel therapy for type 2 diabetes. *Stem Cells International.* 2022 Aug 22; 2022.

García-Lizarribar A, Fernández-Garibay X, Velasco-Mallorquí F, Castaño AG, Samitier J, Ramon-Azcon J. Composite biomaterials as long-lasting scaffolds for 3D bioprinting of highly aligned muscle tissue. *Macromolecular Bioscience.* 2018 Oct; 18(10):1800167.

Gattás-Asfura KM, Stabler CL. Bioorthogonal layer-by-layer encapsulation of pancreatic islets via hyperbranched polymers. *ACS Applied Materials & Interfaces.* 2013 Oct 23; 5(20):9964–74.

George S, Ho SS, Wong ES, Tan TT, Verma NK, Aitken RJ, Riediker M, Cummings C, Yu L, Wang ZM, Zink D. The multi-facets of sustainable nanotechnology–Lessons from a nanosafety symposium. *Nanotoxicology.* 2015 Apr 3; 9(3):404–6.

Ghasemi A, Akbari E, Imani R. An overview of engineered hydrogel-based biomaterials for improved β-cell survival and insulin secretion. *Frontiers in Bioengineering and Biotechnology.* 2021:686.

Guan T, Li J, Chen C, Liu Y. Self-assembling peptide-based hydrogels for wound tissue repair. *Advanced Science.* 2022 Apr; 9(10):2104165.

Haines-Butterick L, Rajagopal K, Branco M, Salick D, Rughani R, Pilarz M, Lamm MS, Pochan DJ, Schneider JP. Controlling hydrogelation kinetics by peptide design for three-dimensional encapsulation and injectable delivery of cells. *Proceedings of the National Academy of Sciences.* 2007 May 8; 104(19):7791–6.

Hu T, Cui X, Zhu M, Wu M, Tian Y, Yao B, Song W, Niu Z, Huang S, Fu X. 3D-printable supramolecular hydrogels with shear-thinning property: Fabricating strength tunable bioink via dual crosslinking. *Bioactive Materials.* 2020 Dec 1; 5(4):808–18.

Huang H, Ding Y, Sun XS, Nguyen TA. Peptide hydrogelation and cell encapsulation for 3D culture of MCF-7 breast cancer cells. *PLoS One.* 2013 Mar 20; 8(3):e59482.

Jansen LE, Amer LD, Chen EY, Nguyen TV, Saleh LS, Emrick T, Liu WF, Bryant SJ, Peyton SR. Zwitterionic PEG-PC hydrogels modulate the foreign body response in a modulus-dependent manner. *Biomacromolecules*. 2018 Apr 26; 19(7):2880–8.

Ji S, Abaci A, Morrison T, Gramlich WM, Guvendiren M. Novel bioinks from UV-responsive norbornene-functionalized carboxymethyl cellulose macromers. *Bioprinting*. 2020 Jun 1; 18:e00083.

Jiang X, Yang X, Yang B, Zhang L, Lu A. Highly self-healable and injectable cellulose hydrogels via rapid hydrazone linkage for drug delivery and 3D cell culture. *Carbohydrate Polymers*. 2021; 273:118547.

Jiang Z, Xia B, McBride R, Oakey J. A microfluidic-based cell encapsulation platform to achieve high long-term cell viability in photopolymerized PEGNB hydrogel microspheres. *Journal of Materials Chemistry B*. 2017; 5(1):173–80.

Khademhosseini A, Eng G, Yeh J, Fukuda J, Blumling J III, Langer R, Burdick JA. Micromolding of photocrosslinkable hyaluronic acid for cell encapsulation and entrapment. *Journal of Biomedical Materials Research Part A: An Official Journal of the Society for Biomaterials, The Japanese Society for Biomaterials, and the Australian Society for Biomaterials and the Korean Society for Biomaterials*. 2006 Dec 1; 79(3):522–32.

Kim E, Seok JM, Bae SB, Park SA, Park WH. Silk fibroin enhances cytocompatibilty and dimensional stability of alginate hydrogels for light-based three-dimensional bioprinting. *Biomacromolecules*. 2021 Apr 12; 22(5):1921–31.

Kim MH, Lee YW, Jung WK, Oh J, Nam SY. Enhanced rheological behaviors of alginate hydrogels with carrageenan for extrusion-based bioprinting. *Journal of the Mechanical Behavior of Biomedical Materials*. 2019 Oct 1; 98:187–94.

Koh WG, Revzin A, Pishko MV. Poly (ethylene glycol) hydrogel microstructures encapsulating living cells. *Langmuir*. 2002 Apr 2; 18(7):2459–62.

Kretsinger JK, Haines LA, Ozbas B, Pochan DJ, Schneider JP. Cytocompatibility of self-assembled β-hairpin peptide hydrogel surfaces. *Biomaterials*. 2005 Sep 1; 26(25):5177–86.

Ku J, Seonwoo H, Park S, Jang KJ, Lee J, Lee M, Lim JW, Kim J, Chung JH. Cell-laden thermosensitive chitosan hydrogel bioinks for 3D bioprinting applications. *Applied Sciences*. 2020 Apr 3; 10(7):2455.

Lanza RP, Kühtreiber WM, Chick WL. Encapsulation technologies. *Tissue Engineering*. 1995 Jun 1; 1(2):181–96.

Lee SY, George JH, Nagel DA, Ye H, Kueberuwa G, Seymour LW. Optogenetic control of iPS cell-derived neurons in 2D and 3D culture systems using channelrhodopsin-2 expression driven by the synapsin-1 and calcium-calmodulin kinase II promoters. *Journal of Tissue Engineering and Regenerative Medicine*. 2019 Mar; 13(3):369–84.

Leslie SK, Cohen DJ, Sedlaczek J, Pinsker EJ, Boyan BD, Schwartz Z. Controlled release of rat adipose-derived stem cells from alginate microbeads. *Biomaterials*. 2013 Nov 1; 34(33):8172–84.

Lewicki J, Bergman J, Kerins C, Hermanson O. Optimization of 3D bioprinting of human neuroblastoma cells using sodium alginate hydrogel. *Bioprinting*. 2019 Dec 1; 16:e00053.

Lewińska D, Chwojnowski A, Wojciechowski C, Kupikowska-Stobba B, Grzeczkowicz M, Weryński A. Electrostatic droplet generator with 3-coaxial-nozzle head for microencapsulation of living cells in hydrogel covered by synthetic polymer membranes. *Separation Science and Technology*. 2012 Feb 1; 47(3):463–9.

Li F, Truong VX, Thissen H, Frith JE, Forsythe JS. Microfluidic encapsulation of human mesenchymal stem cells for articular cartilage tissue regeneration. *ACS Applied Materials & Interfaces*. 2017 Mar 15; 9(10):8589–601.

Li H, Li N, Zhang H, Zhang Y, Suo H, Wang L, Xu M. Three-dimensional bioprinting of perfusable hierarchical microchannels with alginate and silk fibroin double cross-linked network. *3D Printing and Additive Manufacturing*. 2020 Apr 1; 7(2):78–84.

Li H, Tan YJ, Liu S, Li L. Three-dimensional bioprinting of oppositely charged hydrogels with super strong interface bonding. *ACS Applied Materials & Interfaces*. 2018 Mar 8; 10(13):11164–74.

Li W, Lei X, Feng H, Li B, Kong J, Xing M. Layer-by-layer cell encapsulation for drug delivery: The history, technique basis, and applications. *Pharmaceutics*. 2022 Jan 27; 14(2):297.

Li X, Deng Q, Zhuang T, Lu Y, Liu T, Zhao W, Lin B, Luo Y, Zhang X. 3D bioprinted breast tumor model for structure–activity relationship study. *Bio-Design and Manufacturing*. 2020 Dec; 3(4):361–72.

Li Y, Wang F, Liang H, Tang D, Huang M, Zhao J, Yang X, Liu Y, Shu L, Wang J, He Z. Efficacy of mesenchymal stem cell transplantation therapy for type 1 and type 2 diabetes mellitus: A meta-analysis. *Stem Cell Research & Therapy*. 2021 Dec; 12(1):1.

Ling SD, Geng Y, Chen A, Du Y, Xu J. Enhanced single-cell encapsulation in microfluidic devices: From droplet generation to single-cell analysis. *Biomicrofluidics*. 2020 Nov 22; 14(6):061508.

Ling Y, Rubin J, Deng Y, Huang C, Demirci U, Karp JM, Khademhosseini A. A cell-laden microfluidic hydrogel. *Lab on a Chip*. 2007; 7(6):756–62.

Liu J, Gao D, Mao S, Lin JM. A microfluidic photolithography for controlled encapsulation of single cells inside hydrogel microstructures. *Science China Chemistry*. 2012 Apr; 55(4):494–501.

Liu M, Zhou Z, Chai Y, Zhang S, Wu X, Huang S, Su J, Jiang J. Synthesis of cell composite alginate microfibers by microfluidics with the application potential of small diameter vascular grafts. *Biofabrication*. 2017 Jun 7; 9(2):025030.

Liu T, Wang Y, Zhong W, Li B, Mequanint K, Luo G, Xing M. Biomedical applications of layer-by-layer self-assembly for cell encapsulation: Current status and future perspectives. *Advanced Healthcare Materials*. 2019 Jan; 8(1):1800939.

Liu VA, Bhatia SN. Three-dimensional photopatterning of hydrogels containing living cells. *Biomedical Microdevices*. 2002 Dec; 4(4):257–66.

López-Marcial GR, Zeng AY, Osuna C, Dennis J, García JM, O'Connell GD. Agarose-based hydrogels as suitable bioprinting materials for tissue engineering. *ACS Biomaterials Science & Engineering*. 2018 Sep 14; 4(10):3610–6.

Ludwig B, Reichel A, Steffen A, Zimerman B, Schally AV, Block NL, Colton CK, Ludwig S, Kersting S, Bonifacio E, Solimena M. Transplantation of human islets without immunosuppression. *Proceedings of the National Academy of Sciences*. 2013 Nov 19; 10(47):19054–8.

Magli S, Rossi GB, Risi G, Bertini S, Cosentino C, Crippa L, Ballarini E, Cavaletti G, Piazza L, Masseroni E, Nicotra F. Design and synthesis of chitosan—gelatin hybrid hydrogels for 3D printable in vitro models. *Frontiers in Chemistry*. 2020 Jul 14; 8:524.

Mancuso P, Raman S, Glynn A, Barry F, Murphy JM. Mesenchymal stem cell therapy for osteoarthritis: The critical role of the cell secretome. *Frontiers in Bioengineering and Biotechnology*. 2019 Jan 29; 7:9.

Marin E, Tapeinos C, Sarasua JR, Larranaga A. Exploiting the layer-by-layer nanoarchitectonics for the fabrication of polymer capsules: A toolbox to provide multifunctional properties to target complex pathologies. *Advances in Colloid and Interface Science*. 2022 Apr 17:102680.

Maturavongsadit P, Narayanan LK, Chansoria P, Shirwaiker R, Benhabbour SR. Cell-laden nanocellulose/chitosan-based bioinks for 3D bioprinting and enhanced osteogenic cell differentiation. *ACS Applied Bio Materials*. 2021 Feb 17; 4(3):2342–53.

Montheil T, Maumus M, Valot L, Lebrun A, Martinez J, Amblard M, Noël D, Mehdi A, Subra G. Inorganic sol–gel polymerization for hydrogel bioprinting. *ACS Omega*. 2020 Feb 6; 5(6):2640–7.

Montoro SR, de Fátima Medeiros S, Alves GM. Nanostructured hydrogels. In *Nanostructured Polymer Blends* (pp. 325–355). William Andrew Publishing, 2014.

Moyer HR, Kinney RC, Singh KA, Williams JK, Schwartz Z, Boyan BD. Alginate microencapsulation technology for the percutaneous delivery of adipose-derived stem cells. *Annals of Plastic Surgery*. 2010 Nov 1; 65(5):497–503.

Ni T, Liu M, Zhang Y, Cao Y, Pei R. 3D bioprinting of bone marrow mesenchymal stem cell-laden silk fibroin double network scaffolds for cartilage tissue repair. *Bioconjugate Chemistry*. 2020 Jul 9; 31(8):1938–47.

Orive G, Hernandez RM, Gascon AR, Igartua M, Pedraz JL. Survival of different cell lines in alginate-agarose microcapsules. *European Journal of Pharmaceutical Sciences*. 2003 Jan 1; 18(1):23–30.

Ou LC, Zhong S, Ou JS, Tian JW. Application of targeted therapy strategies with nanomedicine delivery for atherosclerosis. *Acta Pharmacologica Sinica*. 2021 Jan; 42(1):10–7.

Panda P, Ali S, Lo E, Chung BG, Hatton TA, Khademhosseini A, Doyle PS. Stop-flow lithography to generate cell-laden microgel particles. *Lab on a Chip*. 2008; 8(7):1056–61.

Pandolfi V, Pereira U, Dufresne M, Legallais C. Alginate-based cell microencapsulation for tissue engineering and regenerative medicine. *Current Pharmaceutical Design*. 2017 Jul 1; 23(26):3833–44.

Pirela MR, Rojas V, Pérez EP, Velásquez CL. Cell encapsulation using chitosan: Chemical aspects and applications. *Avances en Química*. 2021; 16(3):89–103.

Pisani S, Dorati R, Scocozza F, Mariotti C, Chiesa E, Bruni G, Genta I, Auricchio F, Conti M, Conti B. Preliminary investigation on a new natural based poly (gamma-glutamic acid)/Chitosan bioink. *Journal of Biomedical Materials Research Part B: Applied Biomaterials*. 2020 Oct; 108(7):2718–32.

Pokrajac L, Abbas A, Chrzanowski W, Dias GM, Eggleton BJ, Maguire S, Maine E, Malloy T, Nathwani J, Nazar L, Sips A. Nanotechnology for a sustainable future: Addressing global challenges with the international network4sustainable nanotechnology. 2021 Dec.

Puertas-Bartolomé M, Włodarczyk-Biegun MK, Del Campo A, Vázquez-Lasa B, San Román J. 3D printing of a reactive hydrogel bio-ink using a static mixing tool. *Polymers*. 2020 Aug 31; 12(9):1986.

Rainbow RS, Lysaght MJ. *Biomaterials Science: Immunoisolation* (Third Edition). Elsevier, 2013.

Ronzoni FL, Aliberti F, Scocozza F, Benedetti L, Auricchio F, Sampaolesi M, Cusella G, Redwan IN, Ceccarelli G, Conti M. Myoblast 3D bioprinting to burst in vitro skeletal muscle differentiation. *Journal of Tissue Engineering and Regenerative Medicine.* 2022 May; 16(5):484–95.

Rossow T, Heyman JA, Ehrlicher AJ, Langhoff A, Weitz DA, Haag R, Seiffert S. Controlled synthesis of cell-laden microgels by radical-free gelation in droplet microfluidics. *Journal of the American Chemical Society.* 2012 Mar 14; 134(10):4983–9.

Sanandiya ND, Vasudevan J, Das R, Lim CT, Fernandez JG. Stimuli-responsive injectable cellulose thixogel for cell encapsulation. *International Journal of Biological Macromolecules.* 2019 Jun 1; 130:1009–17.

Sarker MD, Naghieh S, McInnes AD, Ning L, Schreyer DJ, Chen X. Bio-fabrication of peptide-modified alginate scaffolds: Printability, mechanical stability and neurite outgrowth assessments. *Bioprinting.* 2019 Jun 1; 14:e00045.

Selimović Š, Oh J, Bae H, Dokmeci M, Khademhosseini A. Microscale strategies for generating cell-encapsulating hydrogels. *Polymers.* 2012 Sep 5; 4(3):1554–79.

Shao F, Yu L, Zhang Y, An C, Zhang H, Zhang Y, Xiong Y, Wang H. Microfluidic encapsulation of single cells by alginate microgels using a trigger-gellified strategy. *Frontiers in Bioengineering and Biotechnology.* 2020 Oct 14; 8:583065.

Shen Y, Tang H, Huang X, Hang R, Zhang X, Wang Y, Yao X. DLP printing photocurable chitosan to build bio-constructs for tissue engineering. *Carbohydrate Polymers.* 2020 May 1; 235:115970.

Soon-Shiong P, Feldman E, Nelson R, Heintz R, Yao Q, Yao Z, Zheng T, Merideth N, Skjak-Braek G, Espevik T. Long-term reversal of diabetes by the injection of immunoprotected islets. *Proceedings of the National Academy of Sciences.* 1993 Jun 15; 90(12):5843–7.

Soon-Shiong P, Feldman E, Nelson R, Komtebedde J, Smidsrod O, Skjak-Braek G, Espevik T, Heintz R, Lee M. Successful reversal of spontaneous diabetes in dogs by intraperitoneal microencapsulated islets. *Transplantation.* 1992 Nov 1; 54(5):769–74.

Soon-Shiong P, Heintz RE, Merideth N, Yao QX, Yao ZH, Zheng TI, Murphy M, Moloney MK, Schmehl M, Harris M. Insulin independence in a type 1 diabetic patient after encapsulated islet transplantation. *Lancet (London, England).* 1994 Apr 1; 343(8903):950–1.

Subramanian V, Semenzin E, Hristozov D, Marcomini A, Linkov I. Sustainable nanotechnology: Defining, measuring and teaching. *Nano Today.* 2014 Feb 1; 9(1):6–9.

Tekin H, Tsinman T, Sanchez JG, Jones BJ, Camci-Unal G, Nichol JW, Langer R, Khademhosseini A. Responsive micromolds for sequential patterning of hydrogel microstructures. *Journal of the American Chemical Society.* 2011 Aug 24; 133(33):12944–7.

Tonda-Turo C, Carmagnola I, Chiappone A, Feng Z, Ciardelli G, Hakkarainen M, Sangermano M. Photocurable chitosan as bioink for cellularized therapies towards personalized scaffold architecture. *Bioprinting.* 2020 Jun 1; 18:e00082.

Trachsel L, Johnbosco C, Lang T, Benetti EM, Zenobi-Wong M. Double-network hydrogels including enzymatically crosslinked poly-(2-alkyl-2-oxazoline) s for 3D bioprinting of cartilage-engineering constructs. *Biomacromolecules.* 2019 Nov 12; 20(12):4502–11.

Tuch BE, Keogh GW, Williams LJ, Wu W, Foster JL, Vaithilingam V, Philips R. Safety and viability of microencapsulated human islets transplanted into diabetic humans. *Diabetes Care.* 2009 Oct 1; 32(10):1887–9.

United Nations (n.d.) Do you know all 17 SDGs? Retrieved September 30, 2022, from https://sdgs.un.org/goals.

Veerabadran NG, Goli PL, Stewart-Clark SS, Lvov YM, Mills DK. Nanoencapsulation of stem cells within polyelectrolyte multilayer shells. *Macromolecular Bioscience.* 2007 Jul 9; 7(7):877–82.

Vossoughi A, Matthew HW. Encapsulation of mesenchymal stem cells in glycosaminoglycans-chitosan polyelectrolyte microcapsules using electrospraying technique: Investigating capsule morphology and cell viability. *Bioengineering & Translational Medicine.* 2018 Sep; 3(3):265–74.

Wang T, Adcock J, Kühtreiber W, Qiang D, Salleng KJ, Trenary I, Williams P. Successful allotransplantation of encapsulated islets in pancreatectomized canines for diabetic management without the use of immunosuppression. *Transplantation.* 2008 Feb 15; 85(3):331–7.

Wang X, Maxwell KG, Wang K, Bowers DT, Flanders JA, Liu W, Wang LH, Liu Q, Liu C, Naji A, Wang Y. A nanofibrous encapsulation device for safe delivery of insulin-producing cells to treat type 1 diabetes. *Science Translational Medicine.* 2021 Jun 2; 13(596):eabb4601.

Wei Z, Zhao J, Chen YM, Zhang P, Zhang Q. Self-healing polysaccharide-based hydrogels as injectable carriers for neural stem cells. *Scientific Reports.* 2016 Nov 29; 6(1):1–2.

Wilkens CA, Rivet CJ, Akentjew TL, Alverio J, Khoury M, Acevedo JP. Layer-by-layer approach for a uniformed fabrication of a cell patterned vessel-like construct. *Biofabrication.* 2016 Dec 1; 9(1):015001.

Wu X, Jiang J, Gu Z, Zhang J, Chen Y, Liu X. Mesenchymal stromal cell therapies: Immunomodulatory properties and clinical progress. *Stem Cell Research & Therapy.* 2020 Dec; 11(1):1–6.

Wu Z, Xie S, Kang Y, Shan X, Li Q, Cai Z. Biocompatibility evaluation of a 3D-bioprinted alginate-GelMA-bacteria nanocellulose (BNC) scaffold laden with oriented-growth RSC96 cells. *Materials Science and Engineering: C.* 2021 Oct 1; 129:112393.

Xu J, Qi G, Sui C, Wang W, Sun X. 3D h9e peptide hydrogel: An advanced three-dimensional cell culture system for anticancer prescreening of chemopreventive phenolic agents. *Toxicology in Vitro.* 2019 Dec 1; 1:104599.

Xu L, Varkey M, Jorgensen A, Ju J, Jin Q, Park JH, Fu Y, Zhang G, Ke D, Zhao W, Hou R. Bioprinting small diameter blood vessel constructs with an endothelial and smooth muscle cell bilayer in a single step. *Biofabrication.* 2020 Jul 28; 12(4):045012.

Yamada Y, Patel NL, Kalen JD, Schneider JP. Design of a peptide-based electronegative hydrogel for the direct encapsulation, 3D culturing, in vivo syringe-based delivery, and long-term tissue engraftment of cells. *ACS Applied Materials & Interfaces.* 2019 Aug 26; 11(38):34688–97.

Yang X, Lu Z, Wu H, Li W, Zheng L, Zhao J. Collagen-alginate as bioink for three-dimensional (3D) cell printing based cartilage tissue engineering. *Materials Science and Engineering: C.* 2018 Feb 1; 83:195–201.

Yao R, Zhang R, Wang X. Design and evaluation of a cell microencapsulating device for cell assembly technology. *Journal of Bioactive and Compatible Polymers.* 2009 May; 24(1 Suppl):48–62.

Yu J, Du KT, Fang Q, Gu Y, Mihardja SS, Sievers RE, Wu JC, Lee RJ. The use of human mesenchymal stem cells encapsulated in RGD modified alginate microspheres in the repair of myocardial infarction in the rat. *Biomaterials.* 2010 Sep 1; 31(27):7012–20.

Zhang Q, Gonelle-Gispert C, Li Y, Geng Z, Gerber S, Wang Y, Bühler LH. Islet encapsulation: New developments for the treatment of type 1 diabetes. *Frontiers in Immunology.* 2022 Apr 14:1540.

Zhao X, Liu S, Yildirimer L, Zhao H, Ding R, Wang H, Cui W, Weitz D. Injectable stem cell-laden photocrosslinkable microspheres fabricated using microfluidics for rapid generation of osteogenic tissue constructs. *Advanced Functional Materials.* 2016 May; 26(17):2809–19.

Zhou T, Yuan Z, Weng J, Pei D, Du X, He C, Lai P. Challenges and advances in clinical applications of mesenchymal stromal cells. *Journal of Hematology & Oncology.* 2021 Dec; 14(1):1–24.

Zhou X, Nowicki M, Sun H, Hann SY, Cui H, Esworthy T, Lee JD, Plesniak M, Zhang LG. 3D bioprinting-tunable small-diameter blood vessels with biomimetic biphasic cell layers. *ACS Applied Materials & Interfaces.* 2020 Oct 2; 12(41):45904–15.

8 Wound Healing and Infection Control Using Nanomaterials

Susanta Kumar Behera, Md. Imran Khan, Suraj Kumar Tripathy, Cecilia Stalsby Lundborg, and Amrita Mishra

CONTENTS

DOI: 10.1201/9781003181422-10

8.1 INTRODUCTION

Infections caused by invading microorganisms during skin injury are a major problem associated with the wound healing process [1]. Normal and intact skin act as a primary defense to control the spread of microbial populations which reside on the outer surface of the skin and also hinder the colonization of essential underlying tissues which can be infected by invading pathogens [2]. Damage to the skin exposes subcutaneous tissues which present a warm, moist, and nutrition-rich environment that promotes the growth, proliferation, and colonization of microbes [3]. The presence of microbial populations in damaged skin can drastically influence the immune system's ability to heal the wound [4]. In particular, bacteria create an environment, or biofilm, through colonization in the open wound that often leads to the formation of a chronic wound, hence reducing tissue regeneration processes and delaying wound healing [5]. For instance, *Staphylococcus aureus*, a gram-positive bacterium, is reported to be the most common cause of infection in all types of wounds (i.e. burns, cuts, diabetic ulcers, etc.) [6]. Thus, antibiotics are the current treatment for bacterial infections due to their cost-effectiveness and controlling effect [7]

However, in recent years the extensive use of antibacterial agents has led to the generation of antimicrobial-resistant (AMR) bacterial strains. Currently, multidrug-resistant (MDR) bacterial strains are becoming a serious concern [8, 9]. Current antibiotics target and inhibit cell wall synthesis processes and disrupt the translation and replication of DNA within bacterial cells [10]. However, MDR strains are formed through different mechanisms, such as the alteration in the expression of the antibiotics degrading enzymes (β-lactamases and aminoglycosides), and alteration of the cell wall and internal cell contents (i.e. ribosomes), which makes them resistant to both vancomycin and tetracycline. Resistance to antibiotics can also result from changes in the expression of the efflux pumps [11]. Hence, MDR is becoming a serious challenge for medicine, and it is important to explore novel approaches through new technologies. The nanotechnology-based approach, which uses the unique size and shape of nanoparticles, has been found to be highly effective against many microbials [12]. For instance, the literature reports that nanoparticles (NPs) have extremely good antimicrobial activity against many bacteria, fungi, yeasts, etc. [13]. Importantly, nanoparticle-based treatments are not expected to promote the microbial-resistant properties of current antibiotics [14].

Nanomaterials can be carbon-based, inorganic-based, organic-based, or a combination of these materials to form composites. Nanomaterials can be synthesized through chemical, physical, and biological routes to produce a wide variety of shapes, with at least one dimension ranging in size from 1 to 100 nm [15]. Numerous studies have shown that many types of nanoparticles (NP) have good antimicrobial properties. For instance, both metal and metal oxide NPs have demonstrated good antibacterial activity toward both gram-positive and gram-negative bacteria. In particular, the growth of *Staphylococcus aureus* has been reported to decrease in the presence of metal oxide NPs like magnesium oxide (MgO) and zinc oxide (ZnO) [16, 17]. The best-known metal with antimicrobial properties is silver (Ag), and Ag NPs have extremely good antibacterial activity against *Escherichia coli*, *Klebsiella pneumoniae*, *Pseudomonas aeruginosa*, and *Staphylococcus aureus* [18]. To date, the antibacterial mechanisms of NPs have not been fully explored or extensively explained. However, a number of probable mechanisms have been suggested. These probable mechanisms include 1) the induction of oxidative stress in the bacteria; 2) the release of metal ions that interact with internal cell components like DNA; 3) reducing cell wall permeability by charging the surface charge of the cell; and 4) inducing intracellular reactive oxygen species (ROS) which impair antioxidant defense systems and damage the cell membrane leading to cell death. Most antimicrobial studies report that after NP treatment there is damage to the cell membrane and increased membrane permeability. This is followed by the induction of ROS and the interaction of metal ions with DNA and proteins [14].

In recent years there has been considerable concern expressed about the toxicity of NPs produced using conventional physical and chemical methods because many of these methods involve the use of toxic chemicals and solvents. Because of this concern, recent interest has focused on using biological methods to synthesize nanomaterials. Biosynthesis is categorized as a green synthesis method since it uses biomass or secondary metabolites of bacteria, fungi, algae, and plant extracts (leaf, fruits, stem, and roots) to produce nanomaterials. Hence, this new and promising field has the potential to deliver biocompatible nanomaterials with unique antimicrobial properties [19]. Importantly, biogenic NPs are produced from biological compounds, not toxic chemical compounds and complex energy-intensive physical and chemical processes. Biogenic NPs are currently used in the biomedical field for applications that require antibacterial, antifungal, and mosquitocidal properties. Biogenic NPs have also been used in applications such as photothermal therapy, magnetic drug delivery, various types of cancer treatments, cosmetic and medical tools, biosensors, cell labeling, and gene delivery. Currently synthesized biogenic NPs with antimicrobial properties include Ag, Au, ZO, MgO, and TiO_2 [20]. In addition, biogenic NPs have been used in bioremediation to eliminate contaminants, in water treatment processes to degrade toxic compounds and microbes, and to assist in the production of clean energy [21].

Nanomaterial-based strategies for wound healing are an attractive alternative since NP-based treatments are not expected to promote microbial-resistant strains currently faced by conventional antibiotics [22]. The skin acts as a barrier to prevent the invasion of microorganisms into the body [23]. However, when the skin is damaged, microorganisms can invade and affect the body. Since ancient times, humans have used dressings to inhibit bleeding and prevent microbial infection [24]. To complete these functions, the dressing must cover the wound immediately after the injury has occurred and then promote wound healing. There are three types of wound dressings currently in use: biological, synthetic, and bio-synthetic [25]. The two types of biological dressing currently used in clinical procedures are AlloSkin and pigskin. Regrettably, both of these dressings suffer from a number of drawbacks, such as having a limited supply of quality material, higher antigenicity, poor adhesiveness, and a high risk of cross-contamination [26]. On the other hand, dressings prepared from synthetic materials offer higher stability, longer shelf life, and minimal risk of cross-contamination and inflammation. Synthetic dressings also prevent the spread of microbes from infected sites and reduce the possibility of community transmission. Combinations of biological and synthetic materials have also been used as a dressing. However, studies and clinical data have shown that all dressings made of these materials have some drawbacks. In general, a dressing must have the following properties: 1) provide an effective barrier against invading microorganisms; 2) the ability to promote wound healing; 3) maintain a moist surrounding; 4) allow the exchange of gases; 5) protect the wound during the healing process; 6) should be biocompatible, non-toxic, and non-allergenic toward mammalian cells; and 7) it should have appropriate adherent so that it can be easily removed without any further trauma to the wound surface [27].

8.2 CLASSIFICATION OF SKIN WOUNDS

Skin is the largest organ of the human body and is made of soft tissues that account for around 15% of total body weight [28]. The main function of the skin is to protect the body from the external environment and prevent external invasion of physical, chemical, and microbial agencies. The skin is also involved in the regulation of metabolic processes and body temperature [29]. The skin is also prone to damage from external stimuli like mechanical impact, burns from thermal sources (i.e. fire and sun exposure), and chemical materials. Each of these stimuli can result in different kinds of wounds. The Wound Healing Society has termed a wound as the result of a change in anatomical structure and function of skin after a disturbance [30]. A wound are categorized as either acute or chronic depending on the type of injury and the period it takes to heal. The following two sections explain the differences between acute and chronic wounds in more detail.

8.2.1 Acute Wounds

An acute wound is the result of an injury but heals in a relatively short period of time with limited scarring [31]. Figure 8.1 summarizes several acute wound types experienced by humans. All acute

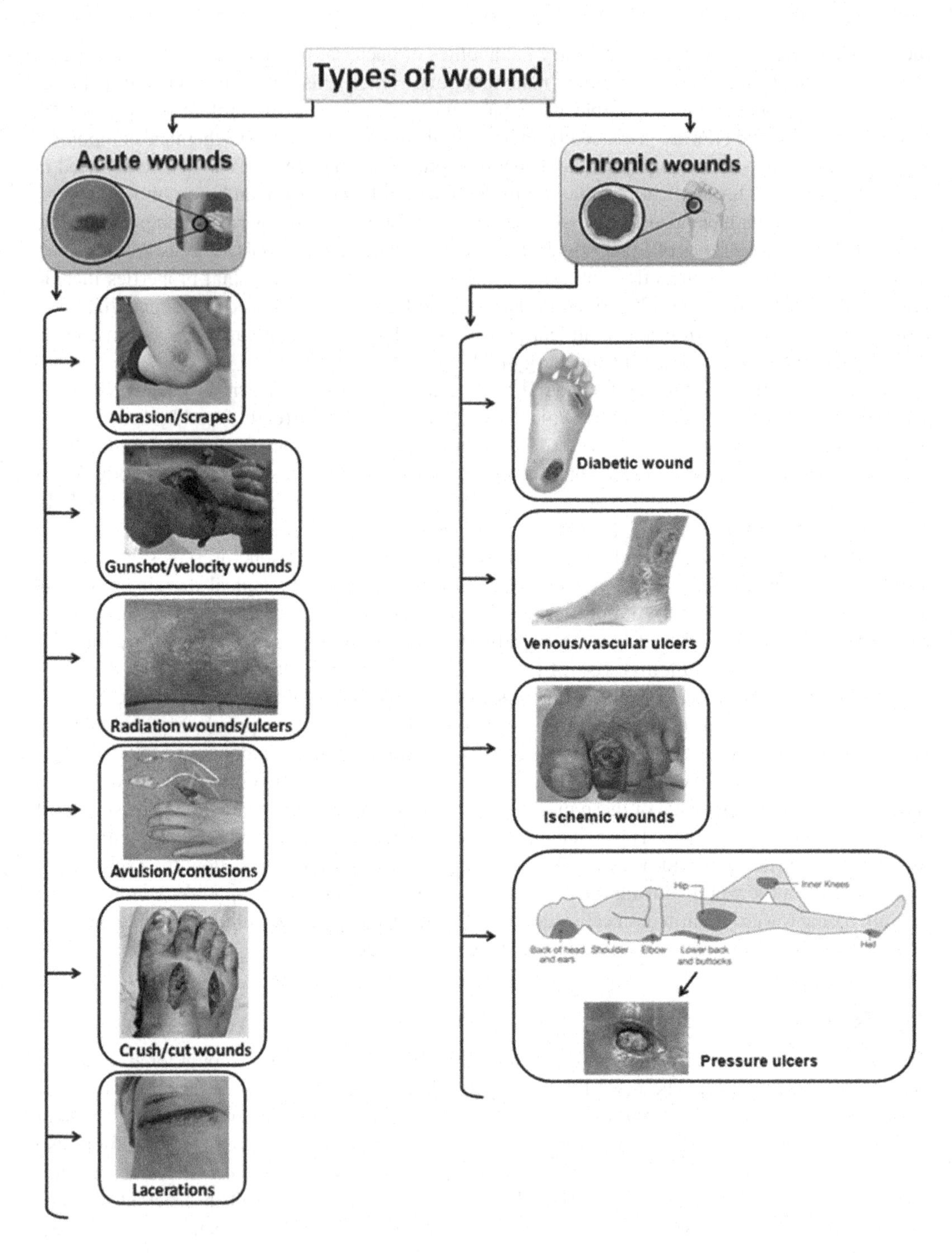

FIGURE 8.1 Schematic presentation of both acute and chronic wounds showing the level of tissue damage resulting from s due to different types of injuries.

wounds result from external environmental factors, for instance, mechanical impacts between hard surfaces where the skin is ripped or scratched. Penetration wounds are caused by knives, gunshot, or even surgical instruments during surgical procedures [32]. Burns are also acute wounds and result from exposure to heat sources such as fire, electricity, or sunburn. Chemical burns can result from a variety of materials in the environment, ranging from domestic cleaning materials to industrial chemicals and solvents. In all cases, acute wounds need immediate treatment from a medical specialist [33].

8.2.2 Chronic Wounds

A chronic wound is the result of an injury that fails to heal within 12 weeks. Healing fails due to disruptions in the normal tissue repair processes. The disruption results from factors such as persistent infections, diabetes, malignancies, and unfavorable primary treatments [34]. Typical chronic wounds include decubitus ulcers, occasionally known as pressure ulcers, bed sores, pressure sores [35], or leg ulcers. In both cases, wound healing is very slow, and even after a year of treatment, the wound may still persist. Typical chronic wound types are presented in Figure 8.1. In general, wounds are also classified by the different skin layers and areas affected by the trauma [36]. For instance, superficial wounds are restricted to the epidermal skin stratum, which includes the epidermis and deeper dermal layers that contain blood vessels, sweat glands, and hair follicles. Full-thickness wounds, on the other hand, involve trauma to deeper skin tissues [37]. Importantly, both acute and chronic wounds are difficult to heal due to the complex nature of wounds and external influences, for instance, due to loss of the outer protective layer (including skin, hair, and glands) and microbial infection. In both cases, there is usually a loss of tissue, impairment of circulation or tissue death, and occurrence of diseases [38].

8.3 WOUND HEALING PHASES

Wounds occur regularly throughout the lifespan of a human. The human body has evolved mechanisms to regenerate damaged skin tissues through definite healing processes. During the wound healing progress, various cellular compounds and growth factors perform mutually to replace lost tissues and restore the integrity of damaged tissues [39, 40]. During this process, there are four overlapping stages that are responsible for the various biochemical and cellular processes [40]. The process stages include hemostasis, inflammation and defense, migration, proliferation, and maturation. The stages are schematically presented in Figure 8.2. Importantly, because wound healing occurs at both the cellular and molecular levels, it is helpful to understand the process, as this understanding will assist in developing new wound dressing materials and wound management protocols [41].

8.3.1 Hemostasis

After encountering an injury, the initial phase of healing begins with bleeding to wash out microbes and antigens from the wound site. During this period the body stimulates the emergency repair system that promotes blood clotting which in turn creates a barrier to prevent infection and prevent further bleeding. During the clotting mechanism, fibrinogen promotes the coagulation of the platelets and blood (without cells) to form a fibrin matrix which creates the clot. Also during this process, both collagen and platelets interact with each other to further solidify the clot. Once dried, the clot forms a scab which strengthens the injured tissues in the wound area, promotes the healing process, and provides a protective barrier from the surrounding environment [43].

8.3.2 Inflammation or Defense Response

The inflammation phase, or defense response, occurs after the hemostasis phase and focuses on destroying and removing bacteria and other debris. The inflammatory phase usually starts within a few minutes and remains active for up to three days. During this period the white blood cells (neutrophils) kill bacteria and remove the debris. Also during this period, histamine and serotonin are released from protein-rich cells and widen blood vessels so that macrophages (phagocytes) can enter the wound to clear the debris by engulfing dead cells (necrotic tissue). Necrotic cells appear as

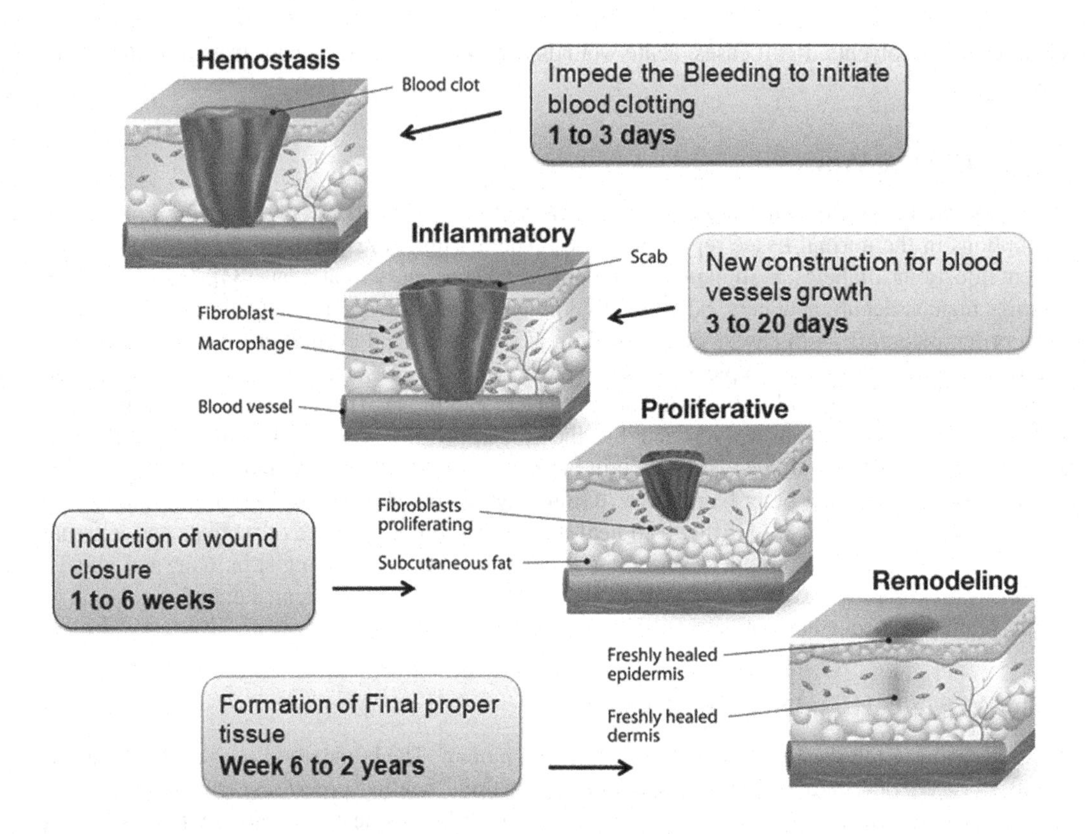

FIGURE 8.2 Schematic representation of the various phases involved with of wound healing, their purpose, and timeline. Image of wound healing phases has been taken from Shield Health Care [42].

a yellow-colored fluid due to enzymatic activity. Also during this period, macrophages secrete proteins and growth factors, which attract migrating cells to the wound. The inflammation phase typically lasts four to six days and is frequently accompanied by edema, erythema, heat, and pain [42].

8.3.3 Proliferation

The proliferation phase occurs immediately after the arrival of migrating cells to the wound area. The phase can take up to 24 days depending on the extent of the wound and occurs in three distinct stages. The stages comprise 1) filling the wound; 2) contraction of the wound margins; and 3) covering the wound through epithelialization. During the first stage, granulation tissue forms and appears shiny and red in color due to the presence of capillary and lymphatic vessels. The optimal level of new granulation tissue and blood vessels typically occurs after around five days. Importantly, the optimal levels of blood vessels promote the synthesis of collagen by fibroblast cells in the wound region. The presence of collagen in the wound strengthens and maintains the appearance of the skin. During the second stage, the margins of the wound migrate toward the center of the wound tissues. Also during this period the thickness of the epithelial layer over the wound increases. During the third stage, fibroblast proliferation rates continue to increase; and higher levels of collagen are produced until a strong structural epithelium layer covers the wound. Generally, this usually takes around two weeks to achieve [43].

8.3.4 Remodeling or Maturation

In the final phase of the healing process, remodeling or maturation takes place. During this period, connective tissues form, and skin flexibility and strength slowly improve while scars slowly break down. Collagen fibers are rearranged, and the tensile strength of the newly formed tissues improves. Overall, wound healing is very complex and is prone to several disruptive factors that can hinder the process. Typical disruptive factors include 1) the extent of the wound; 2) the presence of moisture; 3) levels of microbial infection; and 4) the health of the patient. Because of these factors, wound healing can take anywhere between 21 days to 2 years [43].

8.4 MICROBIAL INFECTIONS AND CURRENT THERAPIES

8.4.1 Microbial Infections and Delays in Wound Healing

Bacterial infection is a serious problem that occurs during wound healing. The growth of colonizing microbes in wounds not only increases infection rates but can also prolong the healing process and even prevent wound closure [44]. Under normal circumstances, natural skin microbiota act as a protective barrier that prevents harmful microbes from colonizing. However, when harmful microbes colonize near a wound, they form a biofilm and disrupt the healing process, as seen in Figure 8.3. Biofilms consist of large numbers of clustering harmful microorganisms (bacteria/fungi) that form on the skin's surface. The microorganisms also secret materials like exopolysaccharides (EPS), DNA, and other compounds to form their surrounding extracellular matrix (ECM). Biofilm formation is considered to be the main reason for wound treatment failure and the key factor in developing antimicrobial-resistant strains. However, the major cause of wound infection is bacteria like *Staphylococcus aureus* (*S. aureus*), methicillin-resistant *S. aureus* (MRSA), *Streptococcus pyogenes* (*S. pyogenes*), and *Pseudomonas aeruginosa* (*P. aeruginosa*). In addition, some reports have suggested the presence of *Escherichia coli (E. coli)* and *Acinetobacter baumani* in the skin and soft tissues also infects open wounds. Other studies have shown that coagulase-negative staphylococci

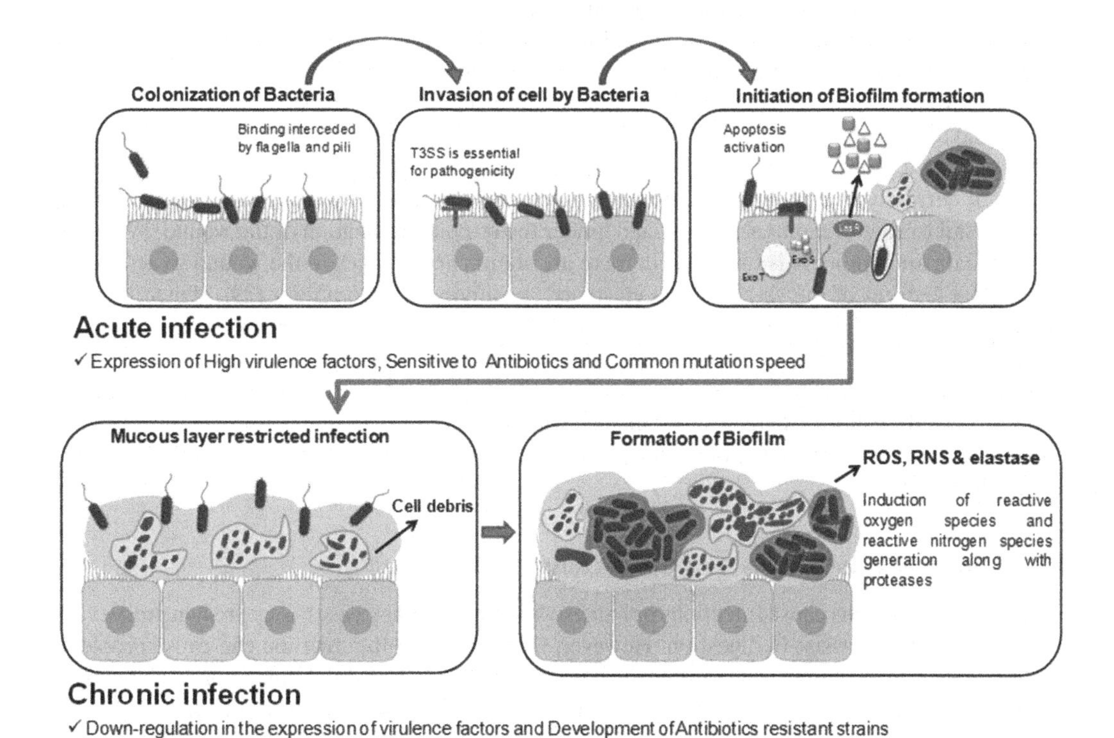

FIGURE 8.3 Schematic presentation of bacterial infection, cross-contamination, and biofilm formation in acute and chronic wounds.

(CNS) such as *Staphylococcus epidermidis* and *Staphylococcus lugdunensis* can cause infections [45, 46]. Moreover, gram-positive bacterial strains like *S. aureus* and MRSA also hinder wound healing. When deeper skin layers get infected by bacteria such as *P. aeruginosa* and *E. coli*, the result is chronic wound infection [47]. Importantly, bacterial infections can delay wound healing and in severe cases threaten the patient's life.

Several microbial species have been observed infecting chronic wounds, and in many cases, these microbes synergistically work together to maintain their environment [48]. For instance, aerobic bacteria consume oxygen for cell division and growth, which creates hypoxic tissue surroundings that promote the growth of anaerobic bacteria. In turn, anaerobic bacteria release short-chain fatty acids that interfere with the activity of host phagocyte cells. As a result, increasing numbers of microbes can colonize the wound area. Furthermore, microorganisms are known to exchange nutrition with each other to sustain growth and promote each other's proliferation within the wound. For instance, studies have reported that both *S. aureus* and *P. aeruginosa* grow simultaneously in co-cultures obtained from chronic wound infections [49]. Opportunistic bacteria like *P. aureuginosa* can readily form biofilms under both biotic and abiotic conditions. It is this ability that has allowed *P. aureuginosa* to develop resistance to both host immune defenses and conventional antibiotic treatments [49]. Crucially, the biofilm, along with the ECM produced by colonizing bacteria, provides protection against host immune responses and from antibacterial treatments [50]. Typical aerobic bacteria strains found in biofilms include *Staphylococcus* spp. and *Enterococcus* spp., while typical anaerobic strains include *Bacteroides* spp., *Prevotella* spp., and *Peptostreptococcus* spp. [48]. In addition, reactive oxygen species (ROS) generated by the biofilm inhibit re-epithelialization of the wound.

The rates of serious microbial infections in acute wounds are generally low, and the healing process begins with fibroblast migration, re-epithelialization, and angiogenesis [51]. However, infections can occur, and colonization of the wound can take place. Infection is signaled by the early development of micro-colonies of microbes and the release of biochemical quorum sensing (QS) components into the wound site. If microbial infection clearance does not take place, the immune system is compromised, and the wound becomes more persistent and severe [52]. Also observed in chronic wounds are increasing levels of inflammation and reduced oxygen levels that hinder the growth and proliferation of fibroblast cells. Both of these effects delay re-epithelialization processes and impair angiogenesis, thus making the treatment of chronic wounds difficult [30].

Not all currently available wound treatment therapies are clinically successful. Many of these therapies fail to control the loss of fluids and hinder the re-epithelialization of the wound. When the re-epithelialization process is interrupted, there are significant changes in the wound structure and histological features of the skin, such as elasticity, sensitivity, and durability [39]. Hence, treating wounds during each phase of the healing process has its challenges. In recent years, nanotechnology-based approaches have been developed to tackle these challenges. The diverse nature and physiochemical properties of nanomaterials offer many opportunities for developing new and more effective wound treatment therapies. For instance, modifying the size, shape, and electric charge of a nanomaterial can adjust its hydrophobicity and its ability to interact with targeted damaged tissues in the wound [53].

8.4.2 Microbial Infections that Compromise Wound Healing

Not all microorganisms cause infections or harm to humans. In many cases, they are beneficial, such as gut microbes that assist in digestion. However, the location of the microbe can cause problems [35]. For instance, *S. aureus* is a gram-positive bacterium that is found in human noses, comprising around 20% of the population. Under normal circumstances, it does not cause infection, but it can become a detrimental agent when it enters wounds. *S. aureus* is found globally, but infections tend to be more frequent in developing countries, and in many cases determining the source of infection can be difficult [55]. *S. aureus* usually infects patients during hospitalization [54]. The main sources of infection result from surgical procedures (by surgical tools) and pressure ulcers (bedsores) that are caused by extended pressure on the skin [56]. Hospital-acquired infections tend to have higher morbidity and mortality levels, and also increase the cost of treating the wound [57].

Several studies have reported that the most frequently found microorganisms in serious wounds are *S. aureus* and *P. aeruginosa*. Both species are effective colonizers and have recently attained resistance to antibiotics [58, 59]. For instance, methicillin-resistant *S. aureus* (MRSA) and several other antibiotic-resistant strains, including nosocomial infections, can enlarge the wound and hinder the healing process [60, 61]. In addition, *S. aureus* is the primary pathogen associated with diabetic foot osteomyelitis, which is a chronic wound that promotes persistent bone infections [62]. Studies have also shown that microorganisms such as *Streptococcus pneumoniae*, *Haemophilus influenzae*, *Corynebacterium* sp., *Enterococcus faecalis*, *Lactobacillus* sp., influenza virus, and *Candida albicans* are found in infected wounds [63]. Several studies have reported that a synergetic relationship exists between *S. aureus* and pathogens such as *E. faecalis*, influenza virus, *H. influenzae*, and *C. albicans*. Other studies have reported an emulative connection between *S. aureus* and pathogens like *Streptococcus pneumoniae*, *P. aeruginosa*, *Corynebacterium* sp., and *Lactobacillus* sp. Interestingly, the interactions between *S. aureus* and other pathogens tend to change *S. aureus*'s behavioral properties, such as protein expression, pathogenicity, colony-forming efficiency, and susceptibility to antibiotics [54].

8.4.3 Current Wound Therapies and Antibiotic-Resistant Strains

Traditionally, wound dressings have been used to provide a protective barrier against external contamination and to prevent bleeding. Typically these traditional dressings are made of cotton or wool and act as a passive barrier only. However, issues with wound complexity and problems associated with the healing process need to be considered, and new improved dressing systems and different therapeutic approaches are needed [64]. Because of these issues, traditional dressings need to be replaced by technically more advanced wound dressing materials and protocols that not only act as protective barriers but also promote the healing process [65].

In recent years, a variety of wound dressing types such as sponges, films, hydrocolloids, hydrogels, and hydrofiber mats have been produced from either natural or synthetic materials, or a combination of both [49]. Dressings made from hydrocolloids, calcium alginate, and semi-permeable polyurethane are designed to be used in moist environments since these materials enhance re-epithelialization in sterile wounds [66], while antibacterial dressing systems have been functionalized with encapsulated antibacterial agents or antibiotics to control and manage infected wounds [49]. Typical antibiotics used in these types of dressing include tetracycline, quinolones, cephalosporin, and aminoglycosides. These antibiotics kill bacteria through various mechanisms, such as modifying cell nucleic acid and disrupting metabolic activity. This leads to changes in cell wall integrity which in turn disrupts cell division [49]. Consequently, the overprescription of antibiotics has led to antibiotic-resistant strains of bacteria being found in wounds. For instance, around 70% of colonizing bacteria found in chronic wounds have acquired antibiotic resistance to at least one antibiotic [67]. Furthermore, oral administration of antibiotics for chronic wound treatment is also ineffective due to poor blood circulation within the wound, which prevents the antibiotic from reaching its targeted site [68]. Hence, oral treatment is not recommended for chronic wounds, since it has the potential to promote the development of antibiotic-resistant strains [69]. Consequently, the administration of non-antibiotic antimicrobials must be used to prevent further antibiotic-resistant strains from developing [70].

8.5 THERAPEUTIC STRATEGIES FOR WOUND TREATMENT

8.5.1 Wound Therapy Using Natural Compounds

Natural compounds are an attractive alternative to current antibiotics due to their advantageous antimicrobial properties and low tendency to cause resistance. Thus, preventing the growth of antibiotic-resistant bacteria can be achieved by combining traditional wound dressing with natural non-antibiotic materials such as essential oils and honey [71]. Essential oils are secondary metabolites derived from plant sources and found to have anti-oxidative, anti-inflammation, and antibacterial properties. The mechanisms involved with inhibiting or killing bacterial cells begin with the essential oil damaging the phospholipids covering the cell membrane. Once inside the cell, the essential oil disrupts the activity of cellular components and changes the cytoplasm pH. The resulting change in pH causes the cytoplasm to erupt from the cell. Because of these advantageous antibacterial mechanisms, it is unlikely that antibacterial-resistant strains will develop. In addition, essential oil-based therapies can be an effective method for treating multidrug-resistant (MDR) bacteria [72]. Another natural product with the ability to fight bacterial and fungal infections is honey. Honey is also a rich source of antioxidants and has good anti-inflammation properties. When honey is used in the treatment of wounds, it actively promotes granulation and re-epithelialization. Due to its anti-inflammation properties, it helps promote the formation of new blood vessels in the wound area [73]. However, the potential use of both natural compounds as viable alternative therapeutics is compromised due to some detrimental material properties. For example, essential oils are hydrophobic and have a volatile nature. Because of these types of shortcomings, research in recent years has focused on developing nanomaterial-based delivery systems. This type of carrier system has the potential to stabilize the natural compound, facilitate its delivery to targeted sites, and release its payload in a controlled manner [74].

8.5.2 Wound Therapy Using Natural and Synthetic Polymers

In recent years, various kinds of natural and synthetic polymer materials have been explored for potential use in hydrogels-based wound dressings. A selection of representative polymeric materials used in wound dressings is presented in Figure 8.4. Typically, hydrogels are made up of long polymer fibers and numerous hydrophilic groups that promote cross-linking between the fibers. The cross-linking process produces a complex three-dimensional matrix that permits the exchange of gases and the ability to trap and store fluids. Thus, hydrogel-based dressing can overcome the two main drawbacks of traditional cloth-based dressings. First, hydrogels, permit the exchange of gases

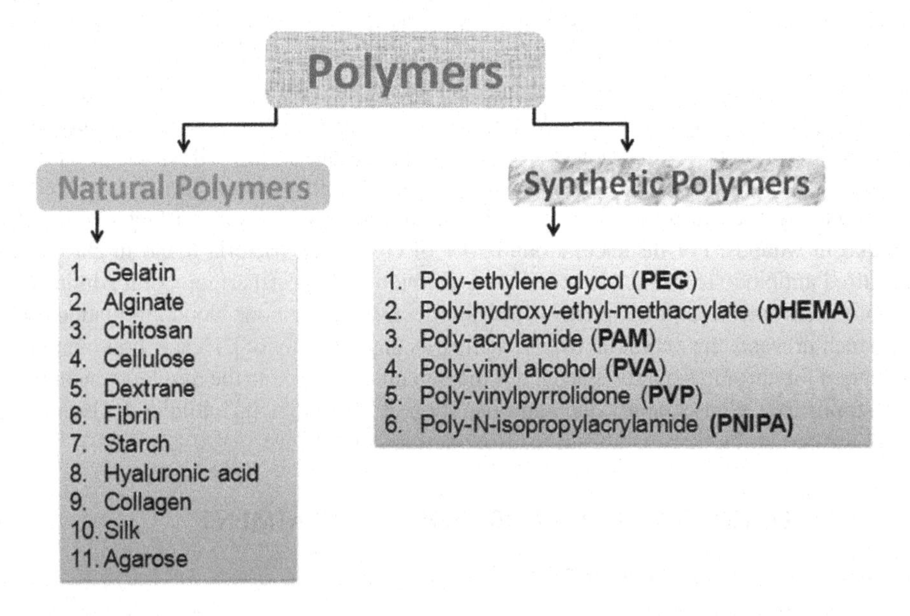

FIGURE 8.4 A selection of representative natural and synthetic polymers used to formulate hydrogel-based wound dressings.

by gas diffusion [75]. Second, hydrogel-based dressings reduce the threat of infection normally associated with traditional cloth-based dressing by absorbing wound fluids [76]. Furthermore, when hydrogel-based dressings are combined with antimicrobial agents, the resulting efficacity can be quite significant. For instance, both silver nitrate and silver sulfadiazine are commonly used to treat burns and chronic wounds, as the release of silver ions displays astonishing antibacterial activity toward the colonizing bacteria in the wound [77].

The antibacterial activity of silver results from the interaction and binding of silver ions with thiol groups in the peptidoglycans present in the cell membrane. The interaction and binding process results in membrane disruption and ultimately leads to cell death [78]. In addition, silver ions interact and inhibit the synthesis of enzymes in cell DNA. Hence, silver-based antibacterials are effective against MDR bacteria and also impede biofilm formation.

Nanoparticle-based therapies using natural and synthetic polymers can also be used to tackle both biofilms and microorganisms resident in wounds. Importantly, properties like size, shape, and surface charge of nanoparticles (NPs) can be tailored to target specific microorganisms and biofilm features. For instance, therapies directed toward biofilm features include 1) water channel penetration through biofilms; 2) targeting high membrane pressure; 3) antimicrobial loading; 4) photothermal effects; 5) photosensitizer loading; and 6) biofilm penetration by contact killing [69, 79].

8.5.3 Wound Therapy Using Growth Factors

Wound healing is complex, and the correct levels of cytokines and growth factors are required at each stage of the healing process. The depletion of growth factors in wounds is the major cause of delayed healing and mostly occurs in chronic wounds [80]. Studies have shown the inclusion of growth factors into wound treatments can promote healing. Currently, REGRANEX gel, a recombinant human platelet-derived growth factor (rHu-PDGF) is the only Food and Drug Authority (FDA, United States of America) approved product available for the treatment of chronic wounds [66, 81]. Studies have reported that PDGF is an important chemo-attractant that enhances the survival of mesenchymal cells [82, 83]. PDGF exists in various dimer forms, such as PDGF-AA, PDGF-AB, and PDGF-BB. The dimers also contain two chains of polypeptides, which are coded as either A or B. Importantly, PDGF is usually released from cells like fibroblasts, keratinocytes, and macrophages during the different stages of wound healing [84].

Another group of growth factors being investigated is the transforming growth factor beta (TGF-β) family. These growth factors are involved in inducing and promoting wound epithelialization [85]. The family consists of three isoforms: TGF-β1, TGF-β2, and TGF-β3. All three isoforms have been shown to play an important role during tissue repair [86]. TGF-β1 has been reported to be effective in the treatment of wounds in animal models [87]. TGF-β3 isoform has been found to assist in repairing the dermal layer and restoring its tensile strength in shorter healing periods [66]. However, implementing the correct levels of growth factors during each stage of the healing process can be difficult. For instance, a TGF-β3-based drug (Avotermin) was found to speed up wound epithelialization and reduce scarring in the early stages of the healing process during clinical trials. However, in the later stages of wound healing, the drug was found to be ineffective [88].

Serine protease-activated protein C (APC) is another important growth factor that acts as an anti-apoptosis and anti-inflammatory to stabilize the endothelial and epithelial tissues. It is also known as protein C and was initially reported in 1960 as an anticoagulation agent [89]. Importantly, APC helps promote the secretion of various important growth factors in the epidermis (keratinocytes), dermis (fibroblasts), and endothelial cells. Typical growth factors secreted include matrix metalloproteinase-2 (MMP-2), vascular endothelial growth factors (VEGF), and membrane cofactor protein-1 (MCP-1) [66]. Overall, APC-mediated growth factors decrease inflammation and induce and promote stable blood vessels (angiogenesis). The growth factors also 1) promote the growth of fibroblasts; 2) promote the growth and migration of keratinocytes; 3) enhance epidermal barrier integrity; and 4) promote epithelialization and tissue granulation [90]. It has also been shown that exogenous APC promotes wound healing, and endogenous APC prevents skin necrosis [91].

From another perspective, vacuum-assisted closure (VAC) therapy is another approach to wound healing. Studies have reported promising results when VAC is applied to wound healing [92]. The therapy involves applying negative pressures to the wound region. Under low pressure, tissue swell is reduced, and the cells experience micromechanical stretching forces that stimulate the cells into secreting transcription growth factors. The procedure also promotes angiogenesis and cell proliferation [92]. Furthermore, VAC can induce the formation of granulation tissues in wounds after infected and damaged tissues are removed by surgery [93].

8.5.4 Wound Dressing Products Currently Available in the Market

Several types of biologically and synthetically engineered skin alternatives have been developed to address dermal injury and promote wound healing. Typical therapies have used autografts (patient's own tissues) and allografts (tissues sourced from another donor) for replacing lost skin tissues and wound healing. To restore dermal integrity, dermal replacements need a similar structure to natural skin tissues, and as a result, many structural matrices are composed of acellular networks [94]. For instance, Alloderm© is an acellular-structured allogenic tissue matrix with regenerative properties. A key feature of this autograft is its use of cadaveric tissue to promote wound healing [95]. Another dermal replacement product is Integra©, which is a skin template that is placed into the pre-cleaned wound. The template is made up of two layers. The first is a bovine-derived collagen matrix rich in chondroitin sulfate, and the second is the silicone cover sheet. The collagen matrix promotes the migration and subsequent proliferation of skin cells in the wound. The adhering silicon sheet covers the wound area during the initial healing period but is replaced by skin grafts after about 14 to 21 days [96]. Similarly, Dermagraft-TC© is a nylon mesh containing fibroblast cells, which is applied onto the wound to accelerate the healing process [97]. Another living cell therapy is done by FDA-approved Apligraf©, which is currently used to treat venous leg ulcers (VLUs) and diabetic foot ulcers (DFUs). Apligraf© uses two cell-type-based approaches. The first approach uses allogeneic neonatal keratinocytes, while the second approach uses collagen and fibroblasts. However, to fully explore the advantages of these various dermal replacement therapies more clinical studies are needed [98].

8.6 PRODUCING NANOMATERIALS FOR WOUND HEALING THERAPIES

8.6.1 Introduction

Nanotechnology is a rapidly developing field that has focused on manipulating and fabricating materials (organic or inorganic) at the atomic and molecular levels. At this scale, materials are designated as nanoparticles (NPs) and range in size from 1 to 100 nm [78]. A unique feature of NPs is their large surface area to volume ratio. In addition, NPs have unique sizes, shapes, and surface chemistries that give them exceptional physiochemical properties and can be categorized on this basis. They can also be categorized on the basis of their composition. For example, there are metal NPs, metal oxide NPs, carbon NPs, ceramic NPs and polymer-based NPs. It is these materials and their unique physiochemical properties that have led to their use in applications like bio-imaging, catalysis, electronics, and new energy-based systems. In medicine, NPs have been used in several therapies that include antimicrobials, anticancer treatments, and wound healing. The following sections discuss the different types of NPs, their synthesis, and their application in wound healing therapies [99].

8.6.2 Nanomaterial Synthesis Methods

There are two approaches for producing nanometer-scale materials. The first is the top-down approach, and the second is the bottom-up approach. The top-down approach basically breaks down bulk materials to form NPs, and the bottom-up approach assembles atoms or molecules to build NPs. The approaches are schematically presented in Figure 8.5 and show the various synthesis methods used to produce NPs. Overall, there are three different synthesis routes used to produce NPs. These include physical, chemical, and biological approaches [100]. The following sections discuss the top-down and bottom-up approaches, the advantages of using the green synthesis approach, and the types of NPs and their application in wound healing therapies.

8.6.2.1 Top-Down Approaches

The top-down approach involves progressively breaking down larger materials to produce smaller materials. Eventually, the breakdown procedure produces the required NPs. Several currently used methods are listed in Figure 8.5. For example, in microfabrication, external energy, force, and a tool are used to break, cut, or mill down a bulk material to produce the desired nanomaterial.

8.6.2.2 Bottom-Up Approaches

The bottom-up approach is based on assembling atoms and molecules to build NPs. Using this approach, NPs can be synthesized to form specific sizes and shapes with a pure homogenous crystal

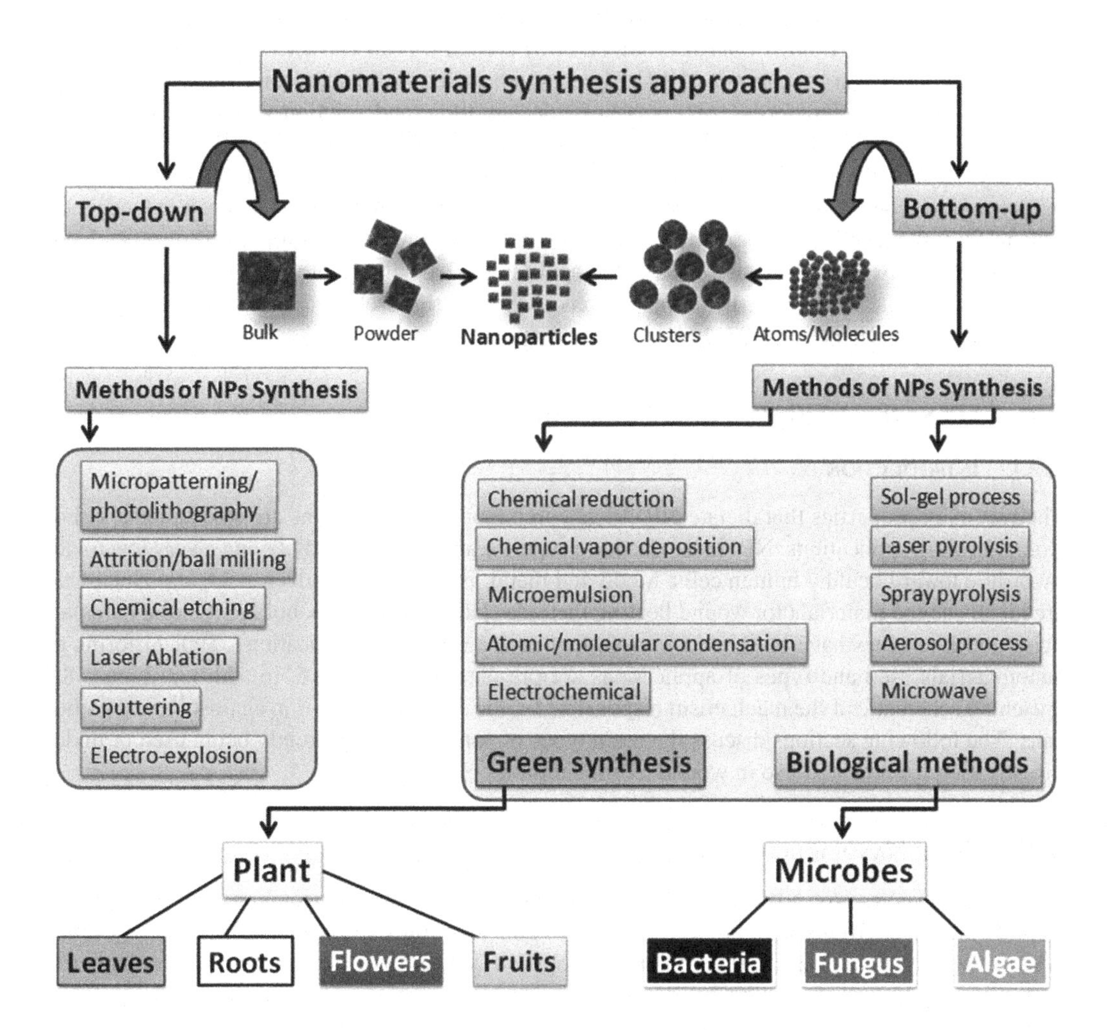

FIGURE 8.5 Schematic of the top-down and bottom-up approaches, with each approach showing the different synthesis methods used to produce nanomaterials.

structure. Typical methods that are routinely used to generate NPs from the bottom up are listed in Figure 8.5.

8.6.3 Advantages of Using Green Synthesis Methods

Biological synthesis, or green synthesis, has emerged as an alternative method for producing NPs. Green synthesis methods are considered to be dependable, reproducible, sustainable, and eco-friendly. Importantly, green synthesis methods do not use toxic chemicals and solvents during the synthesis process. Thus, NPs are not coated with harmful chemicals that are difficult to remove. Surface contamination is a serious issue if the NPs are expected to be used in applications where biocompatibility is important. Several biological entities, including algae, bacteria, fungi, and plant extracts, have been shown to produce NPs. However, NP formation rates vary between the entities. For instance, NP synthesis by microorganisms is slower when compared to plant extracts. In addition, large-scale production of different types of metal and metal oxide NPs by plant extracts is more straightforward when compared to microbial synthesis.

Recently, attention has focused on scaling up the production of NPs synthesized by plant extracts [100]. Typical sources of medicinal and non-medicinal plant extracts that have been investigated include 1) green tea leaf; 2) *Terminalia chebula* fruit extract; 3) *Oolong* and black tea; 4) banana peel and leaf extract from *Colocasia esculenta*; 5) sorghum bran extract; 6) several *eucalyptus* species; and 7) *Tridax procumbens* [101]. Green synthesis is also dependent on different process parameters such as temperature, pH, pressure, concentration, and solution volumes. Importantly, plant diversity must be also considered, since the presence and levels of essential phytochemicals can vary between plants. For instance, plant location, weather conditions, and time of the year can influence phytochemical levels. Typical phytochemicals found in plant leaves that can reduce metal salts to form metal NPs include 1) aldehydes; 2) ketones; 3) flavones; 4) amides; 5) terpenoids, 6) phenols; 7) carboxylic acids; and 8) ascorbic acid [102]. Crucially, biologically synthesized NPs have been evaluated for biomedical applications such as antimicrobials, molecular sensing, diagnostics, and optical imaging [103, 104].

8.7 TYPES OF ANTIBACTERIAL NANOMATERIALS

8.7.1 Introduction

The two main properties that dictate NP efficacy in biological applications are size and shape. For wound healing applications, NPs must also have superior antibacterial activity and a very low toxicity profile toward healthy human cells. Metal and metal oxide NPs like silver, gold, and zinc oxide are ideally suited materials for wound healing therapies [105]. In addition, both polymeric nanoparticles and liposomes have been used in a variety of drug delivery applications. Typical forms of nanomaterials used and types of applications are presented in Figure 8.6. In addition, Figure 8.7 presents a schematic of the mechanism responsible for the antibacterial action against wound pathogens. The following sections discuss the main types of nanoparticles currently being used or under investigation for potential use in wound treatment protocols.

8.7.2 Silver Nanoparticles

Silver nanoparticles (Ag NPs) have been widely used as antibacterial agents to treat a variety of wound infections [106]. Ag NPs are highly active in low concentrations, unlike bulk Ag compounds, which reduces their toxicity to human cells [107]. Their antibacterial activity results in the destruction of bacteria's membrane, disruption of enzyme activity, and compromising the functionality of DNA, which leads to cell death, as seen in Figure 8.7 [106]. Ag NPs also promote the release of anti-inflammatory cytokines, and assist in the relocation and proliferation of fibroblasts

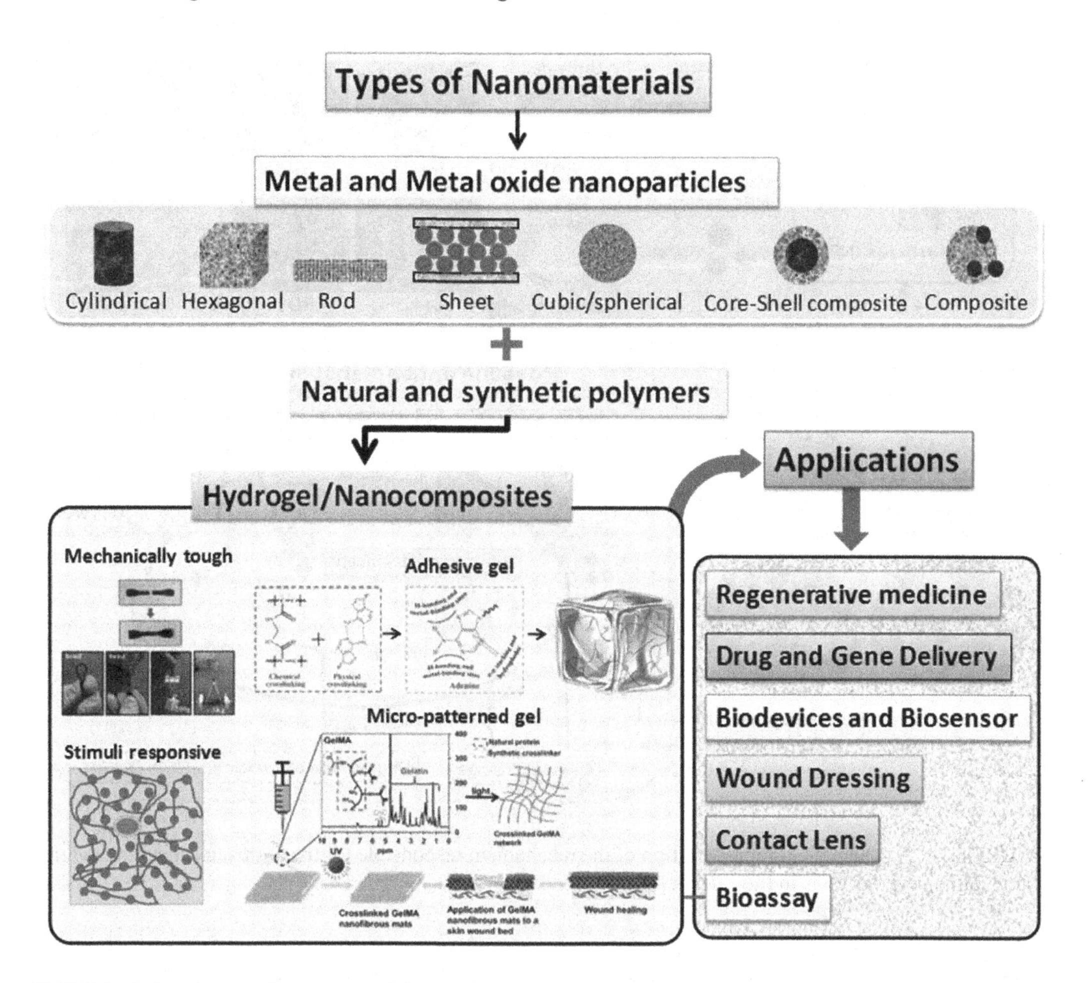

FIGURE 8.6 Types of nanomaterials, different morphologies, and formulations for typical medical procedures.

and keratinocytes, which accelerate the wound healing process [108] to the wound site. The presence of Ag NPs in the wound also promotes epidermal re-epithelialization and encourages wound closure [109].

Because of their superior antibacterial properties, Ag NPs have been incorporated into a variety of wound dressings and hydrogels. Typically, they have been encapsulated in materials like polyvinyl alcohol (PVA), sodium alginate (SA), chitosan, and tissue scaffolds of electrospun polycaprolactone (PCL) [110, 111]. Ag NPs embedded in biocompatible polymeric dressings have shown good antibacterial activity toward pathogens such as *S. aureus*, *S. epidermidis*, *S.* Typhimurium, and *E. coli* [112]. Clinical trials have also shown that Ag NPs can be used to treat burns and chronic wounds [113]. For instance, Acticoat© is a commercially available wound dressing that incorporates 15 nm diameter Ag NPs as the active ingredient. The dressing reduces pain, decreases microbial infection, and promotes wound healing [114].

8.7.3 Gold Nanoparticles

Gold (Au) NPs have attracted considerable scientific interest over recent years due to the material's unique properties and novel applications. The application of Au NPs in biomedical protocols includes the fluorescent labeling of immunoassays and the targeted delivery of therapeutic drugs and genetic materials [115]. In addition, studies have shown that Au NPs have antibacterial

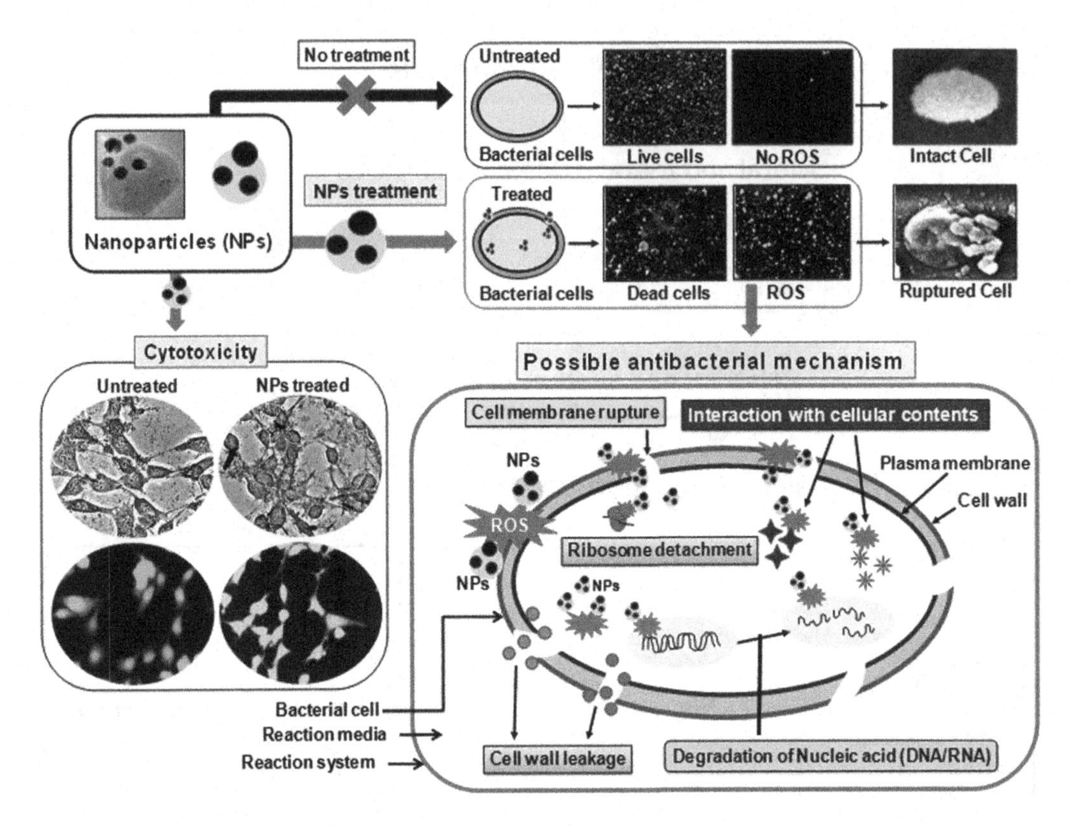

FIGURE 8.7 A schematical representation of the mechanism responsible for the antibacterial action against wound pathogens that ends in their death [20].

properties that can be used in wound therapies [116]. It is believed that the presence of Au NPs on the cell wall creates holes that compromise the cell wall's integrity. In addition, the biosorption and cellular uptake of Au NPs disrupt the functionality of cellular DNA which ultimately leads to cell death [109]. For instance, Au NPs are an effective antibacterial agent against *E. coli*, *P. aeruginosa*, *S. aureus*, and MRSA [70]. In addition, biogenically generated Au NPs produced from the leaf extract of *Gundelia tournefortii* L (GT) are highly effective against bacteria and fungi [117], while Au NPs encapsulated in hydrogels have also been found to be effective. For instance, Poloxamer 407 has been found to be effective against *S. aureus* and *P. aeruginosa* [118].

8.7.4 Zinc Oxides Nanoparticles

Zinc Oxide (ZnO) NPs are biocompatible and biologically safe and have been used in a variety of consumer products and cosmetics. They have also been found to be effective antibacterial agents [119]. The antibacterial activity attributed to ZnO NPs results from the generation of ROS that rupture the cell wall, enter, and then disrupt other cell functions, ultimately leading to cell death. However, the antibacterial properties of ZnO NPs are also size and concentration-dependent. For example, a recent study by Raghupathi et al. showed that a 1 mM concentration of 8 nm-sized ZnO NPs inhibited the growth of *S. aureus* by around 95%, while a 5 mM concentration containing NPs between 50 and 70 nm only inhibited growth to between 40 and 50% [120]. Moreover, ZnO NPs at concentrations between 100 and 1,000 μg/mL have been found to inhibit the formation of biofilms containing *S. aureus* and *P. aeruginosa* [121]. The incorporation of ZnO NPs into various wound dressing materials such as chitosan hydrogels, collagen dressings, and cellulose sheets has been investigated. The incorporation of ZnO NPs into dressings has been found to enhance antibacterial activity and promote wound healing [122].

8.7.5 Calcium Oxide Nanoparticles

Calcium Oxide (CaO) NPs are biocompatible with human cells and also display antibacterial activity toward several pathogens. However, it has been shown that the antibacterial activity of CaO NPs is directly dependent on the synthesis route used. For example, CaO NPs produced from the thermal decomposition of egg shells showed no antibacterial activity against *S. aureus* and MRSA [123], while chemically synthesized CaO NPs were found to have the ability to inhibit the growth of *S. epidermidis*, *P. aeruginosa*, and *Candida tropicalis* [124]. Moreover, the biosynthesis of CaO NPs from *Cissus quadrangularis* stem extracts produces NPs with strong antibacterial activity against both gram-negative and gram-positive bacteria such as *S. typhi*, *Shigella dysenteriae*, *Vibrio cholerae*, *P. aeruginosa*, *K. pneumonia*, *E. coli*, and *S. aureus* [125]. BiSCaO ointments prepared from scallop shell powders (CaO ~ 99.5%), can reduce infection rates of *P. aeruginosa*-infected wounds in hairless rats. The study found that 0.2% (w/w) BiSCaO ointment was an effective treatment for infected wounds [126].

8.7.6 Polymeric Nanoparticles

Because many natural and synthetic polymeric materials are biocompatible, they have received considerable scientific interest. Polymeric NPs have the potential to deliver growth factors, antibacterial peptides, and natural antimicrobial compounds to the wound site [127]. To date, polymeric NPs have been used to deliver drugs to treat neurodegenerative and brain-associated diseases [128]. Importantly, polymeric NPs encapsulate and protect the drug until it arrives at the target site and delivers its payload [129]. Typically, polymeric materials used in medical procedures range in size from 1 to 1000 nm [130]. Importantly, a wide range of chemical compounds and drugs can be encapsulated and delivered by polymeric NPs. For instance, curcumin is an antioxidant and antibacterial compound that can be used to treat wounds. Typically, curcumin has been encapsulated in polymeric NPs composed of materials like polycaprolactone (PCL), chitosan (CS), and poly(lactic-*co*-glycolic) acid (PLGA). In addition, curcumin can also be encapsulated in dressings in the form of membranes, hydrogels, and electrospun fibers [127].

8.7.7 Liposomes

Liposomes are the most extensively used delivery system for pharmaceuticals in the medical field. A liposome is made up of a spherical-shaped lipid bilayer membrane, with a hydrophobic outer surface and hydrophilic inner core. The advantage of using liposome-based delivery systems for wound healing is that they can carry both water-soluble and non-water-soluble drugs and antibiotics [131, 132]. The surface charge of liposomes is positive and is attracted to the negatively charged surfaces of bacterial cells. If the liposomes' surface charge was negative or neutral, it would fail to bind strongly with bacterial cell membranes [133]. Thus, the administration of liposomes results in their strong attraction to the negatively charged bacterial cell membranes. Once on the cell surface, the liposomes transfer antibacterial agents to the inside of the bacteria. The concentration level of antibacterial agents steadily increases until cell death occurs [134, 135]. Additionally, liposomes can be modified or incorporated into dressing materials to improve their performance when treating wounds [136]. For instance, liposomes containing gentamicin have been incorporated into electrospun chitosan meshes. Subsequent application of meshes on wounds found the controlled release of gentamicin was effective against *S. aureus*, *P. aeruginosa*, and *E. coli* for up to 24 hours [132]. Furthermore, to enhance antibacterial release times, improving dressings based on collagen mimetic peptide (CMP) and incorporating liposomes loaded with vancomycin (CMP/Van/Lipo) are being investigated. For instance, the controlled and continuous release of vancomycin from a CMP/Van/Lipo-based dressing over a 48-hour period revealed a strong antibacterial activity against MRSA under both *in vitro* and *in vivo* conditions [136]. The study revealed the controlled release of antibiotics was necessary for the effective treatment of bacteria in the wound area. Importantly, liposomes loaded with antibiotics or other antibacterial compounds are suitable agents to be incorporated into wound dressings to inhibit bacterial growth and prevent biofilm formation.

8.8 CONCLUDING REMARKS

Wound healing is a four-phase process that includes hemostasis, inflammation, proliferation, and remodeling (maturation). Wound healing occurs at both the cellular and molecular levels and can be disrupted by both internal and external influences. The most important is the invasion of microorganisms and the resulting infection. Therefore, understanding the healing process and the detrimental influence of harmful microorganisms will assist in developing new wound dressing materials and wound management protocols. In recent years, this has become a serious concern due to the increased numbers of multidrug-resistant microorganisms. Nanoparticles, due to their superior physiochemical and antimicrobial properties can be used in wound healing therapies. Metal and metal oxide (Ag, Au, ZnO, and CaO) NPs have shown their superior antimicrobial properties, while polymeric NPs and liposomes have the ability to be used as vehicles for delivering antimicrobial agents, thus making NPs ideal antimicrobial agents for incorporation into new and more effective wound dressings. Furthermore, studies have shown that NPs can be used as carriers for pharmaceuticals, growth factors, and genetic compounds, thus providing greater opportunities for developing new and more effective wound healing therapies in the future.

REFERENCES

1. Paladini F, Pollini M. Antimicrobial silver nanoparticles for wound healing application: Progress and future trends. *Materials.* 2019; 12 (16): 2540.
2. Robson MC. Wound infection: A failure of wound healing caused by an imbalance of bacteria. *Surgical Clinics of North America.* 1997; 77 (3): 637–650.
3. Okur ME, Karantas ID, Senyigit Z, Okur NU, Siafaka PI. Recent trends on wound management: New therapeutic choices based on polymeric carriers. *Asian Journal of Pharmaceutical Sciences.* 2020; 15 (6): 661–684.
4. Bhardwaj N, Chouhan D, Mandal B. Tissue engineered skin and wound healing: Current strategies and future directions. *Current Pharmaceutical Design.* 2017; 23 (24): 3455–3482.
5. Percival SL, McCarty SM, Lipsky B. Biofilms and wounds: An overview of the evidence. *Advances in Wound Care.* 2015; 4 (7): 373–381.
6. Brook I. Microbiology and management of soft tissue and muscle infections. *International Journal of Surgery.* 2008; 6 (4): 328–338.
7. Simoens S. Factors affecting the cost effectiveness of antibiotics. *Chemotherapy Research and Practice.* 2011; 2011: 249867.
8. World Health Organization. *Antimicrobial Resistance.* World Health Organization, 2019. (https://www.who.int/news-room/fact-sheets/detail/antimicrobial-resistance)
9. Baloch Z, Aslam B, Yasmeen N, Ali A, Liu Z, Rahaman A, Ma Z. *Antimicrobial Resistance and Global Health: Emergence, Drivers, and Perspectives. Handbook of Global Health,* 2020.
10. Kohanski MA, Dwyer DJ, Collins JJ. How antibiotics kill bacteria: From targets to networks. *Nature Reviews Microbiology.* 2010; 8 (6): 423–435.
11. Munita JM, Arias CA. Mechanisms of antibiotic resistance. *Microbiology Spectrum* 2016; 4 (2): 1–37.
12. Baptista PV, McCusker MP, Carvalho A, Ferreira DA, Mohan NM, Martins M, Fernandes AR. Nano-strategies to fight multidrug resistant bacteria—"A battle of the titans". *Frontiers in Microbiology.* 2018; 9: 1441.
13. Singh A, Gautam PK, Verma A, Singh V, Shivapriya PM, Shivalkar S, Sahoo AK, Samanta SK. Green synthesis of metallic nanoparticles as effective alternatives to treat antibiotics resistant bacterial infections: A review. *Biotechnology Reports.* 2020; 1 (25): e00427.

14. Wang L, Hu C, Shao L. The antimicrobial activity of nanoparticles: Present situation and prospects for the future. *International Journal of Nanomedicine.* 2017; 12: 1227.
15. Jeevanandam J, Barhoum A, Chan YS, Dufresne A, Danquah MK. Review on nanoparticles and nanostructured materials: History, sources, toxicity and regulations. *Beilstein Journal of Nanotechnology.* 2018; 9 (1): 1050–1074.
16. Das B, Moumita S, Ghosh S, Khan MI, Indira D, Jayabalan R, Tripathy SK, Mishra A, Balasubramanian P. Biosynthesis of magnesium oxide (MgO) nanoflakes by using leaf extract of Bauhinia purpurea and evaluation of its antibacterial property against Staphylococcus aureus. *Materials Science and Engineering: C.* 2018; 91: 436–444.
17. Sirelkhatim A, Mahmud S, Seeni A, Kaus NH, Ann LC, Bakhori SK, Hasan H, Mohamad D. Review on zinc oxide nanoparticles: Antibacterial activity and toxicity mechanism. *Nano-Micro Letters.* 2015; 7 (3): 219–242.
18. Sanchez-Lopez E, Gomes D, Esteruelas G, Bonilla L, Lopez-Machado AL, Galindo R, Cano A, Espina M, Ettcheto M, Camins A, Silva AM. Metal-based nanoparticles as antimicrobial agents: An overview. *Nanomaterials.* 2020; 10 (2): 292.
19. Patra JK, Das G, Fraceto LF, Campos EV, del Pilar Rodriguez-Torres M, Acosta-Torres LS, Diaz-Torres LA, Grillo R, Swamy MK, Sharma S, Habtemariam S. Nano based drug delivery systems: Recent developments and future prospects. *Journal of Nanobiotechnology.* 2018; 16 (1): 71.
20. Khan I, Saeed K, Khan I. Nanoparticles: Properties, applications and toxicities. *Arabian Journal of Chemistry.* 2019; 12 (7): 908–931.
21. McNamara K, Tofail SA. Nanoparticles in biomedical applications. *Advances in Physics: X.* 2017; 2 (1): 54–88.
22. Olsson M, Järbrink K, Divakar U, Bajpai R, Upton Z, Schmidtchen A, Car J. The humanistic and economic burden of chronic wounds: A systematic review. *Wound Repair and Regeneration.* 2019; 27 (1): 114–125.
23. Wang Y, Yang Y, Shi Y, Song H, Yu C. Antibiotic-free antibacterial strategies enabled by nanomaterials: Progress and perspectives. *Advanced Materials.* 2020; 32 (18): 1904106.
24. Wuthisuthimethawee P, Lindquist SJ, Sandler N, Clavisi O, Korin S, Watters D, Gruen RL. Wound management in disaster settings. *World Journal of Surgery.* 2015; 39 (4): 842–853.
25. Eyerich S, Eyerich K, Traidl-Hoffmann C, Biedermann T. Cutaneous barriers and skin immunity: Differentiating a connected network. *Trends in Immunology.* 2018; 39 (4): 315–327.
26. Dhivya S, Padma VV, Santhini E. Wound dressings – A review. *BioMedicine.* 2015; 5(4): 22.
27. Zuo H, Song G, Shi W, Jia J, Zhang Y. Observation of viable alloskin vs xenoskin grafted onto subcutaneous tissue wounds after tangential excision in massive burns. *Burns & Trauma.* 2016; 4: 1–10.
28. Kolarsick PA, Kolarsick MA, Goodwin C. Anatomy and physiology of the skin. *Journal of the Dermatology Nurses' Association.* 2011; 3 (4): 203–213.
29. Kanitakis J. Anatomy, histology and immunohistochemistry of normal human skin. *European Journal of Dermatology.* 2002; 12 (4): 390–401.
30. Guo SA, DiPietro LA. Factors affecting wound healing. *Journal of Dental Research.* 2010; 89 (3): 219–229.
31. Saleh K, Sönnergren HH. Control and treatment of infected wounds. In *Wound Healing Biomaterials,* pp. 107–115. Woodhead Publishing, 2016.
32. Boateng J, Catanzano O. Advanced therapeutic dressings for effective wound healing—A review. *Journal of Pharmaceutical Sciences.* 2015; 104 (11): 3653–3680.
33. Abazari M, Ghaffari A, Rashidzadeh H, Badeleh SM, Maleki Y. A systematic review on classification, identification, and healing process of burn wound healing. *The International Journal of Lower Extremity Wounds.* 2020; 21(1): 18–30.
34. Gurtner GC, Werner S, Barrandon Y, Longaker MT. Wound repair and regeneration. *Nature.* 2008; 453 (7193): 314–321.
35. Bhattacharya S, Mishra RK. Pressure ulcers: Current understanding and newer modalities of treatment. *Indian Journal of Plastic Surgery: Official Publication of the Association of Plastic Surgeons of India.* 2015; 48 (1): 416.
36. Loree S, Dompmartin A, Penven K, Harel D, Leroy D. Is vacuum assisted closure a valid technique for debriding chronic leg ulcers? *Journal of Wound Care.* 2004; 13 (6): 249–252.
37. Rittie L. Cellular mechanisms of skin repair in humans and other mammals. *Journal of Cell Communication and Signaling.* 2016; 10 (2): 103–120.
38. Bowler PG, Duerden BI, Armstrong DG. Wound microbiology and associated approaches to wound management. *Clinical Microbiology Reviews.* 2001; 14 (2): 244–269.

39. Eming SA, Martin P, Tomic-Canic M. Wound repair and regeneration: Mechanisms, signaling, and translation. *Science Translational Medicine.* 2014; 6 (265): 265sr6.
40. Krafts KP. Tissue repair: The hidden drama. *Organogenesis.* 2010; 6 (4): 225–233.
41. Schultz GS, Chin GA, Moldawer L, Diegelmann RF. Principles of wound healing. In *Mechanisms of Vascular Disease: A Reference Book for Vascular Specialists.* University of Adelaide Press, 2011.
42. Shield Health Care. (http://www.shieldhealthcare.com)
43. Goldberg SR, Diegelmann RF. Basic science of wound healing. In *Critical Limb Ischemia*, pp. 131–136. Cham: Springer, 2017.
44. DiNubile MJ, Lipsky BA. Complicated infections of skin and skin structures: When the infection is more than skin deep. *Journal of Antimicrobial Chemotherapy.* 2004; 53 (Suppl. 2): 37–50.
45. Martin JM, Zenilman JM, Lazarus GS. Molecular microbiology: New dimensions for cutaneous biology and wound healing. *Journal of Investigative Dermatology.* 2010; 130 (1): 38–48.
46. Percival SL, Emanuel C, Cutting KF, Williams DW. Microbiology of the skin and the role of biofilms in infection. *International Wound Journal.* 2012; 9 (1): 14–32.
47. Mihai MM, Holban AM, Giurcaneanu CĂ, Popa LG, Buzea M, Filipov M, Lazăr VE, Chifiriuc MC, Popa MI. Identification and phenotypic characterization of the most frequent bacterial etiologies in chronic skin ulcers. *Romanian Journal of Morphology and Embryology.* 2014; 55 (4): 1401–1408.
48. Madalina Mihai M, Maria Holban A, Giurcaneanu C, Gabriela Popa L, Mihaela Oanea R, Lazar V, Carmen Chifiriuc M, Popa M, Ioan Popa M. Microbial biofilms: Impact on the pathogenesis of periodontitis, cystic fibrosis, chronic wounds and medical device-related infections. *Current Topics in Medicinal Chemistry.* 2015; 15(16): 1552–1576.
49. Negut I, Grumezescu V, Grumezescu AM. Treatment strategies for infected wounds. *Molecules.* 2018; 23 (9): 2392.
50. Omar A, Wright JB, Schultz G, Burrell R, Nadworny P. Microbial biofilms and chronic wounds. *Microorganisms.* 2017; 5 (1): 9.
51. Demidova-Rice TN, Hamblin MR, Herman IM. Acute and impaired wound healing: Pathophysiology and current methods for drug delivery, part 1: Normal and chronic wounds: Biology, causes, and approaches to care. *Advances in Skin & Wound Care.* 2012; 25 (7): 304.
52. Bjarnsholt T. The role of bacterial biofilms in chronic infections. *Apmis.* 2013; 121: 1–58.
53. Nethi SK, Das S, Patra CR, Mukherjee S. Recent advances in inorganic nanomaterials for wound-healing applications. *Biomaterials Science.* 2019; 7 (7): 2652–2674.
54. Nair N, Biswas R, Götz F, Biswas L. Impact of *Staphylococcus aureus* on pathogenesis in polymicrobial infections. *Infection and Immunity.* 2014; 82 (6): 2162–2169.
55. Ray P, Singh R. Methicillin-resistant Staphylococcus aureus carriage screening in intensive care. *Indian Journal of Critical Care Medicine: Peer-Reviewed, Official Publication of Indian Society of Critical Care Medicine.* 2013; 17 (4): 205.
56. David MZ, Daum RS. Community-associated methicillin-resistant Staphylococcus aureus: Epidemiology and clinical consequences of an emerging epidemic. *Clinical Microbiology Reviews.* 2010; 23 (3): 616–687.
57. Haque M, Sartelli M, McKimm J, Bakar MA. Health care-associated infections–an overview. *Infection and Drug Resistance.* 2018; 11: 2321.
58. Albaugh KW, Biely SA, Cavorsi JP. The effect of a cellulose dressing and topical vancomycin on methicillin-resistant *Staphylococcus aureus* (MRSA) and gram-positive organisms in chronic wounds: A case series. *Ostomy/Wound Management.* 2013; 59 (5): 34.
59. Bessa LJ, Fazii P, Di Giulio M, Cellini L. Bacterial isolates from infected wounds and their antibiotic susceptibility pattern: Some remarks about wound infection. *International Wound Journal.* 2015; 12 (1): 47–52.
60. McKinnell JA, Huang SS, Eells SJ, Cui E, Miller LG. Quantifying the impact of extranasal testing of body sites for methicillin-resistant Staphylococcus aureus colonization at the time of hospital or intensive care unit admission. *Infection Control and Hospital Epidemiology.* 2013; 34 (2): 161–170.
61. Lima AF, Costa LB, Silva JL, Maia MB, Ximenes EC. Interventions for wound healing among diabetic patients infected with *Staphylococcus aureus*: A systematic review. *Sao Paulo Medical Journal.* 2011; 129 (3): 165–170.
62. Boulton AJ, Armstrong DG, Hardman MJ, Malone M, Embil JM, Attinger CE, Lipsky BA, Aragón-Sánchez J, Li HK, Schultz G, Kirsner RS. Diagnosis and management of diabetic foot infections, 2020.
63. Richard JL, Sotto A, Jourdan N, Combescure C, Vannereau D, Rodier M, Lavigne JP, Nîmes University Hospital Working Group on the Diabetic Foot (GP30). Risk factors and healing impact of multidrug-resistant bacteria in diabetic foot ulcers. *Diabetes & Metabolism.* 2008; 34 (4): 363–369.

64. Jones V, Grey JE, Harding KG. Wound dressings. *BMJ*. 2006; 332 (7544): 777–780.
65. Jahromi MA, Zangabad PS, Basri SM, Zangabad KS, Ghamarypour A, Aref AR, Karimi M, Hamblin MR. Nanomedicine and advanced technologies for burns: Preventing infection and facilitating wound healing. *Advanced Drug Delivery Reviews*. 2018; 123: 33–64.
66. Miller MC, Nanchahal J. Advances in the modulation of cutaneous wound healing and scarring. *BioDrugs*. 2005; 19 (6): 363–381.
67. Friedman ND, Temkin E, Carmeli Y. The negative impact of antibiotic resistance. *Clinical Microbiology and Infection*. 2016; 22 (5): 416–422.
68. Mihai MM, Giurcăneanu C, Popa LG, Nitipir C, Popa MI. Controversies and challenges of chronic wound infection diagnosis and treatment. *Modernizing Medicine*. 2015; 22: 375–381.
69. Høiby N, Bjarnsholt T, Moser C, Bassi GL, Coenye T, Donelli G, Hall-Stoodley L, Holá V, Imbert C, Kirketerp-Møller K, Lebeaux D. ESCMID guideline for the diagnosis and treatment of biofilm infections 2014. *Clinical Microbiology and Infection*. 2015; 21 (Suppl. 1): S1–25.
70. Arafa MG, El-Kased RF, Elmazar MM. Thermoresponsive gels containing gold nanoparticles as smart antibacterial and wound healing agents. *Scientific Reports*. 2018; 8 (1): 1–6.
71. Mori HM, Kawanami H, Kawahata H, Aoki M. Wound healing potential of lavender oil by acceleration of granulation and wound contraction through induction of TGF-β in a rat model. *BMC Complementary and Alternative Medicine*. 2016; 16 (1): 144.
72. Orchard A, van Vuuren S. Commercial essential oils as potential antimicrobials to treat skin diseases. *Evidence-Based Complementary and Alternative Medicine*. 2017; 2017: 4517971.
73. Scagnelli AM. Therapeutic review: Manuka honey. *Journal of Exotic Pet Medicine*. 2016; 2 (25): 168–171.
74. Jamil B, Abbasi R, Abbasi S, Imran M, Khan SU, Ihsan A, Javed S, Bokhari H. Encapsulation of cardamom essential oil in chitosan nano-composites: In-vitro efficacy on antibiotic-resistant bacterial pathogens and cytotoxicity studies. *Frontiers in Microbiology*. 2016; 7: 1580.
75. Narayanaswamy R, Torchilin VP. Hydrogels and their applications in targeted drug delivery. *Molecules*. 2019; 24 (3): 603.
76. Sudheesh Kumar PT, Lakshmanan VK, Anilkumar TV, Ramya C, Reshmi P, Unnikrishnan AG, Nair SV, Jayakumar R. Flexible and microporous chitosan hydrogels/nano ZnO composite bandages for wound dressing: In vitro and in vivo evaluation. *ACS Applied Materials & Interfaces*. 2012; 4 (5): 2618–2629.
77. Mehta MA, Shah S, Ranjan V, Sarwade P, Philipose A. Comparative study of silver-sulfadiazine-impregnated collagen dressing versus conventional burn dressings in second-degree burns. *Journal of Family Medicine and Primary Care*. 2019; 8 (1): 215.
78. Niska K, Zielinska E, Radomski MW, Inkielewicz-Stepniak I. Metal nanoparticles in dermatology and cosmetology: Interactions with human skin cells. *Chemico-Biological Interactions*. 2018; 295: 38–51.
79. Hamdan S, Pastar I, Drakulich S, Dikici E, Tomic-Canic M, Deo S, Daunert S. Nanotechnology-driven therapeutic interventions in wound healing: Potential uses and applications. *ACS Central Science*. 2017; 3 (3): 163–175.
80. Rodrigues M, Kosaric N, Bonham CA, Gurtner GC. Wound healing: A cellular perspective. *Physiological Reviews*. 2019; 99 (1): 665–706.
81. Senet P. Becaplermine gel (Regranex® gel). In *Annales de Dermatologie et de Venereologie* 2004; 131 (4): 351–358.
82. Chen PH, Chen X, He X. Platelet-derived growth factors and their receptors: Structural and functional perspectives. *Biochimica Biophysica Acta (BBA)-Proteins and Proteomics*. 2013; 1834 (10): 2176–2186.
83. Hollinger JO, Hart CE, Hirsch SN, Lynch S, Friedlaender GE. Recombinant human platelet-derived growth factor: Biology and clinical applications. *JBJS*. 2008; 90 (Suppl. 1): 48–54.
84. Westermark B, Claesson-Welsh L, Heldin CH. Structural and functional aspects of the receptors for platelet-derived growth factor. *Progress in Growth Factor Research*. 1989; 1 (4): 253–266.
85. Ramirez H, Patel SB, Pastar I. The role of TGFβ signaling in wound epithelialization. *Advances in Wound Care*. 2014; 3 (7): 482–491.
86. Le M, Naridze R, Morrison J, Biggs LC, Rhea L, Schutte BC, Kaartinen V, Dunnwald M. Transforming growth factor Beta 3 is required for excisional wound repair in vivo. *PLoS ONE*. 2012; 7 (10): e48040.
87. Singer AJ, Huang SS, Huang JS, McClain SA, Romanov A, Rooney J, Zimmerman T. A novel TGF-beta antagonist speeds reepithelialization and reduces scarring of partial thickness porcine burns. *Journal of Burn Care & Research*. 2009; 30 (2): 329–334.

88. Ferguson MW, Duncan J, Bond J, Bush J, Durani P, So K, Taylor L, Chantrey J, Mason T, James G, Laverty H. Prophylactic administration of avotermin for improvement of skin scarring: Three double-blind, placebo-controlled, phase I/II studies. *The Lancet.* 2009; 373 (9671): 1264–1274.
89. Mammen EF, Thomas WR, Seegers WH. Activation of purified prothrombin to autoprothrombin I or autoprothrombin II (platelet cofactor II or autoprothrombin II-A). *Thrombosis and Haemostasis.* 1961; 5 (2): 218–249.
90. Johnson KE, Wilgus TA. Vascular endothelial growth factor and angiogenesis in the regulation of cutaneous wound repair. *Advances in Wound Care.* 2014; 3 (10): 647–661.
91. Zhao R, Lin H, Bereza-Malcolm L, Clarke E, Jackson CJ, Xue M. Activated protein c in cutaneous wound healing: from bench to bedside. *International Journal of Molecular Sciences.* 2019; 20 (4): 903.
92. Antony S, Terrazas S. A retrospective study: Clinical experience using vacuum-assisted closure in the treatment of wounds. *Journal of the National Medical Association.* 2004; 96 (8): 1073.
93. Saxena V, Hwang CW, Huang S, Eichbaum Q, Ingber D, Orgill DP. Vacuum-assisted closure: Microdeformations of wounds and cell proliferation. *Plastic and Reconstructive Surgery.* 2004; 114 (5): 1086–1096.
94. Alrubaiy L, Al-Rubaiy KK. Skin substitutes: A brief review of types and clinical applications. *Oman Medical Journal.* 2009; 24 (1): 4.
95. Horch RE, Jeschke MG, Spilker G, Herndon DN, Kopp J. Treatment of second degree facial burns with allografts—Preliminary results. *Burns.* 2005; 31 (5): 597–602.
96. Supp DM, Boyce ST. Engineered skin substitutes: Practices and potentials. *Clinics in Dermatology.* 2005; 23 (4): 403–412.
97. Raguse JD, Gath HJ. The buccal fad pad lined with a metabolic active dermal replacement (Dermagraft) for treatment of defects of the buccal plane. *British Journal of Plastic Surgery.* 2004; 57 (8): 764–768.
98. Curran MP, Plosker GL. Bilayered bioengineered skin substitute (apligraf®). *BioDrugs.* 2002; 16 (6): 439–455.
99. Yildirimer L, Thanh NT, Loizidou M, Seifalian AM. Toxicology and clinical potential of nanoparticles. *Nano Today.* 2011; 6 (6): 585–607.
100. Singh J, Dutta T, Kim KH, Rawat M, Samddar P, Kumar P. 'Green' synthesis of metals and their oxide nanoparticles: Applications for environmental remediation. *Journal of Nanobiotechnology.* 2018; 16 (1): 84.
101. Devatha CP, Thalla AK. Green synthesis of nanomaterials. In *Synthesis of Inorganic Nanomaterials*, pp. 169–184. Woodhead Publishing, 2018.
102. Doble M, Rollins K, Kumar A. *Green Chemistry and Engineering.* Academic Press, 2010.
103. Aguilar Z. *Nanomaterials for Medical Applications.* Newnes, 2012.
104. Lin PC, Lin S, Wang PC, Sridhar R. Techniques for physicochemical characterization of nanomaterials. *Biotechnology Advances.* 2014; 32 (4): 711–726.
105. Rajendran NK, Kumar SS, Houreld NN, Abrahamse H. A review on nanoparticle based treatment for wound healing. *Journal of Drug Delivery Science and Technology.* 2018; 44: 421–430.
106. Burduşel AC, Gherasim O, Grumezescu AM, Mogoantă L, Ficai A, Andronescu E. Biomedical applications of silver nanoparticles: An up-to-date overview. *Nanomaterials.* 2018; 8 (9): 681.
107. Wu Y, Yang Y, Zhang Z, Wang Z, Zhao Y, Sun L. A facile method to prepare size-tunable silver nanoparticles and its antibacterial mechanism. *Advanced Powder Technology.* 2018; 29 (2): 407–415.
108. You C, Li Q, Wang X, Wu P, Ho JK, Jin R, Zhang L, Shao H, Han C. Silver nanoparticle loaded collagen/chitosan scaffolds promote wound healing via regulating fibroblast migration and macrophage activation. *Scientific Reports.* 2017; 7 (1): 1–1.
109. Vijayakumar V, Samal SK, Mohanty S, Nayak SK. Recent advancements in biopolymer and metal nanoparticle-based materials in diabetic wound healing management. *International Journal of Biological Macromolecules.* 2019; 122: 137–148.
110. Kong F, Fan C, Yang Y, Lee BH, Wei K. 5-Hydroxymethylfurfural-embedded poly (vinyl alcohol)/sodium alginate hybrid hydrogels accelerate wound healing. *International Journal of Biological Macromolecules.* 2019; 138: 933–949.
111. Ye H, Cheng J, Yu K. In situ reduction of silver nanoparticles by gelatin to obtain porous silver nanoparticle/chitosan composites with enhanced antimicrobial and wound-healing activity. *International Journal of Biological Macromolecules.* 2019; 121: 633–642.
112. Thanh NT, Hieu MH, Phuong NT, Thuan TD, Thu HN, Do Minh T, Dai HN, Thi HN. Optimization and characterization of electrospun polycaprolactone coated with gelatin-silver nanoparticles for wound healing application. *Materials Science and Engineering: C.* 2018; 91: 318–329.

113. Boroumand Z, Golmakani N, Boroumand S. Clinical trials on silver nanoparticles for wound healing. *Nanomedicine Journal.* 2018; 5 (4):186–191.
114. Fong J, Wood F, Fowler B. A silver coated dressing reduces the incidence of early burn wound cellulitis and associated costs of inpatient treatment: Comparative patient care audits. *Burns.* 2005; 31 (5): 562–567.
115. Lee KX, Shameli K, Yew YP, Teow SY, Jahangirian H, Rafiee-Moghaddam R, Webster TJ. Recent developments in the facile bio-synthesis of gold nanoparticles (AuNPs) and their biomedical applications. *International Journal of Nanomedicine.* 2020; 15: 275.
116. Li Y, Tian Y, Zheng W, Feng Y, Huang R, Shao J, Tang R, Wang P, Jia Y, Zhang J, Zheng W. Composites of bacterial cellulose and small molecule-decorated gold nanoparticles for treating Gram-Negative bacteria-infected wounds. *Small.* 2017; 13 (27): 1700130.
117. Zhaleh M, Zangeneh A, Goorani S, Seydi N, Zangeneh MM, Tahvilian R, Pirabbasi E. In vitro and in vivo evaluation of cytotoxicity, antioxidant, antibacterial, antifungal, and cutaneous wound healing properties of gold nanoparticles produced via a green chemistry synthesis using *Gundelia tournefortii* L. as a capping and reducing agent. *Applied Organometallic Chemistry.* 2019; 33 (9): e5015.
118. Wang S, Yan C, Zhang X, Shi D, Chi L, Luo G, Deng J. Antimicrobial peptide modification enhances the gene delivery and bactericidal efficiency of gold nanoparticles for accelerating diabetic wound healing. *Biomaterials Science.* 2018; 6 (10): 2757–2772.
119. Jones N, Ray B, Ranjit KT, Manna AC. Antibacterial activity of ZnO nanoparticle suspensions on a broad spectrum of microorganisms. *FEMS Microbiology Letters.* 2008; 279 (1): 71–76.
120. Raghupathi KR, Koodali RT, Manna AC. Size-dependent bacterial growth inhibition and mechanism of antibacterial activity of zinc oxide nanoparticles. *Langmuir.* 2011; 27 (7): 4020–4028.
121. Pati R, Mehta RK, Mohanty S, Padhi A, Sengupta M, Vaseeharan B, Goswami C, Sonawane A. Topical application of zinc oxide nanoparticles reduces bacterial skin infection in mice and exhibits antibacterial activity by inducing oxidative stress response and cell membrane disintegration in macrophages. *Nanomedicine: Nanotechnology, Biology and Medicine.* 2014; 10 (6): 1195–1208.
122. Saghazadeh S, Rinoldi C, Schot M, Kashaf SS, Sharifi F, Jalilian E, Nuutila K, Giatsidis G, Mostafalu P, Derakhshandeh H, Yue K. Drug delivery systems and materials for wound healing applications. *Advanced Drug Delivery Reviews.* 2018; 127: 138–166.
123. Khan MI, Mazumdar A, Pathak S, Paul P, Behera SK, Tamhankar AJ, Tripathy SK, Lundborg CS, Mishra A. Biogenic Ag/CaO nanocomposites kill *Staphylococcus aureus* with reduced toxicity towards mammalian cells. *Colloids and Surfaces B: Biointerfaces.* 2020; 189: 110846.
124. Roy A, Gauri SS, Bhattacharya M, Bhattacharya J. Antimicrobial activity of CaO nanoparticles. *Journal of Biomedical Nanotechnology.* 2013; 9 (9): 1570–1578.
125. Marquis G, Ramasamy B, Banwarilal S, Munusamy AP. Evaluation of antibacterial activity of plant mediated CaO nanoparticles using Cissus quadrangularis extract. *Journal of Photochemistry and Photobiology B: Biology.* 2016; 155: 28–33.
126. Takayama T, Ishihara M, Sato Y, Nakamura S, Fukuda K, Murakami K, Yokoe H. Bioshell calcium oxide (BiSCaO) for cleansing and healing *Pseudomonas aeruginosa*–infected wounds in hairless rats. *Bio-Medical Materials and Engineering.* 2020; 31 (2): 95–105.
127. Wu A, Xu C, Akakuru OU, Ma X, Zheng J, Zheng J. Nanoparticle-based wound dressing: Recent progress in the detection and therapy of bacterial infections. *Bioconjugate Chemistry.* 2020; 31 (7): 1708–1723.
128. Calzoni E, Cesaretti A, Polchi A, Di Michele A, Tancini B, Emiliani C. Biocompatible polymer nanoparticles for drug delivery applications in cancer and neurodegenerative disorder therapies. *Journal of Functional Biomaterials.* 2019; 10 (1): 4.
129. Lee JH, Yeo Y. Controlled drug release from pharmaceutical nanocarriers. *Chemical Engineering Science.* 2015; 125: 75–84.
130. Sardoiwala MN, Kaundal B, Choudhury SR. Development of engineered nanoparticles expediting diagnostic and therapeutic applications a cross blood–brain barrier. In *Handbook of Nanomaterials for Industrial Applications*, pp. 696–709. Elsevier, 2018.
131. Scriboni AB, Couto VM, de Morais Ribeiro LN, Freires IA, Groppo FC, De Paula E, Franz-Montan M, Cogo-Müller K. Fusogenic liposomes increase the antimicrobial activity of vancomycin against *Staphylococcus aureus* biofilm. *Frontiers in Pharmacology.* 2019; 10 (Article 1401): 1–11.
132. Monteiro N, Martins M, Martins A, Fonseca NA, Moreira JN, Reis RL, Neves NM. Antibacterial activity of chitosan nanofiber meshes with liposomes immobilized releasing gentamicin. *Acta biomaterialia.* 2015; 18: 196–205.

133. Rukavina Z, Klaric MS, Filipovic-Grcic J, Lovric J, Vanic Z. Azithromycin-loaded liposomes for enhanced topical treatment of methicillin-resistant *Staphyloccocus aureus* (mrsa) infections. *International Journal of Pharmaceutics*. 2018; 553 (1–2):109–119.
134. Sanderson N, Jones M. Encapsulation of vancomycin and gentamicin within cationic liposomes for inhibition of growth of Staphylococcus epidermidis. *Journal of Drug Targeting*. 1996; 4 (3): 181–189.
135. Kim HJ, Gias EL, Jones MN. The adsorption of cationic liposomes to Staphylococcus aureus biofilms. *Colloids and Surfaces A: Physicochemical and Engineering Aspects*. 1999; 149 (1–3): 561–570.
136. Thapa RK, Kiick KL, Sullivan MO. Encapsulation of collagen mimetic peptide-tethered vancomycin liposomes in collagen-based scaffolds for infection control in wounds. *Acta Biomaterialia*. 2020; 103: 115–128.

9 Calcium Carbonate Micro/Nanoparticles as Versatile Carriers for the Controlled Delivery of Pharmaceuticals for Cancer Treatment, Imaging, and Gene Therapy

Gérrard Eddy Jai Poinern, Marjan Nasseh, Triana Wulandari, and Derek Fawcett

CONTENTS

9.1 INTRODUCTION

Cancer is the second most frequent cause of death in the world. For instance, in the United States, 606,520 Americans lost their lives to cancer in 2020, equating to around 1,660 deaths per day [1], while deaths related to antibiotic-resistant infections in the US resulted in only around 35,000 deaths [2]. Similarly, around 150,000 new cancer cases per year were reported in Australia, along with an annual death rate of around 50,000 in 2020 [3], thus indicating the magnitude and impact of

DOI: 10.1201/9781003181422-11

cancer on the community. However, if diagnosed early and dependent on the type of cancer, treatment is possible using various cancer therapy methods. These methods have varying efficacy and include procedures such as surgery, radiation treatment, and drug therapy [4]. Even after more than a century of intense biomedical research and innovations there is still no effective cure for beating cancer. And in many cases, after first defeating cancer, it reappears and begins its aggressive and destructive activities in the patient. Recurrence rates can vary significantly, with cancers with low recurrence rates typically around 5% to 12% and cancers with high recurrence rates being around 36% to 85% [5, 6]. The burden to the community is the extremely large economic cost that runs into several billion dollars annually. For instance, the economic cost resulting from cancers among people in Australia diagnosed during 2009–2013 was estimated to be around AUS$ 6.3 billion or 0.4% of Australia's gross domestic product (GDP), with the largest costs for colorectal cancer ($ 1.1 billion), breast cancer ($ 0.8 billion), lung cancer ($ 0.6 billion), and prostate cancer ($ 0.5 billion) [7].

The efficacy of any pharmaceutical-based treatment is heavily dependent on the delivery system. Factors contributing to successful treatment include targeting cancer cell sites, optimizing drug pharmacokinetics to minimize side effects, effective destruction of cancer cells, and promoting the patients' adherence to the treatment procedure [8, 9]. However, many clinically used anticancer therapeutic drugs have several disadvantages such as low bioavailability, poor solubility, and limited distribution/penetration into targeted cancerous cells and tissues. Other issues also include limited drug stability resulting from variations of temperature and pH in the body and sensitivity to enzymatic activity, thus reducing the half-life of many conventional pharmaceutical-based treatments [10, 11]. Currently, polymer-based carriers such as polyethylene (glycol/oxide) (aka PEO, POE) or poly(lactic acid) PLA are used in commercialized products. Still, many of these products suffer from unfavorable immune system responses, cytotoxicity, and loss of bioactivity in the physiological environment of the body [12]. To overcome these limitations, improvements in drug delivery effectiveness, increased chemotherapeutic outcomes, and new alternative strategies are needed to treat cancer. Recent studies have shown that nanotechnology-based techniques can produce nanometer-scale materials that can be used as effective and efficient drug delivery carriers [13, 14]. In addition, studies have also demonstrated that nanotechnology-based strategies can also be used to assist in the early detection and diagnosis of cancer. This chapter summarizes and discusses drug delivery and targeting, the use of nanotechnology-based medicinal treatments, and dye-doped nanoparticles for the detection and diagnosis of cancer. In particular, the chapter summarizes current research into using pH-sensitive $CaCO_3$ micro/nanometer-scale materials as targeted drug carriers for the delivery of pharmaceuticals to the acidic microenvironment of cancer cells and tissues [15].

9.2 CANCER AND NANOMEDICINE

Cancer is a life-threatening disease and begins when some of the trillions of cells that make up the body experience a series of genetic and epigenetic changes or mutations. These changes can be either inherited or triggered by cancer-causing materials (carcinogens) in the environment. Normally, cells grow and multiply (via a process called cell division), only when they receive signals. Under instruction from these signals, new cells are formed, or cells are told to stop dividing, or when cells become old or damaged, they are told to die (apoptosis). Therefore, under normal circumstances, new cells are formed when the body needs them or terminates them when their function becomes impaired. However, cancer cells not only grow without signal instructions from the body, but they also disregard signals usually directing normal cell division processes and continue to grow and invade nearby tissues. Some cancers not only invade surrounding tissues but also enter the bloodstream and lymphatic system to travel (metastasize) to other regions of the body far from the original source of cancer. Notably, cancer cells hide to evade the immune system, deceive it, and gain its protection. In this environment, many cancers grow to form solid tumors, and at the same time, they direct the growth of blood vessels toward themselves. And by the time they have reached a size range between 150 to 200 micrometers, they have established their vasculature system to supply themselves with nutrients and oxygen, while at the same time removing their waste products [16], thus, ensuring the tumor's ability to grow further, proliferate, and spread. However, if detected early enough and dependent on the type of cancer, treatment by either surgery or radiation therapy or chemotherapy, or a combination of these treatments, a cure is possible, or the respective treatment protocol can prolong the patient's life.

Notably, the tumor's abnormal survival behavior offers several opportunities for developing new anticancer therapies. One such opportunity, discovered by Matsumura and Maeda in 1986, is the enhanced permeability and retention (EPR) effect found in tumors and their associated vasculature systems [17]. The phenomenon promotes greater drug delivery efficiencies to the tumor *via* greater permeability and more significant accumulations of drugs within the tumor site. Thus, the EPR effect offers a new approach to delivering new types of anticancer drugs and developing innovative chemotherapy strategies [18]. Furthermore, a recent report by the European Commission's Technology Platform on Nanomedicine group has stated, "the nanomedicine field is concretely able to design products that overcome critical barriers in conventional medicine in a unique manner" [19], thus recognizing the ability of nanomedicine to address several medical challenges and limitations currently faced by conventional anticancer drug treatment protocols, such as poor targeting, low bioavailability, and organ and systemic toxicity [20, 21]. Nanomedicine can combine nanotechnology-based techniques with medicine to create new and alternative chemotherapy strategies to combat cancer and its effects on patients.

Nanomedicine, a relatively new field, brings together materials engineering at the nanometer-scale level and molecular biology. From the materials engineering perspective, the quantum mechanical effects occurring at the nanometer scale (dimensions smaller than 100 nm) result in the nanomaterial's unique set of properties [22]. In terms of relative size, the human hair fiber is about 0.1 mm wide, and a COVID-19 virus is typically around 100 nanometers (nm) wide, so 1,000 COVID-19 viruses can fit on one human hair fiber. Importantly, a nanomaterial's properties are significantly different from their respective bulk counterparts and have the potential to be tailored for specific applications. These unique properties can include increased strength, different chemical reactivity, and increased thermal or electrical conductivities [23]. On the other hand, the cell is the basic unit of living organisms and is generally in the micrometer-size range. Thus, the size difference between a typical cell and a nanoparticle can be as much as 1;000 times. And because of this significant size difference, nanoparticles can biophysically interact with surface molecules at the cell membrane and/or enter the cell and interact with component structures [24]. Also at the nanometer scale are the molecular building blocks of life, such as carbohydrates, proteins, nucleic acids, and lipids [25]. Importantly, nanoparticles smaller than 50 nm have been shown to directly enter most cells and interact with internal cell structures and molecules with little functionalization, thus making nanoparticles an attractive carrier platform technology for drug delivery procedures [26]. Recent studies have shown that targeted nanoparticle-based drug delivery systems with particle diameters less than 600 nm can take advantage of the EPR effect, thus enhancing therapeutic effectiveness and reducing harmful side effects of therapeutic agents [27, 28].

Nanoparticles can be made from a variety of materials, have varied chemical structures, and come in a wide range of morphologies. They can range in size from 10 nm to a few hundred nanometers and have a large surface area to volume ratio. Because of their unique material properties, nanoparticles have been used in applications like engineering, electronics, catalysis, environmental remediation, and medicine. Nanoparticles can be made from metals, quantum dots, ceramics, lipids (liposomes), polymers, organic materials, and composites. For instance, quantum dots are used in the latest ultra-high-definition television screens, and gold nanoparticles are used in medical imaging and medicinal treatments [29, 30]. From the medical perspective, studies have shown that the release of drugs from targeted nanoparticles can promote better cell penetration and permit slow and controlled release at the cancer cell [31, 32]. Two currently available forms of nanoparticle-based carrier systems used in medicine for delivering pharmaceuticals are liposomal drug carriers and polymers-based drug delivery platforms [33, 34]. The main advantage of these two types of delivery systems is the controlled and sustained release of drugs at specific sites. However, controlling the concentration, plasma half-life, and release profile of the active therapeutic agent are still challenging issues [35]. Furthermore, because of the complex multi-step production methods used to produce many of these delivery systems, cytotoxicity (an unfavorable response from the immune system) and poor bioactivity still limit their effectiveness [36, 37].

Because of the limitations associated with current nanomedicine-based therapies, cancer is still responsible for around 25% of all deaths globally [38, 39]. Most unsuccessful or poor patient responses to current nanomedicine-based treatments are the result of poor tumor targeting and poor plasma half-lives [17]. Despite their current shortcomings, it is expected that further research will improve healthcare outcomes in all phases of cancer therapy. With this objective in mind, recent research has focused on $CaCO_3$ micro/nanoparticles as an alternative medicinal-based platform to overcome current limitations and achieve positive clinical outcomes. The recent scientific interest in $CaCO_3$ micro/nanoparticles arises from their advantageous properties that include 1) biocompatibility; 2) slow biodegradability that allows greater drug retention; 3) in aqueous environments $CaCO_3$ does not swell or change porosity [40]; and 4) pH-responsiveness since cancer cells and tissues tend to have an acidic microenvironment [41]. Because of these advantageous properties, $CaCO_3$-based particle systems have become a drug carrier candidate for the targeted and sustained release of drugs to combat cellular cancer [42].

9.3 CALCIUM CARBONATE ($CaCO_3$) MICRO/NANOPARTICLES

9.3.1 Sources, Polymorphs of Calcium Carbonate, and Its Medical Use

Calcium carbonate ($CaCO_3$) is naturally found in the earth's crust in the form of limestone and chalk. It is also the main inorganic component found in seashells, corals, and eggs shells. Historically, it has been used by humanity as a construction material (limestone) for centuries. Current industrial applications include its use as a viscosity modifier in products like plastics, coatings, rubber, paint, paper, and pigments [43]. Importantly, because $CaCO_3$ is accepted as a food and pharmaceutical additive by the European Food Safety Agency, it is widely used in pharmaceuticals such as in oral antacid formulations and medications for calcium deficiency [44, 45]. In addition, because of $CaCO_3$ biocompatibility, it is used in tissue engineering applications like bone cement, scaffolds, and dental implants [46]. More recently, it has been employed as a drug carrier for the targeted delivery of anticancer formulations [47, 48].

$CaCO_3$ occurs naturally in three different anhydrous polymorphs (aragonite, calcite, and vaterite) and two different hydrated crystalline forms (monohydrocalcite and ikaite) [49, 50]. The needle-like aragonite is also subjected to repeated twinning that results in the formation of pseudo-hexagonal crystals. Aragonite is metastable and readily transforms into the more stable calcite polymorph. The aragonite crystal morphologies are naturally found in most mollusk shells and the endoskeleton of corals. Vaterite has a spherical morphology and occurs naturally in mineral springs, organic tissues, and gallstones [51]. Vaterite is metastable and has poor stability in the presence of water, and slowly dissolves and recrystallizes to form the stable calcite polymorph [52]. Vaterite also has a porous inner structure and a much higher solubility compared to the other two polymorphs. The most thermodynamically stable calcite polymorph can have several morphologies that include prismatic, rhombohedral, and tabular forms [53].

An important property of $CaCO_3$ is its sensitivity to a decreasing pH environment. This property is very advantageous since increasing levels of intracellular hydrogen accompany tumor growth and creates a localized acidic medium around the tumor [54]. For instance, while the normal physiological environment has a pH of around 7.4, tumors can have pH levels of less than 6.5 [55]. Therefore, in the acidic tumor environment, a pH-sensitive $CaCO_3$-based carrier not only targets the tumor but also slowly degrades to release its anticancer drug payload and increase the cell's CO_2 levels. In terms of delivery, micrometer-scale particles (typically between 4 and 10 μm) administered intravenously are usually caught and filtered out of the body and offer few therapeutic benefits [56]. However, smaller nanometer-scale particles (ranging from 10 to 1000 nm) can be administered via several routes (i.e., intravenous, transdermal, subcutaneous) and travel throughout the body, including passage through the blood-brain barrier without being caught and filtered out. Thus, the nanometer-scale particles ultimately accumulate in the targeted tumor sites and deliver their drug payload [57]. Another therapeutic advantage of using $CaCO_3$ nanoparticle-based carriers is their ability to change tumor tissue pH during particle decomposition and hinder further tumor growth [58, 59].

9.3.2 Common Methods for Producing $CaCO_3$ Micro/Nanoparticles

Several methods have been designed and used to produce $CaCO_3$ nanoparticles. These methods include 1) processing raw $CaCO_3$ sources derived from nature; 2) carbonation; 3) solution precipitation; 4) reverse emulsion; and 5) ultrasound-assisted synthesis. These methods a briefly discussed in the following sections.

9.3.2.1 Processing Raw $CaCO_3$ Sources Derived from Nature

Several studies have investigated using biological materials derived from nature. Typical material sources studied include seashells and waste shells from commercial aquacultures, such as bivalve mollusks and pearl oysters, and even terrestrial waste sources like eggshells [60–65]. Generally, after initially cleaning the raw source material, the shell structure is milled into a micron-sized powder and then further processed into a nanometer-scale material by either chemical reduction, ultrasound processing, nano-milling, or by a combination of processes [60, 63, 65]. Typically, mechanical-chemical processing techniques can produce nanoparticles with uniform size and shape with low agglomeration rates [66].

9.3.2.2 Carbonation

Initially, limestone (raw $CaCO_3$) undergoes calcination at around 900 °C to produce CaO and CO_2. The CaO then undergoes hydration to produce $Ca(OH)_2$. The carbonation step follows and involves bubbling CO_2 through the $Ca(OH)_2$ solution, as seen in Equations 9.1 and 9.2 below.

$$CO_2 + H_2O \rightleftarrows H_2CO_3 \tag{9.1}$$

$$H_2CO_3 + Ca(OH)_2 \rightleftarrows CaCO_3\ (\text{nanoparticles}) + 2H_2O \tag{9.2}$$

The carbonation step generally produces nanometer-scale calcite particles. Still, the generated particles' polymorph type, particle size, shape, and physicochemical properties can be controlled by changing the Ca^{2+} concentration, CO_2 flow rate, temperature, and types and amounts of additives [43, 67].

9.3.2.3 Solution Precipitation

The solution precipitation method or mixing method is a well-established laboratory-based method for creating $CaCO_3$ micro/nanoparticles [68]. The method involves the mixing of supersaturated solutions of a carbonate source (i.e., $NaHCO_3$) and calcium source (i.e., $CaCl_2$ or $Ca(NO_3)_2$). Controlling the physiochemical properties of the created $CaCO_3$ nanoparticles is achieved by changing the synthesis variables (i.e., reactant concentration, temperature) and through the addition of additives like surfactants [69, 70]. For instance, particle size can be controlled through reactant concentration, while other properties, including particle size, can be controlled with the help of surfactants [71]. The advantage of this method is that it allows easy control of surfactant concentration, which effectively contains particle size and shape by the degree of capping the particle experiences during formation and growth. However, a disadvantage of many surfactants is their toxicity, which makes them generally unsuitable for biomedical applications.

9.3.2.4 Reverse Emulsion

Synthesizing $CaCO_3$ nanoparticles using the reverse emulsion method involves creating a hydrophobic continuous phase that contains large numbers of dispersed hydrophilic droplets (water in oil or W/O). Surfactants are used to stabilize the dispersed hydrophilic droplets in the organic phase. The dispersed hydrophilic droplets forming the micelles contain either a carbonate source solution (CO_3^{2-}) or a calcium source solution (Ca^{2+}). When the emulsions are mixed, the micelles collide, and their contents combine, forming a micro-reactor that creates the $CaCO_3$ nanoparticles [72]. The advantage of this room temperature-based method is the creation of nanoparticles with a narrow size distribution [43]. The physiochemical properties like the type of polymorph, particle size, and shape of the generated $CaCO_3$ nanoparticles can be regulated by adjusting synthesis parameters. These parameters include adjusting the water-to-surfactant ratio, water-to-oil ratio, temperature, pH, the ratio of CO_3^{2-} to Ca^{2+}, and using different additives in the reaction mixture [73, 74].

9.3.2.5 Ultrasound-Assisted Synthesis

Several researchers have used sonochemical processing (ultrasound irradiation) to assist during solution-based wet chemical synthesis of $CaCO_3$ particles [75, 76]. The advantage of using sonochemical processing stems from the generation of an acoustic cavitation effect in the chemical solution. The cavitation produces the rapid formation and subsequent collapse of bubbles within the chemical solution. During bubble collapse, extremely high temperatures (~ 5000 K) and high pressures (~ 20 MPa) are generated. Under these extreme reaction conditions, novel micro- and nano-scale materials with unique physiochemical properties can be formed [77]. In addition, sonication also facilitates greater mixing, reduces particle agglomeration, and minimizes particle size distribution [78].

9.4 $CaCO_3$-BASED PARTICLE SYSTEMS FOR BIOMEDICAL IMAGING, TARGETED DRUG DELIVERY, AND GENE THERAPY

There are three main medical applications for $CaCO_3$ micro/nanoparticles. These applications include 1) delivery of contrast agents to the tumor for biomedical imaging; 2) targeted drug therapy and controlled release of anticancer therapeutics; and 3) gene therapy. Each of these methods is briefly discussed in the following sections.

9.4.1 Nanoparticle-Based Systems for Biomedical Imaging

9.4.1.1 Biomedical Imaging and Contrast Agents

Biomedical imaging is extensively used for the diagnosis of life-threatening diseases and cancers. Many nanoparticle-based platforms have unique optical, magnetic, and radioactive characteristics that are derived from their nanometer-scale features or agents loaded into their core or incorporated onto their surfaces. Nanoparticle-based platforms are used as imaging contrast agents to determine specific tumor-related information that can be further used for diagnosis. These agents are used in various medical imaging techniques, such as radioisotope imaging, magnetic resonance imaging (MRI), and X-ray computed tomography (CT scans). Optical biological imaging is currently used to perform diagnosis at the cellular level and provides imaging guidance for the intraoperative surgical removal of tumors [79]. Biological imaging uses the contrast generated by different optical properties, such as absorption, light scattering, and luminescence properties such as fluorescence and phosphorescence. For many years, organic dyes and metal complexes were extensively used in luminescent microscopy. However, organic dyes and metal complexes suffer from photo-bleaching during continuous intensive light exposure [80]. To overcome this problem, nanoparticles (e.g., silica-based or polymeric), doped with photo-luminescent organic molecules, and nano-formulations containing semiconductors materials (either quantum dots (QDs) or rods (QRs)) were developed and successfully used in a variety of optical microscopy techniques [81, 82]. In addition, QDs are also incorporated into nanoparticle-based platforms for the delivery of therapeutic agents, with luminescent QDs being used for optical tracking [83].

Similarly, in magnetic resonance imaging (MRI), a few non-toxic super-paramagnetic iron oxide-based (Fe_3O_4) nano-formulations have been used in preclinical and clinical applications [84]. In addition, luminescent QDs can be incorporated with iron oxide nano-formulations for use in a combined MRI and optical imaging procedure [85]. A few nanoparticle formulations incorporating suitable radionuclides have also been used in biomedical-based radioisotope imaging. For instance, silver nanoparticles incorporating ^{125}I were intravenously injected into a mouse model (*in vivo*), and imaging was carried out to determine the full-body distribution pattern of the radionuclide [86]. Correspondingly, the use of nanoparticles as contrast agents for X-ray computed tomography (CT) imaging has also been extensively studied in the past. In particular, gold nanoparticles have attracted considerable interest due to their high X-ray attenuation coefficient, biocompatibility, and longer circulation lifetimes in the body. Further, they can be safely targeted to specific regions of interest within the body. Significantly, it is the beneficial attenuation properties of the targeted gold nanoparticles that enable the production of high-contrast anatomical images of body structures, like blood vessels, tissues and organs, and cancer tumors [87]. In recent years, nanoparticle-based carriers containing two or more contrast elements have also been evaluated for use in CT imaging. For instance, $BaYbF_5$ nanoparticles containing two contrast elements (Ba and Yb) were found to significantly increase contrast enhancement, producing higher-resolution images [88]. Likewise, high-stability bismuth sulfide (Bi_2S_3) nanoparticles have five times more X-ray absorption than conventional contrast agents like iodine [89].

9.4.1.2 Toxicity Concerns of Current Nanoparticle-Based Systems

Despite the many advantages of nanoparticle-based imaging agents, there have also been concerns regarding the potential risks resulting from their interactions with host cells and tissues [90]. The perfect nanoparticle-based system should deliver its payload at the specific target, perform its signaling function and then exit from the body without causing any adverse effects [91]. However, several studies have shown that nanoparticles can produce minor to significant hazardous effects to host cells and tissues. These hazardous effects are dependent on the type of interactions (chemical, physical, or immunological) and the dosage levels [92]. For instance, studies have shown gold nanoparticles smaller than 5 nm are more toxic than those greater than 15 nm [93, 94]. The resulting genotoxicity arises from the smaller-sized (< 5 nm) gold nanoparticles being able to enter the cell nucleus and bind to DNA [92, 95]. Studies have also shown the nanoparticles' composition and surface chemistry, including the release of ions when their matrix breaks down, can produce chemical and immunological interactions with host tissues. For example, the primary source of toxicity arising from CdSe and CdTe quantum dots used in biomedical imaging results from the release of Cd^{2+} ions and the subsequent increase in oxidative stress [96, 97]. Current strategies for reducing Cd^{2+} ion release include coating the quantum dot with a zinc sulfide (ZnS) shell or non-toxic capping agents [98, 99]. However, any long-term protection provided by either of these methods is still largely unknown and needs to be elucidated. Similarly, carbon nanomaterials have also been studied for several biomedical applications, and correspondingly, their respective toxicological studies have attracted considerable interest [90]. These studies have shown that their toxicity can result from physical and structural characteristics of the carbon nanomaterial like size, shape, length, the number of layers, and agglomerates formed [100-105]. For instance, studies have shown the needle-like form of carbon nanotubes (CNTs) can be highly toxic to human tissues [106]. In addition, single-walled CNTs can generate oxidative stress that negatively influences the cell's nuclear activity in human keratinocytes [107].

9.4.1.3 $CaCO_3$ Nanoparticle-Based Systems for Biomedical Imaging

Because of the toxicological issues associated with many nanomaterials, alternative nanoparticle-based platform systems are constantly being studied for potential use as signaling agents in biomedical imaging and diagnostic applications. Fluorescence imaging is emerging as an indispensable tool for surgical navigation and for aiding in optimal tumor excision. This beneficial combination significantly increases the success of surgical outcomes, reduces damage to surrounding tissues, and improves the patient's recovery rate from the procedure [108]. However, despite the variety of fluorescent nanoparticles studied in recent years [109, 110], toxicity concerns have refocused research efforts on finding non-toxic alternatives. Due to their stability at normal physiological pH (7.3 to 7.4), $CaCO_3$ nanoparticles are considered a suitable pH-dependent delivery platform for imaging agents (i.e., fluorescent dyes, gold nanoparticles) and anticancer drugs [42, 111]. Importantly, when $CaCO_3$ nanoparticles are taken up by tumor cells, the tumor's acidic pH environment rapidly decomposes the $CaCO_3$ matrix to release the imaging agents and/or anticancer drugs [15]. For instance, a recent study by Sun et al. found folate-PEG-modified $CaCO_3$ nanoparticles could be loaded with a fluorescent imaging agent for targeted delivery to tumor cells. After the decomposition of the $CaCO_3$ nanoparticles, the imaging agent was quickly released, and its fluorescence was activated under the excitation light source. The results of their study demonstrated a $CaCO_3$ nanoparticle-based imaging agent delivery platform enabled the *in vivo* imaging of tumors and the ability to aid in their surgical treatment [112]. Also, a study by Kiranda et al. investigated the cytotoxicity of spherical conjugated gold-$CaCO_3$ nanoparticles for biomedical applications. Their study found that gold-$CaCO_3$ conjugates were a stable and biocompatible nanoparticle-based carrier for biomedical applications [113]. Similarly, a study by Cheng et al. revealed that gold nanoparticles could be used for molecular imaging of living tissues [114].

This highlights the importance of gold as an imaging agent and $CaCO_3$ as an effective nano-based delivery platform. However, the efficient delivery of hydrophobic drugs and imaging agents in the physiological environment is challenging. The low water solubility of these drugs and agents dramatically reduces their clinical effectiveness and overall performance. Due to this shortcoming, recent studies have focused on developing nanoparticle-based carrier systems that can be loaded with several hydrophobic and hydrophilic drugs and agents for delivery simultaneously. And because $CaCO_3$ nanoparticles have a large surface area to volume ratio and high porosity, combined with their stability in the physiological environment, they have attracted considerable interest in recent years as a carrier system. For instance, Manabe et al. studied the pH response of an ionic probe (Rhodamine B) and its release from the porous core of a $CaCO_3$ matrix. Their study found that the probe was stable and could be encapsulated within the matrix and released by varying the medium pH between 5 and 7.4. The lower the medium pH, the more acidic the tumor environment. Their study also confirmed that it was possible to simultaneously load and release three different probes from within the $CaCO_3$ nanoparticle matrix [115]. Studies have also shown that pH-responsive $CaCO_3$ nanoparticle-based platforms, doped with suitable luminescent imaging agents and anticancer drugs, can be used for real-time drug release monitoring. Thus, enabling a combination therapy with real-time drug release monitoring during treatment [48, 116].

9.4.2 Targeted Drug Therapy and Controlled Release Using $CaCO_3$ Nanoparticles

Research over the years has shown that nanoparticles can be effectively used for cancer diagnosis and therapy [117]. Nanoparticle-based drug carriers have several advantages that enable them to overcome problems generally associated with conventional drug therapies. Many conventional treatments suffer from poor target specificity, low bioavailability, and severe side effects [8]. On the other hand, nanoparticle-based carriers have the potential to solve drug delivery concerns by improving absorption properties, enhancing solubilization, optimizing therapeutic efficacy, increasing specific targeting and monitoring, and reducing side effects [118]. Accordingly, targeted nanoparticle-based carrier systems for the treatment of cancer have received considerable interest in recent years. And much of this interest has focused on $CaCO_3$ nanoparticle-based carriers due to their excellent biocompatibility and low toxicity [119]. The pH sensitivity of $CaCO_3$ nanoparticles means the target location can control their degradation rate. Because the pH environment of tumor cells is acidic (around 4.5 to 6.5), pH-sensitive $CaCO_3$ nanoparticles are an innovative carrier system for delivering anticancer drugs [120]. This innovative nanoparticle-based drug delivery system offers many future opportunities for tackling various types of resistant cancer. [36]. For instance, a study by Shafiu Kamba et al. found that pH-sensitive $CaCO_3$ nanocrystal carriers loaded with doxorubicin (DOX) a potent cancer therapeutic drug released far more DOX in the acidic pH-simulating tumor environments than in the normal physiological pH environment of 7.4. Their study indicated that DOX-$CaCO_3$ nanocrystals could be an effective anticancer therapy for the clinical treatment of osteosarcoma bone cancer [121]. Similarly, a study by Peng et al. found a $CaCO_3$-based carrier loaded with DOX showed a slow-release profile in the normal physiological environment (pH 7.4) and a much faster release rate in acidic tumor environments (pH 4.8). Their study also confirmed the porous $CaCO_3$-based carrier was a promising material for delivering anticancer drugs [122]. In another study, Kim et al. used lipid-coated $CaCO_3$ nanoparticles loaded with the therapeutic peptide EEEEpYFELV (EV) to target lung cancer cells. Their study found the initial encapsulation efficiency of the EV peptide was between 65 and 70%. On delivery, the pH-sensitive nanoparticle matrix was able to sustain the release of the EV peptide. Their study also confirmed the $CaCO_3$ nanoparticle-based carriers could be used to deliver a controlled and sustained release of this therapeutic drug to produce a high retardation tumor growth effect [123].

9.4.3 Gene Therapy Using $CaCO_3$ Nanoparticles

At the cellular level, every cell contains a set of hereditary genes. A gene is considered to act as an instruction book. Each letter sequence of each gene holds a specific piece of data that relates to specific molecules, like hormones and proteins, which are vital for cell growth, well-being, and propagation. Unfortunately, many diseases originate from defective or malfunctioning genes inside the body. Genetic problems are severe and can result in transferrable diseases like tuberculosis and AIDS, and non-transmissible diseases like cancer and diabetes [124]. Importantly, the body's principal defense against invading foreign materials and pathogens is the immune system, which is also controlled by specific genetic protocols, thus emphasizing the importance of genetic materials for signaling and regulating pathways responsible for combating disease within the body [125]. This highlights the importance of gene therapy for treating a wide range of healthcare issues ranging from genetically inherited disorders to acquired diseases and cancer [126]. In recent years, several gene delivery vectors ranging from viral (engineered viruses) to non-viral vectors have been studied as a method of delivering effective gene therapy. However, because viruses suffer from problems like mutagenicity and immunogenicity, nanoparticle-based non-viral carriers are considered a safe alternative [127, 128], hence delivering a nanoparticle-based carrier with an appropriate therapeutic gene, as payload has been extensively studied in recent years [14, 129]. However, the studies have also highlighted several challenges that reduce the effectiveness of this type of gene therapy method. These challenges include 1) preventing enzymatic degradation of the delivered DNA or siRNA; 2) promoting greater cellular uptake to optimal levels; 3) improving the release of DNA or siRNA from the carrier; 4) promoting greater diffusion throughout the cytoplasm and nucleus; 5) improving the efficacy of the gene therapy; and 6) reducing side effects of the therapy [130].

Several non-viral gene therapies have investigated and used liposomes and polymeric materials for gene delivery [131, 132]. In recent years, several types of inorganic-based nanoparticle carriers have also been investigated for potential use in gene therapy applications [133]. Types of inorganic nanoparticles investigated include noble metals like gold, metal oxides, quantum dots, silica-based materials, and calcium phosphates [134-137]. However, these non-viral carrier-based delivery platforms suffer from one or more of the above-mentioned challenging issues, which can significantly reduce their therapeutic effectiveness. Also, many of these nanoparticle-based carriers require surface modifications, which have resulted in unfavorable immune system responses that further hinder their therapeutic effectiveness [138, 139]. In recent years, $CaCO_3$-based materials have attracted considerable interest due to their uniquely desirable properties like biocompatibility, biodegradability, pH sensitivity, and controlled release profile [15, 140]. Studies have shown $CaCO_3$

nanoparticles also have the potential to become a new non-viral carrier system for gene delivery [141]. A study by Chen et al demonstrated that $CaCO_3$ nanoparticles could be used to co-deliver p53 expression plasmid (DNA) and doxorubicin hydrochloride (DOX). During the preparation of the carrier, particle size was kept below 100 nm, and the $CaCO_3$ matrix was found to have a high encapsulation efficiency. The *in vitro* cell inhibition evaluation revealed that simultaneous administration of both gene and drug effectively inhibited the proliferation of HeLa cells, thus demonstrating the $CaCO_3$/DNA/DOX nanoparticle-based carrier could co-deliver a gene and drug to targeted cells and offer a promising cancer treatment [142]. In addition, a study by Zhao et al. reported that a lipid-coated $CaCO_3$ nanoparticle-based carrier could successfully co-deliver miRNA-375 (tumor suppressor) and DOX to treat hepatocellular carcinoma (HCC). The concurrent delivery found the pH-sensitive liberation of DOX and miR-375 improved efficacy, promoted greater anticancer action, and improved therapeutic efficiency [143]. Furthermore, a study by He et al. reported the successful delivery of VEGF-C targeted siRNA using a $CaCO_3$ nanoparticle-based carrier system for *in vitro* cancer models. Their study observed a reduced VEGF-C expression in the SGC-7901 (gastric cancer cell line) cells and a significant suppression in carcinogenesis, thus confirming the effectiveness of the $CaCO_3$ nanoparticle-based carrier and its potential use as an anticancer treatment [144].

9.5 CONCLUDING REMARKS

This chapter has reviewed the current literature and found that $CaCO_3$ nanoparticles have a unique set of properties that make them a promising carrier platform for the delivery of imaging agents used in diagnostics and pharmaceuticals used in gene therapy and cancer therapy. Several chemical methods can be used to produce $CaCO_3$ nanoparticles and engineer their structure and surface interface to protect their drug/agent payload. Studies have also shown the pH-sensitive behavior of $CaCO_3$ nanoparticles allows them to be specifically targeted to tumor sites. However, based on current literature, a few challenges need to be addressed in future studies to fully elucidate the behavior of $CaCO_3$ nanoparticles when used for specific medical applications, for instance, improving and optimizing the loading capacity of specific agents and pharmaceuticals into the nanoparticles matrix and then functionalizing the surface to enhance interface properties. Furthermore, more *in vivo* studies are needed to investigate the degradation mechanism of the carrier platform and then the subsequent elimination of by-products from the body. Thus, further studies would significantly contribute to knowledge in this field and promote $CaCO_3$ nanoparticle-based carriers for future therapeutic applications. This nanotherapeutical use of calcium carbonate nanoparticle shows a viable pathway to synthesise cancer drug platforms and potentially enhance the treatment of this prevalent disease in the community.

REFERENCES

1. American Cancer Society. Cancer facts and figures 2020. https://www.cancer.org/research/cancer-facts-statistics.html.
2. CDC. Antibiotic resistance threats in the United States, Atlanta, GA 2019. DOI: 10.15620/cdc:82532.
3. Australia Institute of Health and Wellbeing. Cancer data in Australia 2020. https://www.aihw.gov.au/reports/cancer/cancer-data-in-australia.
4. Fathia N, Rashidi G, Khodadadi A, Shahi S, Sharifi S. STAT3 and apoptosis challenges in cancer. *International Journal of Biological Macromolecules*. 2018; 117 (1): 993–1001. DOI: 10.1016/j.ijbiomac.2018.05.121.
5. Brookman-May SD, May M, Shariat SF, Novara G, Zigeuner R, Cindolo L, De Cobelli O, De Nunzio C, Pahernik S, Wirth MP, Longo N. Time to recurrence is a significant predictor of cancer-specific survival after recurrence in patients with recurrent renal cell carcinoma–results from a comprehensive multi-centre database (CORONA/SATURN-Project). *BJU International*. 2013; 112 (7): 909–916. DOI: 10.1111/bju.12246.
6. Corrado G, Salutari V, Palluzzi E, Distefano MG, Scambia G, Ferrandina G. Optimizing treatment in recurrent epithelial ovarian cancer. *Expert Review of Anticancer Therapy*. 2017; 17 (12): 1147–1158. DOI: 10.1080/14737140.2017.1398088.

7. Goldsbury DE, Yap S, Weber MF, Veerman L, Rankin N, Banks E, Canfell K, O'Connell DL. Health services costs for cancer care in Australia: Estimates from the 45 and up study. *PLoS ONE*. 2018; 13 (7): e0201552. DOI: 10.1371/journal.pone.0201552.
8. Campbell RB. Tumor physiology and delivery of nanopharmaceuticals. *Anti-Cancer Agents in Medicinal Chemistry*. 2006; 6 (6): 503–512. DOI: 10.2174/187152006778699077.
9. Cho K, Wang X, Nie S, Chen Z, Shin DM. Therapeutic nanoparticles for drug delivery in cancer. *Clinical Cancer Research*. 2008; 14 (5): 1310–1316. DOI: 10.1158/1078-0432.CCR-07-1441.
10. Le Thi X, Poinern GEJ, Subramaniam S, Fawcett D. Applications of nanometre scale particles as pharmaceutical delivery vehicles in medicine. *OJBMR*. 2015; 2 (2): 11–26. http://doi.org/10.12966/ojbmr.04.01.2015.
11. Shi J, Votruba AR, Farokhzad OC, Langer RS. Nanotechnology in drug delivery and tissue engineering: From discovery to application. *Nano Letters*. 2010; 10 (9): 3223–3230. DOI: 10.1021/nl102184c.
12. Anselmo AC, Mitragotri S. An overview of clinical and commercial impact of drug delivery systems. *Journal of Controlled Release*. 2014; 190: 15–28. DOI: 10.1016/j.jconrel.2014.03.053.
13. Emerich DF, Thanos CG. Targeted nanoparticle-based drug delivery and diagnosis. *Journal of Drug Targeting*. 2007; 15: 163–183. DOI: 10.1080/10611860701231810.
14. Dizaj SM, Jafari S, Khosroushahi AY. A sight on the current nanoparticle-based gene delivery vectors. *Nanoscale Research Letters*. 2014; 9 (1): 252, 1–9. DOI: 10.1186/1556-276X-9-252.
15. Dizaj SM, Barzegar-Jalali M, Zarrintan MH, Adibkia K, Lotfipour F. Calcium carbonate nanoparticles as cancer drug delivery system. *Expert Opinion on Drug Delivery*. 2015; 12 (10): 1649–1660. DOI: 10.1517/17425247.2015.1049530.
16. Jasim A, Abdelghany S, Greish K. Chapter 2: Current update on the role of enhanced permeability and retention effect in cancer nanomedicine. In: Mishra V, Kesharwani P, Mohd Amin MCI, Iyer A (Eds.), *Nanotechnology-Based Approaches for Targeting and Delivery of Drugs and Genes*. Academic Press, 2017, pp. 62–109. DOI: 10.1016/B978-0-12-809717-5.00002-6.
17. Maeda H. The 35th anniversary of the discovery of EPR effect: A new wave of nanomedicines for tumor-targeted drug delivery-personal remarks and future prospects. *Journal of Personalized Medicine*. 2021; 11 (229): 1–17. DOI: 10.3390/jpm11030229.
18. Maeda H, Matsumura Y. EPR effect-based drug design and clinical outlook for enhanced cancer chemotherapy. *Advanced Drug Delivery Reviews*. 2011; 63 (3): 129–130. DOI: 10.1016/j.addr.2010.05.001.
19. Germain M, Caputo F, Metcalfe S, Tosi G, Spring K, Åslund AK, Pottier A, Schiffelers R, Ceccaldi A, Schmid R. Delivering the power of nanomedicine to patients today. *Journal of Controlled Release*. 2020; 326: 164–171. DOI: 10.1016/j.jconrel.2020.07.007.
20. Allen TM, Cullis PR. Drug delivery systems: Entering the mainstream. *Science*. 2004; 303 (5665): 1818–1822.
21. Riehemann K, Schneider SW, Luger TA, Godin B, Ferrari M, Fuchs H. Nanomedicine-challenge and perspectives. Angewandte Chemie International Edition. 2009; 48 (5): 872–897.
22. Feynman R. There's plenty of room at the bottom. *Science*. 1991; 254: 1300–1301.
23. Working safely with nanomaterials 2012. https://www.safenano.org/knowledgebase/resources.
24. Cai W, Gao T, Hong H, Sun J. Applications of gold nanoparticles in cancer nanotechnology. *Nanotechnology, Science and Applications*. 2008; 1: 17–32. DOI: 10.2147/NSA.S3788.
25. Hardin J, Bertoni G. *The World of the Cell*. A&S Academic Science; 9th edition, 2017. ISBN-10: 9781292177694.
26. Parak WJ, Gerion D, Pellegrino T, Zanchet D, Micheel C, Williams SC, Boudreau R, Le Gros MA, Larabell CA, Alivisatos AP. Biological applications of colloidal nanocrystals. *Nanotechnology*. 2003; 14: R15–R27.
27. Lee P, Zhang R, Li V, Liu X, Sun RW, Che CM, Wong KK. Enhancement of anticancer efficacy using modified lipophilic nanoparticle drug encapsulation. *International Journal of Nanomedicine*. 2012; 7: 731–737. DOI: 10.2147/IJN.S28783.
28. Eftekhari A, Ahmadian E, Azami A, Johari-Ahar M, Eghbal MA. Protective effects of coenzyme Q10 nanoparticles on dichlorvos-induced hepatotoxicity and mitochondrial/lysosomal injury. *Environmental Toxicology*. 2018; 33 (2): 167–177. DOI: 10.1002/tox.22505.
29. Naito M, Yokoyama T, Hosokawa K, Nogi K. *Nanoparticle Technology Handbook*. 3rd Edition. Elsevier, 2018. ISBN: 9780444641106.
30. Jarvie H, King S, Dobson P. Nanoparticles. *Encyclopaedia Britannica*. https://www.britannica.com/science/nanoparticle.

31. Singh B, Sharma D, Gupta A. In vitro release dynamics of thiram fungicide from starch and poly (methacrylic acid)-based hydrogels. *The Journal of Hazardous Materials*. 2008; 154: 278–286. DOI: 10.1016/j.jhazmat.2007.10.024.
32. Aouada FA, de Moura MR, Orts WJ, Mattoso LHC. Polyacrylamide and methylcellulose hydrogel as delivery vehicle for the controlled release of paraquat pesticide. *Journal of Materials Science*. 2010; 45 (18): 4977–4985. DOI:10.1007/s10853-009-4180-6.
33. Perez-Herrero E, Fernandez-Medarde A. Advanced targeted therapies in cancer: Drug nanocarriers, the future of chemotherapy. *European Journal of Pharmaceutics and Biopharmaceutics*. 2015; 93: 52–79. DOI: 10.1016/j.ejpb.2015.03.018.
34. Faraji AH, Wipf P. Nanoparticles in cellular drug delivery. *Bioorganic & Medicinal Chemistry*. 2009; 15: 2950–2962. DOI: 10.1016/j.bmc.2009.02.043.
35. Barzegar-Jalali M, Adibkia K, Valizadeh H, Shadbad MRS. Kinetic analysis of drug release from nanoparticles. *Journal of Pharmacy & Pharmaceutical Sciences*. 2008; 11 (1): 167–177. DOI: 10.18433/j3d59t.
36. Patra JK, Das G, Fraceto LF, Campos EVR, Acosta-Torres LS, Diaz-Torres LA, Grillo R, Swamy MK, Sharma S, Habtemariam S, Shin H-S. Nano based drug delivery systems: Recent developments and future prospects. *Journal of Nanobiotechnology*. 2018; 16 (71): 1–33. DOI: 10.1186/s12951-018-0392-8.
37. Ryu JH, Lee S, Son S, Kim SH, Leary JF, Choi K, Kwon IC. Theranostic nanoparticles for future personalized medicine. *Journal of Controlled Release*. 2014; 190: 477–484.
38. Park K. The beginning of the end of the nanomedicine hype. *Journal of Controlled Release*. 2019; 305: 221–222. DOI: 10.1016/j.jconrel.2019.05.044.
39. Wilhelm S, Tavares AJ, Dai Q, Ohta S. Analysis of nanoparticle delivery to tumours. *Nature Reviews, Materials*. 2016; 1: 1–12. DOI: 10.1038/natrevmats.2016.14.
40. Yang L, Sheldon BW, Webster TJ. Nanophase ceramics for improved drug delivery. *The American Ceramic Society Bulletin*. 2010; 89: 24–31.
41. Zhang Y, Ma P, Wang Y, Du J, Zhou Q, Zhu Z, Yang X, Yuan J. Biocompatibility of porous spherical calcium carbonate microparticles on Hela cells. *World Journal of Nano Science and Engineering*. 2012; 2 (1): 25–31.
42. Zhou C, Chen T, Wu C, Zhu G, Qiu L, Cui C, Hou W, Tan W. Aptamer $CaCO_3$ nanostructures: A facile, pH responsive, specific platform for targeted anticancer theranostics. *Chemistry: An Asian Journal*. 2014; 10 (1): 166–171. DOI: 10.1002/asia.201403115.
43. Boyjoo Y, Pareek VK, Liu J. Synthesis of micro and nano-sized calcium carbonate particles and their applications. *Journal of Materials Chemistry A*. 2014; 2: 14270–14288. DOI: 10.1039/c4ta02070g.
44. European Food Safety Authority. Scientific opinion on re-evaluation of calcium carbonate (E 170) as a food additive. *EFSA Journal*. 2011; 9 (7): 1–73. DOI: 10.2903/j.efsa.2011.2318.
45. Trushina DB, Bukreeva TV, Kovalchuk MV, Antipina MN. CaCO3 vaterite microparticles for biomedical and personal care applications. *Materials Science and Engineering C*. 2014; 45: 644–658. DOI: 10.1016/j.msec.2014.04.050.
46. Tas AC. Use of vaterite and calcite in forming calcium phosphate cement scaffolds. In Brito M, Case E, Kriven WM, Salem J, Zhu D (Eds.), *Developments in Porous, Biological and Geopolymer Ceramics: Ceramics and Engineering Science Proceedings*. Hoboken: John Wiley & Sons, Inc., 2008; 28: 135–150.
47. Hammadi NI, Abba Y, Hezmee MNM, Razak ISA, Jaji AZ, Isa T, Mahmood SK, Zakaria MZ. Formulation of a sustained release docetaxel loaded cockle shell-derived calcium carbonate nanoparticles against breast cancer. *Pharmaceutical Research*. 2017; 34(6): 1193–1203. DOI: 10.1007/s11095-017-2135-1.
48. Dong Z, Feng L, Zhu W, Sun X, Gao M, Zhao H, Chao Y, Liu Z. $CaCO_3$ nanoparticles as an ultrasensitive tumor-pH-responsive nanoplatform enabling real-time drug release monitoring and cancer combination therapy. *Biomaterials*. 2016; 110: 60–70. DOI: 10.1016/j.biomaterials.2016.09.025.
49. Addadi L, Raz S, Weiner S. Taking advantage of disorder: Amorphous calcium carbonate and its roles in biomineralization. *Advanced Materials*. 2003; 15: 959–970.
50. Bushuev YG, Finney AR, Rodger PM. Stability and structure of hydrated amorphous calcium carbonate. *Crystal Growth & Design*. 2015; 15: 5269–5279. DOI: 10.1021/acs.cgd.5b00771.
51. Kuther J, Seshadri R, Knoll W, Tremel W. Templated growth of calcite, vaterite and aragonite crystals onself-assembled monolayers of substituted alkylthiols on gold. *Journal of Materials Chemistry*. 1998; 8: 641–650.
52. Caruso F, Hyeon T, Rotello VM. Nanomedicine. *Chemical Society Reviews*. 2012; 41 (7): 2537–2538.

53. Griesshaber E, Schmahl WW, Neuser R, Pettke T, Blüm M, Mutterlose J, Brand U. Crystallographic texture and microstructure of terebratulide brachiopod shell calcite: An optimized materials design with hierarchical architecture. American Mineralogist. 2007; 92: 722–734.
54. Neri D, Supuran CT. Interfering with pH regulation in tumours as a therapeutic strategy. *Nature Reviews Drug Discovery.* 2011; 10: 767–777.
55. Som A, Raliya R, Tian L, Akers W, Ippolito JE, Singamaneni S, Biswas P, Achilefu S. Monodispersed calcium carbonate nanoparticles modulate local pH and inhibit tumor growth *in vivo. Nanoscale.* 2016; 8 (25): 12639–12647. DOI: 10.1039/c5nr06162h.
56. Cho K, Wang X, Nie S, Chen Z, Shin DM. Therapeutic nanoparticles for drug delivery in cancer. *Clinical Cancer Research.* 2008; 14 (5): 1310–1316.
57. Lammers T, Hennink WE, Storm G. Tumor-targeted nanomedicine: Principles and practice. British Journal of Cancer. 2008; 99: 392–397.
58. Guo Y, Fang Q, Li H, Shi W, Zhang J, Feng J, Jia W, Yang L. Hollow silica nanospheres coated with insoluble calcium salts for pH-responsive sustained release of anticancer drugs. *Chemical Communications.* 2016; 52 (70): 10652–10655.
59. Davis ME, Chen Z, Shin DM. Nanoparticle therapeutics: An emerging treatment modality for cancer. *Nature Reviews Drug Discovery.* 2008; 7 (9) 771–782.
60. Islam KN, Bakar MZA, Ali ME, Hussein MZB. A novel method for the synthesis of calcium carbonate (aragonite) nanoparticles from cockle shells. *Powder Technology.* 2013; 235: 70–75. DOI: 10.1016/j.powtec.2012.09.041.
61. Mohd Abd Ghafar SL, Hussein MZ, Abu Bakar Zakaria Z. Synthesis and characterization of cockle shell-based calcium carbonate aragonite polymorph nanoparticles with surface functionalization. *Journal of Nanoparticles.* 2017; 2017: 1–12. DOI: 10.1155/2017/8196172.
62. Poinern GEJ, Fawcett D. The manufacture of a novel 3D hydroxyapatite microstructure derived from cuttlefish bones for potential tissue engineering applications. *American Journal of Materials Science.* 2013; 3 (5): 130–135.
63. Poinern GEJ, Brundavanam RK, Fawcett D. Synthesis of a bone like composite material derived from natural pearl oyster shells for potential tissue engineering applications. *International Journal of Research Medical Sciences.* 2017; 5 (6): 2454–2461.
64. Borhade AV, Kale AS. Calcined eggshell as a cost-effective material for removal of dyes from aqueous solution. *Applied Water Sciences.* 2017; 7: 4255–4268. DOI: 10.1007/s13201-017-0558-9.
65. Hassan TA, Rangari VK, Rana RK, Jeelani, S. Sonochemical effect on size reduction of $CaCO_3$ nanoparticles derived from waste eggshells. *Ultrasonics Sonochemistry.* 2013; 20 (5): 1308–1315. https://doi:10.1016/j.ultsonch.2013.01.016.
66. Tsuzuki T, McCormick PG. Mechanochemical synthesis of nanoparticles. *Journal of Materials Science.* 2004; 39 (16–17): 5143–5146.
67. Chuajiw W, Takatori K, Igarashi T, Hara H, Fukushima Y. The influence of aliphatic amines, diamines, and amino acids on the polymorph of calcium carbonate precipitated by the introduction of carbon dioxide gas into calcium hydroxide aqueous suspensions. Journal of Crystal Growth. 2014; 386: 119–127. DOI: 10.1016/j.jcrysgro.2013.10.09.
68. Trushina DB, Bukreeva TV, Antipina MN. Size-controlled synthesis of vaterite calcium carbonate by the mixing method: Aiming for nanosized particles. *Crystal Growth & Design.* 2016; 16: 1311–1319. DOI: 10.1021/acs.cgd.5b01422.
69. Le Thi B, Shi R, Long BD, Ramesh S, Xingling S, Sugiura Y, Ishikawa K. Biological responses of MC3T3-E1 on calcium carbonate coatings fabricated by hydrothermal reaction on titanium. *Biomedical Materials.* 2020; 15 (3): 035004. DOI: 10.1088/1748-605X/ab6939.
70. Nagaraja AT, Pradhan S, McShane MJ. Poly (vinylsulfonic acid) assisted synthesis of aqueous solution stable vaterite calcium carbonate nanoparticles. *Journal of Colloid and Interface Science.* 2014; 418: 366–372. DOI: 10.1016/j.jcis.2013.12.008.
71. Biradar S, Ravichandran P, Gopikrishnan R, Goornavar V, Hall JC, Ramesh V, Baluchamy S, Jeffers RB, Ramesh GT. Calcium carbonate nanoparticles: Synthesis, characterization and biocompatibility. *Journal of Nanoscience and Nanotechnology.* 2011; 11 (8): 6868–6874. DOI: 10.1166/jnn.2011.4251.
72. Liu HS, Chen KA, Tai CY. Droplet stability and product quality in the Higee assisted microemulsion process for preparing $CaCO_3$ particles. *The Chemical Engineering Journal* 2012; 197: 101–109. DOI: 10.1016/j.cej.2012.05.022.
73. Wang C, Sheng Y, Zhao X, Pan Y, Wang Z. Synthesis of hydrophobic $CaCO_3$ nanoparticles. *Materials Letters.* 2006; 60 (6): 854–857.

74. Bodnarchuk MS, Dini D, Heyes DM, Chahine S, Edwards S. Self-assembly of calcium carbonate nanoparticles in water and hydrophobic solvents. *The Journal of Physical Chemistry C*. 2014; 118 (36): 21092–21103.
75. Shimpi NG, Mali AD, Hansora DP, Mishra S. Synthesis and surface modification of calcium carbonate nanoparticles using ultrasound cavitation technique. *Nanoscience and Nanoengineering*. 2015; 3 (1): 8–12.
76. He M, Forssberg E, Wang Y. Ultrasonication-assisted synthesis of calcium carbonate nanoparticles. *Chemical Engineering Communications*. 2005; 192 (11): 1468–1481.
77. Gedanken A. Using sonochemistry for the fabrication of nanomaterials. *Ultrasonics Sonochemistry*. 2007; 11: 47–55.
78. Poinern GEJ, Brundavanam R, Thi Le X, Djordjevic S, Prokic M, Fawcett D. Thermal and ultrasonic influence in the formation of nanometer scale hydroxyapatite bio-ceramic. *International Journal of Nanomedicine*. 2011; 6: 2083–2095.
79. Orringer DA, Koo YE, Chen T, Kopelman R, Sagher O, Philbert MA. Small solutions for big problems: The application of nanoparticles to brain tumor diagnosis and therapy. *Clinical Pharmacology & Therapeutics*. 2009; 85 (5): 531–534.
80. Yeow EKL, Melnikov SM, Bell TDM, De Schryver FC, Hofkens J. Characterizing the fluorescence intermittency and photobleaching kinetics of dye molecules immobilized on a glass surface. *The Journal of Physical Chemistry*. 2006; 110 (5): 1726–1734.
81. Sokolov I, Naik S. Novel fluorescent silica nanoparticles: Towards ultrabright silica nanoparticles. *Small*. 2008; 4 (7): 934–939.
82. Nicolas J, Mura S, Brambilla D, Mackiewicz N, Couvreur P. Design, functionalization strategies and biomedical applications of targeted biodegradable/biocompatible polymer-based nanocarriers for drug delivery. *Chemical Society Reviews*. 2013; 42 (3): 1147–1235.
83. Michalet X, Pinaud FF, Bentolila LA, Tsay JM, Doose SJ, Li JJ, Sundaresan G, Wu AM, Gambhir SS, Weiss S. Quantum dots for live cells, *in vivo* imaging, and diagnostics. *Science*. 2005; 307 (5709): 538–544.
84. Wang YXJ, Hussain SM, Krestin GP. Superparamagnetic iron oxide contrast agents: Physicochemical characteristics and applications in MR imaging. *European Radiology*. 2001; 11 (11): 2319–2331.
85. Erogbogbo F, Yong KT, Hu R, Law WC, Ding H, Chang CW, Prasad PN, Swihart MT. Biocompatible magneto fluorescent probes: Luminescent silicon quantum dots coupled with superparamagnetic iron (III) oxide. *ACS Nano*. 2010; 4 (9): 5131–5138.
86. Chrastina A, Schnitzer JE. Iodine-125 radiolabeling of silver nanoparticles for *in vivo* SPECT imaging. International Journal of Nanomedicine. 2010; 5: 653–659.
87. Popovtzer R, Agrawal A, Kotov NA, Popovtzer A, Balter J, Carey TE, Kopelman R. Targeted gold nanoparticles enable molecular CT imaging of cancer. *Nano Letters*. 2008; 8 (12): 4593–4596.
88. Liu Y, Ai K, Liu J, Yuan Q, He Y, Lu L. Hybrid BaYbF5 nanoparticles: Novel binary contrast agent for high-resolution *in vivo* X-ray computed tomography angiography. *Advanced Healthcare Materials*. 2012; 1 (4): 461–466.
89. Ai K, Liu Y, Liu J, Yuan Q, He Y, Lu L. Large-scale synthesis of Bi_2S_3 nanodots as a contrast agent for *in vivo* X-ray computed tomography imaging. *Advanced Materials*. 2011; 23 (42): 4886–4891.
90. Lewinski N, Colvin V, Drezek R. Cytotoxicity of nanoparticles. *Small*. 2008; 4 (1): 26–49.
91. Soo Choi H, Liu W, Misra P, Tanaka E, Zimmer JP, Itty Ipe B, Bawendi MG, Frangioni JV. Renal clearance of quantum dots. *Nature Biotechnology*. 2007; 25 (10): 1165–1170.
92. Aillon KL, Xie Y, El-Gendy N, Berkland CJ, Forrest M. Effects of nanomaterial physicochemical properties on *in vivo* toxicity. *Advanced Drug Delivery Reviews*. 2009; 61 (6): 457–466.
93. Perrault SD, Chan WCW. Synthesis and surface modification of highly monodispersed, spherical gold nanoparticles of 50–200 nm. *Journal of the American Chemical Society*. 2009; 131 (47): 17042–17043.
94. Dykman L, Khlebtsov N. Gold nanoparticles in biomedical applications: Recent advances and perspectives. *Chemical Society Reviews*. 2012; 41 (6): 2256–2282.
95. Wani MY, Hashim MA, Nabi F, Malik MA. Nanotoxicity: Dimensional and morphological concerns. *Advances in Physical Chemistry*. 2011; 2011: 450912.
96. Rzigalinski BA, Strobl JS. Cadmium-containing nanoparticles: Perspectives on pharmacology and toxicology of quantum dots. *Toxicology Applied Pharmacology*. 2009; 238 (3): 280–288.
97. Li H, Li M, Shih WY, Lelkes PI, Shih WH. Cytotoxicity tests of water soluble ZnS and CdS quantum dots. *Journal of Nanoscience Nanotechnology*. 2011; 11 (4): 3543–3551.
98. Derfus AM, Chan WCW, Bhatia SN. Probing the cytotoxicity of semiconductor quantum dots. *Nano Letters*. 2004; 4 (1): 11–18.

99. Smith AM, Duan H, Mohs AM, Nie S. Bioconjugated quantum dots for *in vivo* molecular and cellular imaging. *Advances in Drug Delivery Reviews*. 2008; 60 (11): 1226–1240.
100. Jia G, Wang H, Yan L, Wang X, Pei R, Yan T, Zhao Y, Guo X. Cytotoxicity of carbon nanomaterials: Single-wall nanotube, multi-wall nanotube, and fullerene. *Environmental Science & Technology*. 2005; 39: 1378–1383.
101. Nagai H, Okazaki Y, Chew S, Misawa N, Yamashita Y, Akatsuka S, Ishihara T, Yamashita K, Yoshikawa Y, Yasui H, Jiang L. Diameter and rigidity of multiwalled carbon nanotubes are critical factors in mesothelial injury and carcinogenesis. *Proceedings of the National Academy of Sciences of the United States of America*. 2011; 108: E1330–E1338.
102. Chen Z, Meng H, Xing G, Chen C, Zhao Y. Toxicological and biological effects of nanomaterials. *International Journal of Nanotechnology*. 2007; 4: 179–196.
103. Poland C, Duffin R, Kinloch I, Maynard A, Wallace WA, Seaton A, Stone V, Brown S, MacNee W, Donaldson K. Carbon nanotubes introduced into the abdominal cavity of mice show asbestos-like pathogenicity in a pilot study. *Nature Nanotechnology*. 2008; 3: 423–428.
104. Yamashita K, Yoshioka Y, Higashisaka K, Morishita Y, Yoshida T, Fujimura M, Kayamuro H, Nabeshi H, Yamashita T, Nagano K, Abe Y. Carbon nanotubes elicit DNA damage and inflammatory response relative to their size and shape. *Inflammation*. 2010; 33: 276–280.
105. Wick P, Manser P, Limbach L, Dettlaff-Weglikowska U, Krumeich F, Roth S, Stark WJ, Bruinink A. The degree and kind of agglomeration affect carbon nanotube cytotoxicity. *Toxicology Letters*. 2007; 168: 121–131.
106. Lui Y, Zhao Y, Sun B, Chen C. Understanding the toxicity of carbon nanotubes. *Accounts of Chemical Research*. 2013; 46(3): 702–713.
107. Manna SK, Sarkar S, Barr J, Wise K, Barrera EV, Jejelowo O, Rice-Ficht AC, Ramesh GT. Single-walled carbon nanotube induces oxidative stress and activates nuclear transcription factor-kappa B in human keratinocytes. *Nano Letters*. 2005; 5 (9): 1676–1684.
108. Bu L, Shen B, Cheng Z. Fluorescent imaging of cancerous tissues for targeted surgery. *Advanced Drug Delivery Reviews*. 2014; 76: 21–38. DOI: 10.1016/j.addr.2014.07.008.
109. Gonda K, Watanabe M, Tada H, Miyashita M, Takahashi-Aoyama Y, Kamei T, Ishida T, Usami S, Hirakawa H, Kakugawa Y, Hamanaka Y. Quantitative diagnostic imaging of cancer tissues by using phosphor-integrated dots with ultrahigh brightness. *Science Reports*. 2017; 7(1): 7509. DOI: 10.1038/s41598-017-06534-z.
110. Dai Z, Ma H, Tian L, Song B, Tan M, Zheng X, Yuan J. Construction of a multifunctional nanoprobe for tumor-targeted time-gated luminescence and magnetic resonance imaging *in vitro* and *in vivo*. *Nanoscale*. 2018; 10 (24): 11597–11603. doi: 10.1039/c8nr03085e.
111. He X, Li J, An S, Jiang C. pH-sensitive drug-delivery systems for tumor targeting. *Therapy Delivery*. 2013; 4 (12): 1499–1510. DOI: 10.4155/tde.13.120.
112. Sun N, Wang D, Yao G, Li X, Mei T, Zhou X, Wong KY, Jiang B, Fang Z. pH-dependent and cathepsin B activable $CaCO_3$ nanoprobe for targeted *in vivo* tumor imaging. *International Journal of Nanomedicine*. 2019; 14: 4309–4317.
113. Kiranda HK, Mahmud R, Abubakar D, Zakaria ZA. Fabrication, characterization and cytotoxicity of spherical-shaped conjugated Gold-Cockle shell derived calcium carbonate nanoparticles for biomedical applications. *Nanoscale Research Letters*. 2018; 13 (1): 1–10. DOI: 10.1186/s11671-017-2411-3.
114. Cheng K, Kothapalli SR, Liu H, Koh AL, Jokerst JV, Jiang H, Yang M, Li J, Levi J, Wu JC, Gambhir SS. Construction and validation of nano gold tripods for molecular imaging of living subjects. *Journal of the American Chemical Society*. 2014; 136: 3560–3571.
115. Manabe K, Oniszczuk J, Michely L, Belbekhouche S. pH- and redox-responsive hybrid porous $CaCO_3$ microparticles based on cyclodextrin for loading three probes all at once. *Colloids and Surfaces A: Physicochemical and Engineering Aspects*. 2020; 602: 125072. DOI: 10.1016/j.colsurfa.2020.125072.
116. Lin Y, Chan CM. Calcium carbonate nanocomposites. In: Gao F (Ed.), *Advances in Polymer Nanocomposites*. Woodhead Publishing, 2012, pp. 55–90. DOI: 10.1533/9780857096241.1.55.
117. Koo YEL, Reddy GR, Bhojani M, Schneider R, Philbert MA, Rehemtulla A, Ross BD, Kopelman R. Brain cancer diagnosis and therapy with nanoplatforms. *Advances in Drug Delivery Reviews*. 2006; 58 (14): 1556–1577.
118. Hu CM, Aryal S, Zhang L. Nanoparticle-assisted combination therapies for effective cancer treatment. *Therapeutic Delivery*. 2010; 1 (2): 323–334.
119. Wang CQ, Gong MQ, Wu JL, Zhuo RX, Cheng SX. Dual-functionalized calcium carbonate based gene delivery system for efficient gene delivery. *RSC Advances*. 2014; 4: 38623–38629.

120. Zhou H, Wei J, Dai Q, Wang L, Luo J, Cheang T, Wang S. $CaCO_3$/CaIP6 composite nanoparticles effectively deliver AKT1 small interfering RNA to inhibit human breast cancer growth. *International Journal of Nanomedicine*. 2015; 10: 4255.
121. Shafiu Kamba A, Ismail M, Tengku Ibrahim TA, Zakaria ZA. A pH-sensitive, biobased calcium carbonate aragonite nanocrystal as a novel anticancer delivery system. *BioMed Research International*. 2013; 2013: 1–10.
122. Peng C, Zhao Q, Gao C. Sustained delivery of doxorubicin by porous $CaCO_3$ and chitosan/alginate multilayers-coated $CaCO_3$ microparticles. *Colloids and Surfaces A: Physicochemical and Engineering Aspects*. 2010; 353 (2–3): 132–139.
123. Kim SK, Foote MB, Huang L. Targeted delivery of EV peptide to tumor cell cytoplasm using lipid coated calcium carbonate nanoparticles. *Cancer Letters*. 2013; 334 (2): 311–318.
124. Culliton BJ. (1991). Gene therapy on the move. *Nature (London)*. 1991; 354 (6353): 429–429. DOI: 10.1038/354429a0.
125. Verma IM, Somia N. Gene therapy - Promises, problems and prospects. *Nature*. 1997; 389 (6648): 239–242.
126. Biffi S, Voltan R, Rampazzo E, Prodi L, Zauli G, Secchiero P. Applications of nanoparticles in cancer medicine and beyond: Optical and multimodal *in vivo* imaging, tissue targeting and drug delivery. *Expert Opinion on Drug Delivery*. 2015; 12 (12): 1837–1849.
127. Thomas CE, Ehrhardt A, Kay MA. Progress and problems with the use of viral vectors for gene therapy. *Nature Reviews Genetics*. 2003; 4 (5): 346–358.
128. Whitehead KA, Langer R, Anderson DG. Knocking down barriers: Advances in siRNA delivery. *Nature Reviews Drug Discovery*. 2009; 8 (2): 129–138.
129. Mintzer MA, Simanek EE. Nonviral vectors for gene delivery. *Chemical Reviews*. 2009; 109 (2): 259–302.
130. Atkinson H, Chalmers R. Delivering the goods: Viral and non-viral gene therapy systems and the inherent limits on cargo DNA and internal sequences. *Genetica*. 2010; 138 (5): 485–498.
131. Balazs DA, Godbey W. Liposomes for use in gene delivery. *Journal of Drug Delivery*. 2011; 2011: 326497.
132. Lungwitz U, Breunig M, Blunk T, Gopferich A. Polyethylenimine-based non-viral gene delivery systems. *European Journal of Pharmaceutics and Biopharmaceutics*. 2005; 60 (2): 247–266.
133. Chen C, Roy I, Yang C, Prasad PN. Nanochemistry and nanomedicine for nanoparticle-based diagnostics and therapy. *Chemical Review*. 2016; 116: 2826–2885.
134. Bonoiu AC, Mahajan SD, Ding H, Roy I, Yong KT, Kumar R, Hu R, Bergey EJ, Schwartz SA, Prasad PN. Nanotechnology approach for drug addiction therapy: Gene silencing using delivery of gold nanorod-siRNA nanoplex in dopaminergic neurons. *Proceedings of the National Academy of Sciences USA*. 2009; 106 (14): 5546–5550.
135. Arsianti M, Lim M, Marquis CP, Amal R. Assembly of polyethylenimine-based magnetic iron oxide vectors: Insights into gene delivery. *Langmuir*. 2010; 26 (10): 7314–7326.
136. Yezhelyev MV, Qi L, O'Regan RM, Nie, S, Gao X. Proton-sponge coated quantum dots for siRNA delivery and intracellular imaging. *Journal of the American Chemical Society*. 2008; 130 (28): 9006–9012.
137. Bisht S, Bhakta G, Mitra S, Maitra A. pDNA loaded calcium phosphate nanoparticles: Highly efficient non-viral vector for gene delivery. *International Journal of Pharmaceutics*. 2005; 288 (1): 157–168.
138. Burnett JC, Rossi JJ, Tiemann K. Current progress of siRNA/shRNA therapeutics in clinical trials. *Biotechnology Journal*. 2011; 6 (9): 1130–1146.
139. Coelho T, Adams D, Silva A, Lozeron P, Hawkins PN, Mant T, Perez J, Chiesa J, Warrington S, Tranter E, Munisamy M. Safety and efficacy of RNAi therapy for transthyretin amyloidosis. *New England Journal of Medicine*. 2013; 369 (9): 819–829.
140. Kong X, Xu S, Wang X, Cui F, Yao J. Calcium carbonate microparticles used as a gene vector for delivering p53 gene into cancer cells. *Journal of Biomedical Materials Research Part A*. 2012; 100 (9): 2312–2318.
141. Chen S, Li F, Zhuo RX, Cheng SX. Efficient non-viral gene delivery mediated by nanostructured calcium carbonate in solution-based transfection and solid phase transfection. *Molecular Bio Systems*. 2011; 7: 2841–2847.
142. Chen S, Zhao D, Li F, Zhuo RX, Cheng SX. Co-delivery of genes and drugs with nanostructured calcium carbonate for cancer therapy. *RSC Advances*. 2012; 2: 1820–1826.

143. Zhao P, Wu S, Cheng Y, You J, Chen Y, Li M, He C, Zhang X, Yang T, Lu Y, Lee RJ. MiR-375 delivered by lipid-coated doxorubicin calcium carbonate nanoparticles overcomes chemo resistance in hepatocellular carcinoma. *Nanomedicine: Nanotechnology, Biology and Medicine.* 2017; 13 (8): 2507–2516.
144. He XW, Liu T, Chen YX, Cheng DJ, Li XR, Xiao Y, Feng YL. Calcium carbonate nanoparticle delivering vascular endothelial growth factor-C siRNA effectively inhibits lymphangiogenesis and growth of gastric cancer *in vivo. Cancer Gene Therapy.* 2008; 15 (3): 193–202.

10 Ultrasonically Engineered Silicon Substituted Nanometer-Scale Hydroxyapatite Nanoparticles for Dental and Bone Restorative Procedures

Synthesis, Characterization, and Property Evaluation

Supriya Rattan, Derek Fawcett, and Gérrard Eddy Jai Poinern

CONTENTS

10.1 INTRODUCTION

The human skeletal system, which is made up of hard tissues and cartilaginous materials, has evolved over millions of years and its versatile design incorporates the support of organs and soft tissues and the movement of the human body. Hard tissues like bone and teeth are composite materials composed of a biological (organic) component and a mineral (inorganic) component. Bone is composed of rod-like nano-hydroxyapatite (nano-HAP) particles ranging in size from 25 to 50 nm in length that are embedded in collagen fibrils [1, 2]. Pure nano-HAP is composed of calcium phosphate groups, with the general formula of $Ca_{10}(OH)_2(PO_4)_6$, which form a unit cell with a hexagonal structure [3]. However, the natural form of nano-HAP found in bone is not pure. Instead, it exhibits deficiencies in calcium (Ca), hydroxyl (OH^-), and phosphorus (P), which are naturally replaced by a variety of different ionic substitutions of varying concentrations. The presence of ionic substitutions like magnesium (Mg), sodium (Na), silicon (Si), and strontium (Sr) modifies the surface chemistry, charge, and structure of bone and teeth, thus influencing the interaction between hard tissues and their physiological environments [4]. Importantly, both crystallographic and chemical studies have

DOI: 10.1201/9781003181422-12

shown the close similarity between synthetically produced nano-HAP and nano-HAP found naturally in bone. Because of this phenomenon, synthesized micro-HAP and nano-HAP-based materials are widely used in biomaterial products for repairing both bone and teeth [5–7]. Although synthetic nano-HAP offers good osteoconductivity and osteoinductivity, its slow biodegradability tends to hinder tissue regeneration [8]. Because of this shortcoming, recent research has focused on improving synthetic nano-HAPs *in vivo* biodegradability and bioactivity, thus promoting greater bone formation and calcification. Early *in vitro* and *in vivo* studies by Carlisle detected silicon in active growth areas like the osteoid regions in the tibiae of young mice and rats, thus revealing the importance of silicon in calcification processes and for promoting the growth and development of hard tissues in general [9, 10]. Interestingly, several studies following the work of Carlisle have also shown that the addition of small quantities of silicon to synthetic nano-HAP can improve its bioactivity and its dissolution rate [11, 12]. However, studies have found that only small amounts of silicon can be added for property improvement. Typically, amounts ranging from 0.5 to 3% by weight can significantly improve *in vivo* bioactivity and promote bone formation [13, 14]. Studies have also shown the crystallographic, mechanical, and physiochemical properties of synthetic nano-HAP are also influenced by the synthesis processes used during powder manufacture [15, 16].

The present study used a two-step powder manufacturing process that was followed by a powder annealing step. The first step involves sonochemical synthesis. During this acoustic cavitation step, highly energetic bubbles are generated and subsequently collapse within the reaction medium. During bubble collapse, there are extreme reaction conditions generated. These conditions include extremely high temperatures (~ 5000 K) and intense pressures (~ 20 MPa). In this extreme bubble environment, it is possible to generate novel nanometer-scale materials with unique and distinctive physiochemical properties [17]. The second step involves microwave heating the slurry produced by the first step. The microwave heating step has the advantage of volumetrically generating thermal energy throughout the slurry. This process is fundamentally different from the conventional thermal conduction that takes place in a conventional furnace [18]. After the microwave treatment, the raw powders were then ground into ultrafine powders and then subjected to annealing. After processing, all ultrafine powders were then subjected to property evaluation using several advanced characterization techniques. These techniques included X-ray diffraction (XRD) spectroscopy, Fourier transform infrared spectroscopy (FT-IR), energy dispersive spectroscopy (EDS), and field emission scanning electron microscopy (FESEM). XRD spectroscopy was used to identify Miller indices and study the effects of peak broadening in the respective ultrafine powder samples. The XRD data was subsequently used in the Williamson-Hall (W-H) analysis methods to determine crystalline parameters like size, lattice strain, surface stress, and energy density. Fourier transform infrared spectroscopy (FT-IR) was used to identify functional groups and their respective vibration modes present in the respective ultrafine powder samples. Elemental analysis of the respective ultrafine powders was carried out by EDS, and field emission scanning electron microscopy (FESEM) imagery was used to determine both mean particle size and morphology.

10.2 MATERIALS AND METHODS

10.2.1 Materials

All chemicals used in synthesizing the samples were supplied by Chem. Supply Pty Ltd (Gillman, South Australia, Australia) and Sigma Aldrich (United States of America). All aqueous-based solutions were prepared using Milli-Q® water (18.3 M Ω cm^{-1}) produced by an ultrapure water system (Barnstead Ultrapure Water System D11931) supplied by Thermo Scientific Australia.

10.2.2 Experimental Methods: Preparation of Ultrafine Powders

Pure nano-HAP powder synthesis began by pouring a 40 ml solution of 0.32M $Ca(NO_3)_2{\cdot}4H_2O$ into a small glass beaker. The solution was then subjected to ultrasonic processing for 20 minutes at maximum amplitude and operating at 200 W. The processor used was a UP400S supplied by Hielscher Ultrasound Technology (Teltow, Germany) and fitted with a 22 mm diameter Sonotrode operating at 24 kHz. During the 20 minutes, solution pH was stabilized at 9 using 5 ml of NH_4OH that was added drop-wise during the procedure. After the first 20 minutes, a 60 ml solution of 0.19M [KH_2PO_4] was added drop-wise to the solution during a second 20-minute processing period. During the second period, the solution pH was sustained at 9, and the Ca/P ratio of 1.67 was maintained. After ultrasonic processing, the mixture underwent centrifugation (~2000 g), which resulted in the production of the precipitate. The Si-doped nano-HAP ultrafine powders were synthesized using the abovementioned procedure, except for the drop-wise addition of the sodium silicate dopant in parallel with the addition of the KH_2PO_4. Two Si-doped nano-HAP powders were prepared with the 0.19M $Na_2Si_2O_5$ dopant. The first Si-based powder consisted of adding a 1 ml solution of dopant to the reaction mixture, which equated to a 3 % (At. %) Si nano-HAP powder. The second consisted of adding a 5 ml solution of dopant to the reaction mixture, which equated to a 7% (At. %) Si nano-HAP powder. In both cases, elemental analysis was carried out to verify the inclusion of the At. % of Si into the lattice structure of the respective sample. After sonochemical processing, the respective slurry samples were collected and then individually subjected to microwave heating. The heating treatment was carried out in a microwave oven (Model TMOSS25, operated at 900 W at 2450 MHz and 240V and 50Hz) at full power for a 10-minute period. After the microwave treatment, the raw powder samples were individually ground to the consistency of an ultrafine powder using a mortar and pestle. The respective powders were then subjected to annealing at either 400 °C or 800 °C. All powders were then studied using advanced characterization techniques.

10.2.3 Powder Characterization

Particle shape and size range was determined from image analysis derived from electron microscopy. Initially, a JEOL JCM-6000 electron microscope was used to undertake preliminary studies. Prior to imaging, samples were dried and then deposited onto carbon adhesive tape-covered holders. Samples were then sputter coated (Cressington 208HR) with a 2 nm layer of platinum to prevent charge build-up. Elemental analysis of each powder sample was also carried out at this stage using the energy dispersive spectroscopy (EDS) attachment of the microscope. Higher resolution images were obtained via a FESEM: FEI-Verios 460 operating at 5 kV, with 0.10 nA current and operating under secondary electron mode (Microscopy Centre at University of Western Australia). Fourier transform infrared spectroscopy (FT-IR) was used to identify functional groups and vibrational modes present in the powder samples. A PerkinElmer FT-IR / NIR Spectrometer Frontier fitted with a universal signal bounce Diamond ATR attachment was used to collect spectra. The spectra were recorded over the wavenumber interval ranging from 400 to 4000 cm^{-1} with a resolution step of 1 cm^{-1}. X-ray diffraction (XRD) spectroscopy was carried out using a GBC® eMMA X-Ray Powder Diffractometer (Cu $K\alpha$ = 1.54056 Å radiation source) operating at 35 kV and 28 mA. The diffraction spectra were collected over a 2θ range starting at 10° and ending at 80°. The incremental step size over the 2θ range was 0.02°, and the acquisition speed used was 2° min^{-1}. Diffraction peaks in the pattern were identified using the ICDD (International Centre for Diffraction Data) and JCPDS (Joint Committee on Powder Diffraction Standards) databases. After peak identification, the Miller indices were assigned, and peak broadening analysis was carried out using Williamson-Hall (W-H) methods to determine crystalline parameters like size, strain, stress, and energy density.

10.3 RESULTS AND DISCUSSIONS

10.3.1 Williamson-Hall Analysis of XRD Peak Broadening

Representative XRD patterns for pure nano-HAP and Si-doped nano-HAP ultrafine powders are presented in Figure 10.1. All patterns display diffraction peaks and confirm the sample powders are polycrystalline materials. All diffraction peak positions were identified and indexed with respect to

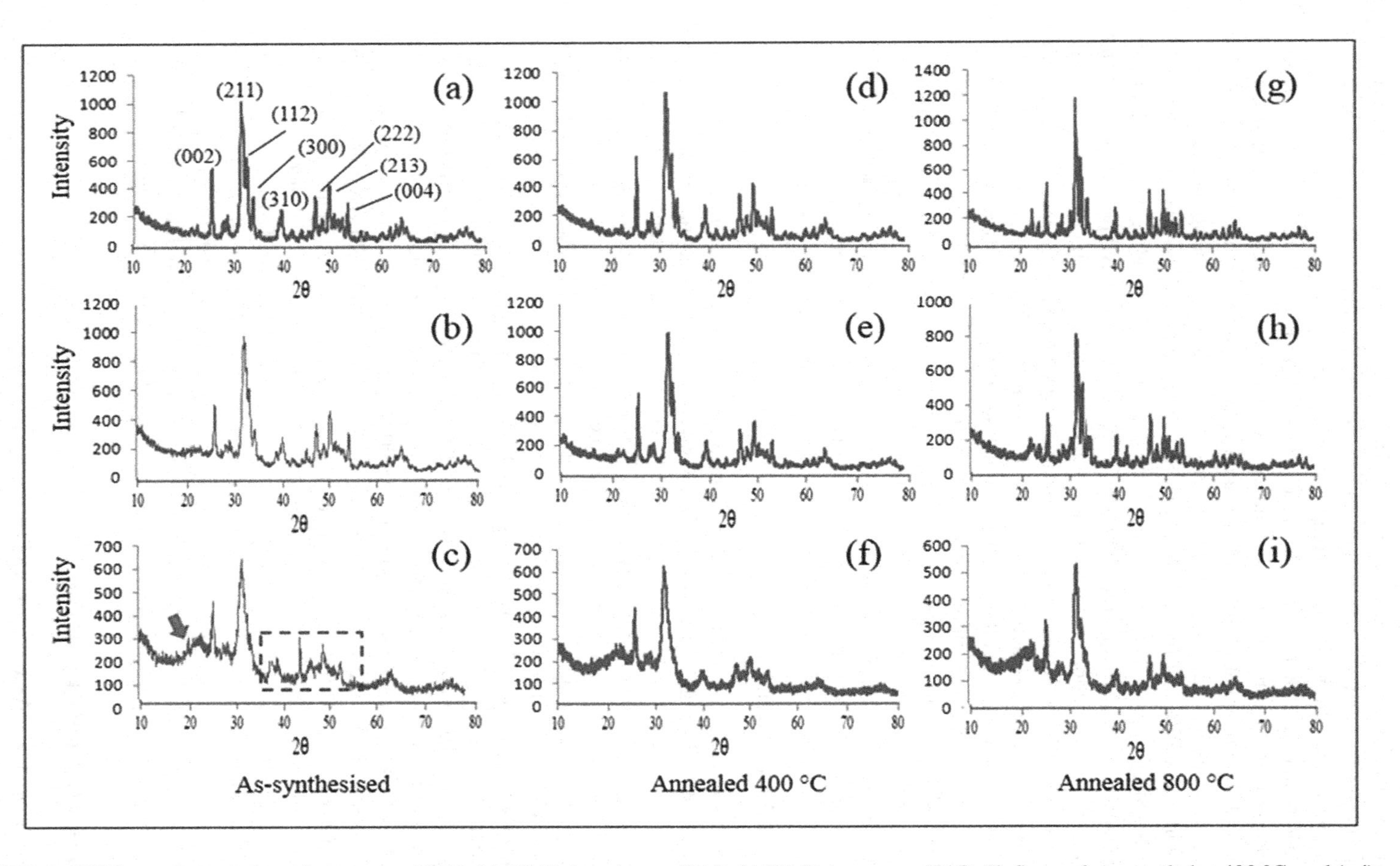

FIGURE 10.1 XRD powder samples: (a) pure nano-HAP; (b) 3% Si-doped nano-HAP; (c) 7% Si dope nano-HAP; (d–f) samples annealed at 400 °C; and (g-i) samples annealed at 800 °C.

the Joint Committee on Powder Diffraction Standards for pure hexagonal crystalline hydroxyapatite (JCPDS No. 09–0432).

Analysis of XRD data failed to detect the presence of any impurities; if any were present in the samples they were below the detection limit of the X-ray equipment. Figure 10.1 (a) presents a representative pure nano-HAP ultrafine powder sample. Figure 10.1 (d) shows a pure sample annealed at 400 °C, and Figure 10.1 (g) shows a sample annealed at 800 °C. Examination of these patterns reveals the annealing process increases both peak intensity and sharpness compared to the non-annealed sample. The annealing temperature of 800 °C produced the largest increase in peak sharpness and indicated enhanced crystallinity and increased crystallite size. XRD patterns for the 400 °C annealing temperature were found to be similar to those of the non-annealed samples. Similarly, other studies have reported improvements in powder crystallinity when annealing temperatures are typically around 800 °C [19, 20]. Figure 10.1 (b) presents a representative 3% Si-doped nano-HAP ultrafine powder pattern. Inspection reveals that its pattern is similar to the pure powder sample presented in Figure 10.1 (a). However, the representative 7% Si-doped nano-HAP ultrafine powder pattern presented in Figure 10.1 (c) shows significant differences from both Figure 10.1 (a) and (b). The major difference between the pure and 7% Si-doped nano-HAP ultrafine powder patterns is the significant decrease in peak intensity and broadening of several peaks, as highlighted in the red square box seen in Figure 10.1 (c). In addition, the appearance of a very broad peak indicated by the red arrow in Figure 10.1 (c) signifies a decrease in crystallinity. Furthermore, when both Si-doped nano-HAP ultrafine powders were annealed they showed increased peak sharpening. However, their respective peak intensities and sharpness were less pronounced than in the annealed pure powder sample.

In the present work, only the graphical analysis of a pure nano-HAP and a 3% Si-doped nano-HAP ultrafine powder are presented in Figures 10.2 and 10.3. The graphical analysis for the 7% Si-doped nano-HAP is not presented, but the results are presented in Table 10.1 for completeness. In addition, because there was little difference between non-annealed and samples annealed at 400 °C, only the analysis of the 800 °C sample is presented. The crystallite size of the respective ultrafine powders was determined using two techniques. The first analysis technique used was the Scherrer ($D_{S\,(hkl)}$) Equation (10.1) and the second was the Williamson-Hall ($D_{\,(hkl)}$) Equation (10.2). The Williamson-Hall analysis was also used to estimate crystalline lattice properties like strain (ε), crystalline stress (σ), and anisotropic energy density (u):

$$D_{s\,(hkl)} = \frac{k\lambda}{\beta_{(hkl)}\cos\theta_{(hkl)}} \tag{10.1}$$

$$\beta_{(hkl)}\cos\theta = \frac{k\lambda}{D_{(hkl)}} + 4\varepsilon\sin\theta \tag{10.2}$$

where λ is the wavelength of the monochromatic X-ray beam, and k is the crystallite shape which is 0.9 for spherical crystals with cubic unit cells. β is the full width at half maximum (FWHM) of the peak at the maximum intensity, $\theta_{\,(hkl)}$ is the peak diffraction angle that satisfies Bragg's law for the ($h\ k\ l$) plane, ε is strain experienced by the lattice, and $D_{s\,(hkl)}$ and $D_{(hkl)}$ are the respective crystallite sizes. Equation (10.2) above is known as the uniform deformation model (UDM) and assumes uniform strain in all crystallographic directions. It also assumes material properties of the crystalline structure are independent of direction. Thus, plotting 4 sin θ along the x-axis and $\beta_{\,(hkl)}$ cos θ along the y-axis facilitates the calculation of line slope (ε) and the (y-intercept) which aids in estimating the crystallite size ($D_{(hkl)}$). Representative graphs of the pure and 3% doped Si-doped nano-HAP ultrafine powders are presented in Figures 10.2 and 10.3, while the estimated values of ($D_{(hkl)}$) and ε are presented in Table 10.1.

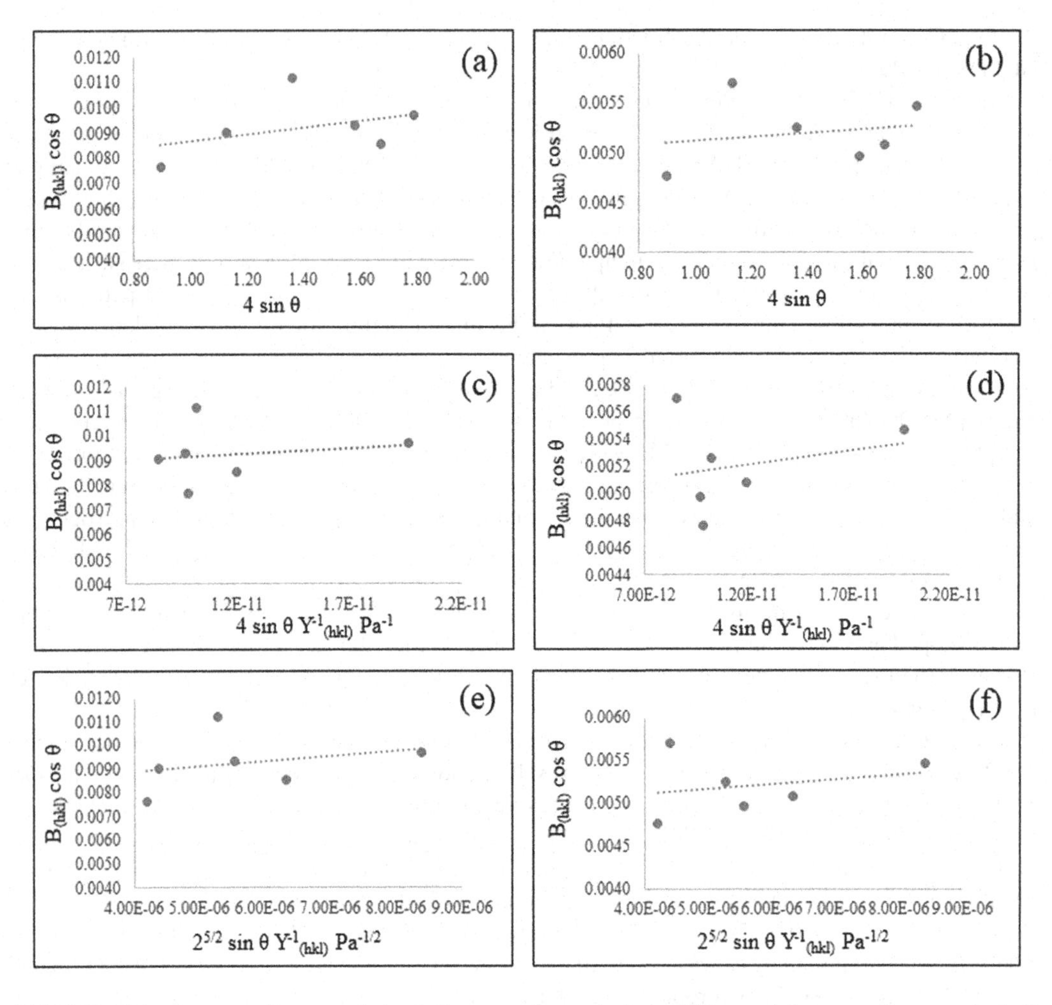

FIGURE 10.2 $\beta_{(hkl)}$ cos θ versus 4 sin θ; $\beta_{(hkl)}$ cos θ versus 4 sin $\theta/Y_{(hkl)}$ and $\beta_{(hkl)}$ cos θ versus 4 sin θ $(2u/Y_{(hkl)})^{1/2}$ for as-synthesized pure nano-HAP sample (a, c, e) and annealed (800 °C) sample (b, d, f).

The crystallite surface stress was calculated using the uniform stress deformation mode (USDM) which assumes linear proportionality between stress and strain (Hooke's law) within the lattice structure:

$$Y_{(hkl)} = \frac{\sigma}{\varepsilon} \tag{10.3}$$

Therefore, when the modulus of elasticity ($Y_{(hkl)}$) or Young's modulus is substituted into Equation (10.2) and rearranged, it produces Equation (10.4) below:

$$\beta_{(hkl)} \cos\theta = \frac{k\lambda}{D_{(hkl)}} + \frac{4\sigma \sin\theta}{Y_{(hkl)}} \tag{10.4}$$

In hexagonal crystallites like hydroxyapatite the elastic modulus $Y_{(hkl)}$ is dependent on the crystallographic direction and is expressed by Equation (10.5):

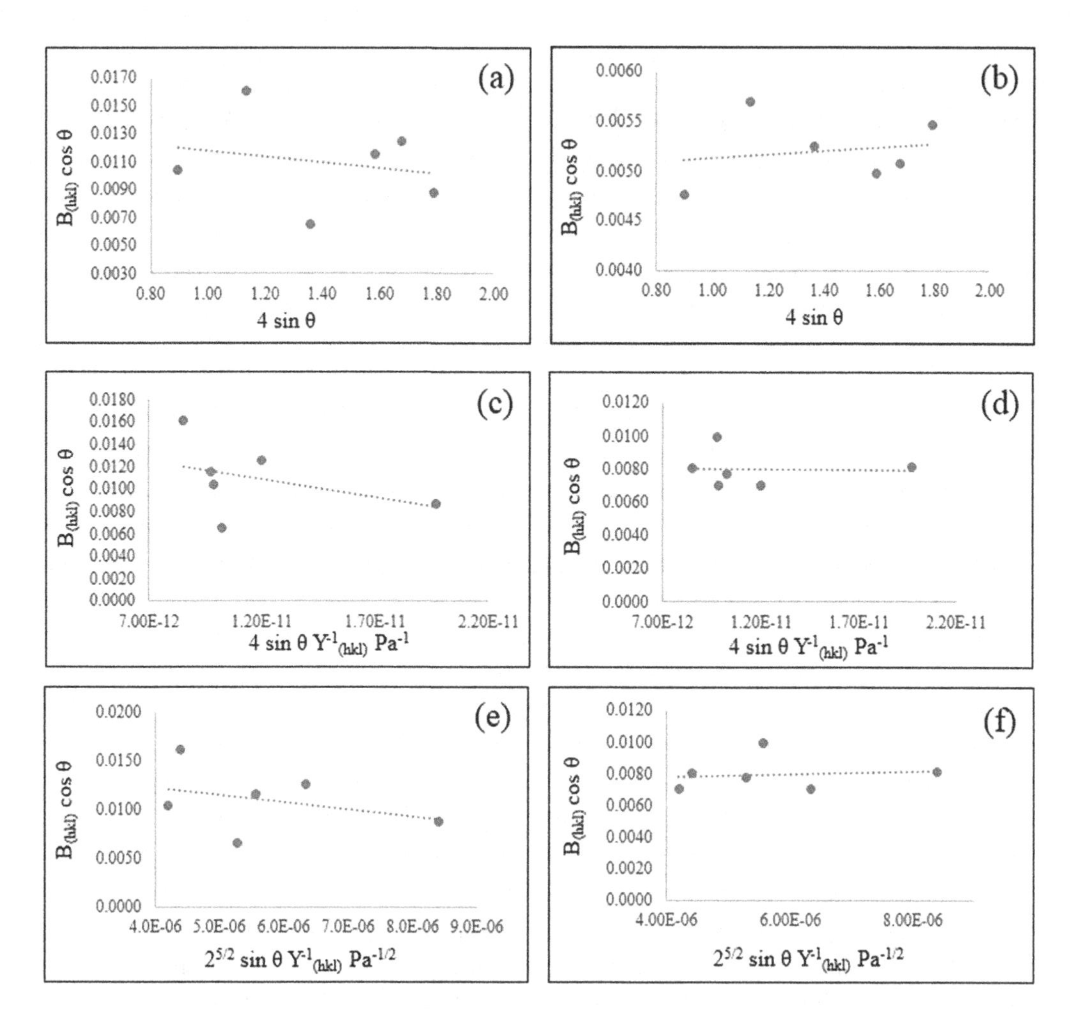

FIGURE 10.3 $\beta_{(hkl)}$ cos θ versus 4 sin θ; $\beta_{(hkl)}$ cos θ versus 4 sin $\theta/Y_{(hkl)}$ and $\beta_{(hkl)}$ cos θ versus 4 sin θ $(2u/Y_{(hkl)})^{1/2}$ for as-synthesized 3% doped nano-HAP sample (a, c, e) and annealed (800 °C) sample (b, d, f).

$$Y_{hkl} = \left(\frac{\left[h^2 + \frac{(h+2k)^2}{3} + \frac{(al)^2}{c} \right]^2}{S_{11}\left(h^2 + \frac{(h+2k)^2}{3}\right)^2 + S_{33}\left(\frac{al}{c}\right)^4 + (2S_{13}+S_{44})\left(h^2 + \frac{(h+2k)^2}{3} + \left(\frac{al}{c}\right)^2\right)} \right) \quad (10.5)$$

The elastic compliances for the various crystallographic orientations are $S_{11} = 7.49 \times 10^{-12}$, $S_{33} = 10.9 \times 10^{-12}$, $S_{44} = 15.1 \times 10^{-12}$, and $S_{13} = -4.0 \times 10^{-12}$ m^2N^{-1} [22]. Therefore, plotting 4 sin $\theta/Y_{(hkl)}$ along the x-axis and $\beta_{(hkl)}$ cos θ along the y-axis enables the slope (σ) to be calculated and from the y-intercept estimate the crystallite size ($D_{(hkl)}$). Representative graphs of the pure nano-HAP and 3% Si-doped nano-HAP powders are presented in Figures 10.2 and 10.3, while the estimated ($D_{(hkl)}$) and σ values are presented in Table 10.1.

The anisotropic energy density (u) within the crystalline lattice structure was determined from the uniform deformation energy density model (UDEDM). This model is derived by substituting the

TABLE 10.1
XRD peak analysis results of pure and Si-doped nano-HAP ultrafine powder samples

		Williamson-Hall methods					
		UDM		USDM		UDEDM	
Sample	**Scherrer (002) (nm)**	**$D_{(hkl)}$ (nm)**	**ε x 10-3**	**$D_{(hkl)}$ (nm)**	**σ (M Pa)**	**$D_{(hkl)}$ (nm)**	**u (KJ/m³)**
Pure nano-HAP	19.87	18.74	1.310	15.94	47.82	17.35	14.76
Annealed 800 °C	29.11	28.29	0.181	27.93	20.16	28.36	7.40
3% Si-nano-HAP	13.36	9.99	2.065	9.40	324.16	9.07	27.53
Annealed 800 °C	19.88	21.62	1.098	17.14	11.19	18.56	9.25
7% Si-nano-HAP	11.32	11.37	3.066	11.71	400.32	12.86	31.72
Annealed 800 °C	18.22	19.32	1.204	20.98	15.63	26.37	12.08

strain energy for an elastic crystallite (Hooke's law) represented by Equation (10.6) into Equation (10.2) to give Equation (10.7) below:

$$u = \frac{\varepsilon^2 Y_{(hkl)}}{2} \tag{10.6}$$

$$\beta_{(hkl)} \cos\theta = \frac{k\lambda}{D_{(hkl)}} + 4 \sin\theta \left(\frac{2u}{Y_{(hkl)}} \right)^{\frac{1}{2}} \tag{10.7}$$

Thus, using Equation (10.7) and plotting $\beta_{(hkl)} \cos \theta$ along the y-axis and $4 \sin \theta \, (2u/Y_{(hkl)})^{1/2}$ along the x-axis, it is possible to determine the anisotropic energy density u from the slope of the graph and determine crystallite size from the y-intercept. The respective graphs for the pure nano-HAP powder sample and the 3% Si-doped nano-HAP powder sample are presented in Figures 10.2 and 10.3, respectively.

Also determined from the XRD data were lattice constants (a, c) and unit cell volumes (V) of the various samples using Equations (10.8) and (10.9), respectively:

$$\frac{1}{d^2} = \frac{4}{3}\left(\frac{h^2 + hk + k^2}{a^2} \right) + \frac{l^2}{c^2} \tag{10.8}$$

$$V = \frac{\sqrt{3}\, a^2 c}{2} = 0.866 a^2 c \tag{10.9}$$

The calculated lattice constants and unit cell volumes of the non-annealed and annealed samples are presented in Table 10.2.

The XRD studies revealed Si was incorporated into the lattice structure of HAP, thus allowing Si doping of nano-HAP powders to take place. Analysis of the XRD data revealed the 3% Si-doped nano-HAP powder maintained its original HAP structure. A similar study by Kim et al. also reported that the substitution of small quantities of Si (~ 2 % wt.) does not have a major impact

TABLE 10.2
Lattice parameters and unit cell volumes of pure and Si-doped nano-HAP ultrafine powder samples

	Lattice parameters		Unit cell volume
Sample	a (Å)	c (Å)	V (Å³)
Pure nano-HAP	9.4252	6.8832	529.53
Annealed 800 °C	9.4248	6.8753	528.88
3% Si-nano-HAP	9.4264	6.8870	529.94
Annealed 800 °C	9.4254	6.8821	529.47
7% Si-nano-HAP	9.4279	6.8911	530.44
Annealed 800 °C	9.4261	6.8865	529.88

on the original HAP lattice structure. In addition, their study also found only a single Si-HAP phase formed, and no other phases (silicon oxide or other calcium phosphates) were present [23]. Similarly, a study by Gibson et al. also found that the addition of small amounts of Si (1.6 wt.) only produced a single phase with only small changes in the lattice constants [24]. Recently, a study by Jamil et al. also found that the addition of small quantities of Si produced only small changes in the lattice constants [25]. This result was also seen in the present study as seen in Table 10.2. For instance, in the non-annealed powder samples for the Si (3 % wt.)/doped and pure nano-HAP powder samples, there was only a slight increase in the unit cell volume (0.077 %) of the Si-doped powder sample. In terms of the lattice parameters (Table 2), there were slight increases in both the (a) constant (0.013%) and the (c) constant (0.055 %). A similar study by Gomes et al. also reported similar changes in the lattice parameters and unit cell volumes with Si doping [26]. Figure 10.4 presents the influence of increasing Si (At. %) content on lattice constants and unit cell volumes for powder samples investigated in the present study.

The study also found increasing Si content in the powders resulted in increasing values of ε, σ, and u for the non-annealed samples, as seen in Table 10.1. For instance, the lattice strain (ε) for the pure nano-HAP powder sample was 1.310×10^{-3}, while the 3% Si-doped nano-HAP powder had a higher value of 2.065×10^{-3}, and the 7% had a higher value of 3.066×10^{-3}. Similarly, there were also an increase in both the surface stress (σ) and the anisotropic energy density with increasing Si content, as seen in Table 10.1. On the other hand, powder samples annealed at 800 °C showed increases in crystallite size and decreasing levels of strain, stress, and anisotropic energy density, as seen in Table 10.1. Interestingly, the XRD patterns for the 7% Si-doped nano-HAP powder sample (Figure 1) revealed a significant reduction in peak intensity and peak broadening. Moreover, the non-annealed powder sample had a large broad peak (2θ around 22°), which is indicated by the red arrow as shown in Figure 10.1. The presence of this feature suggests another phase (amorphous SiO_2) was formed due to the increased Si doping level. A similar result was reported by Nakata et al. for Si-doped HAP powders with increasing levels of Si content [27]. In addition, prior to annealing, the 7% Si-doped nano-HAP powder sample displayed much higher strain levels than the pure nano-HAP powder, thus indicating that the presence of Si was subjecting the lattice structure to distortion. However, after annealing, the stress levels were dramatically reduced.

10.3.2 FT-IR Spectroscopy Analysis

FT-IR spectra showing major bands for pure, 3% Si-doped and 7% doped nano-HAP powder samples are presented in Figure 10.5. FT-IR spectra presented in Figure 10.5 contains a non-annealed

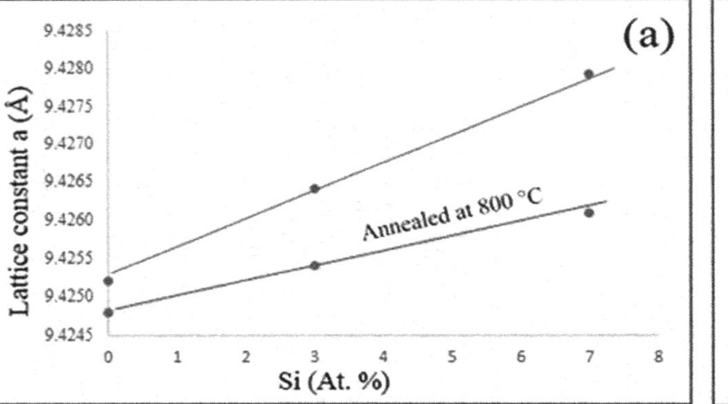

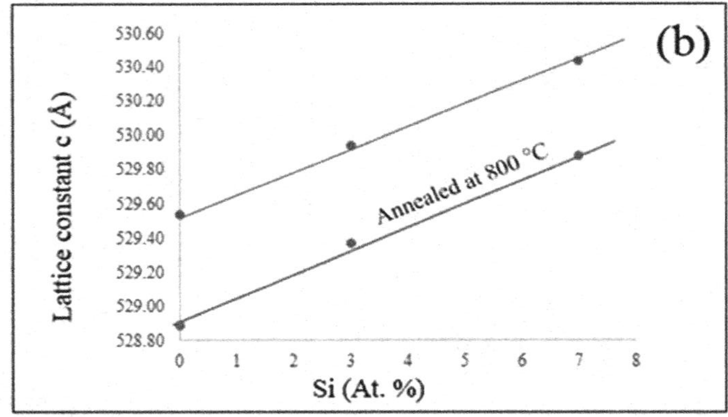

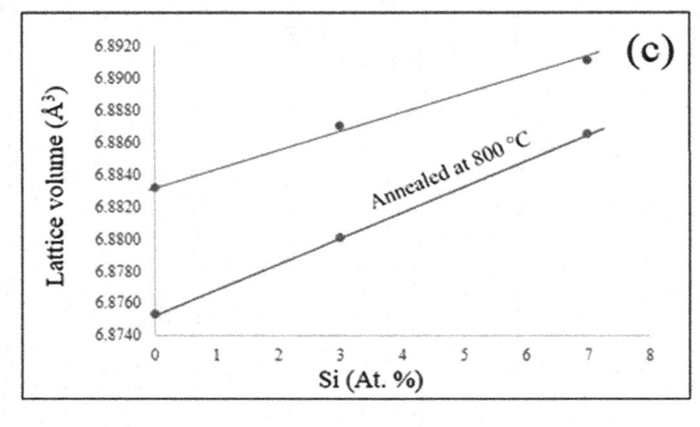

FIGURE 10.4 The influence of Si (At. %) content on lattice constants and unit cell volumes.

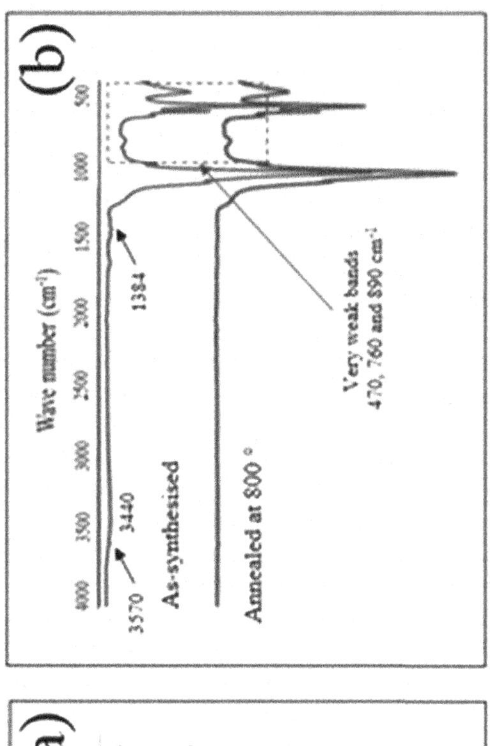

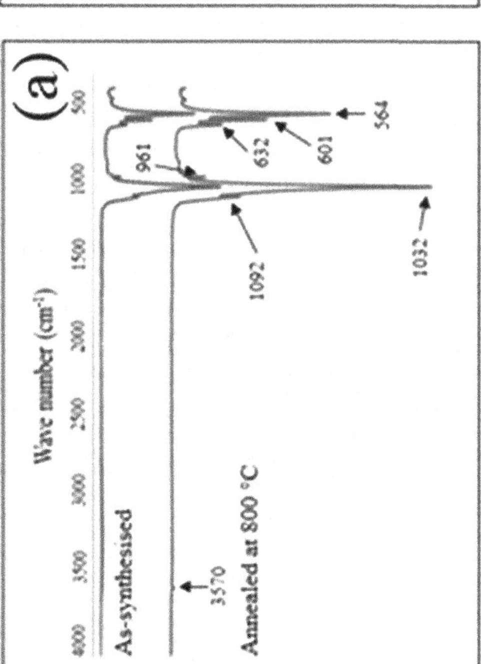

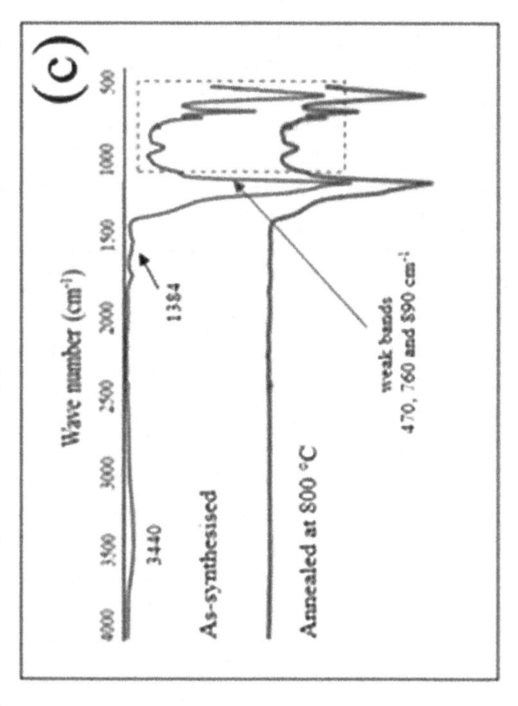

FIGURE 10.5 FT-IR spectra of non-annealed and annealed (800 °C) nano-HAP powders: (a) pure (b) Si (3%) doped and (c) Si (7%) doped.

sample and a sample annealed at 800 °C. Spectra for samples annealed at 400 °C showed no difference from those that were not annealed. Figure 10.5 (a) presents spectra for non-annealed and annealed pure nano-HAP powder samples. Starting from the left-hand side of Figure 10.5 (a), there is a very weak band located at 3570 cm^{-1} in both spectra that corresponds to the O-H bond vibrations. However, in both cases, there are no bands around 3440 cm^{-1} (OH^- ion vibrations) that are normally associated with adsorbed water in the lattice structure. The absence of these bands is due to the microwave-based heat treatment which significantly reduces the presence of adsorbed water and has also been reported in similar studies [28]. Moving further to the right, there are two strong bands and one weak band located at 1092 cm^{-1}, 1032 cm^{-1}, and 961 cm^{-1} respectively. These bands correspond to the vibrational modes of phosphate groups $(PO_4)^{3-}$ present in the samples [29]. The weaker band located at 632 cm^{-1} indicates a hydroxyl vibrational mode. Bands located further to the right at 564 cm^{-1} and 601 cm^{-1} are associated with O-P-O vibrational modes. Comparing the spectra reveals that both are similar in shape, except the annealed sample has sharper bands compared to the non-annealed sample. This confirms that the annealing process improved crystallinity and corroborates the results of the XRD data.

FT-IR spectra presented in Figure 10.5 (b) contain a 3% Si-doped non-annealed sample and a 3% Si-doped sample annealed at 800 °C. Si in the spectra confirms the results of the XRD analysis and the results of elemental analysis derived from energy dispersive spectroscopy. Inspecting Figure 10.5 (b) reveals the overall nano-HAP powder profile is still present but with a few additional bands. Like the pure nano-HAP powder sample presented in Figure 10.5 (a), there is a wide and weak band located around 3440 cm^{-1} in the non-annealed sample that is associated with OH^- ion vibrations and indicates the presence of adsorbed water in the lattice structure. However, this band is not seen in the annealed sample and indicates the annealing process has removed adsorbed water from the lattice structure. Also in the non-annealed sample is a very weak band located at 1384 cm^{-1} that is associated with the presence of carbonate ions $(CO_3)^{2-}$ in the sample. This band is also not seen in the annealed sample. In addition, there are three new low-intensity bands located at around 470, 760, and 890 cm^{-1} that were not seen in the pure nano-HAP samples. These new bands were attributed to the presence of $(SiO_4)^{4-}$ groups present in the lattice structure. Similar studies have also reported the presence of these weak bands in HAP structures [25, 30]. In addition, the presence of Si in the powder samples was found to broaden the bands, indicating a decrease in crystallinity. The broadening seen in both FT-IR bands and XRD peaks (as discussed above) indicates decreasing crystallinity, which has also been reported by other researchers for similar studies [31, 32]. FT-IR spectra presented in Figure 10.5 (c) contain a 7% Si-doped non-annealed sample and a 7% Si-doped sample annealed at 800 °C. The spectra presented in Figure 10.5 (c) are similar to the abovementioned 3% Si-doped sample presented in Figure 10.5 (b). However, the bands in the 7% doped samples are much broader. In addition, the bands indicating $(SiO_4)^{4-}$ groups located at 470, 760, and 890 cm^{-1} have increased in intensity, while the intensity of the $(PO_4)^{3-}$ bands located at 1092 cm^{-1}, 1032 cm^{-1}, and 961 cm^{-1} have slightly decreased. The FT-IR results not only confirm the presence of Si in all the Si-doped samples but also indicate that $(PO_4)^{3-}$ tetrahedrons in the lattice structure are being replaced by $(SiO_4)^{4-}$ tetrahedrons. The exchange of $(PO_4)^{3-}$ tetrahedrons for $(SiO_4)^{4-}$ tetrahedrons in the HAP lattice structure has also been reported in similar Si doping studies [33, 34].

10.3.3 Elemental Analysis and Electron Microscopy Evaluation

Representative results of elemental analysis for a 3% Si-doped nano-HAP powder sample are presented in Figure 10.6. The analysis shows the widespread distribution of Ca, P, and Si within the sample. Figure 10.6 (b) presents the elemental content of Ca, P, and Si in the sample and reveals that the 2.87% Si content is slightly less than the theoretically predicted value of 3%. However, XRD analysis failed to detect the presence of any secondary phases resulting from this slight difference. From another perspective, several studies have suggested that Si doping levels up to 5% can be easily incorporated into the HAP lattice structure [13]. The present study also suggests that 5% is more

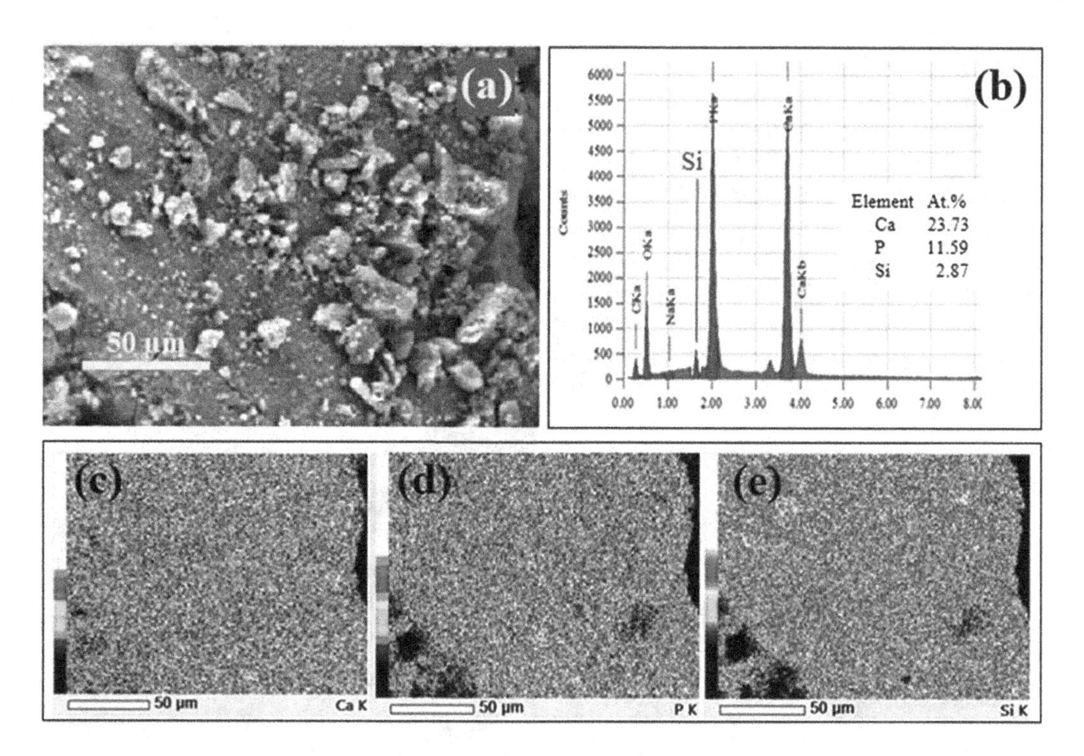

FIGURE 10.6 (a) SEM image of a 3% doped Si nano-HAP powder sample, (b) typical elemental analysis spectra images (c), (d), and (e) typical elemental distributions for Ca, P and Si.

likely to be the upper limit because significant changes were seen in XRD patterns for 7% Si-doped samples.

A FESEM investigation was also carried out on as-synthesized and annealed (800 °C) samples. Representative images of the various sample types are presented in Figure 10.7. Pure nano-HAP powder images of both as-synthesized (non-annealed) and annealed samples are presented in Figures 10.7 (a) and (b). The as-synthesized (non-annealed) nanoparticles making up the nano-HAP powder are ellipsoidal in shape with an average aspect ratio of around 1:7. Nanoparticle diameters range from 18 to 30 nm, and lengths range from 130 nm to 200 nm. In addition, the nanoparticles are found to cluster together to form large agglomerations, as seen in the images presented in Figure 10.7 (a). Similarly, the annealed nanoparticles also tend to agglomerate. However, in the annealed case, the nanoparticles are found to have a different particle size range and morphology, as seen in image Figure 7 (b). Inspection of Figure 7 (b) reveals that the nanoparticles range in size from 80 to 150 nm. In addition, nanoparticle morphology is granular and has a rough irregular rectangular appearance.

Representative images of 3% Si-doped nano-HAP powders are presented in Figures 10.7 (c) as-synthesized (non-annealed) and (d) annealed at 800 °C. Inspecting the respective images reveals significant changes in both nanoparticle size and morphology. In the as-synthesized (non-annealed) sample, Figure 10.7 (c), two morphologies can be seen. The two shapes seen are granular and ellipsoidal. However, the ellipsoidal shapes have smaller dimensions than those present in the pure nano-HAP powders samples. The image reveals the ellipsoidal shapes have similar diameters ranging from 18 to 35 nm but have smaller lengths ranging from 50 to 80 nm, with a typical aspect ratio of around 3. Moreover, there are also large numbers of granular nanoparticles present that range in size from 45 nm to 80 nm. However, in the annealed sample, the nanoparticles are predominantly granular in nature and range in size from 27 to 135 nm. Representative images of the 7% Si-doped nano-HAP powder sample are presented in Figures 10.7 (e) and (f) and also display different particle size ranges and morphologies. The as-synthesized (non-annealed) sample presented in Figure 10.7

FIGURE 10.7 Pure nano-HAP powders: (a) as-synthesized and (b) annealed at 800 °C, 3% Si-doped nano-HAP powders: (c) as-synthesized and (d) annealed at 800 °C, and 7% Si-doped nano-HAP powders: (e) as-synthesized and (f) annealed at 800 °C

(e) shows both granular and ellipsoidal shapes. In this case, the granular shapes range in size from 20 to 40 nm, while the ellipsoidal shapes have diameters ranging from 14 to 28 nm and lengths ranging from 40 to 75 nm. The annealed sample presented in Figure 10.7 (f) has nanoparticles with similar granular morphology to those seen in Figure 10.7 (d) for the 3% Si-doped annealed sample. The only difference between the two annealed samples is a larger particle size range, with the 7% Si-doped sample having a larger granular particle size range between 25 and 80 nm. Image analysis confirms that the presence of Si in the respective samples does influence particle size. The presence of Si in the respective samples was confirmed by both EDS and FT-IR. In addition, both XRD analysis and FT-IR analysis confirm that Si doping is inhibiting crystallite size growth, and $(SiO_4)^{4-}$ ions are being substituted for $(PO_4)^{3-}$ ions in the HAP lattice structure [27]. Similar studies have also reported decreasing grain growth with increasing Si content [24]. Thus, the present study has confirmed that Si doping acts as an inhibitor to grain growth during the manufacture of ultrasonically engineered and microwave-treated Si-doped nano-HAP ultrafine powders.

10.4 CONCLUSION

The present study investigated the influence of Si doping on ultrasonically engineered and microwave-treated Si-doped nano-HAP ultrafine powders. Advanced characterization techniques like XRD, FT-IR, EDS, and FESEM all confirmed the presence of Si in the crystalline structure of the nano-HAP-based ultrafine powder. XRD analysis of the Si-doped nano-HAP powders revealed a single, crystalline phase, with a crystalline structure similar to pure nano-HAP. In addition, both FT-IR and EDS analysis indicated that $(SiO_4)^{4-}$ ions were being substituted for $(PO_4)^{3-}$ ions in the nano-HAP lattice structure. X-ray peak broadening analysis using Williamson-Hall methods revealed that increasing amounts of Si distorted the HAP lattice structure and resulted in increased levels of strain, stress, and anisotropic energy density. The XRD analysis also revealed small increases in lattice constants and the unit cell volume with increasing Si doping levels. Annealing at 800 °C was found to be an effective method for reducing levels of strain, stress, and anisotropic energy density in the Si-doped samples. In addition, the presence of Si tended to inhibit grain growth and reduce crystallinity in the Si-doped nano-HAP ultrafine powders. This ultrasonic pathway to create Si-doped nanoHAP powders can thus be harnessed in the treatment of dental and bone related injured tissues for the benefit of the community.

REFERENCES

1. Zhou H, Lee J. Nanoscale hydroxyapatite particles for bone tissue engineering. *Acta Biomater.* 2011; 7: 2769–2781.
2. Poinern GJE, Brundavanam R, Le X, Djordjevic S, Prokic M, Fawcett D. Thermal and ultrasonic influence in the formation of nanometre scale hydroxyapatite bio-ceramic. *Int. J. Nanomed.* 2011; 6: 2083–2095.
3. Kalita SJ, Bhardwaj A, Bhatt HA. Nanocrystalline calcium phosphate ceramics in biomedical engineering. *Mater. Sci. Eng. C* 2007; 27(3): 441–449.
4. LeGeros RZ. Calcium phosphates in oral biology and medicine. *Mon. Oral Sci.* 1991; 15: 1–201.
5. Silva RV, Camilli JA, Bertran JA, Moreira NH. The use of hydroxyapatite and autogenous cancellous bone grafts to repair bone defects in rats. *Int. J. Oral Maxillofac. Surg.* 2005; 34(2): 178–184.
6. Hutmacher DW, Schantz JT, Lam CXF, Tan KC, Lim TC. State of the art and future directions of scaffold-based bone engineering from a biomaterials perspective. *J. Tissue Eng. Regen. Med.* 2007; 1: 245–260.
7. Stoch A, Jastrzebski W, Dlugon E, Lejda W, Trybalska B, Stoch GJ, Adamczyk A. Sol-gel derived hydroxyapatite coatings on titanium and its alloy Ti6 Al4V. *J. Mol. Struct.* 2005; 744–747: 633–640.
8. Habibovic P, de Groot K. Osteoinductive biomaterials-properties and relevance in bone repair. *J. Tissue Eng. Regen. Med.* 2007; 1(1): 25–32.
9. Carlisle EM. Silicon: A possible factor in bone calcification. *Science* 1970; 167: 179–280.
10. Carlisle EM. Silicon—A requirement in bone-formation independent of vitamin-D1. *Calcif. Tissue Int.* 1981; 33: 27–34.
11. Hing KA, Revell PA, Smith N, Buckland T. Effect of silicon level on rate, quality and progression of bone healing within silicate-substituted porous hydroxyapatite scaffolds. *Biomaterials* 2006; 27: 5014–5026.
12. Porter AE, Patel N, Skepper JN, Best SM, Bonfield W. Comparison of in vivo dissolution processes in hydroxyapatite and silicon-substituted hydroxyapatite bioceramics. *Biomaterials* 2003; 24: 4609–4620.
13. Balas, F., Perez-Pariente, J., Vallet-Regı, M. In vitro bioactivity of silicon- substituted hydroxyapatites introduction. *J. Biomed. Mater. Res.* 2003; 66A: 364–375.
14. Vallet-Regi M, Arcos D. Silicon substituted hydroxyapatites: A method to upgrade calcium phosphate based implants. *J. Mater. Chem.* 2005; 15: 1509–1516.
15. Patel N, Best SM, Bonfield W, Gibson IR, Hing KA, Damien E, Revell PA. A comparative study on the in vivo behaviour of hydroxyapatite and silicon substituted hydroxyapatite granules. *J. Mater. Sci. Mater. Med.* 2002; 13: 1199–1206.
16. Gasqueres G, Bonhomme C, Maquet J, Babonneau F, Hayakawa S, Kanaya T, Osaka A. Revisiting silicate substituted hydroxyapatite by solid-state. *NMR Magn. Reson. Chem.* 2008; 46: 342–346.
17. Gedanken A. Using sonochemistry for the fabrication of nanomaterials. *Ultrason. Sonochem.* 2007; 11: 47–55.

18. Vollmer M. Physics of the microwave oven. *Phys. Ed.* 2004; 39(1): 74–81.
19. Zhou H, Lee J. Nanoscale hydroxyapatite particles for bone tissue engineering. *Acta Biomater.* 2011; 7: 2769–2781.
20. Venkateswarlu K, Chandra Bose A, Rameshbabu N. X-ray peak broadening studies of nanocrystalline hydroxyapatite by Williamson–Hall analysis. *Phys. B.* 2010; 405: 4256–4261.
21. Kim SR, Lee JH, Kim YT, Riu DH, Jung SJ, Lee YJ, Chung SC, Kim YH. Synthesis of Si, Mg substituted hydroxyapatites and their sintering behaviours. *J. Biomater.* 2003; 24: 1389–1398.
22. Gibson IR, Best SM, Bonfield W. Chemical characterization of silicon substituted hydroxyapatite. *J. Biomed. Mater. Res.* 1999; 4: 422–428.
23. Jamil M, Elouatli B, Khallok H, Elouahli A, Gourri E, Ezzahmouly M, Abida F, Hatim Z. Silicon substituted hydroxyapatite: Preparation with solid-state reaction, characterization and dissolution properties. *J. Mater. Environ. Sci.* 2018; 9(8): 2322–2327.
24. Gomes S, Renaudin G, Mesbah A, Jallot E, Bonhomme C, Babonneau F, Nedelec JM. Thorough analysis of silicon substitution in biphasic calcium phosphate bioceramics: A multi-technique study. *Acta Biomater.* 2010; 115: 114707–114713.
25. Nakata K, Kubo T, Numako C, Onoki T, Nakahira A. Synthesis and characterization of silicon-doped hydroxyapatite. *Mater. Trans.* 2009; 50(5): 1046–1049.
26. Liangzhi G, Weibin Z, Yuhui S. Magnesium substituted hydroxyapatite whiskers: Synthesis, characterization and bioactivity evaluation. *RSC Adv.* 2016; 115: 114707–114713.
27. Lijuan X, Liuyun J, Lixin J, Chengdong X. Synthesis of Mg-substituted hydroxyapatite nanopowders: Effect of two different magnesium sources. *Mater. Lett.* 2013; 106: 246–249.
28. Aminian A, Solati-Hashjin M, Samadikuchaksaraei A, Bakhshi F, Gorjipour F, Farzadi A, Moztarzadeh F, Schmucker M. Synthesis of silicon-substituted hydroxyapatite by a hydrothermal method with two different phosphorous sources. *Ceramics Int.* 2011; 37: 1219–1229.
29. Tang XL, Xiao XF, Liu RF. Structural characterization of silicon-substituted hydroxyapatite synthesized by a hydrothermal method. *J. Mater. Lett.* 2005; 59: 3841–3846.
30. Botelho CM, Lopes MA, Gibson IR, Best SM, Santos JD. Structural analysis of Si–substituted hydroxyapatite: Zeta potential and X-ray photoelectron spectroscopy. *J. Mater. Sci. Mater. Med.* 2002; 13: 1123–1127.
31. Solla EL, Malz F, Gonzalez P, Serra J, Jaeger C, Leon B. The role of Si substitution into hydroxyapatite coatings. *J. Key Eng. Mater.* 2008; 361–363: 175–178.
32. Tian T, Jiang D, Zhang J, Lin Q. Synthesis of Si-substituted hydroxyapatite by a wet mechanichemical method. *J. Mater. Sci. Eng. C.* 2008; 28: 57–63.
33. Hadden DJ, Skakle JMS, Gibson IR. Optimisation of the aqueous precipitation synthesis of silicate substituted hydroxyapatite. *Key Eng. Mater.* 2008; 361–363: 55–58.
34. Li XW, Yasuda HY, Umakoshi Y. Bioactive ceramic compositions sintered from hydroxyapatite and silica at 1200°C: Preparation, microstructure and in vitro bone-like layer growth. *J. Mater. Sci. Mater. Med.* 2006; 17: 573–581.

Theme 3

Advanced Materials for Agriculture and Environmental Concerns

11 Nanoparticles for Agriculture

Owen Horoch and David Henry

CONTENTS

11.1 INTRODUCTION

Nanoparticles for agricultural applications need to possess a number of important characteristics including biocompatibility, high efficacy for the specific application, environmental stability, and no inherent or residual toxicity to crops, animals, or humans, while ideally providing a cost-effective alternative to bulk materials [1]. Silicon nanoparticles (Si-NPs) are well-known for displaying many of these characteristics and have been developed for a variety of uses in agriculture, as recently cataloged by Rastogi et al. [2]. For example, Si-NPs have been used as vehicles for fertilizers, herbicides, pesticides, soil monitoring, and water retention. They have been shown to interact with plants positively by improving plant growth, preventing infection, and increasing yield. Si-NPs have also been shown to guard plants against certain environmental stresses such as salinity and UVB irradiation, in addition to offering some levels of defense against heavy metal toxicity and dehydration. However, in some cases, they have also negatively impacted plants by altering soil pH and causing toxic conditions.

As pesticides, they can either be used directly or as carriers for other pesticides which allow slow release for heightened efficiency and durability, and are commonly applied directly to the plant [2–6]. Alternatively, Si-NPs can carry fertilizers and offer the benefit of being small enough

DOI: 10.1201/9781003181422-14

to penetrate plant cells to deliver these chemicals. Furthermore, Si-NPs can carry nucleotides, proteins, and zeolites for improving soil conditions for plants by aiding in water availability. They are also used extensively as nano-sensors for the purpose of detecting either harmful or beneficial species within plants or soil.

However, nanomaterial applications in agriculture are not restricted to Si-NPs. For example, NPs formed from other materials such as TiO_2 have seen use as fertilizers, bio-fortifiers, or pesticides against cotton leafworm [7, 8]. ZnO has also been used as a fertilizer or pesticide and can be produced by green synthesis [7, 9]. Silver is a well-known antibacterial agent and sees the same use in agriculture. However, depending on the form applied to the crops, it can have positive or detrimental effects [10]. CuO has been shown to protect lettuce crops and possesses cytotoxicity against bacteria, but there is growing concern about its toxicity toward humans and certain plants [11–13]. Chitosan NPs have seen extensive use as a delivery system for herbicides, pesticides, micronutrients, and biosensors [14].

11.2 APPLICATION OF NANOMATERIALS IN AGRICULTURE

Sustainability in agriculture is an important goal for ensuring continued food production for a growing population while conserving the environment. Nanomaterials can assist in this endeavor by allowing for efficient solutions to problems such as delivering water and nutrients to plants, feeding humans and animals, and reducing negative impacts on ecosystems and the environment.

11.2.1 Nano-Fertilizers to Improve Crop Yields

The most common agricultural application of nanoparticles involves the application of fertilizers or growth stimulants. One of the main features of this application is the slow release of nutrients. The outcome is that less fertilizer is used overall, which drastically reduces leaching into water tables and surrounding ecosystems. NPs used as carriers often either encapsulate or adsorb the desired fertilizer (such as urea) and release their payload during a pH change within a plant or from the enzymatic breakdown of the carrier NP [15].

Nanoparticles can also function as a standalone growth stimulant, as demonstrated by Tymoszuk and Wojnarowicz, where zinc addition from ZnO NPs was essentially providing the plants with an added micronutrient that could work in tandem with a fertilizer [16].

Batsmanova et al. investigated the effect of a colloidal solution of metal nanoparticles containing Mn, Cu, Zn, Ag, and Fe on the productivity of soybeans [17]. The colloidal solution used was patented and thus details were not revealed within the study including the percent composition of each metal. Results indicated that treatment through foliar and seed uptake gave the best results when the obscure "$N_{60}P_{60}K_{60}$" fertilizer was applied with assistance from the colloidal solution at a concentration of 240 mgL^{-1}, resulting in an 18% increase in productivity over fertilizer treatment alone.

Compounds that are used to offset crop stresses caused by biotic and abiotic factors on crops can sometimes leave toxic remnants which may be consumed by humans. A possible non-toxic and environmentally compatible solution might be an extract of quinoa (*Chenopodium quinoa* Willd). Quinoa husk is an unutilized biomass, which possesses saponins and polyphenolic compounds that have been shown to be important in plant metabolism and have exhibited efficacy in treating a range of plant pests including nematodes, insects, and fungi. The saponins from quinoa husk have also been used in the synthesis of antibacterial silver nanoparticles. Segura et al. investigated the phytostimulant properties of quinoa husk extract (QE) and quinoa extract silver nanoparticles (QEAgNPs) using a radish seed germination assay [18]. Importantly, both QE and QEAgNPs were found to have no phytotoxic effects on the radish seeds. However, QE did present as inhibiting root growth. The optimal concentration of QEAgNPs for improving root growth and germination was also determined (500 µg/mL). Furthermore, the germination index (GI) obtained with the optimal QEAgNPs solution was higher than for QE alone, indicating synergy between the silver nanoparticles and quinoa extract. The results demonstrated that QEAgNPs have a strong potential to be used as plant growth bio-promoters.

Zinc oxide nanoparticles (ZnO NPs) are some of the most widely manufactured and used NPs across the world. It is for this reason that Tymoszuk and Wojnarowicz investigated the effects of the material on onion seedling (*Allium cepa* L) growth [16]. Given the small size of the nanoparticles, they could very easily be absorbed into plants via roots and foliar transfer, and may have both advantageous and detrimental effects. The researchers investigated the effect of both ZnO NPs and ZnO submicron particles (SMPs) upon the plants at different concentrations. The best germination rates were observed for 800 mg/L of ZnO NPs and SMPs, which were almost double the control. However, all concentrations ranging between 50–1600 mg/L also gave positive effects upon plant germination. Zinc is a constituent of an enzyme that influences the release of indole acetic acid (IAA), a phytohormone that regulates plant growth, which might explain the positive effect observed [16]. However, the outcome of the Tymoszuk and Wojnarowicz study differed from some literature, as only germination percent and time were altered. In previous comparative studies, length, fresh weight, dry weight, and shoot and root length were also studied, where ZnO NPs had a positive effect on biomass production of sunflower, mung, and gram seedlings, in addition to higher germination rates of onion seeds [19–23]. Tymoszuk and Wojnarowicz ascribe these differences in outcome to plant genotype, chemical site deposition, environment, NP characteristics, and exposure time, which have been contributing factors mentioned in previous research [16, 24]. Further experimentation is required with variations in growth media (*in vitro*/*in vivo*), plant genotype, and a greater level of replication. The lack of difference between ZnO NPs and SMPs in the study was likely due to the overall release levels of Zn^{2+} being similar over time and the size difference between particles only affecting the initial solubility.

11.2.2 Delivery of Essential Nutrients for Bio-Fortification

Bio-fortification is the process of enriching crops with essential nutrients (such as P, K, S, Ca, and Mg) for the purpose of increasing their nutritional value of foods produced from these crops. These and other trace elements are needed for optimal human health but often also improve the health of the crops. Each has a crucial role, such as assisting in the transfer of energy from sunlight, stimulating plant growth, or increasing disease resistance. Bio-fortification is especially important in agricultural areas where crops lack nutrients due to deficits within the soil or areas of the world where micronutrient malnourishment is common.

Delivery of NPs for bio-fortification focuses upon physical application. Foliar application is a common method of delivering NPs to plants, where nanoparticles in suspension are either sprayed or painted onto leaves [25–28]. Fortis-Hernández et al. used cylindrical atomizers to deliver their Cu NPs, with the addition of an adherent in an aqueous solution. In another study, Read et al. placed 3 and 5 μL droplets on the leaves with the addition of 0.05% surfactant [25–28]. Seed priming is another method wherein seeds are pre-soaked in an NP solution before being sown [29]. The final method of delivery is through the soil, where nanoparticle uptake happens through the root systems of crops [30].

11.3 NANO-PESTICIDES TO PATHOGENIC DISEASES

Up to USD 2 trillion is lost globally per year due to agricultural setbacks such as pests [31]. Nanoparticles provide the means for the efficient delivery of pesticides by controlling the destination of chemicals and reducing their persistence in the environment. Nanoparticles for pesticidal use function in a similar manner to NPs for fertilizers and growth stimulants in that they can deploy a payload or work as a standalone solution.

11.3.1 NP/Mechanism for Disease Control

There are numerous agricultural plant diseases that have shown susceptibility to treatment by nanoparticles. Some of the most commonly used nanoparticles for treating plant diseases are Ag, Cu, CuO, TiO_2, and SiO_2, as well as Se and ZnO to a lesser extent [32]. Many of these can be prepared using extracts from different plants such as lemons (*Citrus limon*), olive leaves (*Olea europaea*), wormwood leaves (*Artemisia absinthium*), and cloves (*Syzygium aromaticum*) [33–36].

In addition to nanoparticles working independently to combat diseases, it is also possible to load some nanoparticles with antibacterial compounds or fungicides as an effective treatment for plant diseases [37, 38].

The treatment of bacteria that cause diseases in plants is largely reliant on breaching or degrading the cell membrane, allowing access to the vulnerable inner sections of the targeted bacterial cells. Sreelatha et al. used chitosan NPs as a carrier for the antibacterial agent thymol, creating thymol-loaded chitosan nanoparticles (TCNPs). These TCNPs showed effective bactericidal outcomes when tested against black rot-causing bacteria (*Xanthomonas campestris* pv. *Campestris* (*Xcc*)) [37]. The cytotoxicity of TCNPs has been attributed to their suppression of xanthomonadin and exopolysaccharide (EPS) production and inhibiting the formation of biofilms, which drastically reduces virulence and cellular integrity of the bacteria cells as well as enhancing reactive oxygen species (ROS) formation within cells. Both of these modes result in the disruption of membrane integrity and reduction of cell viability.

Another example of this disruption mechanism was provided by silver nanoparticles (Ag NPs) produced by Hirpara and Gajera via a green synthesis method using exometabolites of a biocontrol fungus (Trichoderma). These nanoparticles were found to be effective at degrading outer cell walls and greatly reducing cell respiration of *Sclerotium rolfsii*, a fungal pathogen that causes stem rot disease in ground nuts [39].

Okra (*Abelmoschus esculentus* L.) is a useful plant found in Africa, Asia, Europe, and America and possesses anti-diabetic, anti-cancer, anti-ulcer, and anti-fungal properties. Keerthana et al. synthesized ZnO nanoparticles from citron (*Citrus medica*) peel and zinc acetate for the purpose of improving the growth of okra [40]. The antimicrobial capabilities of the ZnO NPs were tested for gram-positive and gram-negative bacteria, and fungal strains. These ZnO NPs outperformed plant extract, zinc acetate, and gentamycin and were competitive with nystatin, a common anti-fungal agent. The researchers ascribe the anti-fungal effects to reactive oxygen species (ROS) generated by ZnO NPs in water suspensions [40, 41]. The ZnO NPs were also considered to act as a nano-fertilizer, providing Zn as a micronutrient to the plants. Pot trials revealed 20 mg/L dosages of the nano-fertilizer improved the number of branches, root length, leaf area, fresh shoot weight, seed germination, shoot length, dry weight, and number of pods in okra plants. The authors conclude that the safe, non-toxic, environmentally compatible ZnO NPs demonstrated excellent antimicrobial and nano-fertilizer abilities and are suitable for agriculture applications.

11.3.2 Nanomaterial-Based Insecticides and Herbicides

The negative effects of inefficient pesticides are seen the world over in agricultural deployment, caused by excessive application, which results in resistant populations of pests and, in turn, an increased persistence within the environment that may lead to environmental damage. Additionally, pesticides have been known to accidentally affect non-target organisms, and residual quantities can negatively impact human health.

Nanoparticles can be utilized to augment the delivery of chemicals for practical purposes, thus becoming classified as nano-carriers. They can be modified for targeted delivery under specific conditions and have been used previously with the chemical avermectin to treat worms and other pests [42, 43]. Nano-carriers have also been developed for the stimulated release of chemicals by environmental triggers such as specific pH, enzyme presence, temperature, and irradiation by near-infrared (NIR) light [44–46].

Zein is a protein that has seen application in a number of fields including the pharmaceutical and food industries due to its biodegradability, biocompatibility, and low cost [47, 48]. For these reasons, Monteiro et al. developed a corn-protein nanoparticle loaded with botanical pesticides for the purpose of efficient, environmentally friendly insect control [49]. Limonene (LIM) and carvacrol (CVC) are naturally derived compounds that can be used as insecticides, antifungals, antimicrobials, and pesticides. The drawbacks of these chemicals are that they possess low persistency, high volatility, photosensitivity, and low solubility in aqueous solutions. These researchers investigated a way to deploy these chemicals in biodegradable NPs that would break down when exposed to enzymes within insects. This would ensure that the release of the insecticide occurred only at the intended time and location. Trypsin, an enzyme produced by pests such as cotton bollworms (*Helicoverpa armigera*) is able to break down corn proteins such as zein. Therefore, when the bollworms ingest zein particles loaded with a specific pesticide, the enzyme breaks down the protein and releases the pesticide. Results indicated that the zein NPs were stable in storage, possessed high encapsulation efficiency, and were not phytotoxic. The zein NPs also increased tolerances to oxidative stress in plants by increasing carotenoid content. Unfortunately, when subjected to trypsin, only 12% of the zein particles demonstrated effective breakdown after 120 minutes. However, when the loaded particles were applied to the fall armyworm larvae (*Spodoptera frugiperda*), mortality rates of $30 \pm 12\%$ were recorded, which was much higher than individual mortality rates of carvacrol and limonene (4.3% and 16.2%, respectively) [50]. Fluorescent probing conducted upon the specimens found treatment with NPs resulted in increased concentrations within midgut and feces, indicating high transport within larvae. The authors concluded the discovery was a "major contribution to sustainable agriculture", as the NPs provided controlled release within the desired organisms without negatively affecting the surrounding environment [49].

Copper oxide nanoparticles have been used in agriculture as pesticides to act against fungus and cotton leafworm [51, 52]. Biosynthesis of CuO nanoparticles from copper sulfate and neem leaf extract is a viable method through phytosynthesis [51]. Additionally, CuO nanoparticles have been described by Ayoub et al. to be an eco-friendly pesticide because they are an alternative treatment to methomyl, chlorfluazuron, and flufenoxuron [52–54].

11.4 NANO-BIOSENSORS FOR SOIL-PLANT SYSTEMS

Biosensors possess sensing elements that are biologically active and can be used to measure biological processes or physical changes [55]. Nanoparticles have also been developed for biosensor applications. However, the nanoparticle sections of the biosensor do not necessarily need to be of biological origin and commonly consist of semiconductor, oxide, or metal NPs [56].

Persistent organic pollutants (POPs), including polychlorinated biphenyls (PCBs), pose a significant threat to the health and safety of humans. POPs have been identified in polluted soils and the environment as a whole [57]. Therefore, having a reliable and accurate method for the detection of these compounds in soils prior to establishing food crops is an important safety requirement. Surface-enhanced Raman scattering (SERS) is a technique that utilizes the adsorption of organic molecules on the surface of nanoparticles to improve detection limits. One group investigated the use of gold nanoparticles (Au NPs) with cyclodextrin (ß-CD) functionalized on the surface of the AuNP. Cyclodextrin possesses a hydrophobic internal cavity, which can facilitate the formation of a cage and channel complex with targeted molecules, in this case, POPs [57]. The study demonstrated that trace amounts of PCBs could be detected with Raman when extracted from the soil with acetone. Unfortunately, the researchers did not quantify the amount of PCB present within contaminated soil samples or report detection limits.

There is an increasing need to determine heavy metal levels in the environment, as they have become increasingly prevalent globally. Heavy metals are of concern because they can alter biochemical cycles, cause cancer, and are not biodegradable [58]. Analytical instruments for the detection of heavy metals are often not field-deployable and have high costs associated with use and sample preparation. Swivedi et al. utilized selenite-reducing rhizospheric bacteria *Stenotrophomonas* acidaminaphila, a soil bacteria, to make selenium nanoparticles [59]. The Se NPs were synthesized by combining the bacteria with sodium selenite (Na_2SeO_3) and incubating the samples. The resulting red solution can be characterized by fluorescence to determine the concentration of Se NPs. However, the bio-reduction process is inhibited by the presence of heavy metals, and this forms the basis of the biosensor. By measuring decreases in fluorescence, decreases in enzymatic activity could be determined, and the presence of heavy metals could be quantified with concentrations of pollutants as low as 5 μM.

11.5 NANOMATERIALS FOR SOIL REMEDIATION

There is a multitude of different soil contaminants found throughout the world in agricultural areas, including metals such as As, Cd, Cu, Zn, and Pb as well as polycyclic aromatic hydrocarbons, herbicides, insecticides, excess and remnant fertilizers, petroleum products, and phthalate esters [60, 61]. Contamination creates negative effects on agricultural outcomes, such as reducing available nutrients and crop growth, decreasing soil quality, and in some cases poisoning crops.

11.5.1 Nano-Assisted Abiotic Remediation of Contaminated Soils

Most soil remediation techniques are abiotic and do not use biological methods of ameliorating contamination. For example, iron NPs can be used to remove petroleum contamination in water and soils by reduction [62]. Adsorption is another common method for remediation, especially for arsenic, where titanium, iron, manganese, and copper NPs have seen use, allowing for separation after adsorption [63]. Organochlorides are still creating issues in many countries even many years after their use as pesticides was banned due to their long half-lives [64]. However photocatalytic degradation of organochlorines is possible using TiO_2 or Fe NPs.

Iron nanoparticles and iron oxide nanoparticles have been used as decontamination agents, specifically for diesel-contaminated soils as reported by Karthick et al. [65]. However, Zhou et al. investigated the effects of nanoparticles composed of metallic iron, Fe(II), and Fe(II) oxides on soil physiochemical properties in red and Wushan soil (RS and WS, respectively) and found metallic iron reduced P availability [66]. In comparison, nitrogen and dissolved organic carbon availability increased. Therefore, the use of these nanoparticles to overcome one soil constraint (contamination) may introduce an alternate constraint (reduced fertility).

11.5.2 Nano-Assisted Bioremediation of Contaminated Soils

Bioremediation involves using a biological system for the purpose of removing, capturing, or neutralizing pollutants to aid in the decontamination of the affected area. Bioremediation tends toward being environmentally friendly and offers cost savings when compared to other classes of soil remediation, including physical and chemical approaches [67].

Mosmeri et al. used CaO_2 NPs to improve the bioremediation of groundwater contaminated with aromatic compounds [68]. Benzene, toluene, ethylbenzene, and xylene (BTEX) are known groundwater contaminants, which can be degraded by various microorganisms such as *Pseudomonas*. CaO_2 NPs were chosen because they can stimulate microorganism growth by the release of O_2 and can also directly degrade the aromatic structures of these contaminants by the release of OH•. The addition of Fe^{3+} to the CaO_2 NPs was expected to improve degradation via the promotion of a modified Fenton process, but this was not successful. However, the CaO_2 NPs did increase populations of contaminant biodegrading microorganisms without considerable drawbacks.

Another approach to bioremediation is to produce the microorganisms initially instead of deploying an amendment where they naturally occur [69]. Li et al. did just that with pickles, by synthesizing exopolysaccharide (EPS) NPs from the bacterial strain *L. plantarum-605*. EPS is a polymer secreted by microorganisms and has been utilized previously as a bio-adsorbent and antibiotic. The 88 nm EPS-605 NPs reported in the study were self-assembling and demonstrated excellent adsorption of lead, copper, and cadmium ions, as well as organic compounds like methylene blue.

11.5.3 Nanomaterials for Soil Health

Nanoparticles can be used for soil treatment and to support favorable growing conditions for plants by ensuring water availability and removal or entrapment of contaminants in soil.

Alsharef et al. investigated the use of carbon nanomaterials for the purpose of stabilizing the physical properties of clayey sand soil (UKM) [70]. Multiwall carbon nanotubes (MWCNTs) and carbon nano-fibers (CNFs) were tested at loadings of less than 0.2%. Enhanced hydraulic conductivity was observed, and it was suggested that water could more easily pass through the soil. The carbon nanomaterials also led to a reduction in soil cracks, which might otherwise have led to a loss of nutrients and interruption of plant growth. This study demonstrated that the nano-carbon amendments were effective in improving soil texture. As stated by Singh et al., improving soil texture is "the most important factor that governs the fertility of soil" [71].

Bayat et al. tested Fe and Mg nano-oxide amendments in calcareous loamy soil [72]. The effects of Fe_3O_4 and MgO additions (0–5%) on soil bulk density, compression, soil void, and soil tensile strength were assessed. An improved root growth environment was observed as the void ratio was boosted, and bulk density reduced with the addition of the Mg nano-oxide. The authors also reported enhanced soil structure and improved porosity with the Mg amendment and augmented tensile strength of the aggregates with Fe nano-oxide addition.

11.6 EFFECT OF NANOMATERIALS ON SOIL AND PLANT SYSTEMS

Understanding how nanomaterials affect soils and plant systems is imperative before utilizing them for agricultural purposes. Factors such as nanoparticle residence time, how they interact with organic matter and microbes, if they penetrate farm crops, and ultimately where they end up must be considered.

11.6.1 Effect of Nanomaterials on Soil Organic Matter

Very little to no research on how nanomaterials affect soil organic matter has been conducted, but the inverse has been researched. Soil interacting with nanoparticles can induce many changes in the chemistry of the NP species including mobility, dissolution, and redox properties, which can have different effects on ecosystems.

Zehlike et al. identified that nanoparticle mobility and retention within the soil are dependent on colloidal stability, which can be affected by composition, size, surface characteristics, and aggregation state [73]. Furthermore, dissolved organic matter (DOM) characteristics such as mass distribution, structure, composition, and concentration also affect the mobility of NPs in soil, as does the presence of multivalent cations [73]. The authors conclude that DOM interaction upon NP stability is unlikely to be consistent and suggest that when attempting to assess the environmental impact that NPs have, the entirety of the impacted area needs to be assessed. For these reasons, several different nanoparticles have been studied in different soil conditions.

For example, Hortin et al. investigated CuO NP dissolution and proposed that it is controlled by DOM [74]. The authors noted that differences reported for the toxicity to plants of CuO NPs under acidic conditions may be attributed to differences in the dissolution of the NPs. At high pH levels, CuO is expected to have low solubility. However, dissolved non-organic material (DNOM) and aqueous carbonates could lead to copper dissolving due to complexation, in addition to metabolites also affecting dissolution. In fact, Hortin et al. conclusively demonstrated that CuO NPs have long residence times in alkaline/calcareous soil and that crops, soil organic matter, and microbes can vastly affect CuO NP dissolution and increase their bioavailability [74]. These findings are in line with the conclusions made by Zehlike et al. that DOM plays a part in how NPs react within the environment [73].

Li et al. investigated the speciation and distribution of Ag NPs in paddy soil [75]. The authors noted that due to the unique conditions in paddy soils, silver NP transformations may differ from those in oxic soils [75]. The authors noted that "soil organic matter (OM) was found to increase Ag retention in the soil solids and inhibit the release of dissolved Ag, while low redox conditions tended to increase Ag sulfidation and also decrease dissolved Ag levels" [75]. Furthermore, the majority (> 73%) of silver ended up as Ag_2S regardless of the form of silver added. However, different pathways were observed between NP and ionic forms due to the formation of AgCl, Ag_3PO_4, and Ag_2O [75]. These findings indicate that the prediction of the environmental impacts of silver and silver NPs should be considered on a case-by-case basis, which is in agreement with previous research [73].

11.6.2 Effect of Nanomaterials on Soil Microbes

There are a plethora of microorganisms within soil, which vary by location, climate, soil type, and composition, all of which play varying roles within their environments. Functioning ecosystems are said to be reliant upon soil microbes, and because of this, factors which influence these microbes are vital to study [76].

CuO NPs are known to cause a reduction in soil microbial biomass, enzyme activities, microbial community composition, and biodiversity. In addition to inducing stress upon microbes, CuO NPs affect nutrient bioavailability [77]. TiO_2 is also known to have a complicated effect on enzymes and has been shown to slightly impact microbial biomass and stress [77].

Silver NPs have known antibacterial characteristics, therefore it is no surprise that even at low concentrations Ag NPs have been shown to reduce biological activity in soil and impact bacterial and fungal community structures, whereas Al_2O_3 and SiO_2 NPs have little effect [78, 79].

McGee et al. found similar results to Schlich and Hund-Rinke in that ammonia-oxidizing microbes showed sensitivity to Ag NPs, which implies that nitrogen cycling in soil may be adversely affected [78, 80]. Schlich and Hund-Rinke go on to explain that Ag NP toxicity within soils relies upon multiple parameters and is influenced by the same parameters that affect conventional chemicals such as sand content, pH, and organic carbon content [80]. The factors that influence NP toxicity to microbes partly depend upon the environment that they are found in as well as the type, form, and composition of the NPs. This is similar to the aforementioned conclusion with respect to how NPs affect soil organic matter. Specific studies that consider all these aspects and assess the entirety of the impacted area are required to accurately assess how NPs will interact with soil and the environment as a whole. McGee et al. recommend that future research investigate the thresholds at which NPs will negatively affect key functional groups and biological activity within ecosystems [78]. This would assist in establishing improved agricultural and environmental practices that utilize nanoparticles by determining levels at which they are safe and levels beyond which they represent a potential hazard.

11.6.3 Nanomaterials in Plants

11.6.3.1 Uptake and Translocation Mechanism

Understanding how nanoparticles enter and move around in plants is important, as it assists in understanding how they impact plant growth and health. Kapoor et al. described the uptake of nanoparticles through plant leaves as occurring via two pathways, depending on the nature of the particles. These can be summarized as hydrophobic/lipophilic particles entering through the cuticle or polar/ionic particles entering through the stomata [30]. Furthermore, they described uptake through plant roots as requiring particles to traverse the root cuticle, epidermis, cortex, endodermis, and then the Casparian strip [30]. In this case, NPs initially adhere to the root surface and then penetrate via carrier proteins, ion channels, aquaporins, or most commonly endocytosis. Once particles penetrate the plant, the xylem and phloem allow for transport away from and toward the root system, respectively. The efficacy of uptake depends upon the size, stability, surface functionality, and chemistry of the particles as well as the characteristics of the plant species. Surface area to charge and volume ratios of the nanoparticles also impact this absorption and transport within the plant.

11.6.3.2 Influence of Nanomaterials on Plants

There are many ways that nanoparticles can interact with plants, some of which have a positive impact, and in other cases, they have a negative impact. For example luminescent carbon dots (CDs) increase enzyme activity to enhance photosynthesis through dual emission, which has a positive impact on plant growth [81]. Luminescent up-conversion nanoparticles (UCNPs) assist in plant growth in low doses and inhibit growth in high doses [81]. In comparison, luminescent quantum dots (QDs) have been shown to generate ROS in cells, which can induce lipid membrane peroxidation, DNA damage, oxidative stress, protein damage, early flowering, Fe accumulation in roots, and reduction of photosynthetic ability and promote leaf senescence, all of which are negative to healthy plant growth [81]. C_{60} fullerenes have been found to reduce biomass in corn and soybeans [82]. Alternatively, fullerenol increased biomass yield, water content, fruit length, fruit number, and fruit fresh weight, and increased two anti-cancer phytomedicines, cucurbitacin-B and lycopene, and two anti-diabetic phytomedicines, charantin and insulin [83, 84]. These are just a few examples of the effects that different types of nanoparticles can have on plant and crop growth, which reinforces the need to assess the full impact of NPs for specific applications.

11.7 FATE OF NANOMATERIALS IN SOIL AND ENVIRONMENT

When considering the fate of nanoparticles, it is necessary to consider the movement of the particles in the environment, their stability, accumulation, potential transformations, and eventual or immediate toxicity. This information is necessary to understand how applied quantities of nanoparticles will impact the environment long term.

In 2013, production estimates for engineered nanoparticles (ENPs), i.e. man-made nanoparticles for a specific purpose, ranged between 220,000–320,000 tons per year [85]. From this significant amount of material, 17% was predicted to end up in soil, 21% in water, 2.5% in air, and the remaining in landfill [85]. As already noted, the interaction of NPs with the environment depends upon characteristics such as size, surface properties, and ambient environmental conditions and characteristics. This can make it difficult to precisely predict the fate of nanoparticles in the environment. Short of testing all types of nanoparticles, the best that can currently be achieved is identifying patterns of reactivity to estimate present and future outcomes. Some general patterns of NP behavior in the environment include aggregation, transformation by oxidation or sulfidation, complexation with natural organic material (NOM), and dissolution with time [86–90].

When NPs are released into the air during crop or weed spraying, their residence time is short, and often less is released compared to applications to water or soil [85]. NPs within the air can be removed by either wet or dry deposition. Dry deposition occurs at air-surface interfaces with the formation of bigger aggregates occurring at higher rates than small particles [91, 92]. With wet deposition, the aerosol NPs are either removed via interactions with clouds or precipitation depending on particle size [93].

When NPs are released into water, their fate is strongly dependent on the environment, as the quality of the water (fresh, storm, sea, or ground) affects aggregation, sedimentation, and dissolution. Both homogeneous and heterogeneous aggregation can occur, with the former more common than the latter. Nevertheless, aggregation and deposition of nanoparticles are key factors that influence their transport in the environment [94–97]. However, depending on the characteristics of the particle, environmental conditions, and initial concentration, some NPs can de-aggregate naturally [98, 99]. Most NPs are stable in fresh- and stormwater but unstable in ground- and seawater [100]. To best describe NP aggregation, the Derjaguin–Landau–Verwey–Overbeek (DLVO) theory is used to evaluate the stability of the isolated and aggregated particles [86, 101, 102]. Aggregation corresponds to sedimentation, which tends to occur over time [103]. When sedimentation occurs quickly, a lower residence time is achieved in the water column, and this equates to lower exposure times to living organisms [104]. Sedimentation of nanoparticles generally takes place more rapidly in seawater than in other water bodies [100]. In comparison, the dissolution of nanoparticles in water causes a release of dissolved ions from the NPs. The rate of dissolution is closely linked to surface properties such as the surface area-to-size ratio and surface charge. Surface modifications not only determine the aqueous solubility of nanoparticles but also influence dissolution characteristics [105]. Transformations of NPs occur when chemical and physical properties are altered, which tends to lead to less reactive particles [100].

Following NP release into soils is complex. Nevertheless, most NPs act in a similar fashion in soil as in water and tend to aggregate, adsorb/desorb, sediment, or dissolve [106–108]. In particular, most NPs released in non-saturated soil become trapped within air-water interfaces of soil particles [109]. Transport of nanoparticles in soil occurs by three mechanisms: direct interaction with soil, sedimentation, or diffusion [106, 110]. The fate of the nanoparticles is dependent upon primary particle size, aggregate particle size, environmental conditions, and surface charge [108, 111, 112]. The surface charge also affects particle–particle interaction and particle–soil interaction [108, 113, 114]. For example, a low zeta potential facilitates aggregation and adhesion to soil particles. Furthermore, the particle and pore size of the soil affects transportation [100]. pH often affects the surface charge of nanoparticles and therefore is also linked to aggregation and mobility, which has been specifically observed with TiO_2 and Cu NPs [106, 115]. Other factors that affect mobility in soil include the presence of electrolyte species, NP concentration, and organic matter characteristics, which can stabilize or destabilize NPs, and the flow rate of groundwater [116, 120].

11.8 HAZARDS OF NANOMATERIALS IN SOIL/PLANT/ENVIRONMENT

As NPs are so small, it is very easy for them to pass through barriers such as plant cell walls and ultimately end up in food. For example, nanoparticles applied to plants can increase the metal content in crops that humans consume. Even though in many cases NPs contain micronutrients that increase growth in plants, overexposure can also cause stress or damage to plants.

Copper oxide nanoparticles (CuO NPs) are commonly found in agricultural chemicals such as growth regulators and fertilizers used in urban agriculture (UA). Xiong et al. found these CuO NPs could have adverse effects on crops [121]. In particular, they investigated the effects of daily pollutant intake and maximal daily intake of copper on the leaves of lettuce and cabbage plants. Throughout the investigation, it was noted that stimulation of growth occurred when small amounts of copper were introduced to lettuce due to Cu being a vital micronutrient. However, at larger doses, phytotoxicity occurred within the plants. At higher levels, copper caused necrosis, stunting, chlorosis, and inhibition of photosynthesis, as well as blockages to stomata, which in turn limited H_2O, CO_2, and O_2 exchange. Plants were exposed to up to 45 times the tolerable daily intake (TDI) as set by the USEPA, which was representative of levels observed in field studies in some countries [122, 123].

Some pesticides have been known to be residual in soil and to subsequently find their way into crops [124]. Synthetic pyrethroid pesticides used in agriculture are of particular concern, specifically bifenthrin which is the most detected pesticide of that type in the US[125, 126]. Bifenthrin is also suspected of causing harm to wildlife [127–129]. The presence of residual nanoparticles may also have effects on crops [130, 131]. For example, 181 tons of copper and copper oxide nanoparticles have been produced annually since 2013 [132]. CuO NPs are used in agriculture as pesticides but have also been proven to be toxic to crops [133–135].

It has been suggested that the combination of residual pesticides and residual nanoparticles may have synergistic effects on crops. For example, Li et al. investigated the translocation of soil pollutants to vegetables to assess the risk of bifenthrin and Cu bioaccumulation in rapeseed [136].

Given the average size of CuO NP (46.1 nm), they can easily be absorbed by diffusion into plant roots. This finding corroborated previous work that demonstrated that CuO NPs did not dissolve as Cu^{2+} in soil but in fact were absorbed by plants as intact NPs [137]. Data indicated that while copper content in contaminated samples was several-fold higher than in controls, most of it tended to stay within the root as opposed to being transported into the leaves. Plant damage can occur with increased copper content through the generation of reactive oxygen species (ROS), which causes lipid peroxidation and is detrimental to the fluidity and integrity of the cell membrane. It was observed that the antioxidant (SOD), designed to ward against plant damage, was produced at higher rates when plants were exposed to higher levels of copper. Increased MDA content was also present in addition to electrolyte leakage, indicating that the permeability of the root cell was compromised. The aforementioned factors show that plant root damage occurred, and metabolic functions were jeopardized. Furthermore, large quantities of bifenthrin were detected in plants exposed to both CuO and the pesticide, which the authors propose is facilitated by the root damage caused by the CuO. Previous studies have demonstrated a lack of bifenthrin adsorption onto CuO, which removes the possibility of it acting as a carrier [137]. Given the evidence discovered, the authors warn that the synergistic influence of nanoparticles must be considered with respect to the detrimental effects of pesticides.

Nanoscale pesticides have not been thoroughly investigated for detrimental effects on humans and the environment. Furthermore, reports of the detection and characterization of nanoparticles in agricultural produce are limited.

Zhang et al. used a range of characterization techniques to measure the concentration of silver nanoparticles on and in pears [138]. The characterization techniques were chosen to take into consideration the difficulty in extracting nanoparticles from food matrixes especially when present in low concentrations.

Transmission electron microscopy (TEM) revealed that Ag NPs penetrated into the pear skin through gaps in skin tissues and were often firmly attached, which the authors ascribe to the high surface area of the NPs. Ag NPs were also found within the pulp (inside the fruit), and therefore could not simply be removed by washing with water. SEM was advantageous for the identification of NPs, as the backscattering of electrons made them highly visible. ICP-OES was reliable in quantifying nanoparticle contents. The authors believe Ag NPs present in the skin are not of great importance because "usually the pear skin is peeled off before consumption"; however, this is not a global practice [138]. Smaller-sized silver nanoparticles (20 nm) were found in higher concentrations than larger ones (70 nm) which the authors ascribe to a higher penetrant ability of the smaller nanoparticles [139, 140]. Thus, smaller NPs have a higher potential to be harmful to consumers, and nano-silver pesticides should be carefully regulated when applied to agricultural crops.

11.9 LIMITATIONS, KNOWLEDGE GAP, AND FUTURE PERSPECTIVES

11.9.1 Nanomaterial-Based Systems for Sustainable Agriculture

There are significant gaps in research to determine what nanoparticles are safe and useful for the cultivation of crops and for consumption by humans and livestock. Furthermore, research into controls and methods to prevent the spread of unwanted nanoparticles into and out of agricultural land needs to be investigated. Nevertheless, mesoporous silica nanoparticles are relatively benign and have been approved for medical applications by the Food and Drug Administration (FDA) in the United States.

11.10 CONCLUDING REMARKS

Nanomaterials are a powerful tool for enriching agriculture and have great potential to enhance the sector further. Nanomaterials can be utilized as fertilizers, biosensors, and pesticides, while assisting in disease control and soil remediation, with undoubtedly more applications to follow. However, the warnings in the literature should be heeded, and relevant studies should be performed to determine the fate and potential hazards of nanomaterials as their application to agriculture is extended.

REFERENCES

1. Vert, M.; Doi, Y.; Hellwich, K.-H.; Hess, M.; Hodge, P.; Kubisa, P.; Rinaudo, M.; Schue, F. Terminology for Biorelated Polymers and Applications (IUPAC Recommendations 2012). *Pure and Applied Chemistry* 2012, 84 (2), 377–410. https://doi.org/10.1351/PAC-REC-10-12-04.
2. Rastogi, A.; Tripathi, D. K.; Yadav, S.; Chauhan, D. K.; Zivcak, M.; Ghorbanpour, M.; El-Sheery, N. I.; Brestic, M. Application of Silicon Nanoparticles in Agriculture. *Biotechnology* 2019, 9 (3), 90. https://doi.org/10.1007/s13205-019-1626-7.
3. El-bendary, H. M.; El-Helaly, A. A. *First Record Nanotechnology in Agricultural: Silica Nano-Particles a Potential New Insecticide for Pest Control.* 6.
4. Rouhani, M.; Samih, M. A.; Kalantari, S. Insecticidal Effect of Silica and Silver Nanoparticles on the Cowpea Seed Beetle, Callosobruchus Maculatus F. (Col.: Bruchidae). 10.
5. Magda, S. Determinations of the Effect of Using Silca Gel and Nano-Silica Gel Against Tuta Absoluta (Lepidoptera: Gelechiidae) in Tomato Fields. 2016, 7.
6. Ziaee, M.; Ganji, Z. Insecticidal Efficacy of Silica Nanoparticles Against Rhyzopertha Dominica F. and Tribolium Confusum Jacquelin Du Val. *Journal of Plant Protection Research* 2016. https://doi.org/10.1515/jppr-2016-0037.
7. Šebesta, M.; Kolenčík, M.; Sunil, B. R.; Illa, R.; Mosnáček, J.; Ingle, A. P.; Urík, M. Field Application of ZnO and TiO_2 Nanoparticles on Agricultural Plants. *Agronomy* 2021, 11 (11), 2281. https://doi.org/10.3390/agronomy11112281.
8. Shaker, A. M.; Zaki, A. H.; Abdel-Rahim, E. F. M.; Khedr, M. H. TiO_2 Nanoparticles as an Effective Nanopesticide for Cotton Leaf Worm. 2017, 8.
9. Sabir, S.; Arshad, M.; Chaudhari, S. K. Zinc Oxide Nanoparticles for Revolutionizing Agriculture: Synthesis and Applications. *The Scientific World Journal* 2014, 2014, 1–8. https://doi.org/10.1155/2014/925494.
10. Savassa, S. M.; Castillo-Michel, H.; Pradas del Real, A. E.; Reyes-Herrera, J.; Marques, J. P. R.; de Carvalho, H. W. P. Ag Nanoparticles Enhancing Phaseolus Vulgaris Seedling Development: Understanding Nanoparticle Migration and Chemical Transformation Across the Seed Coat. *Environmental Science: Nano* 2021, 8 (2), 493–501. https://doi.org/10.1039/D0EN00959H.
11. Kohatsu, M. Y.; Lange, C. N.; Pelegrino, M. T.; Pieretti, J. C.; Tortella, G.; Rubilar, O.; Batista, B. L.; Seabra, A. B.; Jesus, T. A. de. Foliar Spraying of Biogenic CuO Nanoparticles Protects the Defence System and Photosynthetic Pigments of Lettuce (*Lactuca Sativa*). *Journal of Cleaner Production* 2021, 324, 129264. https://doi.org/10.1016/j.jclepro.2021.129264.
12. Concha-Guerrero, S. I.; Brito, E. M. S.; Piñón-Castillo, H. A.; Tarango-Rivero, S. H.; Caretta, C. A.; Luna-Velasco, A.; Duran, R.; Orrantia-Borunda, E. Effect of CuO Nanoparticles Over Isolated Bacterial Strains From Agricultural Soil. *Journal of Nanomaterials* 2014, 2014, 1–13. https://doi.org/10.1155/2014/148743.
13. Rajput, V.; Minkina, T.; Sushkova, S.; Behal, A.; Maksimov, A.; Blicharska, E.; Ghazaryan, K.; Movsesyan, H.; Barsova, N. ZnO and CuO Nanoparticles: A Threat to Soil Organisms, Plants, and Human Health. *Environmental Geochemical Health* 2020, 42 (1), 147–158. https://doi.org/10.1007/s10653-019-00317-3.
14. Kashyap, P. L.; Xiang, X.; Heiden, P. Chitosan Nanoparticle Based Delivery Systems for Sustainable Agriculture. *International Journal of Biological Macromolecules* 2015, 77, 36–51. https://doi.org/10.1016/j.ijbiomac.2015.02.039.
15. Salimi, M.; Motamedi, E.; Motesharezedeh, B.; Hosseini, H. M.; Alikhani, H. A. Starch-g-Poly(Acrylic Acid-Co-Acrylamide) Composites Reinforced With Natural Char Nanoparticles Toward Environmentally Benign Slow-Release Urea Fertilizers. *Journal of Environmental Chemical Engineering* 2020, 8 (3), 103765. https://doi.org/10.1016/j.jece.2020.103765.

16. Tymoszuk, A.; Wojnarowicz, J. Zinc Oxide and Zinc Oxide Nanoparticles Impact on In Vitro Germination and Seedling Growth in *Allium cepa* L. *Materials* 2020, 13 (12), 2784. https://doi.org/10.3390/ma13122784.
17. Batsmanova, L.; Taran, N.; Konotop, Y.; Kalenska, S.; Novytska, N. Use of a Colloidal Solution of Metal and Metal Oxide-Containing Nanoparticles as Fertilizer for Increasing Soybean Productivity. *Journal of Central European Agriculture* 2020, 21 (2), 311–319. https://doi.org/10.5513/JCEA01/21.2.2414.
18. Segura, R.; Vásquez, G.; Colson, E.; Gerbaux, P.; Frischmon, C.; Nesic, A.; García, D. E.; Cabrera-Barjas, G. Phytostimulant Properties of Highly Stable Silver Nanoparticles Obtained With Saponin Extract From *Chenopodium quinoa*. *Journal of Science Food Agriculture* 2020, 100 (13), 4987–4994. https://doi.org/10.1002/jsfa.10529.
19. Torabian, S.; Zahedi, M.; Khoshgoftarmanesh, A. Effect of Foliar Spray of Zinc Oxide on Some Antioxidant Enzymes Activity of Sunflower Under Salt Stress. 12.
20. Mahajan, P.; Dhoke, S. K.; Khanna, A. S. Effect of Nano-ZnO Particle Suspension on Growth of Mung (*Vigna radiata*) and Gram (*Cicer arietinum*) Seedlings Using Plant Agar Method. *Journal of Nanotechnology* 2011, 2011, 1–7. https://doi.org/10.1155/2011/696535.
21. Raskar, S. V.; Laware, S. L. Original Research Article Effect of Zinc Oxide Nanoparticles on Cytology and Seed Germination in Onion. 2014, 8.
22. Sedghi, M.; Hadi, M.; Toluie, S. G. Effect of Nano Zinc Oxide on the Germination Parameters of Soybean Seeds Under Drought Stress. 2013, 6.
23. Faizan, M.; Faraz, A.; Yusuf, M.; Khan, S. T.; Hayat, S. Zinc Oxide Nanoparticle-Mediated Changes in Photosynthetic Efficiency and Antioxidant System of Tomato Plants. *Photosynthesis* 2018, 56 (2), 678–686. https://doi.org/10.1007/s11099-017-0717-0.
24. Nalci, O. B.; Nadaroglu, H.; Pour, A. H.; Gungor, A. A.; Haliloglu, K. Effects of ZnO, CuO and γ-Fe_3O_4 Nanoparticles on Mature Embryo Culture of Wheat (*Triticum aestivum* L.). *Plant Cell Tissue Organ Culture* 2019, 136 (2), 269–277. https://doi.org/10.1007/s11240-018-1512-8.
25. Fortis Hernández, M.; Ortiz Lopez, J.; Preciado Rangel, P.; Trejo Valencia, R.; Lagunes Fortiz, E.; Andrade-Sifuentes, A.; Rueda Puente, E. O. Biofortification With Copper Nanoparticles (Nps Cu) and Its Effect on the Physical and Nutraceutical Quality of Hydroponic Melon Fruits. *Notulae Botanicae Horti Agrobotanici Cluj-Napoca* 2022, 50 (1), 12568. https://doi.org/10.15835/nbha50112568.
26. Palacio-Marquez, A.; Ramírez-Estrada, C. A.; Sanchez, E.; Ojeda-Barrios, D. L.; Chavez-Mendoza, C.; Sida-Arreola, J. P. Biofortification With Nanoparticles and Zinc Nitrate Plus Chitosan in Green Beans: Effects on Yield and Mineral Content. *Notulae Botanicae Horti Agrobotanici Cluj-Napoca* 2022, 50 (2), 12672–12672. https://doi.org/10.15835/nbha50212672.
27. Golubkina, N. A.; Folmanis, G. E.; Tananaev, I. G.; Krivenkov, L. V.; Kosheleva, O. V.; Soldatenko, A. V. Comparative Evaluation of Spinach Biofortification With Selenium Nanoparticles and Ionic Forms of the Element. *Nanotechnology Russia* 2017, 12 (9–10), 569–576. https://doi.org/10.1134/S1995078017050032.
28. Read, T. L.; Doolette, C. L.; Cresswell, T.; Howell, N. R.; Aughterson, R.; Karatchevtseva, I.; Donner, E.; Kopittke, P. M.; Schjoerring, J. K.; Lombi, E. Investigating the Foliar Uptake of Zinc From Conventional and Nano-Formulations: A Methodological Study. *Environmental Chemistry* 2019, 16 (6), 459. https://doi.org/10.1071/EN19019.
29. Sundaria, N.; Singh, M.; Upreti, P.; Chauhan, R. P.; Jaiswal, J. P.; Kumar, A. Seed Priming With Iron Oxide Nanoparticles Triggers Iron Acquisition and Biofortification in Wheat (*Triticum aestivum* L.) Grains. *The Journal of Plant Growth Regulation* 2019, 38 (1), 122–131. https://doi.org/10.1007/s00344-018-9818-7.
30. Kapoor, P.; Dhaka, R. K.; Sihag, P.; Mehla, S.; Sagwal, V.; Singh, Y.; Langaya, S.; Balyan, P.; Singh, K. P.; Xing, B.; White, J. C.; Dhankher, O. P.; Kumar, U. Nanotechnology-Enabled Biofortification Strategies for Micronutrients Enrichment of Food Crops: Current Understanding and Future Scope. *NanoImpact* 2022, 26, 100407. https://doi.org/10.1016/j.impact.2022.100407.
31. Peshin, R.; Bandral, R. S.; Zhang, W. Integrated Pest Management: A Global Overview of History, Programs and Adoption. *Integrated Pest Management: Innovation-Development Process*, 2009.
32. Kumar, A.; Choudhary, A.; Kaur, H.; Guha, S.; Mehta, S.; Husen, A. Potential Applications of Engineered Nanoparticles in Plant Disease Management: A Critical Update. *Chemosphere* 2022, 295, 133798. https://doi.org/10.1016/j.chemosphere.2022.133798.
33. Hossain, A.; Abdallah, Y.; Ali, M. A.; Masum, M. M. I.; Li, B.; Sun, G.; Meng, Y.; Wang, Y.; An, Q. Lemon-Fruit-Based Green Synthesis of Zinc Oxide Nanoparticles and Titanium Dioxide Nanoparticles Against Soft Rot Bacterial Pathogen Dickeya Dadantii. *Biomolecules* 2019, 9 (12), 863. https://doi.org/10.3390/biom9120863.

34. Ogunyemi, S. O.; Abdallah, Y.; Zhang, M.; Fouad, H.; Hong, X.; Ibrahim, E.; Masum, Md. M. I.; Hossain, A.; Mo, J.; Li, B. Green Synthesis of Zinc Oxide Nanoparticles Using Different Plant Extracts and Their Antibacterial Activity Against *Xanthomonas Oryzae Pv. Oryzae*. *Artificial Cells, Nanomedicine, and Biotechnology* 2019, 47 (1), 341–352. https://doi.org/10.1080/21691401.2018.1557671.
35. Ali, M.; Kim, B.; Belfield, K. D.; Norman, D.; Brennan, M.; Ali, G. S. Inhibition of Phytophthora Parasitica and *P. Capsici* by Silver Nanoparticles Synthesized Using Aqueous Extract of Artemisia Absinthium. *Phytopathology* 2015, 105 (9), 1183–1190. https://doi.org/10.1094/PHYTO-01-15-0006-R.
36. Rajesh, K. M.; Ajitha, B.; Reddy, Y. A. K.; Suneetha, Y.; Reddy, P. S. Assisted Green Synthesis of Copper Nanoparticles Using Syzygium Aromaticum Bud Extract: Physical, Optical and Antimicrobial Properties. *Optik* 2018, 154, 593–600. https://doi.org/10.1016/j.ijleo.2017.10.074.
37. Sreelatha, S.; Kumar, N.; Yin, T. S.; Rajani, S. Evaluating the Antibacterial Activity and Mode of Action of Thymol-Loaded Chitosan Nanoparticles Against Plant Bacterial Pathogen Xanthomonas Campestris Pv. Campestris. *Frontiers in Microbiology* 2022, 12, 792737. https://doi.org/10.3389/fmicb.2021.792737.
38. Campos, E. V. R.; Oliveira, J. L. de; da Silva, C. M. G.; Pascoli, M.; Pasquoto, T.; Lima, R.; Abhilash, P. C.; Fernandes Fraceto, L. Polymeric and Solid Lipid Nanoparticles for Sustained Release of Carbendazim and Tebuconazole in Agricultural Applications. *Science Report* 2015, 5 (1), 13809. https://doi.org/10.1038/srep13809.
39. Hirpara, D. G.; Gajera, H.; Green, P. Synthesis and Antifungal Mechanism of Silver Nanoparticles Derived From Chitin-Induced Exometabolites of Trichoderma Interfusant. *Applied Organometallic Chemistry* 2020, 34 (3), e5407. https://doi.org/10.1002/aoc.5407.
40. Keerthana, P.; Vijayakumar, S.; Vidhya, E.; Punitha, V. N.; Nilavukkarasi, M., Praseetha, P. K. Biogenesis of ZnO Nanoparticles for Revolutionizing Agriculture: A Step Towards Anti-Infection and Growth Promotion in Plants. *Industrial Crops and Products* 2021, 170, 113762. https://doi.org/10.1016/j.indcrop.2021.113762.
41. Doan Thi, T. U.; Nguyen, T. T.; Thi, Y. D.; Ta Thi, K. H.; Phan, B. T.; Pham, K. N. Green Synthesis of ZnO Nanoparticles Using Orange Fruit Peel Extract for Antibacterial Activities. *RSC Advances* 2020, 10 (40), 23899–23907. https://doi.org/10.1039/D0RA04926C.
42. Smith, A. M.; Gilbertson, L. M. Rational Ligand Design to Improve Agrochemical Delivery Efficiency and Advance Agriculture Sustainability. *ACS Sustainable Chemical Engineering* 2018, 6 (11), 13599–13610. https://doi.org/10.1021/acssuschemeng.8b03457.
43. Liang, Y.; Gao, Y.; Wang, W.; Dong, H.; Tang, R.; Yang, J.; Niu, J.; Zhou, Z.; Jiang, N.; Cao, Y. Fabrication of Smart Stimuli-Responsive Mesoporous Organosilica Nano-Vehicles for Targeted Pesticide Delivery. *Journal of Hazardous Materials* 2020, 389, 122075. https://doi.org/10.1016/j.jhazmat.2020.122075.
44. Xu, X.; Bai, B.; Wang, H.; Suo, Y. A Near-Infrared and Temperature-Responsive Pesticide Release Platform Through Core–Shell Polydopamine@PNIPAm Nanocomposites. *ACS Applied Material Interfaces* 2017, 9 (7), 6424–6432. https://doi.org/10.1021/acsami.6b15393.
45. Zhang, Y.; Chen, W.; Jing, M.; Liu, S.; Feng, J.; Wu, H.; Zhou, Y.; Zhang, X.; Ma, Z. Self-Assembled Mixed Micelle Loaded With Natural Pyrethrins as an Intelligent Nano-Insecticide With a Novel Temperature-Responsive Release Mode. *Chemical Engineering Journal* 2019, 361, 1381–1391. https://doi.org/10.1016/j.cej.2018.10.132.
46. Zhu, H.; Shen, Y.; Cui, J.; Wang, A.; Li, N.; Wang, C.; Cui, B.; Sun, C.; Zhao, X.; Wang, C.; Gao, F.; Zhan, S.; Guo, L.; Zhang, L.; Zeng, Z.; Wang, Y.; Cui, H. Avermectin Loaded Carboxymethyl Cellulose Nanoparticles With Stimuli-Responsive and Controlled Release Properties. *Industrial Crops and Products* 2020, 152, 112497. https://doi.org/10.1016/j.indcrop.2020.112497.
47. Anderson, T. J.; Lamsal, B. P. Review: Zein Extraction From Corn, Corn Products, and Coproducts and Modifications for Various Applications: A Review. *Cereal Chemistry* 2011, 88 (2), 159–173. https://doi.org/10.1094/CCHEM-06-10-0091.
48. Berardi, A.; Bisharat, L.; AlKhatib, H. S.; Cespi, M. Zein as a Pharmaceutical Excipient in Oral Solid Dosage Forms: State of the Art and Future Perspectives. *AAPS PharmSciTech* 2018, 19 (5), 2009–2022. https://doi.org/10.1208/s12249-018-1035-y.
49. Monteiro, R. A.; Camara, M. C.; de Oliveira, J. L.; Campos, E. V. R.; Carvalho, L. B.; Proença, P. L. de F.; Guilger-Casagrande, M.; Lima, R.; do Nascimento, J.; Gonçalves, K. C.; Polanczyk, R. A.; Fraceto, L. F. Zein Based-Nanoparticles Loaded Botanical Pesticides in Pest Control: An Enzyme Stimuli-Responsive Approach Aiming Sustainable Agriculture. *Journal of Hazardous Materials* 2021, 417, 126004. https://doi.org/10.1016/j.jhazmat.2021.126004.
50. Pavela, R. Acute, Synergistic and Antagonistic Effects of Some Aromatic Compounds on the Spodoptera Littoralis Boisd. (Lep., Noctuidae) Larvae. *Industrial Crops and Products* 2014, 60, 247–258. https://doi.org/10.1016/j.indcrop.2014.06.030.

51. Ahmad, H.; Venugopal, K.; Bhat, A. H.; Kavitha, K.; Ramanan, A.; Rajagopal, K.; Srinivasan, R.; Manikandan, E. Enhanced Biosynthesis Synthesis of Copper Oxide Nanoparticles (CuO-NPs) for Their Antifungal Activity Toxicity Against Major Phyto-Pathogens of Apple Orchards. *Pharmaceutical Research* 2020, 37 (12), 246. https://doi.org/10.1007/s11095-020-02966-x.
52. Ayoub, H. A.; Khairy, M.; Elsaid, S.; Rashwan, F. A.; Abdel-Hafez, H. F. Pesticidal Activity of Nanostructured Metal Oxides for Generation of Alternative Pesticide Formulations. *Journal of Agricultural and Food Chemistry* 2018, 66 (22), 5491–5498. https://doi.org/10.1021/acs.jafc.8b01600.
53. Shaurub, E.-S. H.; Zohdy, N. Z.; Abdel-Aal, A. E.; Emara, S. A. Effect of Chlorfluazuron and Flufenoxuron on Development and Reproductive Performance of the Black Cutworm, Agrotis Ipsilon (Hufnagel) (Lepidoptera: Noctuidae). *Invertebrate Reproduction & Development* 2018, 62 (1), 27–34. https://doi.org/10.1080/07924259.2017.1384407.
54. America, United States Environmental Protection Agency, Office of Drinking Water. (1987). Methomyl Health Advisory, Office Of Drinking Water US Environmental Protection Agency (pp. 1–19).
55. Li, Y.; Schluesener, H. J.; Xu, S. Gold Nanoparticle-Based Biosensors. *Gold Bulletin* 2010, 43 (1), 29–41. https://doi.org/10.1007/BF03214964.
56. Cai, H.; Wang, Y.; He, P.; Fang, Y. Electrochemical Detection of DNA Hybridization Based on Silver-Enhanced Gold Nanoparticle Label. *Analytica Chimica Acta* 2002, 469 (2), 165–172. https://doi.org/10.1016/S0003-2670(02)00670-0.
57. Jency, D. A.; Umadevi, M.; Sathe, G. V. SERS Detection of Polychlorinated Biphenyls Using β-Cyclodextrin Functionalized Gold Nanoparticles on Agriculture Land Soil. *Journal of Raman Spectroscopy* 2015, 46 (4), 377–383. https://doi.org/10.1002/jrs.4654.
58. Ahmed, F.; Dwivedi, S.; Shaalan, N. M.; Kumar, S.; Arshi, N.; Alshoaibi, A.; Husain, F. M. Development of Selenium Nanoparticle Based Agriculture Sensor for Heavy Metal Toxicity Detection. Agriculture 2020, 10 (12), 610. https://doi.org/10.3390/agriculture10120610.
59. Dwivedi, S.; AlKhedhairy, A. A.; Ahamed, M.; Musarrat, J. Biomimetic Synthesis of Selenium Nanospheres by Bacterial Strain JS-11 and Its Role as a Biosensor for Nanotoxicity Assessment: A Novel Se-Bioassay. *PLoS One* 2013, 8 (3), e57404. http://dx.doi.org.libproxy.murdoch.edu.au/10.1371/journal.pone.0057404.
60. Kibblewhite, M. G. Contamination of Agricultural Soil by Urban and Peri-Urban Highways: An Overlooked Priority? *Environmental Pollution* 2018, 242, 1331–1336. https://doi.org/10.1016/j.envpol.2018.08.008.
61. Zeng, F.; Cui, K.; Xie, Z.; Wu, L.; Liu, M.; Sun, G.; Lin, Y.; Luo, D.; Zeng, Z. Phthalate Esters (PAEs): Emerging Organic Contaminants in Agricultural Soils in Peri-Urban Areas around Guangzhou, China. *Environmental Pollution* 2008, 156 (2), 425–434. https://doi.org/10.1016/j.envpol.2008.01.045.
62. Murgueitio, E.; Cumbal, L.; Abril, M.; Izquierdo, A.; Debut, A.; Tinoco, O. Green Synthesis of Iron Nanoparticles: Application on the Removal of Petroleum Oil From Contaminated Water and Soils. *Journal of Nanotechnology* 2018, 2018, 1–8. https://doi.org/10.1155/2018/4184769.
63. Habuda-Stanić, M.; Stjepanović, M. Arsenic Removal by Nanoparticles: A Review. *Environmental Science and Pollution Research International* 2015, 22, 8094–8123. https://doi.org/10.1007/s11356-015-4307-z.
64. Rani, M.; Shanker, U.; Jassal, V. Recent Strategies for Removal and Degradation of Persistent & Toxic Organochlorine Pesticides Using Nanoparticles: A Review. *Journal of Environmental Management* 2017, 190, 208–222. https://doi.org/10.1016/j.jenvman.2016.12.068.
65. Karthick, A.; Roy, B.; Chattopadhyay, P. Comparison of Zero-Valent Iron and Iron Oxide Nanoparticle Stabilized Alkyl Polyglucoside Phosphate Foams for Remediation of Diesel-Contaminated Soils. *Journal of Environmental Management* 2019, 240, 93–107. https://doi.org/10.1016/j.jenvman.2019.03.088.
66. Zhou, D.-M.; Jin, S.-Y.; Wang, Y.-J.; Wang, P.; Weng, N.-Y.; Wang, Y. Assessing the Impact of Iron-Based Nanoparticles on PH, Dissolved Organic Carbon, and Nutrient Availability in Soils. *Soil and Sediment Contamination: An International Journal* 2012, 21 (1), 101–114. https://doi.org/10.1080/15320383.2012.636778.
67. Azubuike, C. C.; Chikere, C. B.; Okpokwasili, G. C. Bioremediation Techniques – Classification Based on Site of Application: Principles, Advantages, Limitations and Prospects. *World Journal of Microbiology Biotechnology* 2016, 32 (11), 180. https://doi.org/10.1007/s11274-016-2137-x.
68. Mosmeri, H.; Gholami, F.; Shavandi, M.; Dastgheib, S. M. M.; Alaie, E. Bioremediation of Benzene-Contaminated Groundwater by Calcium Peroxide (CaO_2) Nanoparticles: Continuous-Flow and Biodiversity Studies. *Journal of Hazardous Materials* 2019, 371, 183–190. https://doi.org/10.1016/j.jhazmat.2019.02.071.

69. Li, C.; Zhou, L.; Yang, H.; Lv, R.; Tian, P.; Li, X.; Zhang, Y.; Chen, Z.; Lin, F. Self-Assembled Exopolysaccharide Nanoparticles for Bioremediation and Green Synthesis of Noble Metal Nanoparticles. *ACS Applied Material Interfaces* 2017, 9 (27), 22808–22818. https://doi.org/10.1021/acsami.7b02908.
70. Alsharef, J. M. A.; Taha, M. R.; Firoozi, A. A.; Govindasamy, P. Potential of Using Nanocarbons to Stabilize Weak Soils. *Applied and Environmental Soil Science* 2016, 2016, e5060531. https://doi.org/10.1155/2016/5060531.
71. Singh, R. P.; Handa, R.; Manchanda, G. Nanoparticles in Sustainable Agriculture: An Emerging Opportunity. *Journal of Controlled Release* 2021, 329, 1234–1248. https://doi.org/10.1016/j.jconrel.2020.10.051.
72. Bayat, H.; Kolahchi, Z.; Valaey, S.; Rastgou, M.; Mahdavi, S. Novel Impacts of Nanoparticles on Soil Properties: Tensile Strength of Aggregates and Compression Characteristics of Soil. *Archives of Agronomy and Soil Science* 2018, 64 (6), 776–789. https://doi.org/10.1080/03650340.2017.1393527.
73. Zehlike, L.; Peters, A.; Ellerbrock, R. H.; Degenkolb, L.; Klitzke, S. Aggregation of TiO_2 and Ag Nanoparticles in Soil Solution – Effects of Primary Nanoparticle Size and Dissolved Organic Matter Characteristics. *Science of The Total Environment* 2019, 688, 288–298. https://doi.org/10.1016/j.scitotenv.2019.06.020.
74. Hortin, J. M.; Anderson, A. J.; Britt, D. W.; Jacobson, A. R.; McLean, J. E. Copper Oxide Nanoparticle Dissolution at Alkaline PH is Controlled by Dissolved Organic Matter: Influence of Soil-Derived Organic Matter, Wheat, Bacteria, and Nanoparticle Coating. *Environmental Science: Nano* 2020, 7 (9), 2618–2631. https://doi.org/10.1039/D0EN00574F.
75. Li, M.; Wang, P.; Dang, F.; Zhou, D.-M. The Transformation and Fate of Silver Nanoparticles in Paddy Soil: Effects of Soil Organic Matter and Redox Conditions. *Environmental Science: Nano* 2017, 4 (4), 919–928. https://doi.org/10.1039/C6EN00682E.
76. Gajjar, P.; Pettee, B.; Britt, D. W.; Huang, W.; Johnson, W. P.; Anderson, A. J. Antimicrobial Activities of Commercial Nanoparticles against an Environmental Soil Microbe, Pseudomonas Putida KT2440. *Journal of Biological Engineering* 2009, 3 (1), 9. https://doi.org/10.1186/1754-1611-3-9.
77. Xu, C.; Peng, C.; Sun, L.; Zhang, S.; Huang, H.; Chen, Y.; Shi, J. Distinctive Effects of TiO_2 and CuO Nanoparticles on Soil Microbes and Their Community Structures in Flooded Paddy Soil. *Soil Biology and Biochemistry* 2015, 86, 24–33. https://doi.org/10.1016/j.soilbio.2015.03.011.
78. McGee, C. F. Soil Microbial Community Responses to Contamination With Silver, Aluminium Oxide and Silicon Dioxide Nanoparticles. 10.
79. Sharma, S.; Sanpui, P.; Chattopadhyay, A.; Ghosh, S. S. Fabrication of Antibacterial Silver Nanoparticle—Sodium Alginate–Chitosan Composite Films. *RSC Advances* 2012, 2 (13), 5837. https://doi.org/10.1039/c2ra00006g.
80. Schlich, K.; Hund-Rinke, K. Influence of Soil Properties on the Effect of Silver Nanomaterials on Microbial Activity in Five Soils. *Environmental Pollution* 2015, 196, 321–330. https://doi.org/10.1016/j.envpol.2014.10.021.
81. Yu, J.; Yu, H.; Li, L.; Ni, X.; Song, K.; Wang, L. Influence of Luminescent Nanomaterials on Plant Growth and Development. *ChemNanoMat* 2021, 7 (8), 859–872. https://doi.org/10.1002/cnma.202100138.
82. De La Torre-Roche, R.; Hawthorne, J.; Deng, Y.; Xing, B.; Cai, W.; Newman, L. A.; Wang, Q.; Ma, X.; Hamdi, H.; White, J. C. Multiwalled Carbon Nanotubes and C 60 Fullerenes Differentially Impact the Accumulation of Weathered Pesticides in Four Agricultural Plants. *Environmental Science Technology* 2013, 47 (21), 12539–12547. https://doi.org/10.1021/es4034809.
83. Kole, C.; Kole, P.; Randunu, K. M.; Choudhary, P.; Podila, R.; Ke, P. C.; Rao, A. M.; Marcus, R. K. Nanobiotechnology Can Boost Crop Production and Quality: First Evidence From Increased Plant Biomass, Fruit Yield and Phytomedicine Content in Bitter Melon (*Momordica charantia*). *BMC Biotechnology* 2013, 13 (1), 37. https://doi.org/10.1186/1472-6750-13-37.
84. Husen, A.; Siddiqi, K. Carbon and Fullerene Nanomaterials in Plant System. *Journal of Nanobiotechnology* 2014, 12 (1), 16. https://doi.org/10.1186/1477-3155-12-16.
85. Keller, A. A.; Lazareva, A. Predicted Releases of Engineered Nanomaterials: From Global to Regional to Local. *Environmental Science Technology Letters* 2014, 1 (1), 65–70. https://doi.org/10.1021/ez400106t.
86. Stebounova, L. V.; Guio, E.; Grassian, V. H. Silver Nanoparticles in Simulated Biological Media: A Study of Aggregation, Sedimentation, and Dissolution. *Journal of Nanoparticle Research* 2011, 13 (1), 233–244. https://doi.org/10.1007/s11051-010-0022-3.
87. Praetorius, A.; Scheringer, M.; Hungerbühler, K. Development of Environmental Fate Models for Engineered Nanoparticles - A Case Study of TiO_2 Nanoparticles in the Rhine River. *Environmental Science Technology* 2012, 46 (12), 6705–6713. https://doi.org/10.1021/es204530n.

88. Lowry, G. V.; Espinasse, B. P.; Badireddy, A. R.; Richardson, C. J.; Reinsch, B. C.; Bryant, L. D.; Bone, A. J.; Deonarine, A.; Chae, S.; Therezien, M.; Colman, B. P.; Hsu- Kim, H.; Bernhardt, E. S.; Matson, C. W.; Wiesner, M. R. Long-Term Transformation and Fate of Manufactured Ag Nanoparticles in a Simulated Large Scale Freshwater Emergent Wetland. *Environmental Science Technology* 2012, 46 (13), 7027–7036. https://doi.org/10.1021/es204608d.
89. Lowry, G. V.; Gregory, K. B.; Apte, S. C.; Lead, J. R. Transformations of Nanomaterials in the Environment. *Environmental Science Technology* 2012, 46 (13), 6893–6899. https://doi.org/10.1021/es300839e.
90. Arvidsson, R.; Molander, S.; Sandén, B. A.; Hassellöv, M. Challenges in Exposure Modeling of Nanoparticles in Aquatic Environments. *Human and Ecological Risk Assessment: An International Journal* 2011, 17 (1), 245–262. https://doi.org/10.1080/10807039.2011.538639.
91. Friedlander, S. K.; Pui, D. Y. H. Emerging Issues in Nanoparticle Aerosol Science and Technology. *Journal of Nanoparticle Research* 2004, 6 (2/3), 313–320. https://doi.org/10.1023/B:NANO.0000034725.89027.6b.
92. Gong, L.; Xu, B.; Zhu, Y. Ultrafine Particles Deposition Inside Passenger Vehicles. *Aerosol Science and Technology* 2009, 43 (6), 544–553. https://doi.org/10.1080/02786820902791901.
93. Laakso, L. Ultrafine Particle Scavenging Coefficients Calculated From 6 Years Field Measurements. *Atmospheric Environment* 2003, 37 (25), 3605–3613. https://doi.org/10.1016/S1352-2310(03)00326-1.
94. Blaser, S. A.; Scheringer, M.; MacLeod, M.; Hungerbühler, K. Estimation of Cumulative Aquatic Exposure and Risk Due to Silver: Contribution of Nano-Functionalized Plastics and Textiles. *Science of The Total Environment* 2008, 390 (2–3), 396–409. https://doi.org/10.1016/j.scitotenv.2007.10.010.
95. Petosa, A. R.; Jaisi, D. P.; Quevedo, I. R.; Elimelech, M.; Tufenkji, N. Aggregation and Deposition of Engineered Nanomaterials in Aquatic Environments: Role of Physicochemical Interactions. *Environmental Science Technology* 2010, 44 (17), 6532–6549. https://doi.org/10.1021/es100598h.
96. Liu, H. H.; Cohen, Y. Multimedia Environmental Distribution of Engineered Nanomaterials. *Environmental Science Technology* 2014, 48 (6), 3281–3292. https://doi.org/10.1021/es405132z.
97. Quik, J. T. K.; Velzeboer, I.; Wouterse, M.; Koelmans, A. A.; van de Meent, D. Heteroaggregation and Sedimentation Rates for Nanomaterials in Natural Waters. *Water Research* 2014, 48, 269–279. https://doi.org/10.1016/j.watres.2013.09.036.
98. Zhou, D.; Abdel-Fattah, A. I.; Keller, A. A. Clay Particles Destabilize Engineered Nanoparticles in Aqueous Environments. *Environmental Science Technology* 2012, 46 (14), 7520–7526. https://doi.org/10.1021/es3004427.
99. Zhou, D.; Bennett, S. W.; Keller, A. A. Increased Mobility of Metal Oxide Nanoparticles Due to Photo and Thermal Induced Disagglomeration. *PLoS ONE* 2012, 7 (5), e37363. https://doi.org/10.1371/journal.pone.0037363.
100. Garner, K. L.; Keller, A. A. Emerging Patterns for Engineered Nanomaterials in the Environment: A Review of Fate and Toxicity Studies. *Journal of Nanoparticle Research* 2014, 16 (8), 2503. https://doi.org/10.1007/s11051-014-2503-2.
101. Elimelech, M.; Gregory, J.; Jia, X.; Williams, R. Particle Deposition and Aggregation, Measurement, Modeling and Simulation. *Colloids and Surfaces A: Physicochemical and Engineering Aspects* 1997, 1 (125), 93–94.
102. Wang, P.; Keller, A. A. Natural and Engineered Nano and Colloidal Transport: Role of Zeta Potential in Prediction of Particle Deposition. *Langmuir* 2009, 25 (12), 6856–6862. https://doi.org/10.1021/la900134f.
103. Elzey, S.; Grassian, V. H. Agglomeration, Isolation and Dissolution of Commercially Manufactured Silver Nanoparticles in Aqueous Environments. *Journal of Nanoparticle Research* 2010, 12 (5), 1945–1958. https://doi.org/10.1007/s11051-009-9783-y.
104. Klaine, S. J.; Alvarez, P. J. J.; Batley, G. E.; Fernandes, T. F.; Handy, R. D.; Lyon, D. Y.; Mahendra, S.; McLaughlin, M. J.; Lead, J. R. Nanomaterials in the Environment: Behavior, Fate, Bioavailability, and Effects. *Environmental Toxicology and Chemistry* 2008, 27 (9), 1825–1851. https://doi.org/10.1897/08-090.1.
105. Verma, A.; Stellacci, F. Effect of Surface Properties on Nanoparticle–Cell Interactions. *Small* 2010, 6 (1), 12–21. https://doi.org/10.1002/smll.200901158.
106. Dunphy Guzman, K. A.; Finnegan, M. P.; Banfield, J. F. Influence of Surface Potential on Aggregation and Transport of Titania Nanoparticles. *Environmental Science Technology* 2006, 40 (24), 7688–7693. https://doi.org/10.1021/es060847g.

107. Franklin, N. M.; Rogers, N. J.; Apte, S. C.; Batley, G. E.; Gadd, G. E.; Casey, P. S. Comparative Toxicity of Nanoparticulate ZnO, Bulk ZnO, and $ZnCl_2$ to a Freshwater Microalga (*Pseudokirchneriella subcapitata*): The Importance of Particle Solubility. *Environmental Science Technology* 2007, 41 (24), 8484–8490. https://doi.org/10.1021/es071445r.
108. Tourinho, P. S.; van Gestel, C. A. M.; Lofts, S.; Svendsen, C.; Soares, A. M. V. M.; Loureiro, S. Metal-Based Nanoparticles in Soil: Fate, Behavior, and Effects on Soil Invertebrates. *Environmental Toxicology and Chemistry* 2012, 31 (8), 1679–1692. https://doi.org/10.1002/etc.1880.
109. Sirivithayapakorn, S.; Keller, A. Transport of Colloids in Saturated Porous Media: A Pore-Scale Observation of the Size Exclusion Effect and Colloid Acceleration. *Water Resources Research* 2003, 39 (4). https://doi.org/10.1029/2002WR001583.
110. Fang, J.; Shan, X.; Wen, B.; Lin, J.; Owens, G. Stability of Titania Nanoparticles in Soil Suspensions and Transport in Saturated Homogeneous Soil Columns. *Environmental Pollution* 2009, 157 (4), 1101–1109. https://doi.org/10.1016/j.envpol.2008.11.006.
111. Darlington, T. K.; Neigh, A. M.; Spencer, M. T.; Guyen, O. T. N.; Oldenburg, S. J. Nanoparticle Characteristics Affecting Environmental Fate and Transport Through Soil. *Environmental Toxicology and Chemistry* 2009, 28 (6), 1191–1199. https://doi.org/10.1897/08-341.1.
112. Tufenkji, N.; Elimelech, M. Correlation Equation for Predicting Single-Collector Efficiency in Physicochemical Filtration in Saturated Porous Media. *Environmental Science Technology* 2004, 38 (2), 529–536. https://doi.org/10.1021/es034049r.
113. Saleh, N. B.; Pfefferle, L. D.; Elimelech, M. Aggregation Kinetics of Multiwalled Carbon Nanotubes in Aquatic Systems: Measurements and Environmental Implications. *Environmental Science Technology* 2008, 42 (21), 7963–7969. https://doi.org/10.1021/es801251c.
114. Saleh, N.; Kim, H.-J.; Phenrat, T.; Matyjaszewski, K.; Tilton, R. D.; Lowry, G. V. Ionic Strength and Composition Affect the Mobility of Surface-Modified FeO Nanoparticles in Water-Saturated Sand Columns. *Environmental Science Technology* 2008, 42 (9), 3349–3355. https://doi.org/10.1021/es071936b.
115. Jones, E. H.; Su, C. Fate and Transport of Elemental Copper (Cu0) Nanoparticles Through Saturated Porous Media in the Presence of Organic Materials. *Water Research* 2012, 46 (7), 2445–2456. https://doi.org/10.1016/j.watres.2012.02.022.
116. Espinasse, B.; Hotze, E. M.; Wiesner, M. R. Transport and Retention of Colloidal Aggregates of C 60 in Porous Media: Effects of Organic Macromolecules, Ionic Composition, and Preparation Method. *Environmental Science Technology* 2007, 41 (21), 7396–7402. https://doi.org/10.1021/es0708767.
117. Jaisi, D. P.; Elimelech, M. Single-Walled Carbon Nanotubes Exhibit Limited Transport in Soil Columns. *Environmental Science Technology* 2009, 43 (24), 9161–9166. https://doi.org/10.1021/es901927y.
118. Wang, P.; Shi, Q.; Liang, H.; Steuerman, D. W.; Stucky, G. D.; Keller, A. A. Enhanced Environmental Mobility of Carbon Nanotubes in the Presence of Humic Acid and Their Removal From Aqueous Solution. *Small* 2008, 4 (12), 2166–2170. https://doi.org/10.1002/smll.200800753.
119. Wang, Y.; Li, Y.; Pennell, K. D. Influence of Electrolyte Species and Concentration on the Aggregation and Transport of Fullerene Nanoparticles in Quartz Sands. *Environmental Toxicology and Chemistry* 2008, 27 (9), 1860–1867. https://doi.org/10.1897/08-039.1.
120. Ben-Moshe, T.; Dror, I.; Berkowitz, B. Transport of Metal Oxide Nanoparticles in Saturated Porous Media. *Chemosphere* 2010, 81 (3), 387–393. https://doi.org/10.1016/j.chemosphere.2010.07.007.
121. Xiong, T.; Dumat, C.; Dappe, V.; Schreck, E.; Shahid, M.; Pierart, A.; Sobanska, S. Copper Oxide Nanoparticle Foliar Uptake, Phytotoxicity, and Consequences for Sustainable Urban Agriculture. *Environmental Science Technology* 2017, 10.
122. Sharma, R. K.; Agrawal, M.; Marshall, F. M. Heavy Metal (Cu, Zn, Cd and Pb) Contamination of Vegetables in Urban India: A Case Study in Varanasi. *Environmental Pollution* 2008, 154 (2), 254–263. https://doi.org/10.1016/j.envpol.2007.10.010.
123. Liu, H.; Probst, A.; Liao, B. Metal Contamination of Soils and Crops Affected by the Chenzhou Lead/Zinc Mine Spill (Hunan, China). *Science of the Total Environment* 2005, 339 (1), 153–166. https://doi.org/10.1016/j.scitotenv.2004.07.030.
124. Cámara, M. A.; Cermeño, S.; Martínez, G.; Oliva, J. Removal Residues of Pesticides in Apricot, Peach and Orange Processed and Dietary Exposure Assessment. *Food Chemistry* 2020, 325, 126936. https://doi.org/10.1016/j.foodchem.2020.126936.
125. Dou, R.; Sun, J.; Deng, F.; Wang, P.; Zhou, H.; Wei, Z.; Chen, M.; He, Z.; Lai, M.; Ye, T.; Zhu, L. Contamination of Pyrethroids and Atrazine in Greenhouse and Open-Field Agricultural Soils in China. *Science of the Total Environment* 2020, 701, 134916. https://doi.org/10.1016/j.scitotenv.2019.134916.

126. Huff Hartz, K. E.; Nutile, S. A.; Fung, C. Y.; Sinche, F. L.; Moran, P. W.; Van Metre, P. C.; Nowell, L. H.; Lydy, M. J. Survey of Bioaccessible Pyrethroid Insecticides and Sediment Toxicity in Urban Streams of the Northeast United States. *Environmental Pollution* 2019, 254, 112931. https://doi.org/10.1016/j.envpol.2019.07.099.
127. Bertotto, L. B.; Bruce, R.; Li, S.; Richards, J.; Sikder, R.; Baljkas, L.; Giroux, M.; Gan, J.; Schlenk, D. Effects of Bifenthrin on Sex Differentiation in Japanese Medaka (*Oryzias latipes*). *Environmental Research* 2019, 177, 108564. https://doi.org/10.1016/j.envres.2019.108564.
128. Magnuson, J. T.; Cryder, Z.; Andrzejczyk, N. E.; Harraka, G.; Wolf, D. C.; Gan, J.; Schlenk, D. Metabolomic Profiles in the Brains of Juvenile Steelhead (*Oncorhynchus mykiss*) Following Bifenthrin Treatment. *Environmental Science Technology* 2020, 54 (19), 12245–12253. https://doi.org/10.1021/acs.est.0c04847.
129. Anderson, B. S.; Phillips, B. M.; Voorhees, J. P.; Petersen, M. A.; Jennings, L. L.; Fojut, T. L.; Vasquez, M. E.; Siegler, C.; Tjeerdema, R. S. Relative Toxicity of Bifenthrin to Hyalella Azteca in 10 Day versus 28 Day Exposures. *Integrated Environmental Assessment and Management* 2015, 11 (2), 319–328. https://doi.org/10.1002/ieam.1609.
130. Gogos, A.; Knauer, K.; Bucheli, T. D. Nanomaterials in Plant Protection and Fertilization: Current State, Foreseen Applications, and Research Priorities. *Journal of Agricultural and Food Chemistry* 2012, 60 (39), 9781–9792. https://doi.org/10.1021/jf302154y.
131. Wang, Y.; Jiang, F.; Ma, C.; Rui, Y.; Tsang, D. C. W.; Xing, B. Effect of Metal Oxide Nanoparticles on Amino Acids in Wheat Grains (*Triticum aestivum*) in a Life Cycle Study. *Journal of Environmental Management* 2019, 241, 319–327. https://doi.org/10.1016/j.jenvman.2019.04.041.
132. Keller, A. A.; McFerran, S.; Lazareva, A.; Suh, S. Global Life Cycle Releases of Engineered Nanomaterials. *Journal of Nanoparticle Research* 2013, 15 (6), 1692. https://doi.org/10.1007/s11051-013-1692-4.
133. Ma, J.; Chen, Q.-L.; O'Connor, P.; Sheng, G. D. Does Soil CuO Nanoparticles Pollution Alter the Gut Microbiota and Resistome of Enchytraeus Crypticus? *Environmental Pollution* 2020, 256, 113463. https://doi.org/10.1016/j.envpol.2019.113463.
134. Gao, X.; Avellan, A.; Laughton, S.; Vaidya, R.; Rodrigues, S. M.; Casman, E. A.; Lowry, G. V. CuO Nanoparticle Dissolution and Toxicity to Wheat (*Triticum aestivum*) in Rhizosphere Soil. *Environmental Science Technology* 2018, 52 (5), 2888–2897. https://doi.org/10.1021/acs.est.7b05816.
135. Pu, S.; Yan, C.; Huang, H.; Liu, S.; Deng, D. Toxicity of Nano-CuO Particles to Maize and Microbial Community Largely Depends on Its Bioavailable Fractions. *Environmental Pollution* 2019, 255, 113248. https://doi.org/10.1016/j.envpol.2019.113248.
136. Li, M.; Xu, G.; Huang, F.; Hou, S.; Liu, B.; Yu, Y. Influence of Nano CuO on Uptake and Translocation of Bifenthrin in Rape (*Brassica napus* L.). *Food Control* 2021, 130, 108333. https://doi.org/10.1016/j.foodcont.2021.108333.
137. Li, M.; Xu, G.; Yang, X.; Zeng, Y.; Yu, Y. Metal Oxide Nanoparticles Facilitate the Accumulation of Bifenthrin in Earthworms by Causing Damage to Body Cavity. *Environmental Pollution* 2020, 263, 114629. https://doi.org/10.1016/j.envpol.2020.114629.
138. Zhang, Z.; Kong, F.; Vardhanabhuti, B.; Mustapha, A.; Lin, M. Detection of Engineered Silver Nanoparticle Contamination in Pears. *Journal of Agricultural and Food Chemistry* 2012, 60 (43), 10762–10767. https://doi.org/10.1021/jf303423q.
139. Sondi, I.; Salopek-Sondi, B. Silver Nanoparticles as Antimicrobial Agent: A Case Study on *E. coli* as a Model for Gram-Negative Bacteria. *Journal of Colloid and Interface Science* 2004, 275 (1), 177–182. https://doi.org/10.1016/j.jcis.2004.02.012.
140. Pal, S.; Tak, Y. K.; Song, J. M. Does the Antibacterial Activity of Silver Nanoparticles Depend on the Shape of the Nanoparticle? A Study of the Gram-Negative Bacterium *Escherichia coli*. *Applied and Environmental Microbiology* 2007, 73 (6), 1712–1720. https://doi.org/10.1128/AEM.02218-06.

12 Designing Composite Nano-Systems for Photocatalytic Water Treatment: Opportunities for Off-Grid Applications

Ananyo Jyoti Misra, Amrita Mishra, Cecilia Stalsby Lundborg, and Suraj Kumar Tripathy

CONTENTS

DOI: 10.1201/9781003181422-15

12.1 INTRODUCTION

Human life is dependent on clean water. A society without an adequate source and supply of clean drinking water is unsustainable. The provision of safe drinking water and adequate sanitation standards has been an essential prerequisite for individual health and prosperity since the dawn of human civilization. Because of the importance of water to humanity, the United Nations has recognized water as one of its sustainable development goals [1]. Currently, 1.2 billion people lack access to clean drinking water, and around 3,900 children die every day as a result of waterborne infections. This problem continues despite recent efforts to improve access to clean drinking water and sanitation [2, 3]. According to recent research by the World Health Organization, waterborne illnesses are responsible for nearly 80% of all diseases worldwide, and according to a World Bank analysis, such infections are responsible for 73 million days of lost labor in India alone [4]. Most states belonging to the eastern part of India are solely dependent on water that is sourced from nearby natural waterbodies. However, incessant dumping of untreated sewage from nearby treatment plants and industries often leads to the outbreak of waterborne illnesses in the surrounding population. There have been numerous instances of this being reported in eastern states [5, 6] and in northern regions of India [7, 8]. Globally, portable water is often supplied by a centralized treatment plant via a piped distribution network in developed countries and in many parts of the developing world [9]. However, in low- and middle-income countries, this is not the case, with much of the drinking water being contaminated and not being treated. Even effluents discharged from healthcare facilities to wastewater networks have little or no pre-treatment [10, 11]. This serious problem leads to the transfer of waterborne nosocomial and commensal multidrug-resistant (MDR) bacteria to the wider environment, exposing humans and animals well away from healthcare facilities. Recently, preventing healthcare-associated waterborne infections became a high priority due to the sudden increase in the number of immunologically weakened patients. Thus, any technical process designed to remove MDR bacteria is considered fundamental to treating unclean water and wastewater. However, conventional methods for treating wastewater, such as chlorination, ozone treatments, and UV irradiation, have also shown serious shortcomings. These shortcomings include the formation of potentially hazardous disinfection by-products (DBPs) and questionable disinfection ability [12–14].

Current estimates indicate that MDR bacteria are projected to pose a serious global public health hazard and potentially revert human civilization back to the pre-antibiotic age [15, 16]. According to a recent study based on data derived from the European Antimicrobial Resistance Surveillance Network (EARS-Net), approximately 33,000 people die each year in EU/EEA countries as a direct result of infections caused by MDR bacteria or as a result of associated clinical complications [17]. The threat of MDR bacteria in the environment is increasing in developing countries like India, where there is an incessant and uncontrolled usage of antibiotics to treat a wide variety of ailments. Ultimately, antibiotic residues find their way into waterbodies, and these residues aid in the promotion of antimicrobial resistance in bacteria. This is a serious public health concern in both the developing and developed worlds. [18].

The advent of various pathogens as a result of MDR bacteria in recent years has made supplying clean drinking water even more difficult. For instance, MDR bacteria not only contaminate community water supplies, but they also have the potential to pass their resistance on to unresisting bacteria *via* horizontal gene transfer. Importantly, MDR bacteria have genes that assist in the repair of DNA and bacteria to resist conventional treatment protocols. As a result, the therapeutic potential of antibiotics against bacterial infections may be severely reduced, posing a serious threat to public health [19]. The spread of MDR pathogens through waterbodies, owing to improper disinfection procedures, has significantly contributed to the spread of various illnesses among local populations. In particular, enteric pathogens have become a serious problem. For instance, ingestion of water infected with enteric pathogens is the most common cause of gastrointestinal disease. *Enterobacter sp.*, a motile gram-negative bacteria classified as a facultative anaerobe, has been identified as an increasingly important opportunistic pathogen that can cause eye and skin infections, meningitis, bacteremia (bacterial blood infection), pneumonia, and urinary tract infections [20, 21]. *Enterobacter sp.* are common in nature and can be found in soil, water, and sewage, and due to their prevalence in nature are also found in the intestinal tracts of animals [19]. A similar bacterial that is widely associated with waterborne diseases is *Salmonella sp.*, or *Salmonella enterica* serovar Typhimurium, which is a common gastrointestinal pathogen that can infect humans and animals alike. Typically, infection occurs with the ingestion of contaminated water or food, which allows salmonellae to infect the intestinal epithelium and cause gastrointestinal illness. Salmonellosis, a predominant disease that infects the intestinal tract, is generally found in children and immunocompromised persons, and is one of the most common infections found worldwide [22]. Another important species of bacteria that often spreads from nosocomial settings is *Staphylococcus sp.*, or *Staphylococcus haemolyticus* (*S. haemolyticus*), a coagulase-negative staphylococcus (CoNS) that is the second most often isolated species from human blood cultures [23] and is characterized by its antimicrobial resistance to several commonly used in several antimicrobial agents [24]. This gram-positive bacterium, like sepsis and leukocytosis, can cause infection in neutropenic patients, especially when venous catheters are used [25]. A substantial number of insertion gene sequences have been detected, and suggest that this species has unusual genomic plasticity which may have contributed to its potential to evolve antibiotic resistance. *S. haemolyticus* is likely to be the reservoir of resistance genes for other staphylococci, including *S. aureus*, as demonstrated by interspecies transfer of SCCmec cassettes [22]. Thus, this species has become a major agent for spreading nosocomial infections due to its high adaptability and capacity to survive in hospital environments, notably in hospital wastewater [26]. Other noteworthy etiological agents in contaminated water are *Vibrio cholera, Shigella sp*, and *Escherichia coli*, which are known to cause cholera, bacterial dysentery, and acute diarrhea respectively. According to WHO, globally more than 400,000 deaths result from cholera, typhoid, and other waterborne diseases every year [27].

Typically, a wastewater treatment plant (WWTP) consists of four to five treatment modules or unit operations to handle the pollutant load. However, efficacy is highly dependent on factors such as water type, treatment type, and so on. Microbial pollutants that persist after secondary treatment are usually dealt with in the tertiary treatment stage, as seen in Figure 12.3. Typical tertiary treatments used for disinfection procedures include chlorination, ozonation, and ultraviolet (UV) irradiation. UV light causes nucleic acid mutations that render the microorganism inactive. Chlorination and ozonation use the high oxidative potential of free chlorine and ozone to render the microorganisms inactive [28, 29]. Microbes, particularly bacteria, have developed mechanisms to circumvent the efficiency of these treatments. More crucially, during bacteria's escape from water disinfection, they also develop mechanisms to increase their resistance to antibiotics by using their oxidative stress response [30]. As a result, water has become a reservoir for antibiotic-resistant bacteria (ARB) and, as a result, has become a vehicle for the spread of antibiotic-resistant genes (ARG) to various habitats. Importantly, because bacteria continue to develop resistance against antibiotics, the presence of both interacting in wastewater has created a new level of complexity that must be immediately addressed.

12.2 MOTIVATION FOR DEVELOPING PHOTOCATALYTIC-BASED NANO-SYSTEMS FOR WATER TREATMENTS

Current methods used for water disinfection, namely chlorination, ozonation, and UV irradiation, are not completely effective. In many cases, target microbes are "inactivated temporarily" rather than "disinfected completely" [31, 32]. Moreover, microorganisms, mainly bacteria, under oxidative stress conditions often engage mechanisms like complete dormancy, which makes them extremely difficult to monitor their state under normal culture techniques. Such a state is known as a "viable but non-cultural state" or VBNC [33, 34]. This state makes monitoring the state and activity of bacteria in current water treatment procedures difficult. Furthermore, current water treatment procedures, such as adsorption or coagulation, just concentrate the pollutants and do not completely eliminate or destroy them. In addition, processes like sedimentation, filtration, chemical, and membrane technologies are traditional water treatment methods that have high operating costs and can introduce harmful secondary pollutants into the environment [35]. These harmful secondary pollutants tend to accumulate and have become a source of concern around the world as environmental awareness and regulatory procedures have become more stringent. For instance, chlorination has been widely used globally for many years to disinfect wastewater. However, the chlorination process produces harmful by-products like trihalomethanes (THM), halogenated acetic acids, and halogenated acetonitriles that are mutagenic and carcinogenic to humans [36–38], while ozonation, another important water disinfection method, produces harmful by-products like haloacetic acid and trihalomethanes. In addition, ozone gas used in the disinfectant process is not stable and tends to revert back to its more stable oxygen form; hence, ozone requires additional maintenance, handling, and monitoring to prevent reversion back to oxygen [39, 40]. UV irradiation is another widely used water disinfection technique because of its selectivity toward microorganisms. However, UV treatments are unable to provide complete disinfection [41]. Because of the abovementioned issues, recent research has focused on developing advanced oxidation processes (AOPs) as an alternative disinfection method. The AOP generates highly reactive transient species (i.e. OH, H_2O_2, O_3, O_2^-,) *in situ* for disinfecting organic compounds and water pathogens [42]. The AOP process makes use of heterogeneous semiconductor photocatalysts (i.e. TiO_2, ZnO, Fe_2O_3, CdS, GaP, and ZnS) to assist in the degradations of a wide range of organic compounds and water pathogens to produce readily biodegradable molecules.

Photocatalysis is the process by which a photoreaction is accelerated in the presence of a semiconductor photocatalyst. The photocatalyst is generally a heterogeneous semiconductor in the form of a nanometer-scale material of a nanoparticle. When the photocatalyst is stimulated by light with higher energy than its band gap, electrons (e^-) from the valance band of a semiconductor material are energized to reach the conduction band [43]. This results in a photo-generated hole (h^+) in the valance band, while the e^- is involved in the reduction of dissolved oxygen to superoxide radical (O_2^-), and the h^+ is involved in the oxidation of water to create OH radicals. Thus, the process results in the formation of energy-rich electron-hole pairs that can be exploited in redox processes, as seen in Figure 12.1. Furthermore, the O_2^- can be protonated to generate a hydroperoxyl radical that can then be converted to H_2O_2. However, because the e^- is highly unstable in the conduction band, it tends to revert to its starting state, which prevents ROS production [43, 44]. The most commonly used photocatalyst is titanium dioxide (TiO_2), and its photocatalysis process is schematically presented in Figure 12.1.

TiO_2 is an efficient photocatalyst material since it has a stable and tunable structure. However, it has a band gap that is poorly matched to the solar spectrum and, as a result, needs a specific source of energy for its application. In addition, TO_2 is also difficult to recover from the slurry systems after water treatment. Because of these issues, research has focused on using both metal dopants (Ag, Cu, Ni, Pd) and non-metal dopants (N, C, F, S) to aid in prolonging the charge separation and hence yield a better photocatalytic activity and efficiency [45, 46]. To improve TiO_2 recovery from treatment slurries, several immobilizers such as nanowires, nanorods, and nanotubes [47, 48], and mesoporous clay like bentonite, montmorillonite, and zeolite [49], have been used. Clays are catalytically inert and have high surface areas and high adsorption capacities that promote greater surface contact for the generation of reactive oxygen species (ROS). Moreover, clays are naturally biocompatible and have been used for many years in cosmetic products [50]. To date, the full potential of utilizing clays as supporting materials for semiconductor nanomaterials is yet to be fully explored. Because of the increased interest and innovation in this field, other photocatalysts are being developed. Other photocatalysts used in AOP treatments include ZnO, Fe_2O_3, CdS, GaP, and ZnS). These materials have been found to be highly effective in decomposing organic matter. Furthermore, these photocatalysts can be improved by additional modifications. For instance the use of clay-based ZnO photocatalysts

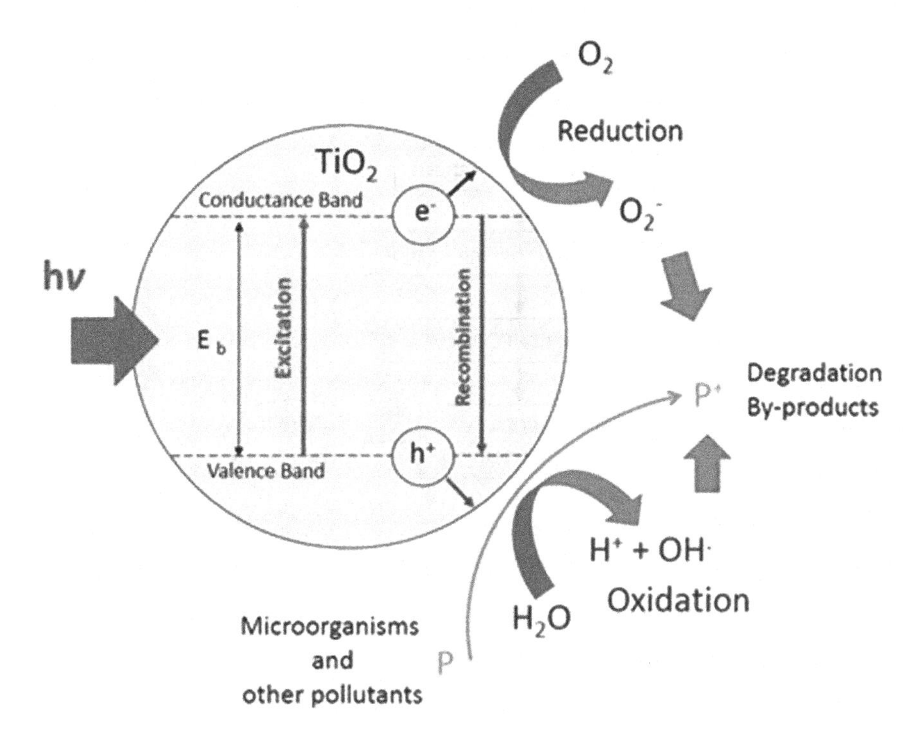

FIGURE 12.1 Schematic representation of TiO_2 mediated photocatalysis.

for use against multidrug-resistant strains of *Enterobacter*, *Salmonella*, and *Staphylococcus sp.* under a visible light source [51]. From our perspective, developing a working prototype for point-of-use (POU) water disinfection under visible light using reusable photocatalytic materials was investigated. In our analysis, we examined three major aspects, these included mechanism, safety, and operational capability with real wastewater conditions. A schematic representation of the POU water disinfection strategy is presented in Figure 12.2. As part of this strategy, the efficacy was not only evaluated for the disinfection of common waterborne bacteria such as *Escherichia coli*, *Staphylococcus haemolyticus*, and *Shigella dysenteriae* but also for the disinfection of total coliform in surface water samples. The water samples were collected from four major rivers in Odisha (India) as part of real-world testing. In addition, the *in-vivo* toxicity of the various treated water samples was assessed using a mouse model. Thus, the analysis of data generated by the POU water disinfection strategy of a proposed photocatalytic material will be used in the future for developing a real-world disinfection module in a wastewater treatment facility.

12.3 LITERATURE REVIEW OF TRADITIONAL WASTEWATER TREATMENTS

The global demand for clean water sources has rapidly increased with population growth, industrialization, and droughts. With the increased demand, numerous approaches, such as rainwater collecting and increasing catchment volumes, have been utilized to improve the supply of clean water [52]. Current estimates indicate that billions of people around the world will have little or no access to clean water, and millions of people may contract severe waterborne diseases and infections as a result of consuming contaminated water supplies. Thus, advancements in economical and highly efficient wastewater treatment (WWT) technologies are needed to clean poor-quality water supplies and overcome impending shortages. Two approaches currently considered for agricultural and industrial uses are the potential on-site reuse of rural wastewater or treated city wastewater from WWTPs. Unfortunately, wastewater reuse and recycling are frequently linked to the existence of dissolved organic components, suspended particles, and fecal coliforms, all of which

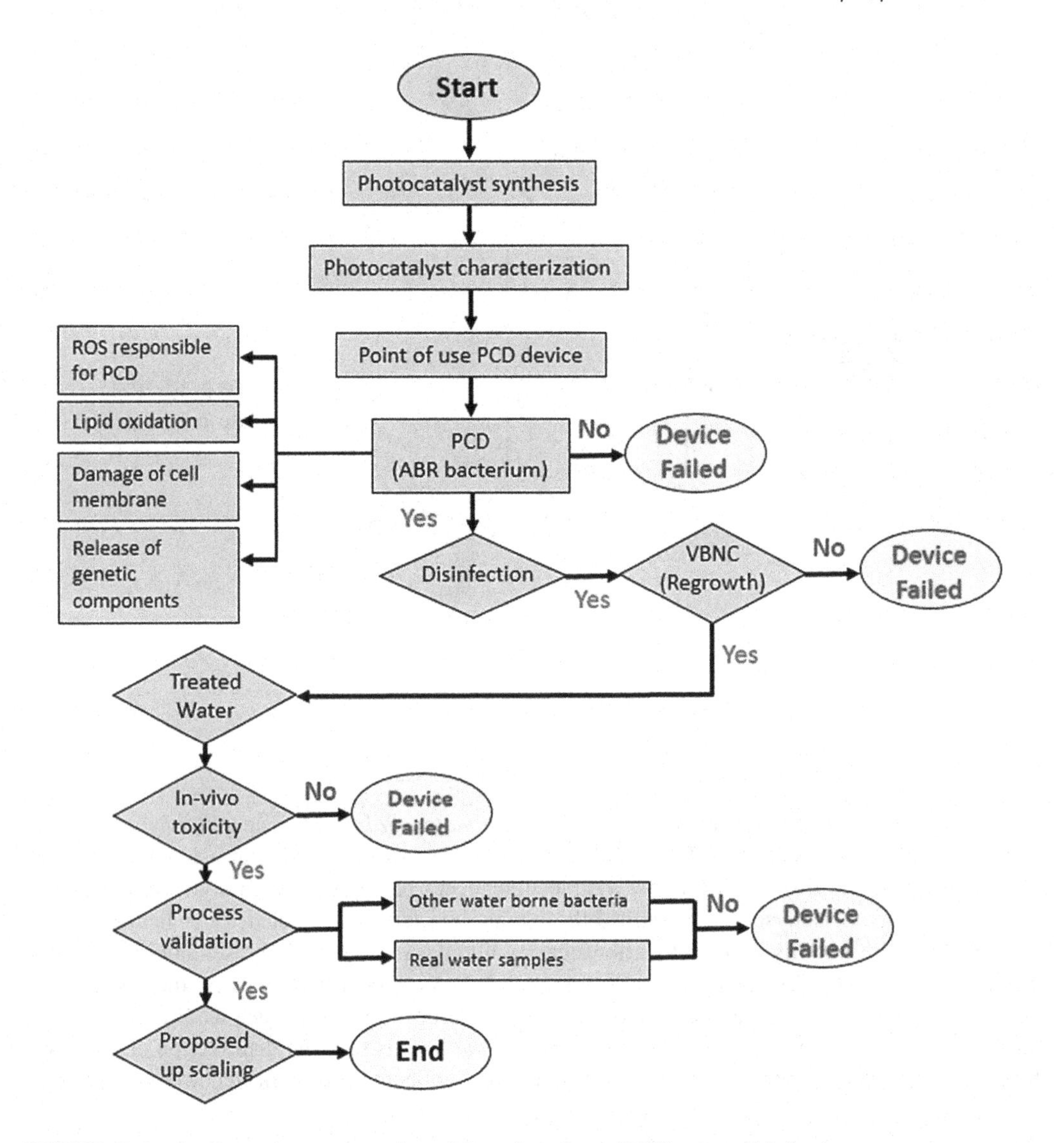

FIGURE 12.2 A schematic representation of the point-of-use (POU) water disinfection strategy.

are undesirable and costly to remove [53]. Moreover, the main component of wastewater, apart from chemical pollutants, is the presence of detrimental microbial like bacteria, fungi, viruses, etc. Therefore, efficient water treatment strategies are needed to combat these problems and prevent the spread of waterborne diseases and remove pollutants from water supplies. The following sections present a selection of traditional wastewater treatment methods and disinfection processes used to reduce or eliminate pollutants and microorganisms present in wastewater.

12.3.1 Conventional Wastewater Treatments

Traditional wastewater treatment processes are designed to remove solids and organic debris. In some circumstances, the removal of contaminants and nutrients using a combination of physical, chemical, and biological processes and activities. The wastewater treatment process consists of several stages that include 1) preliminary; 2) primary; 3) secondary; 4) tertiary; and/or 5) advanced wastewater treatments [54]. In many countries, the fifth stage consists of a disinfection process to remove pathogens. Figure 12.3 presents a schematic representation of a typical wastewater treatment

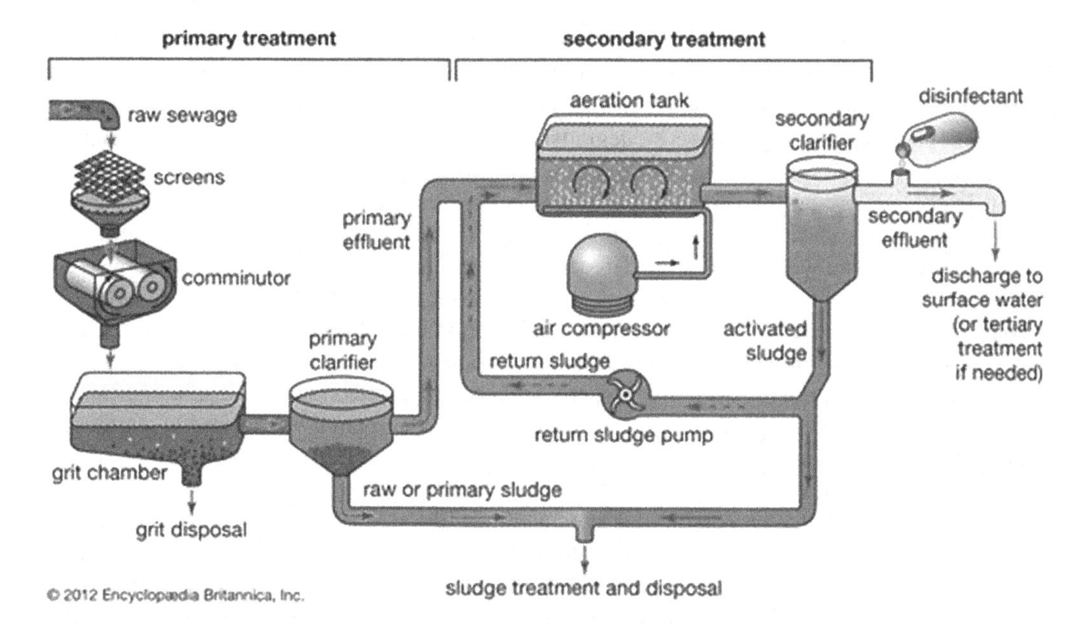

FIGURE 12.3 A schematic representation of a typical wastewater treatment facility.

facility. The schematic shows the various stages involved in treating wastewater. These stages are discussed in the following sections.

12.3.1.1 Preliminary Treatment Stage

The removal of large particles and other big items commonly found in raw wastewater is the goal of preliminary treatment. These larger materials must be eliminated in order to optimize the working and maintenance of subsequent treatment units that are designed to handle smaller-sized materials. Preliminary treatment procedures include coarse screening, grit removal, and, in exceptional cases, the comminution of large materials; see Figure 12.3. Most organic materials do not settle in grit chambers because the velocity of the moving water is kept high, or air is used to prevent settling. Generally, grit removal is not a standard treatment step in most small wastewater treatment plants. In addition, comminutors are sometimes used in conjunction with coarse screening to reduce the size of large materials, so that sludge may be removed in later stages of the treatment process [55]. Importantly, because the water flows need to be high, flow measuring equipment (i.e. standing-wave flumes, which are commonly used) is used to monitor flow rates preliminary treatment stage.

12.3.1.2 Primary Treatment Stage

During the primary treatment process, both sedimentation and skimming are used to remove floatable organic and inorganic contaminants. The primary treatment eliminates between 25 and 50% of incoming biochemical oxygen demand, 55 to 65% of total suspended solids (SS), and 65% of greases and oils. Also removed during the primary stage are some organic forms of nitrogen, organic phosphorus, and solids containing heavy metals. The wastewater generated by the primary treatment stage is known as primary effluent. In many industrialized nations, the primary treatment stage is the bare minimum of pre-treatment required for wastewater irrigation. In these cases, the wastewater is used to irrigate crops that are not consumed directly by humans, such as orchards, vineyards, and some processed food crops. However, even for non-food crop irrigation, further wastewater treatment is often necessary for many countries to minimize the inconvenience caused by storing or using flow-equalizing reservoirs [56]. In some cases, it may be possible to at least use a portion of primary effluent for irrigation if off-grid storage is available.

12.3.1.3 Secondary Treatment

The purpose of secondary treatment is to eliminate any leftover organics or suspended particles that may have escaped processing during the primary treatment. Secondary treatment employs aerobic-based biological methods to remove biodegradable dissolved and colloidal organic materials. In the presence of oxygen, aerobic microorganisms (mainly bacteria) metabolize organic materials present in the wastewater. During this stage, more microbes are produced, along with inorganic end-products such as NH_3, CO_2, and H_2O. A variety of aerobic biological processes are used in secondary treatments, with the main differences between the processes being the way oxygen is delivered to the microorganisms and the rate at which the organisms digest the organic matter [57]. In comparison to low-rate biological processes, high-rate biological processes are characterized by limited reactor volumes and high microbe densities. As a result of the well-controlled environment, the growth rate of new organisms is significantly higher in high-rate systems. To generate clarified secondary effluent, microorganisms must be removed from the treated wastewater by using sedimentation. Thus, secondary or biological sludge is made up of biological materials extracted during secondary sedimentation and sludge from the primary treatment stage. The combined sludge then undergoes high-rate sludge processing that uses methods like sludge activation, trickling filters or bio-filters, oxidation ditches, and rotating biological contactors (RBC). Generally, a combination of the different methods is used in series to treat municipal wastewater with high concentrations of organic materials derived from industrial sources. Typically, bio-filters are used first and are then followed by activated sludge methods. Each of these sludge treatment methods is discussed in the following sections.

12.3.1.3.1 Activated Sludge

An aeration tank or basin holding a suspension of wastewater and microorganisms, known as the mixed liquor, serves as the dispersed-growth reactor in the activated sludge process. The contents of the aeration tank are aggressively mixed while oxygen is supplied to the biological suspension via aeration devices. Aeration devices include submerged diffusers that release compressed air and mechanical surface aerators that stir the liquid surface. Hydraulic retention times in the aeration tanks are usually between three and eight hours but may be longer for wastewaters with high levels of microorganisms and low levels of oxygen.

12.3.1.3.2 Trickling Filters

A trickling filter, often called a bio-filter, is a basin or tower containing a support medium such as stones, plastic forms, or wooden slats. On a random or continuous basis, wastewater is sprayed on the support medium. Microorganisms adhere to the medium, generating a biological layer or a long-lasting bio-film. The organic material in the wastewater then diffuses into the film, where it is digested. The natural flow of air, up or down through the medium, provides oxygen to the bio-film. Diffusion of oxygen through the medium and bio-film is dependent on the relative temperatures of the wastewater and ambient air. Forced air can also be provided by blowers; however, this is rarely necessary. The bio-film grows thicker with the growth of more microorganisms. The effluent and bio-film fragments are then further treated, as seen in Figure 12.3.

12.3.1.3.3 Rotating Biological Contactors

With a similar operating principle to bio-filters, rotating biological contactors (RBCs) are fixed bio-film reactors with microorganisms attached to the support medium. RBCs consist of slowly rotating disks that are partially immersed in the reactor's flowing effluent. The rotating disks serve as the support medium. The surface turbulence induced by the rotating disks delivers oxygen to the waste-water. Oxygen is transferred to the bio-film from the air when the bio-film is out of the water and from the liquid when the bio-film is submerged. Bio-film fragments that slued are then eliminated in a similar way to bio-filter fragments mentioned above.

12.3.1.4 Tertiary Treatment Stage

Tertiary or advanced wastewater treatments are utilized when specific wastewater constituents cannot be removed by the secondary treatment stage. Because advanced treatments follow the secondary treatment stage, they are often referred to as the tertiary treatment stage. The tertiary treatment stage is designed to remove specific wastewater constituents such as nitrogen compounds, phosphorus, organics, heavy metals/particles, and other dissolved materials. However, advanced treatment processes are occasionally used in conjunction with primary or secondary treatments where specific materials need to be removed from the wastewater stream. Chemical treatments are added to primary clarifiers or aeration basins to remove phosphorus. Similarly, adding chemical treatments to the secondary treatment stage removes specific chemical compounds [58].

12.4 WATER DISINFECTION PROCESSES

The number of pathogens present in water determines its microbiological quality. Although the majority of fecal coliforms and other pathogens are removed in conventional WWTPs, there still remains a significant number of pathogens in treated wastewater. The treated wastewater is then subjected to a further disinfection stage designed to eradicate the residual pathogens. Disinfection methods are designed to remove any chance of humans coming into contact with harmful pathogens. The three main disinfection methods commonly used are chlorination, UV treatment, and ozonation. The following sections discuss these methods.

12.4.1 Chlorination

The most widely used disinfection method used to combat waterborne infections is chlorination. The chlorine used for disinfection is usually in the form of sodium or calcium hypochlorite and chloramines, and in some cases, it is used as a gas. When chlorine comes into contact with water, it generates hypochlorous acid and hypochlorite ions. Both products are disinfectants and are collectively referred to as free chlorine. The former disinfectant is more powerful than the latter. Importantly, the generation of free chlorine is influenced by water pH, which in turn influences the efficacy of disinfectant. Hypochlorous acid formation is greater at a lower pH, so the efficacy of chlorination is quite high in lower pHs [59]. Importantly, when pathogens come into contact with chlorine, chlorine replaces one or more hydrogen atoms present in the pathogen's chemical structure. As a result, the entire molecular structure of the pathogen changes, and, as a result, the pathogen is unable to survive [60].

Although chlorine has been widely used, there are some issues associated with the generation of harmful disinfection by-products. Typical by-products resulting from chlorination are trihalomethanes, haloacetic acids, chlorophenols, chloramines, and halofuranones. In particular, both trihalomethanes and haloacetic acids have been found to be carcinogenic in nature. Thus, strict regulation of the chlorination process is required to control and manage these hazardous by-products [61]. Another issue that arises from chlorination is the advancement of antibiotic resistance in wastewater, promoted by horizontal gene transfer from chlorine-resistant bacterial strains. For instance, the number of antibiotic-resistant bacteria such as *Pseudomonas* and *Acidovorax* in wastewater have steadily increased in recent years [62]. In addition, studies have also reported that some bacteria, such as chloramphenicol, penicillin, and ampicillin, which are resistant to antibiotics, have become resistant to chlorination. Thus, these types of bacteria are not fully disinfected and have shown extremely high rates of reactivation after treatment. This trend in resistance indicates that some bacteria are slowly acquiring resistance to chlorination [63]. In recent years, there has been increasing anxiety over the long-term effectiveness of chlorine as a useful water disinfectant.

12.4.2 Ultraviolet Light Irradiation

In recent years, ultraviolet (UV) irradiation as a method for disinfecting water has been extensively used. UV irradiation at wavelengths of 254 nm is known to have bactericidal properties that neutralize microorganisms by forming pyrimidine dimers in their RNA and DNA [64]. The damage caused to microorganism RNA and DNA changes the transcription and replication mechanisms that ultimately render them inactive. Importantly, compared to other disinfection methods, UV irradiation produces no significant levels of by-product, which makes this method a safer option for end users [65]. However, in spite of this advantage, UV irradiation has a number of disadvantages that have limited its widespread use as a "stand-alone" disinfection technique. For instance, some microorganisms have the ability to repair damage caused by exposure to UV irradiation. The repair mechanism, known as "dark repair" or photo-reactivation, even after exposure to high doses of UV allows microorganisms to recover from the treatment [66]. In addition, UV treatment is selective toward microorganisms and is ineffective against the cysts of *Giardia* and *Cryptosporidium* [65]. Furthermore, the UV treatment is not entirely effective in neutralizing the threat of antibiotic-resistant bacteria and antibiotic-resistant genes. There is also a high probability that inactivated and non-inactivated bacteria might transfer their resistance through horizontal gene transfer when they are released into the environment [67]. These factors make the success of UV treatment heavily dependent on the types of microorganisms present in the wastewater and their subsequent ability to execute the repair of UV damage through the photo-reactivation mechanism. Thus, further advancements in this field are necessary before UV irradiation can be considered an effective stand-alone wastewater disinfection method.

12.4.3 Advanced Oxidation Process

Advanced oxidation processes (AOPs) are new cutting-edge technologies that speed up the oxidation and degradation of a wide variety of organic and inorganic chemicals that were previously resistant to treatment. The AOPs create, *in situ*, highly oxidizing radical species like $OH^{\bullet}$ using a wide range of energy sources such as chemical, natural light (solar), and artificial light energy. AOPs were first employed to degrade chemical pollutants before being extended to include microbial inactivation. Ozonation, photocatalysis, and the photo-Fenton reaction are some of the most common forms AOPs utilized for disinfecting water. AOPs operate by producing reactive oxygen species (ROS) that attack microorganisms and make their survival problematic [43]. Among the various types of AOPS, both ozonation and photocatalysis have proven to be effective water disinfection methods. Both processes are discussed further in the following sections.

12.4.4 Ozonation

Ozone (O_3) is a potent oxidizing agent that has been used to clean and improve the quality of wastewater for many years. The three oxygen atoms of the O_3 molecule are very reactive and can easily undergo chemical reactions with chemical species. In the case of wastewater, there are two reaction mechanisms. The first is the direct oxidation of water compounds with ozone, and the second is the indirect oxidation of water compounds by $OH^{\bullet}$ radical generation. Importantly, during indirect oxidation, the free radical reacts with virtually all organic pollutants. In addition, when ozone combines with organic materials, the resultant production of free radicals increases disinfection efficacy. In addition, changing wastewater pH or adding hydrogen peroxide or catalysts can promote radical production, and combined with UV irradiation can greatly increase its disinfectant properties [68]. Studies have shown that ozonation is more effective than chlorination for certain types of bacteria. For instance, ozonation effectively neutralizes several bacterial strains that have developed resistance to chlorination. Even organic pollutants, resistant to chlorination, are effectively neutralized by ozonation [69].

Ozonation works by generating oxidative stress that ultimately dismantles and damages the microorganism's membrane fluidity. However, there are several issues associated with ozonation as a water disinfectant method. These issues include 1) the generation of harmful disinfection by-products, like bromates, that are potentially carcinogenic for humans; 2) ozone is water soluble, as a result, high concentrations are needed to make it an effective disinfection [70]; 3) ozone is not effective against some microorganisms, such as spores, cysts, and viruses [71]; and 4) ozonation suffers from the regrowth of bacterial strains after treatment, which raises questions about its suitability as an effective disinfectant. Therefore, further studies are needed to evaluate and address issues like bacterial regrowth and the production of harmful by-products.

12.4.5 Photocatalysis

Photocatalysis is the acceleration of a photoreaction in the presence of a semiconductor photocatalyst. This method uses a semiconductor, which, when stimulated by light with a higher energy level than its band gap, causes the migration of an electron from its valence band to its conduction band, as seen in Figure 12.4. The net result of this electron transfer is the formation of electron-hole pairs that can be employed in redox reactions. The mechanism needs an illumination source, with photonic energy (h), which is greater than or equal to the band-gap energy of a metal oxide semiconductor. When the surface of the semiconductor is illuminated, photo-excitation takes place, and an electron (e^-) from the valence band moves to the vacant positions in the conduction band (CB). Photo-excitation takes place within femtoseconds and leaves a hole (h^+) in the valance band (VB) of the semiconductor. The charge carrier pair (e^- and h^+) is delivered to the surface of the photocatalyst and contributes to both reduction and oxidation processes. The photocatalysts are heterogeneous semiconductor metal oxides such as TiO_2, ZnO, CdS, GaP, ZnS, and Fe_2O_3. Each of these semiconductor materials has been found to promote the breakdown of a wide variety of organic and inorganic materials. Importantly, because TiO_2 and ZnO can be photo-excited by UV light and visible light, they have attracted considerable interest and are among the most studied photocatalysts [72]. In addition, both materials have good and stable thermal, chemical, and mechanical properties, which makes them suitable candidates for photocatalytic processes. Figure 12.4 presents a schematic representation of a metal oxide-based semiconductor being used as a photocatalyst to generate ROSs.

Photocatalysis has several features that make it a suitable method for disinfecting water. These include that 1) the process can operate at room temperature and pressure; 2) it can effectively mineralize refractory materials and persistent organic pollutants (POP) without generating hazardous secondary pollutants; and 3) it has lower operating costs compared to other processes. ROS produced by photocatalysis have been shown to have effective disinfection properties toward a wide variety of microorganisms and are capable of breaking down POPs without creating secondary

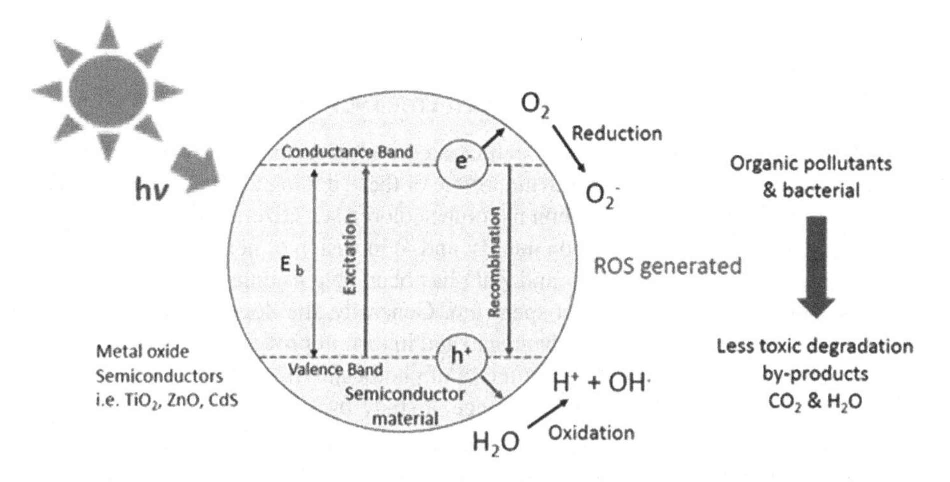

FIGURE 12.4 Schematic representation of semiconductor metal oxide being used to generate redox reactions for the degradation of organic pollutants and bacteria.

pollutants. However, semiconductor-based metal oxide photocatalysts have some technical issues. Some of these issues include 1) the photocatalyst's post-reaction separation after water treatment; 2) their large surface area and surface energy tend to promote agglomeration during application; and 3) developing photocatalysts that are active across a wider spectrum of light while taking into account that sunlight periods around 12 hours.

12.5 CURRENT INNOVATIONS IN PHOTOCATALYST-BASED TECHNOLOGIES

The TiO_2 electrode was originally utilized for water-splitting applications in 1972. Since then, TiO_2 electrodes have been manufactured in a variety of sizes, shapes, and physical and chemical properties in order for them to be used in a variety of photocatalytic applications [73]. Nanometer-scale TiO_2 has been shown to have a much higher oxidation competency than bulkier TiO_2 due to its light opaque qualities [74, 75]. However, the major issue for using TiO_2 in large-scale photocatalytic applications and post-reaction is separation. In particular, their shape, particle size, and sensitivity in hostile chemical environments make separation from wastewater streams difficult. The most significant disadvantage of TiO_2 for photocatalysis is the mismatch between its band-gap energy and the solar spectrum. Because the TiO_2 band gap only overlaps with the UV (320–400 nm) ranges, it can only benefit from around 5% of the solar spectrum reaching the Earth's surface, thus making it less likely to be exploited for real-time photocatalytic applications. This mismatch has had a big impact on photocatalyst research, which is now focusing on making better use of the visible solar spectrum. Because of this issue, the viability of utilizing ZnO and CdS photocatalysts has been re-evaluated in light of recent nanotechnology-based advancements. ZnO, because of its remarkable optical, photovoltaic, photocatalytic, and high photochemical activity has received considerable interest in recent years. It has a band-gap energy of 3.37 eV and a large excitation binding energy of up to 60 m eV. However, a major problem with all photocatalysts is their immobilization in inert substrates or matrices because this can dramatically lower the number of active sites on the catalyst. Thus, lower light penetration and reduced photocatalyst activation create a number of challenges in developing an effective photocatalytic process. To address this issue, integrating photocatalysis with membrane technologies or membrane-based filtration technologies that entrap the photo

catalyst particles during the fabrication process has helped reduce this problem. Examples of this are TiO_2 based-composites, where TiO_2 is supported on polymeric, metal, and ceramic (Al_2O_3) membranes [76]. However, issues such as membrane fouling, obstruction, regeneration, and back-washing difficulties, as well as limited photocatalytic potential and time-dependent catalyst layer degradation, still persist. The following sections discuss recent advances made to improving photocatalyst performance.

12.5.1 Doping Photocatalysts for Increased Efficiency

Various material-doping techniques have been created and used to extend the photo response of TiO_2 catalysts to operate in the solar spectrum. Some of these doping techniques developed include 1) incorporating photocatalysts with carbon nanotube colored sensitizers; 2) inclusion of noble metals or metal ions; 3) inclusion of transition metals; and 4) inclusion of non-metals. Typically, metallic and non-metallic ion doping of TiO_2 and ZnO has been able to improve the photo response of these semiconductors in the visible light spectrum. Generally, the dopant introduces an impurity into the semiconductor that decreases its band gap and in turn improves the semiconductor's capacity to capture more of the solar spectrum. In terms of photocatalytic doping, nitrogen (N) has been found to be a promising dopant [46]. For instance, a study by Rizzo et al. investigated the influence of N-doped TiO_2 on the performance of a photocatalytic device (PCD) toward the multidrug-resistant (MDR) *Escherichia coli* isolated from a WWTP. Their study found N-doped TiO_2 had greater disinfection rates than un-doped TiO_2 [77]. In addition, studies of visible-light-driven PCDs consisting of N-doped TiO_2 mesoporous thin films have found them effective against gram-positive and gram-negative bacteria like *Bacillus amyloliquefaciens* and *Pseudomonas aeruginosa*, respectively [78, 79]. In addition, sulfur (S) and other non-metallic dopants have also been reported as effective dopants in the literature. For instance, S-doped and un-doped TiO_2 synthesized by the sol-gel method found that S-doped TiO_2, when exposed to visible light, was an effective disinfectant against *Micrococcus lylae* [80].

Transition metals such as iron (Fe) have also been found to be effective dopants. For example, Fe-doped ZnO nanoparticles, when used in a solar-photocatalysis reactor, were found to be an effective disinfectant against MDR *Escherichia coli* [81]. In addition, the study revealed that the electronegativity of the dopant, as well as the difference between the dopant and ionic radius of Zn, determined its efficiency. Importantly, Fe, which is chemically stable and exists in two potential oxidation states, Fe^{2+} (ionic radii 0.78) and Fe^{3+} (ionic radii 0.64), is able to easily penetrate through the ZnO lattice without disrupting its crystal structure. This results in more charge carriers, which in turn improves the photocatalytic potency and conductivity of the ZnO [82]. In addition, as the concentration of Fe dopant was increased, the optical band gap of the ZnO gradually decreased [83]. When noble metals such as silver (Ag) are utilized as dopants, they transform n-type ZnO semiconductors into p-type semiconductors. The transformation also increases their optoelectronic and photocatalytic potential. For instance, the photocatalytic inactivation of *Escherichia coli* by ZnO-Ag nanoparticles was found to be effective under solar radiation [84]. As well as single elemental doping, multiple element co-doping has also shown promise in improving the photocatalytic activity of semiconductor materials. For instance, Er-Al co-doped ZnO nanoparticles were found to have enhanced photocatalytic performance that enabled them to break down ethyl orange under visible light [85]. Furthermore, a study by Feilizadeh et al. found Fe-Cd co-doped TiO_2 photocatalysts were able to disinfect *Escherichia coli* bacteria within around 45 minutes when exposed to visible light irradiation [86]. Another study by Wang et al. also found that co-doping with boron (B) and nickel (Ni) can significantly improve photocatalytic activity against *Escherichia coli* when exposed to visible light [87].

12.5.2 Immobilization of Photocatalysts for Improved Efficiency

The most widely used photocatalytic material is TiO_2. It is generally used in powder form and is directly placed into the reaction slurry. However, the introduction of TiO_2 also means there is an additional step of recovery after the treatment stage is over. Because of the need to recover TiO_2, a number of strategies have developed to immobilize the photocatalytic in support materials like photocatalytic membranes, activated carbon, mesoporous clays, and beads to assist in the recovery stage. These strategies are discussed in the following sections.

12.5.2.1 Immobilization in Photocatalytic Membranes

The use of photocatalytic membranes for immobilization has grown in popularity in recent years due to treated water being continuously discharged without losing the embedded photocatalytic material from the membrane surface. A number of organic and inorganic membrane materials have been investigated. For instance, several polymers and ceramic materials like Al_2O_3 membranes have been developed and evaluated. The composite photocatalytic membrane structure consists of the base membrane, with the photocatalytic material, usually TiO_2, embedded in the membrane's surface structure. For example, photocatalytic membrane reactors have been successfully used to degrade azo dyes, toluene, and antibiotics such as diclofenac and oxytetracycline [88–90]. However, photocatalytic membranes suffer from one major operational issue. The issue results from the deterioration of the membrane structure and the subsequent reduction in photocatalytic activity. The deterioration results in poor disinfection of microorganism targets in the wastewater [91, 92].

12.5.2.2 Immobilization in Fiber-Based Materials and Activated Carbon

The photo-oxidation of organic pollutants for water purification has also investigated the use of support materials like glass, carbon, titanate, woven textile fibers, and activated carbon. The majority of fiber materials used generally have an elongated rod shape or longitudinal morphology. Significant improvements in increased surface area and TiO_2 coverage can be achieved using nanofibers, nanowires, or nanorods because of their thin longitudinal shape [93]. However, utilizing less durable immobilizer fibers, such as glass or woven cloths, can result in limited durability because the implanted TiO_2 anatase crystals tend to wear away over time and lose their photocatalytic activity. Moreover, the use of immobilizer fibers tends to increase the flow pressure needed to pump wastewater through the reactor system. On the other hand, commercially successful nano-fiber-based materials have been made into microfiltration membranes (MF), ultrafiltration membranes (UF), and photocatalytic membranes (PMs). MF membranes are especially intriguing because they have high pollutant removal rates while maintaining low trans-membrane pressures (usually less than 300 k Pa) [48]. The photocatalytic activity of TiO_2 was significantly increased when it was immobilized in activated carbon. For instance, the degradation of pharmaceuticals like amoxicillin, ampicillin, diclofenac and paracetamol, dyes, and toluene is greatly enhanced when TiO_2 is immobilized in activated carbon [94]. In addition, immobilized cerium-doped zinc oxide has been successfully used as an effective photocatalytic material for the degradation of antibiotics and the inactivation of antibiotic-resistant bacteria [95]. However, more research is needed to ensure the prolonged activity and increased efficiency of these types of immobilizers. Currently, they are not totally effective as disinfectants against all microorganism targets.

12.5.2.3 Immobilization in Clays

Clays are catalytically inactive materials. However, they have large surface areas and adsorption capacities that can assist a photocatalytic material and subsequently promote reactions to generate large numbers of ROS, which in turn improves the overall efficiency of the photocatalytic process. Furthermore, the abundance of clays on the Earth's surface, their high porosity, their thermal stability, and their bio-compatibility make them suitable candidates for the immobilization of photocatalytic materials [96]. Because of their material properties, several clay types have been investigated in recent years. In particular, clays like kaolinite, sepiolite, bentonite, and montmorillonite have attracted considerable interest. For instance, kaolinite impregnated with ZnO has proven to be an effective photocatalytic composite suitable for the disinfection of MDR bacterial strains under visible light stimulation [97]. Recent studies have also shown that kaolinite, montmorillonite, and bentonite are antibacterial clay minerals and have been found to be effective against waterborne bacteria such as *Escherichia coli* and *Staphylococcus aureus* [98]. In addition, palygorskite and sepiolite clays can be used as photo-disinfectants against *P. putida* in soils [99]. Furthermore, the discovery of medicinal clays also offers an alternative approach to combating antibiotic-resistant bacteria, thus making them an ideal candidate for use in wastewater treatment processes [100]. The most important aspect of having a clay-based immobilizer is that it actively works with the photocatalytic material to disinfect wastewater-borne microorganisms.

12.5.2.4 Immobilization in Beads

Immobilization of photocatalytic materials in beads has gained considerable interest in recent years due to the easier recovery of the beads after the treatment process. Typical materials investigated include glass, ceramic, chitosan, and sodium alginate. These materials have been used for the degradation of a variety of contaminants such as azo dyes, methyl orange, dichlorophenoxyacetic acid, antibiotics, and other pollutants [101–104]. For instance, ZnO supported on bentonite embedded in sodium alginate beads has been successfully used to photocatalytically inactivate *Staphylococcus aureus* [105]. Misra et al. used gypsum clay nanocomposites with sodium alginate beads for the photocatalytic disinfection of *Salmonella sp.* [97]. In addition, kaolinite clay nanocomposites immobilized with sodium alginate beads were found to be effective in killing nosocomial MDR strains isolated from a hospital setting. These findings indicated that metal-doped clay-based nanocomposites, when used with a suitable bead material, have the potential to be effective disinfectants toward waterborne contaminants [106].

12.6 DESIGN AND DEVELOPMENT OF A POINT-OF-USE PHOTOCATALYTIC REACTOR, USING BIOCOMPATIBLE CLAY-BASED NANOCOMPOSITES

Unfortunately, because of the high capital cost and high levels of energy usage, conventional wastewater treatment facilities are unaffordable for rural populations or isolated communities [106–109]. The problem is further compounded by the lack of constant electricity, which makes traditional wastewater treatment plants unviable. In addition, in recent years traditional wastewater treatment facilities have been heavily criticized due to the number of environmental concerns about their operational functionality. For instance, the formation of disinfection by-products can have unfavorable environmental consequences [110]. In this context, a decentralized, off-grid, and point-of-use (POU) wastewater disinfection system would be the most cost-effective method for remote and vulnerable individuals and rural populations. Thus, on-site POU systems should be able to operate without having to rely on grid energy and should be able to deliver high-quality drinking water [111]. To this end, several technologies such as 1) coagulation-flocculation [112, 113]; 2) filtration with silver-coated composites [114–118]; 3) solar disinfection (SODIS) [111, 119–121]; and UV irradiation [122–124] have all been investigated for off-grid POU systems. For instance, Lui et al. reported using a POU water treatment system using UV LED to disinfect *Escherichia coli* and *Enterococcus faecalis* [125].

Recently, Carlson et al. used a photo-electro-catalytic POU system to disinfect *Escherichia coli* CN13 using immobilized black TiO_2 nanotubes under solar irradiation [126]. Similarly, a photo-electro-catalytic POU system was used by Montenegro-Ayo et al. to disinfect *Escherichia coli* using UV light/TiO_2 nanotube photo-anodes [127]. However, these methods have encountered several technical difficulties. These difficulties include 1) poor output rates; 2) bacterial reactivation after treatment; 3) safety issues related to the treated water; and 4) insufficient efficacy against antibiotic-resistant bacteria (ABR) [67, 128–130].

However, recent research suggests that designing POU systems using visible light-assisted photocatalytic disinfection (PCD) has the potential to effectively disinfect microorganisms and, in particular, disinfect problematic antibiotic-resistant bacteria [81, 131–133]. On the other hand, in spite of the availability of several new photocatalytic materials and composites as mentioned above, there are a number of issues restricting progress in this field. These issues include 1) attracting effective policies that promote its broader development; 2) unserviceable photocatalytic materials that are chemically unstable, costly, and composed of biologically toxic components that might have wider safety issues regarding their use; 3) the recovery of powder photocatalytic materials from slurries after treatment, where their subsequent reusability is technically challenging and economically unviable; 4) limited information regarding the safety of PCD treated water toward both human and animal health; 5) information regarding bacterial reactivation after treatment due to DNA repair mechanisms is not fully understood; and 5) the high cost of manufacturing POU facilities in remote locations [134–137].

12.7 CONCLUSION

This chapter has summarized recent research developments into photocatalysis and photocatalytic materials designed for water treatment processes. Initially, the chapter discussed conventional wastewater treatments and outlined a number of concerns related to their operation. For instance, some bacteria resistant to antibiotics such as chloramphenicol, penicillin, and ampicillin have also gained resistance to chlorination. Thus, these bacteria have shown extremely high rates of reactivation after treatment. In addition, chlorination is known to produce harmful disinfection by-products. On the other hand, photocatalysis has the potential to be an efficient and sustainable oxidation technology for treating wastewater without the problems normally associated with conventional wastewater treatments. The photocatalytic oxidation process was discussed and in particular described the importance of charge excitation, separation, transport, adsorption, and surface reactions of the semiconductor photocatalytic materials. It was pointed out that each of these factors has a significant impact on the performance of the photocatalytic material. In addition, the synergistic effect of composite components like clay and immobilization to improve photocatalytic performance was also pointed out. Importantly, it was pointed out that photocatalysis could be used as an off-grid and point-of-use wastewater disinfection system for remote and rural populations because a photocatalysis-based system avoids the high capital cost and high levels of energy usage normally associated with conventional wastewater treatment facilities. However, in spite of significant advances made in recent years, there are still a number of issues that need to be addressed. In particular, particle aggregation and difficulties in separating and recovering photocatalytic materials from treated water after processing are challenging. A number of strategies for dealing with these problems were discussed. However, further research is needed to develop more effective industrial-scale photocatalytic-based systems that optimize photocatalytic material performance, improve immobilization and separation, and in turn improve recovery rates and promote increased material recycling efficiencies.

REFERENCES

1. World Health Organization (WHO). *Progress on Drinking Water, Sanitation and Hygiene: 2017 Update and SDG Baselines.* World Health Organization, 2017.
2. The Lancet. On the question of water: A matter of life and death, 2019.
3. The Lancet. Water and sanitation: Addressing inequalities, 2014.
4. Agapitova N, Navarrete Moreno C. *Waterlife: Improving Access to Safe Drinking Water in India.* World Bank, 2017.
5. Khuntia KH, Samal SK, Kar SK, Pal BB. An Ogawa cholera outbreak 6 months after the Inaba cholera outbreaks in India 2006. *Journal of Microbiology, Immunology and Infection.* 2010; 43: 133–137.
6. Khuntia HK, Pal BB, Meher PK, Chhotray GP. Environmental *Vibrio cholerae* O139 may be the progenitor of outbreak of cholera in coastal area of Orissa, Eastern India, 2000: Molecular evidence. *The American Journal of Tropical Medicine and Hygiene.* 2008; 78: 819–822.
7. Paul D. Research on heavy metal pollution of river Ganga: A review. *Annals of Agrarian Science.* 2015; 15: 278–286.
8. Chakraborti D, Singh SK, Rahman MM, Dutta RN, Mukherjee SC, Pati S, Kar PB. Groundwater arsenic contamination in the ganga river basin: A future health danger. *International Journal of Environmental Research and Public Health.* 2018; 15: 180.
9. Hunter PR, MacDonald AM, Carter RC. Water supply and health. *PLoS Medicine.* 2010; 7: e1000361.
10. Rizzo L, Manaia C, Merlin C, Schwartz T, Dagot C, Ploy MC, Michael I, Fatta-Kassinos D. Urban wastewater treatment plants as hotspots for antibiotic resistant bacteria and genes spread into the environment: A review. *Science of the Total Environment.* 2013; 447: 345–360.
11. Pazda M, Kumirska J, Stepnowski P, Mulkiewicz E. Antibiotic resistance genes identified in wastewater treatment plant systems – A review, *Science of the Total Environment.* 2019; 697: 134023.
12. Xu J, Xu Y, Wang H, Guo C, Qiu H, He Y, Zhang Y, Li X, Meng W. Occurrence of antibiotics and antibiotic resistance genes in a sewage treatment plant and its effluent-receiving river. *Chemosphere.* 2015; 119: 1379–1385.
13. Garcia S, Wade B, Bauer C, Craig C, Nakaoka K, Lorowitz W. The effect of wastewater treatment on antibiotic resistance in *Escherichia coli* and *Enterococcus sp. Water Environment Research.* 2007; 79: 2387–2395.
14. Xi C, Zhang Y, Marrs CF, Ye W, Simon C, Foxman B, Nriagu J. Prevalence of antibiotic resistance in drinking water treatment and distribution systems. *Applied Environmental Microbiology.* 2009; 75: 5714–5718.

15. Ardal C, Balasegaram M, Laxminarayan R, McAdams D, Outterson K, Rex JH, Sumpradit N. Antibiotic development-economic, regulatory and societal challenges. *Nature Reviews Microbiology*. 2020; 18: 1–8.
16. Holmes AH, Moore LS, Sundsfjord A, Steinbakk M, Regmi S, Karkey A, Guerin PJ, Piddock LJ. Understanding the mechanisms and drivers of antimicrobial resistance. *The Lancet*. 2016; 387: 176–187.
17. Nunez-Nunez M, Navarro MD, Palomo V, Rajendran NB, Del Toro MD, Voss A, Sharland M, Sifakis F, Tacconelli E, Rodríguez-Baño J, Burkert F. The methodology of surveillance for antimicrobial resistance and healthcare-associated infections in Europe (SUSPIRE): A systematic review of publicly available information. *Clinical Microbiology and Infection*. 2018; 24: 105–109.
18. Hofer U. The cost of antimicrobial resistance. *Nature Reviews Microbiology*. 2019; 17: 3–3.
19. Pitout JD, Laupland KB. Extended-spectrum β-lactamase producing Enterobacteriaceae: An emerging public-health concern. *The Lancet Infectious Diseases*. 2008; 8: 159–166.
20. Kang CI, Kim SH, Park WB, Lee KD, Kim HB, Oh MD, Kim EC, Choe KW. Bloodstream infections caused by Enterobacter species: Predictors of 30-day mortality rate and impact of broad-spectrum cephalosporin resistance on outcome. *Clinical Infectious Diseases*. 2004; 39: 812–818.
21. Sanders WE Jr, Sanders CC. Enterobacter spp.: Pathogens poised to flourish at the turn of the century. *Clinical Microbiology Reviews*. 1997; 10: 220–241.
22. Fabrega A, Vila J. Salmonella enterica serovar Typhimurium skills to succeed in the host: Virulence and regulation. *Clinical Microbiology Reviews*. 2013; 26: 308–341.
23. Becker K, Heilmann C, Peters G. Coagulase-negative staphylococci. *Clinical Microbiology Reviews*. 2014; 27: 870–926.
24. Czekaj T, Ciszewski M, Szewczyk EM. Staphylococcus haemolyticus – An emerging threat in the twilight of the antibiotics age. *Microbiology*. 2015; 161: 2061–2068.
25. Froggatt JW, Johnston JL, Galetto DW, Archer GL. Antimicrobial resistance in nosocomial isolates of *Staphylococcus haemolyticus*. *Antimicrobial Agents and Chemotherapy*. 1989; 33: 460–466.
26. Agvald-Ohman C, Lund B, Edlund C. Multiresistant coagulase-negative staphylococci disseminate frequently between intubated patients in a multidisciplinary intensive care unit. *Critical Care*. 2003; 8: R42.
27. Cabral JPS. Water microbiology: Bacterial pathogens and water. *International Journal of Environmental Research and Public Health*. 2010; 7: 3657–3703.
28. Collivignarelli MC, Abba A, Benigna I, Sorlini S, Torretta V. Overview of the main disinfection processes for wastewater and drinking water treatment plants. *Sustainability*. 2018; 10: 86.
29. Mc Carlie S, Boucher CE, Bragg RR. Molecular basis of bacterial disinfectant resistance. *Drug Resistance Updates*. 2020; 48: 100672.
30. Jin M, Liu L, Wang DN, Yang D, Liu WL, Yin J, Yang ZW, Wang HR, Qiu ZG, Shen ZQ, Shi DY. Chlorine disinfection promotes the exchange of antibiotic resistance genes across bacterial genera by natural transformation. *The ISME Journal*. 2020; 14: 1847–1856.
31. Li D, Zeng S, Gu AZ, He M, Shi H. Inactivation, reactivation and regrowth of indigenous bacteria in reclaimed water after chlorine disinfection of a municipal wastewater treatment plant. *Journal of Environmental Sciences*. 2013; 25: 1319–1325.
32. Kollu K, Ormeci B. Regrowth potential of bacteria after ultraviolet disinfection in the absence of light and dark repair. *Journal of Environmental Engineering*. 2015; 141: 04014069.
33. Orta de Velasquez MT, Yanez Noguez I, Casasola Rodriguez B, Roman PI. Effects of ozone and chlorine disinfection on VBNC *Helicobacter pylori* by molecular techniques and FESEM images. *Environmental Technology*. 2017; 38: 744–753.
34. Chen S, Li X, Wang Y, Zeng J, Ye C, Li X, Guo L, Zhang S, Yu X. Induction of *Escherichia coli* into a VBNC state through chlorination/chloramination and differences in characteristics of the bacterium between states. *Water Research*. 2018; 142: 279–288.
35. Gaya UI, Abdullah AH. Heterogeneous photocatalytic degradation of organic contaminants over titanium dioxide: A review of fundamentals, progress and problems. *Journal of Photochemistry and Photobiology C: Photochemistry Reviews*. 2008; 9: 1–12.
36. Yang H, Cheng H. Controlling nitrite level in drinking water by chlorination and chloramination. *Separation and Purification Technology*. 2007; 56: 392–396.
37. Lu J, Zhang T, Ma J, Chen Z. Evaluation of disinfection by-products formation during chlorination and chloramination of dissolved natural organic matter fractions isolated from a filtered river water. *Journal of Hazardous Materials*. 2009; 162: 140–145.
38. Coleman H, Marquis C, Scott J, Chin SS, Amal R. Bactericidal effects of titanium dioxide-based photocatalysts. *Chemical Engineering Journal*. 2005; 113: 55–63.

39. Laflamme O, Serodes JB, Simard S, Legay C, Dorea C, Rodriguez MJ. Occurrence and fate of ozonation disinfection by-products in two Canadian drinking water systems. *Chemosphere.* 2020; 260: 127660.
40. Qi W, Zhang H, Hu C, Liu H, Qu J. Effect of ozonation on the characteristics of effluent organic matter fractions and subsequent associations with disinfection by-products formation. *Science of the Total Environment.* 2018; 610: 1057–1064.
41. Pullerits K, Ahlinder J, Holmer L, Salomonsson E, Öhrman C, Jacobsson K, Dryselius R, Forsman M, Paul CJ, Rådström P. Impact of UV irradiation at full scale on bacterial communities in drinking water. *NPJ Clean Water.* 2020; 3: 1–10.
42. Esplugas S, Gimenez J, Contreras S, Pascual E, Rodríguez M. Comparison of different advanced oxidation processes for phenol degradation. *Water Research.* 2002; 36: 1034–1042.
43. Chong MN, Jin B, Chow CW, Saint C. Recent developments in photocatalytic water treatment technology: A review. *Water Research.* 2010; 44: 2997–3027.
44. He J, Kumar A, Khan M, Lo IM. Critical review of photocatalytic disinfection of bacteria: from noble metals-and carbon nanomaterials-TiO_2 composites to challenges of water characteristics and strategic solutions. *Science of the Total Environment.* 2020; 758: 143953.
45. Ni M, Leung MK, Leung DY, Sumathy K. A review and recent developments in photocatalytic water-splitting using TiO_2 for hydrogen production. *Renewable and Sustainable Energy Reviews.* 2007; 11: 401–425.
46. Fujishima A, Zhang X, Tryk DA. TiO_2 photocatalysis and related surface phenomena. *Surface Science Reports.* 2008; 63: 515–582.
47. Pozzo RL, Giombi JL, Baltanas MA, Cassano AE. The performance in a fluidized bed reactor of photocatalysts immobilized onto inert supports. *Catalysis Today.* 2000; 62: 175–187.
48. Zhang X, Du AJ, Lee P, Sun DD, Leckie JO. Grafted multifunctional titanium dioxide nanotube membrane: Separation and photodegradation of aquatic pollutant. *Applied Catalysis B: Environmental.* 2008; 84: 262–267.
49. Sun Z, Chen Y, Ke Q, Yang Y, Yuan J. Photocatalytic degradation of a cationic azo dye by TiO_2/bentonite nanocomposite. *Journal of Photochemistry and Photobiology A: Chemistry.* 2002; 149: 169–174.
50. Kun R, Mogyorosi K, Dekany I. Synthesis and structural and photocatalytic properties of TiO_2/montmorillonite nanocomposites. *Applied Clay Science.* 2006; 32: 99–110.
51. Ong CB, Ng LY, Mohammad AW. A review of ZnO nanoparticles as solar photocatalysts: Synthesis, mechanisms and applications. *Renewable and Sustainable Energy Reviews.* 2018; 81: 536–551.
52. Rahman S, Khan M, Akib S, Din NB, Biswas SK, Shirazi SM. Sustainability of rainwater harvesting system in terms of water quality. *The Scientific World Journal.* 2014.
53. Padmanabhan P, Sreekumar K, Thiyagarajan T, Satpute RU, Bhanumurthy K, Sengupta P, Dey GK, Warrier KG. Nano-crystalline titanium dioxide formed by reactive plasma synthesis. *Vacuum.* 2006; 80: 1252–1255.
54. Clara M, Strenn B, Gans O, Martinez E, Kreuzinger N, Kroiss H. Removal of selected pharmaceuticals, fragrances and endocrine disrupting compounds in a membrane bioreactor and conventional wastewater treatment plants. *Water Research.* 2005; 39: 4797–4807.
55. Sonune A, Ghate R. Developments in wastewater treatment methods. *Desalination.* 2004; 167: 55–63.
56. Taboada-Santos A, Rivadulla E, Paredes L, Carballa M, Romalde J, Lema JM. Comprehensive comparison of chemically enhanced primary treatment and high-rate activated sludge in novel wastewater treatment plant configurations. *Water Research.* 2020; 169: 115258.
57. FAO. Wastewater treatment processes. http://www.fao.org/3/t0551e/t0551e05.htm.
58. Rout PR, Zhang TC, Bhunia P, Surampalli RY. Treatment technologies for emerging contaminants in wastewater treatment plants: A review. *Science of the Total Environment.* 2021; 753: 141990.
59. Ghernaout D. Water treatment chlorination: An updated mechanistic insight review. *Chemistry Research Journal.* 2017; 2: 125–138.
60. Virto R, Manas P, Alvarez I, Condon S, Raso J. Membrane damage and microbial inactivation by chlorine in the absence and presence of a chlorine-demanding substrate. *Applied and Environmental Microbiology.* 2005; 71: 5022–5028.
61. Mazhar MA, Khan NA, Ahmed S, Khan AH, Hussain A, Changani F, Yousefi M, Ahmadi S, Vambol V. Chlorination disinfection by-products in municipal drinking water – A review. *Journal of Cleaner Production.* 2020; 273: 123159.
62. Jia S, Shi P, Hu Q, Li B, Zhang T, Zhang XX. Bacterial community shift drives antibiotic resistance promotion during drinking water chlorination. *Environmental Science & Technology.* 2015; 49: 12271–12279.

63. Huang JJ, Hu HY, Tang F, Li Y, Lu SQ, Lu Y. Inactivation and reactivation of antibiotic-resistant bacteria by chlorination in secondary effluents of a municipal wastewater treatment plant. *Water Research.* 2011; 45: 2775–2781.
64. Cutler TD, Zimmerman JJ. Ultraviolet irradiation and the mechanisms underlying its inactivation of infectious agents. *Animal Health Research Reviews* 2011; 12: 15–23.
65. Pichel N, Vivar M, Fuentes M. The problem of drinking water access: A review of disinfection technologies with an emphasis on solar treatment methods. *Chemosphere.* 2019; 218: 1014–1030.
66. Jungfer C, Schwartz T, Obst U. UV-induced dark repair mechanisms in bacteria associated with drinking water. *Water Research.* 2007; 41: 188–196.
67. Guo MT, Kong C. Antibiotic resistant bacteria survived from UV disinfection: Safety concerns on genes dissemination. *Chemosphere.* 2019; 224: 827–832.
68. Buffle MO, von Gunten U. Phenols and amine induced HO• generation during the initial phase of natural water ozonation. *Environmental Science & Technology.* 2006; 40: 3057–3063.
69. Ding W, Jin W, Cao S, Zhou X, Wang C, Jiang Q, Huang H, Tu R, Han SF, Wang Q. Ozone disinfection of chlorine-resistant bacteria in drinking water. *Water Research.* 2019; 160: 339–349.
70. Agus E, Voutchkov N, Sedlak DL. Disinfection by-products and their potential impact on the quality of water produced by desalination systems: A literature review. *Desalination.* 2009; 237: 214–237.
71. Iakovides I, Michael-Kordatou I, Moreira NF, Ribeiro AR, Fernandes T, Pereira MF, Nunes OC, Manaia CM, Silva AM, Fatta-Kassinos D. Continuous ozonation of urban wastewater: Removal of antibiotics, antibiotic-resistant *Escherichia coli* and antibiotic resistance genes and phytotoxicity. *Water Research.* 2019; 159: 333–347.
72. Ibhadon AO, Fitzpatrick P. Heterogeneous photocatalysis: Recent advances and applications. *Catalysts.* 2013; 3: 189–218.
73. Fujishima A, Honda K. Electrochemical photolysis of water at a semiconductor electrode. *Nature.* 1972; 238: 37–38.
74. Kondo Y, Yoshikawa H, Awaga K, Murayama M, Mori T, Sunada K, Bandow S, Iijima S. Preparation, photocatalytic activities, and dye- sensitized solar-cell performance of submicron-scale TiO_2 hollow spheres. *Langmuir.* 2008; 24: 547–550.
75. Hosono E, Fujihara S, Kakiuchi K, Imai H. Growth of submicrometer-scale rectangular parallelepiped rutile TiO_2 films in aqueous $TiCl_3$ solutions under hydrothermal conditions. *Journal of the American Chemical Society.* 2004; 126: 7790–7791.
76. Artale MA, Augugliaro V, Drioli E, Golemme G, Grande C, Loddo V, Molinari R, Palmisano L, Schiavello M. Preparation and characterization of membranes with entrapped TiO_2 and preliminary photocatalytic tests. *Annali di Chimica.* 2001; 91: 127–136.
77. Rizzo L, Sannino D, Vaiano V, Sacco O, Scarpa A, Pietrogiacomi D. Effect of solar simulated N-doped TiO_2 photocatalysis on the inactivation and antibiotic resistance of an E. coli strain in biologically treated urban wastewater. *Applied Catalysis B: Environmental.* 2014; 144: 369–378.
78. Soni S, Dave G, Henderson M, Gibaud A. Visible light induced cell damage of Gram positive bacteria by N-doped TiO_2 mesoporous thin films. *Thin Solid Films.* 2013; 531: 559–565.
79. Mamane H, Horovitz I, Lozzi L, Di Camillo D, Avisar D. The role of physical and operational parameters in photocatalysis by N-doped TiO_2 sol–gel thin films. *Chemical Engineering Journal.* 2014; 257: 159–169.
80. Yu JC, Ho W, Yu J, Yip H, Wong PK, Zhao J. Efficient visible-light-induced photocatalytic disinfection on sulfur-doped nanocrystalline titania. *Environmental Science & Technology.* 2005; 39: 1175–1179.
81. Das S, Sinha S, Das B, Jayabalan R, Suar M, Mishra A, Tamhankar AJ, Stålsby Lundborg C, Tripathy SK. Disinfection of multidrug resistant *Escherichia coli* by solar-photocatalysis using Fe-doped ZnO nanoparticles. *Scientific Reports.* 2017; 7: 1–14.
82. Srinivasulu T, Saritha K, Reddy KR. Synthesis and characterization of Fe-doped ZnO thin films deposited by chemical spray pyrolysis. *Modern Electronic Materials.* 2017; 3: 76–85.
83. Kafle B, Acharya S, Thapa S, Poudel S. Structural and optical properties of Fe-doped ZnO transparent thin films. *Ceramics International.* 2016; 42: 1133–1139.
84. Adhikari S, Banerjee A, Eswar NK, Sarkar D, Madras G. Photocatalytic inactivation of *E. coli* by ZnO–Ag nanoparticles under solar radiation. *RSC Advances.* 2015; 5: 51067–51077.
85. Zhang X, Dong S, Zhou X, Yan L, Chen G, Dong S, Zhou D. A facile one-pot synthesis of Er-Al co-doped ZnO nanoparticles with enhanced photocatalytic performance under visible light. *Materials Letters.* 2015; 143: 312–314.
86. Feilizadeh M, Mul G, Vossoughi M. *E. coli* inactivation by visible light irradiation using a Fe–Cd/TiO_2 photocatalyst: Statistical analysis and optimization of operating parameters. *Applied Catalysis B: Environmental.* 2015; 168: 441–447.

87. Wang W, Huang G, Jimmy CY, Wong PK. Advances in photocatalytic disinfection of bacteria: Development of photocatalysts and mechanisms. *Journal of Environmental Sciences*. 2015; 34: 232–247.
88. Espindola JC, Cristovao RO, Mendes A, Boaventura RA, Vilar VJ. Photocatalytic membrane reactor performance towards oxytetracycline removal from synthetic and real matrices: Suspended vs immobilized TiO_2-P25. *Chemical Engineering Journal*. 2019; 378: 122114.
89. Wei Z, He Y, Huang Z, Xiao X, Li B, Ming S, Cheng X. Photocatalytic membrane combined with biodegradation for toluene oxidation. *Ecotoxicology and Environmental Safety*. 2019; 184: 109618.
90. Nguyen VH, Tran QB, Nguyen XC, Ho TT, Shokouhimehr M, Vo DV, Lam SS, Nguyen HP, Hoang CT, Ly QV, Peng W. Submerged photocatalytic membrane reactor with suspended and immobilized N-doped TiO_2 under visible irradiation for diclofenac removal from wastewater. *Process Safety and Environmental Protection*. 2020; 142: 229–237.
91. Cheng R, Shen L, Wang Q, Xiang S, Shi L, Zheng X, Lv W. Photocatalytic membrane reactor (PMR) for virus removal in drinking water: Effect of humic acid. *Catalysts*. 2018; 8: 284.
92. Adan C, Marugan J, Mesones S, Casado C, Van Grieken R. Bacterial inactivation and degradation of organic molecules by titanium dioxide supported on porous stainless steel photocatalytic membranes. *Chemical Engineering Journal*. 2017; 318: 29–38.
93. Pozzo RL, Baltanas MA, Cassano AE. Supported titanium oxide as photocatalyst in water decontamination: State of the art. *Catalysis Today*. 1997; 39: 219–231.
94. Alalm MG, Tawfik A, Ookawara S. Enhancement of photocatalytic activity of TiO_2 by immobilization on activated carbon for degradation of pharmaceuticals. *Journal of Environmental Chemical Engineering*. 2016; 4: 1929–1937.
95. Zammit I, Vaiano V, Ribeiro AR, Silva AM, Manaia CM, Rizzo L. Immobilised cerium-doped zinc oxide as a photocatalyst for the degradation of antibiotics and the inactivation of antibiotic-resistant bacteria. *Catalysts*. 2019; 9: 222.
96. Chong MN, Jin B, Saint CP. Bacterial inactivation kinetics of a photo-disinfection system using novel titania-impregnated kaolinite photocatalyst. *Chemical Engineering Journal*. 2011; 171: 16–23.
97. Misra AJ, Das S, Rahman AH, Das B, Jayabalan R, Behera SK, Suar M, Tamhankar AJ, Mishra A, Lundborg CS, Tripathy SK. Doped ZnO nanoparticles impregnated on Kaolinite (Clay): A reusable nanocomposite for photocatalytic disinfection of multidrug resistant Enterobacter sp. under visible light. *Journal of Colloid and Interface Science*. 2018; 530: 610–623.
98. Unuabonah EI, Ugwuja CG, Omorogie MO, Adewuyi A, Oladoja NA. Clays for efficient disinfection of bacteria in water. *Applied Clay Science*. 2018; 151: 211–223.
99. Tavanaee M, Shirvani M, Bakhtiary S. Adhesion of *Pseudomonas putida* onto palygorskite and sepiolite clay minerals. *Geomicrobiology Journal*. 2017; 34: 677–686.
100. Morrison KD, Misra R, Williams LB. Unearthing the antibacterial mechanism of medicinal clay: A geochemical approach to combating antibiotic resistance. *Scientific Reports*. 2016; 6: 1–13.
101. Daneshvar N, Salari D, Niaei A, Rasoulifard M, Khataee A. Immobilization of TiO_2 nanopowder on glass beads for the photocatalytic decolorization of an azo dye CI Direct Red 23. *Journal of Environmental Science and Health: Part A*. 2005; 40: 1605–1617.
102. Khalilian H, Behpour M, Atouf V, Hosseini SN. Immobilization of S, N-codoped TiO_2 nanoparticles on glass beads for photocatalytic degradation of methyl orange by fixed bed photoreactor under visible and sunlight irradiation. *Solar Energy*. 2015; 112: 239–245.
103. Sraw A, Kaur T, Pandey Y, Sobti A, Wanchoo RK, Toor AP. Fixed bed recirculation type photocatalytic reactor with TiO_2 immobilized clay beads for the degradation of pesticide polluted water. *Journal of Environmental Chemical Engineering*. 2018; 6: 7035–7043.
104. Mehmood CT, Zhong Z, Zhou H, Zhang C, Xiao Y. Immobilizing a visible light- responsive photocatalyst on a recyclable polymeric composite for floating and suspended applications in water treatment. *RSC Advances*. 2020; 10: 36349–36362.
105. Motshekga SC, Ray SS, Maity A. Synthesis and characterization of alginate beads encapsulated zinc oxide nanoparticles for bacteria disinfection in water. *Journal of Colloid and Interface Science*. 2018; 512: 686–692.
106. Shannon MA, Bohn PW, Elimelech M, Georgiadis JG, Marinas BJ, Mayes AM. Science and technology for water purification in the coming decades. *Nature*. 2008; 452: 301–310.
107. Roefs I, Meulman B, Vreeburg JH, Spiller M. Centralised, decentralised or hybrid sanitation systems? Economic evaluation under urban development uncertainty and phased expansion. *Water Research*. 2017; 109: 274–286.
108. Zodrow KR, Li Q, Buono RM, Chen W, Daigger G, Dueñas-Osorio L, Elimelech M, Huang X, Jiang G, Kim JH, Logan BE. Advanced materials, technologies, and complex systems analyses: Emerging opportunities to enhance urban water security. *Environmental Science & Technology*. 2017; 51: 10274–10281.

109. National Research. *Drinking Water Distribution Systems: Assessing and Reducing Risks*. Washington, DC: The National Academies Press, 2006.
110. Loeb S, Hofmann R, Kim JH. Beyond the pipeline: Assessing the efficiency limits of advanced technologies for solar water disinfection. *Environmental Science & Technology Letters*. 2016; 3: 73–80.
111. Pooi CK, Ng HY. Review of low-cost point-of-use water treatment systems for developing communities. *NPJ Clean Water*. 2018; 1: 1–8.
112. Zhang S, Zheng H, Tang X, Zhao C, Zheng C, Gao B. Sterilization by flocculants in drinking water treatment. *Chemical Engineering Journal*. 2020; 382: 122961.
113. Shaarani S, Azizan SNF, Md Akhir FN, Muhammad Yuzir MA, Othman NA, Zakaria Z, Mohd Noor MJ, Hara H. Removal efficiency of gram-positive and gram-negative bacteria using a natural coagulant during coagulation, flocculation, and sedimentation processes. *Water Science and Technology*. 2019; 80: 1787–1795.
114. Sankar MU, Aigal S, Maliyekkal SM, Chaudhary A, Anshup Kumar AA, Chaudhari K, Pradeep T. Biopolymer-reinforced synthetic granular nanocomposites for affordable point-of-use water purification. *Proceedings of the National Academy of Sciences*. 2013; 110: 8459–8464.
115. Zeng X, McCarthy DT, Deletic A, Zhang X. Silver/reduced graphene oxide hydrogel as novel bactericidal filter for point-of-use water disinfection. *Advanced Functional Materials*. 2015; 25: 4344–4351.
116. Fan M, Gong L, Sun J, Wang D, Bi F, Gong Z. Killing two birds with one stone: Coating Ag NPs embedded filter paper with chitosan for better and durable point-of-use water disinfection. *ACS Applied Materials & Interfaces*. 2018; 10: 38239–38245.
117. Chen W, Jiang J, Zhang W, Wang T, Zhou J, Huang CH, Xie X. Silver nanowire-modified filter with controllable silver ion release for point-of-use disinfection. *Environmental Science & Technology*. 2019; 53: 7504–7512.
118. Dhiman NK, Agnihotri S. Hierarchically aligned nano silver/chitosan–PVA hydrogel for point-of-use water disinfection: Contact-active mechanism revealed. *Environmental Science: Nano*. 2020; 7: 2337–2350.
119. McGuigan KG, Conroy RM, Mosler HJ, du Preez M, Ubomba-Jaswa E, Fernandez-Ibanez P. Solar water disinfection (SODIS): A review from bench-top to roof-top. *Journal of Hazardous Materials*. 2012; 235: 29–46.
120. Moreno-SanSegundo J, Giannakis S, Samoili S, Farinelli G, McGuigan KG, Pulgarín C, Marugán J. SODIS potential: A novel parameter to assess the suitability of solar water disinfection worldwide. *Chemical Engineering Journal*. 2021; 419: 129889.
121. Pooi C, Ng H. Review of low-cost point-of-use water treatment systems for developing communities. *NPJ Clean Water*. 2018; 1: 11.
122. Li X, Cai M, Wang L, Niu F, Yang D, Zhang G. Evaluation survey of microbial disinfection methods in UV-LED water treatment systems. *Science of the Total Environment*. 2019; 659: 1415–1427.
123. Shen L, Griffith TM, Nyangaresi PO, Qin Y, Pang X, Chen G, Li M, Lu Y, Zhang B. Efficacy of UVC-LED in water disinfection on Bacillus species with consideration of antibiotic resistance issue. *Journal of Hazardous Materials*. 2020; 386: 121968.
124. Chen J, Loeb S, Kim JH. LED revolution: Fundamentals and prospects for UV disinfection applications. *Environmental Science: Water Research & Technology*. 2017; 3: 188–202.
125. Lui GY, Roser D, Corkish R, Ashbolt NJ, Stuetz R. Point-of-use water disinfection using ultraviolet and visible light-emitting diodes. *Science of the Total Environment*. 2016; 553: 626–635.
126. Carlson K, Elliott C, Walker S, Misra M, Mohanty S. An effective, point-of-use water disinfection device using immobilized black TiO_2 nanotubes as an electrocatalyst. *Journal of the Electrochemical Society*. 2016; 163: H395.
127. Montenegro-Ayo R, Barrios AC, Mondal I, Bhagat K, Morales-Gomero JC, Abbaszadegan M, Westerhoff P, Perreault F, Garcia-Segura S. Portable point-of-use photoelectrocatalytic device provides rapid water disinfection. *Science of the Total Environment*. 2020; 737: 140044.
128. Deshmukh S, Patil S, Mullani S, Delekar S. Silver nanoparticles as an effective disinfectant: A review. *Materials Science and Engineering: C*. 2019; 97: 954–965.
129. Cowie BE, Porley V, Robertson N. Solar disinfection (SODIS) provides a much underexploited opportunity for researchers in photocatalytic water treatment (PWT). *ACS Catalysis*. 2020; 10: 11779–11782.
130. Bai X, Ma X, Xu F, Li J, Zhang H, Xiao X. The drinking water treatment process as a potential source of affecting the bacterial antibiotic resistance. *Science of the Total Environment*. 2015; 533: 24–31.
131. Karaolia P, Michael-Kordatou I, Hapeshi E, Drosou C, Bertakis Y, Christofilos D, Armatas GS, Sygellou L, Schwartz T, Xekoukoulotakis NP, Fatta-Kassinos D. Removal of antibiotics, antibiotic-resistant bacteria and their associated genes by graphene-based TiO_2 composite photocatalysts under solar radiation in urban wastewaters. *Applied Catalysis B: Environmental*. 2018; 224: 810–824.

132. Xiong P, Hu J. Inactivation/reactivation of antibiotic-resistant bacteria by a novel UVA/LED/TiO_2 system. *Water Research.* 2013; 47: 4547–4555.
133. Das S, Misra AJ, Rahman AH, Das B, Jayabalan R, Tamhankar AJ, Mishra A, Lundborg CS, Tripathy SK. Ag@ SnO_2@ ZnO core-shell nanocomposites assisted solar-photocatalysis downregulates multi-drug resistance in *Bacillus* sp.: A catalytic approach to impede antibiotic resistance. *Applied Catalysis B: Environmental.* 2019; 259: 118065.
134. Falinski M, Turley R, Kidd J, Lounsbury AW, Lanzarini-Lopes M, Backhaus A, Rudel HE, Lane MK, Fausey CL, Barrios AC, Loyo-Rosales JE. Doing nano-enabled water treatment right: Sustainability considerations from design and research through development and implementation. *Environmental Science: Nano.* 2020; 7: 3255–3278.
135. You J, Guo Y, Guo R, Liu X, A review of visible light-active photocatalysts for water disinfection: Features and prospects. *Chemical Engineering Journal.* 2019; 373: 624–641.
136. Chu C, Ryberg EC, Loeb SK, Suh MJ, Kim JH. Water disinfection in rural areas demands unconventional solar technologies. *Accounts of Chemical Research.* 2019; 52: 1187–1195.
137. Hodges BC, Cates EL, Kim JH. Challenges and prospects of advanced oxidation water treatment processes using catalytic nanomaterials. *Nature Nanotechnology.* 2018; 13: 642–650.

13 Applications of Carbon-Based Heterogeneous Nanomaterials for Industrial Waste Treatment

Sankha Chakrabortty, Jayato Nayak, Shirsendu Banerjee, and Suraj Kumar Tripathy

CONTENTS

13.1 INTRODUCTION

In recent years of environmental protection, researchers have examined the ramifications of using heterogeneous carbon-based nanomaterials on ecosystems around the world. Their studies are trying to determine whether or not nanomaterials have the ability to become pollutants. On the other hand, placing one's primary emphasis on possible effects has the potential to obfuscate the multiple ways in which nanomaterials may be utilized to improve environmental outcomes. It is possible to develop nanomaterials that are safe to use or have very little or no impact on the environment. This chapter looks at possible carbon-based materials that have the potential to make a real positive impact in treating wastewater and reducing the adverse effects of wastewater in the environment. Importantly, the physical, chemical, and electrical properties of carbon-based nanomaterials can be modified, so that they can be used in innovative solutions and solve environmental issues. For instance, in the medical field, carbonaceous nanomaterials have been used in a number of applications such as using functionalized nanotubes as synthetic trans-membrane pores for the transport and delivery of pharmaceuticals across thick tissues and for targeting specific malignant cells. Thus, similar new carbon-based nanomaterials can be developed for new membrane-based technologies for wastewater treatment.

DOI: 10.1201/9781003181422-16

The present chapter evaluates the benefits and drawbacks of different types of carbon-based heterogeneous nanomaterials, such as fullerenes, carbon nanotubes, graphene, etc. The authors believe the scientific community in this field would benefit from the concepts and information presented in this work. This is of particular importance since we found only a small number of studies that have attempted to summarize the various nanomaterials currently employed in water purification [1–5]. In addition, there are relatively few studies evaluating the beneficial effects of nanomaterial-based adsorbent materials for the removal of dyes, heavy metals, pharmaceutical drugs, and organic contaminants from wastewater. Thus, the authors also believe the present work will assist in the search for even more cutting-edge nanomaterials and their application for wastewater treatment.

13.2 DIFFERENT TYPES OF CARBON-BASED HETEROGENEOUS NANOMATERIALS

To fully take advantage of the unique features that occur at the nanometer scale, molecular manipulation is used to assemble carbon atoms to build distinctive structures, as seen in Figure 13.1. When compared to their bulk forms, the physical, mechanical, and electrical properties of carbon-based fullerenes and nanotubes composed of carbon, boron, molybdenum disulfide (MoS_2), kaolinite, and various other additives display novel and greatly enhanced properties [6]. To date, the full potential of manipulating carbon-based nanostructures has not been fully realized, especially when we consider carbon's unique hybridization capabilities and sensitivity during synthesis. While research into new inorganic nanomaterials holds great promise for the future, current research into carbonaceous nanostructures, as well as their applications, is making great progress. In this context, the following sections provide an overview of the composition and properties of carbon-based nanomaterials that makes them suitable candidates for water purification and wastewater treatments. Importantly, it's the carbon atom's structural conformation and, by extension, its hybridization state, that play a significant role in determining the physicochemical and electrical properties of carbon-based nanomaterials [7, 8].

Graphite sheets are more thermodynamically stable when they are arranged in three-dimensional arrays at the nanometer scale. The strain energy produced as a result of the planar curvature of the graphite is balanced out by the decreased number of unfavorable dangling bonds [10]. As a consequence, fullerenes and nanotubes are produced, both of which are similar to graphite in a number of respects. However, they also display some unique and modifiable properties as a consequence of quantum phenomena at the nanometer scale. Among these unique properties are 1) an increase in the proportion of sp^3 bonds; 2) the quantum restriction of wave functions in one or more dimensions; and 3) a closed topology. Consequently, carbonaceous nanostructures have the same

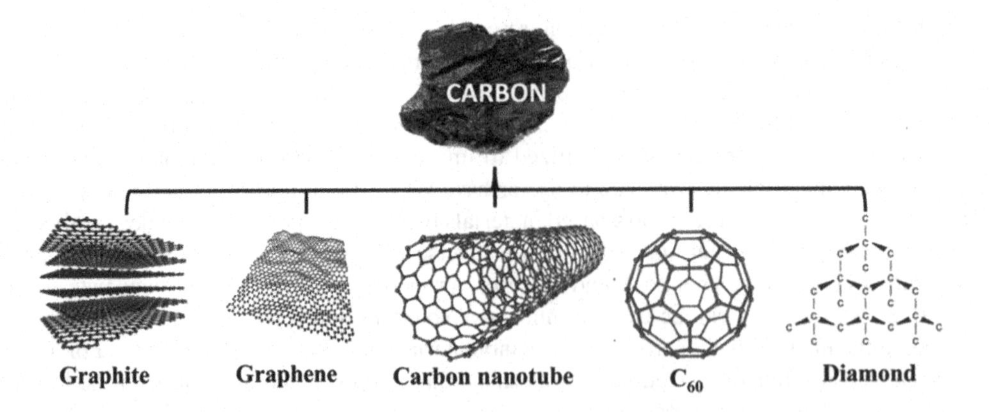

FIGURE 13.1 Schematic illustration of several nanometer-scale carbon-based nanomaterials. (Adapted and reproduced with permission from reference [9].)

bonding arrangements as macroscopic carbon structures. However, their characteristics and shape are dictated by the durability of particular resonance structures rather than the bulk averages of their crystalline forms.

13.2.1 Fullerenes

Fullerenes are distinguished by their one-of-a-kind chemical, optical, and structural properties, all of which are determined by their electronic structures. In their perfect form, fullerenes take the form of a cage-like structure composed of hexagons, as seen in Figure 13.1. The sp3 bonding character, strain energy, and the number of reactive carbon sites are all greater in structures with fewer hexagons compared to structures with more hexagons. Resonance structures delocalize bonds over the fullerene structure, leading to a decrease in the stability and relative abundance of isomers with contiguous pentagons compared to those with isolated pentagons [11, 12]. C-60's chemical behavior falls somewhere between an aromatic molecule and a straight-chained alkene. However, there is substantial debate over the degree to which charging becomes decentralized [13].

To distinguish itself from the degenerate C-20 fullerenes and the inert planar graphite, C-60, or Buckminsterfullerene as it is often known, strikes a balance between the two differing stabilities and reactivities. C-60 is a fullerene with an icosahedral symmetry [14], and its stability comes from resonance structures that keep each carbon atom in the same electrical state and bonding geometry [15]. C-60's relative stability, granted by its symmetry, makes it a stand-alone starting material for chemical processes. Thus, molecular manipulation and the synthesis of polymeric materials designed for environmental applications are made possible through the use of C-60's covalent, supramolecular, and endohedral transformations. In addition, even though C-60 has a very strong resistance to oxidation, it is still possible to fit up to six electrons in orbitals with free slots. In addition, fullerenes can be used as the backbone of reactive adducts, and covalent chemistry is possible [16]. Supramolecular approaches like molecular self-assembly rely on fullerene- or reactant-specific non-covalent van der Waals, electrostatic, and hydrophobic interactions. Applications in the realms of biomedicine and environmental remediation need supramolecular techniques that increase the solubility of hydrophobic fullerene molecules and reduce their inclination to cluster [17]. Thus, the variety of customized structures that can be created using a combination of covalent and supramolecular approaches has been significantly broadened in recent years as a result [18].

Many of the one-of-a-kind traits that fullerenes and other carbonaceous nanomaterials exhibit at the nanometer scale have their origins in the electric and conductive capabilities they possess. As a result, modifying these electric characteristics through either endohedral doping or single substitution has garnered a significant amount of attention from researchers. The photosensitivity and binding energies of a fullerene molecule are significantly influenced by even the smallest amount of atom substitutions within the fullerene structure [19].

Several types of fullerene configurations have been discovered thanks to recent studies of carbonaceous nanomaterials. For instance, there are carbon onions, which consist of a series of concentric fullerenes, and those that consist of very large spherical fullerenes [20, 21]. The symmetry and geometry of the bonds in these molecular structures account for their exceptional reactivity [22]. One theory proposes that extremely large fullerenes and carbon onions have a structure with several faces. However, direct imaging data, on the other hand, indicates that these molecules maintain the familiar C-60 spherical form. This discrepancy between theory and experiment has led to the conclusion that Stone–Wales flaws are an intrinsic feature of these bonding structures [23]. The rearranging of four, six-member rings, into a conformation with numbers 5-7-7-5 is known as the "Stone–Wales defect" [24]. The results of Stone–Wales flaws are larger strain energies and lower resonances. This makes the structure of these types of fullerenes more susceptible to nucleophilic assault. This indicates that this modified structure is the cause of the reported increased sensitivity.

13.2.2 Carbon Nanotubes

With similar molecules to fullerene, but with an extra dimension, carbon nanotubes (CNTs) are a cylindrical structure composed of carbon atoms, as seen in Figure 13.1. A nanotube is a micrometer-scale graphene sheet rolled into a nanometer-scale cylindrical tube. CNT structures with remarkable aspect ratios can be manufactured, with diameters in the nanometer scale to lengths ranging from a few micrometers up to the centimeter scale. There are several in-depth reviews on the synthesis of CNTs and their structural integrity [25–27]. The present work provides a concise overview of CNTs in terms of where they fall on the carbon hybridization spectrum and how useful their features are from an ecological standpoint.

The sidewall of CNTs has the same hexagonal sp^2 structure as graphene, which is also similar to spherical fullerenes. Nanotubes, on the other hand, can only have one degree of curvature at a time, in contrast to their zero-dimensional analogs, which can have any number of degrees. In addition, because of the strong sp^3 character, the strain energy on the sidewall carbons is reduced. Due to this feature, the CNT is more stable than spherical fullerene molecules and is less prone to chemical alteration and rearrangement. But unlike graphene planes, nanotubes can be distinguished by two important features: 1) quantum confinement in the radial and circumferential directions is brought about by the shape of the tubule, which only has a single dimension [10] and 2) limiting graphenes' hexagonal carbon arrangement from two dimensions to one introduces the possibility of structural indeterminacy. For instance, graphene is a two-dimensional material, but the direction in which the rolled graphene sheet is oriented has a significant bearing on the electrical characteristics of the CNT. Thus, CNT properties are dependent on the orientation of the hexagonal lattice in the graphene sheet. For instance, small-diameter, isolated single-walled nanotubes (SWNTs) can exhibit metallic, semi-metallic, or semiconducting characteristics, while graphite in its bulk form acts as a semiconductor [28]. Most nanotubes take on a metallic armchair form during high-yield production processes. However, due to issues with purifying mono-chiral nanotubes, SWNTs have limited economic value and few environmental remediation applications.

Carbon onions and their one-dimensional equivalents are known as double-walled or multi-walled nanotubes (MWNTs). Since there is only a weak link between the 0.34 nm interlayer distances of MWNTs, they have many of the same features as SWNTs [29]. This is also the case with concentric fullerenes. In contrast to SWNTs, MWNTs have the same semiconducting properties as bulk graphite and do not display the distinctive metallic qualities of SWNTs. Another feature of CNTs is the strong attraction forces that exist between nanotubes which makes purification and manipulation more difficult. The strong attraction is also responsible for the tightly packed conformation of CNTs, as well as their limited dispersion in polar as well as apolar solvents [30]. Thus, in order to unbundle nanotube sheaths, it is frequently essential to utilize a physical dispersion method such as exfoliation and sonication. Apart from chemical exfoliation, there are also materials like natural poly-electrolytes [31], aromatic polymers [32], and DNA that can be used for exfoliation [33]. There are also supramolecular de-bundling pathways that can be utilized. These pathways involve steric interference with the bundling process brought about by interactions between aromatic components of the exfoliating agent and the orbitals of the nanotube sidewall [31]. As efficient as supramolecular methods are at breaking up CNT bundles, they have limited applicability in systems that purely rely on the characteristics of unmodified nanotubes. In particular, dispersion issues complicate the covalent and supramolecular modification procedures that have been devised. Despite the fact that numerous functionalization and supramolecular modifications have been achieved [34–38], the scope and control of molecular imprinting for nanotubes are more difficult. For instance, carbon's reactivity is decreased further by the sidewall hexagons' lower strain energy, necessitating the application of more powerful reagents. There are a number of reviews that go into detail about the chemistry involved in the functionalization of fullerenes and how it relates to the functionalization of SWNTs [39].

13.2.3 Graphene

Graphene has attracted a considerable amount of scientific interest due to the material's two-dimensional structure being one atom thick, as seen in Figure 13.2. Graphene sheets have a honey-comb-like crystal structure and a thin coating of conjugated groupings of carbon molecules (sp^2). Graphenes' unique characteristics are a result of this surface layer [40, 41]. Graphenes' sheet-like structure is so advantageous because of the sp^2 nature of its carbon bonds, which permits electron delocalization [42]. This remarkable material has the potential to be used in a wide variety of fascinating applications. For instance, its thin sheet-like structure makes it very transparent, while it absorbs only around 2.3% of the visible light that strikes it, and it also has an exceptional thermal conductivity of 5000 $Wm^{-1}K^{-1}$. In terms of mechanical properties, it has a Young's modulus of 1 TPa and a strength of 130 GPa. In addition, graphenes' electronic conduction is impressive, and it may be possible to increase this electron transport across its sheet-like structure by including various types of metallic nanoparticles. In particular, graphene hybridized with metal nanoparticles is being considered for use in applications like catalytic systems and optoelectronic devices [42]. Because of these unrivaled structural characteristics and exceptional properties, graphene has attracted significant interest from both the scientific and applied engineering fields.

13.2.4 Graphitic Carbon Nitride ($g-C_3N_4$)

Graphitic carbon nitride, also known as $g-C_3N_4$, is a type of polymeric material that is composed of carbon (C), nitrogen (N), and a trace amount of hydrogen (H). These atoms are connected to one another using tris-triazine-based patterns. Because it contains nitrogen and hydrogen atoms, carbon possesses electron-rich features, as well as primitive surface functions and bonding motifs. Not only is $g-C_3N_4$ the most stable allotrope of the carbon nitrides in the atmosphere, but it also exhibits vast surface area features that are striking for a number of applications like catalysis [43]. Because of these characteristics, it can be used in a wide variety of applications. The p state of the idealized $g-C_3N_4$ molecule possesses a remarkable collection of C-N bonds that are devoid of electron localization [44]. Furthermore, $g-C_3N_4$ synthesized by poly-condensation has many hydrogen atoms in the form of primary and secondary amine groups on its terminal edges. And during poly-condensation, hydrogen is incorporated into the cyanamide molecule. Since hydrogen species are present, there is a modest decrease in the amount of true $g-C_3N_4$ and a proliferation of surface

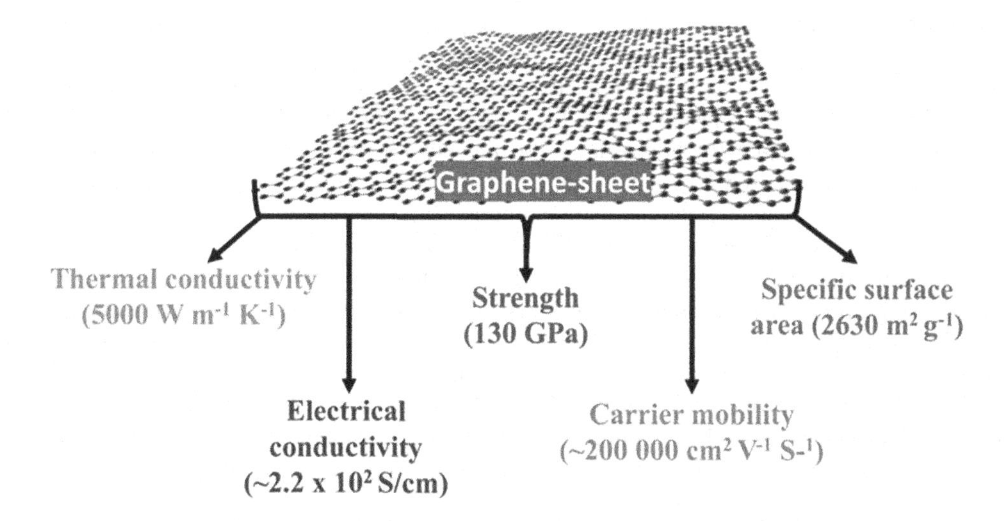

FIGURE 13.2 Layer-by-layer structure and typical property characteristics of graphene sheet. (Adapted and reproduced with permission from reference [9].)

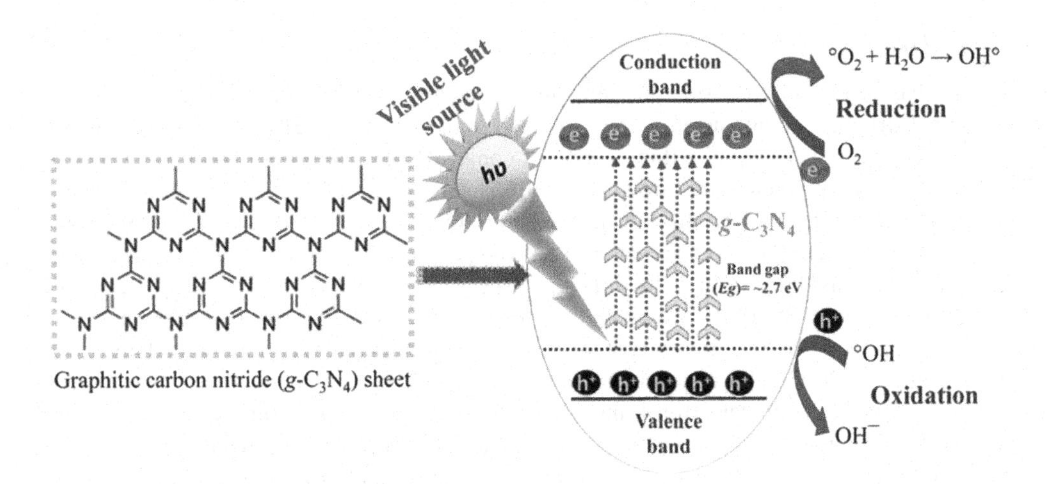

FIGURE 13.3 Graphitic carbon nitride: fundamental steps involved in photocatalysis under visible light irradiation. (Adopted and reproduced with permission from reference [9].)

defects. As mentioned above, these surface defects could be beneficial in catalysis [44], as seen in Figure 13.3. In addition, the material possesses high thermal stability which enables it to function in either liquid or gaseous environments at elevated temperatures, thus improving the operational range of g-C_3N_4 for catalysis applications [44].

13.3 UNIQUE PROPERTIES OF CARBON-BASED NANOMATERIALS

Carbonaceous nanoparticles form a new category of materials. This is due to the unique properties of the sp^2-conjugated carbon bonds and the features of nanometer-scale physics and chemistry. The size, shape, and surface area of carbonaceous nanomaterials are their most outstanding physical features. However, carbonaceous nanomaterials have also attracted considerable interest due to features such as 1) their molecular interactions and sorption properties; 2) unique electronic and optical properties; and 3) their exceptional mechanical and thermal properties. It is these distinctive characteristics and properties that are most frequently mentioned in connection with their use in environmental applications.

The term "molecular manipulation" refers to the process of exerting control over the structure and conformation of a substance. Nanomaterials can have their structure, purity, and physical orientation optimized for particular applications by adjusting their synthesis conditions. Typical synthesis conditions that can be adjusted include temperature, pressure, catalyst selection, electron field, and process gases used [45, 46]. For instance, the diameter of SWNTs is highly sensitive to the type of synthesis technique used for their manufacture. Typically, nanotubes with diameters between 0.7 and 1 nm can be synthesized by HiPCO synthesis, while nanotubes with sizes between 1 and 2 nm can be produced by techniques such as laser ablation and other graphitic-based processes [47]. In addition, the diameter of fullerenes and cylindrical nanostructures is a crucial factor in defining their respective properties and applications. The number of carbon atoms contained within a fullerene spherical molecule is used as a measurement of its diameter. The inner diameter of C-60, the smallest fullerene that abides by the stabilizing pentagon isolation criterion, approaches 0.7 nanometers. This is also the lower size limit for SWNT diameter dimensions, while typical diameters of 1.4 nm have been reported for a number of commercial production processes [48]. In fact, the smaller the diameter of the SWNT, the greater the strain energy that is induced in its structure, as well as the re-hybridization of the electron orbitals.

The characteristic qualities that are dependent on nanotube diameter are supplemented by the physical size exclusion and their capillary behavior. For instance, because of their small inner diameter, nanotubes have been used in a variety of innovative molding, separation, and size exclusion procedures, such as nanowire synthesis and membrane filtration. Their use in membrane filtration is of particular interest since these types of membranes can be developed for water purification [49]. However, the smaller inner diameters have made it more difficult to purify carbon nanotubes, i.e. the removal of growth catalysts and other materials from the nanotubes, which has in turn hampered the use of CNT for particular applications [50]. Typically, nanotubular structures are characterized by small diameters and narrow lengthy tubules, thus giving these structures extraordinary aspect ratios. Additional information on the characteristics and potential uses of high aspect ratio nanomaterials can be found in referenced reviews [27, 34, 51].

Nanomaterials are distinguishable from their micrometer-scale or macro-scale analogs by their high surface area to volume ratios. The symbol 'G' represents the free energy differential between the bulk material and nanometer-scale materials. Thus, the $G_{surface} / G_{volume}$ ratio increases as atom clusters approach the nanometer scale [52]. In the case of nanotubes, greater numbers of atoms are located at the surface, which increases their ability to adsorb molecules and also provides a useful scaffold for molecules depositing on the surface [53]. In addition, the size, shape, and surface area of carbonaceous nanoparticles are highly reliant on the aggregation (bundling) state as well as the chemistry of the solvent used in their manufacture. Furthermore, carbonaceous nanomaterials are prone to contamination by materials such as gases, biomolecules, and metals. This can lead to secondary structures of nanomaterial aggregates, which are often credited with a wide variety of physicochemical properties. However, these properties are often difficult to quantify. The resolution of these issues is crucial for the widespread deployment of carbonaceous nanomaterials.

13.4 APPLICATION OF HETEROGENEOUS CARBON-BASED NANOMATERIALS IN WASTEWATER TREATMENT

13.4.1 Adsorption of Dyes from Wastewater

In 2004, Fugetsu et al. initiated research into the use of CNTs for the removal of dyes from contaminated water. Their study evaluated the adsorption of common dyes such as eosin blue, ethidium bromide, acridine orange, and orange G onto carbon nanotubes [54]. Similarly, a study by Wang et al. demonstrated that the cationic dye methylene blue had a high adsorption ability in comparison to the anionic acid red 183 when multi-walled CNTs were used as the absorbent and conducted at a temperature of 298 K. The highest adsorption capacities of methylene blue and acid red 183 were 59.70 and 45.20 mg/g, respectively. The experimental setup only accommodated a single colorant analysis, and all studies were conducted with solution pH held constant at 6.0. However, a significant reduction of acid red 183 (10.0 mg/L) was shown to increase the adsorption of both dyes onto the multi-walled CNTs in a system containing two dyes, thus suggesting that a synergistic effect due to electrical interaction between the two dyes had taken place. An electrical pull between the two dyes was found to produce this phenomenon. The study revealed the methylene blues chromophore planar structure supplied a face-to-face conformation that was advantageous for π-π bond interactions to take place with the CNTs. As well as the surface charge, the structure of the dyes was found to have a major impact on the CNT's ability to adsorb a wide variety of dyes. The study found the separation of anionic dyes with a planar structure was best achieved with CNTs [55].

In order to facilitate more effective separation, a number of researchers have experimented with creating a magnetic nanocomposite based on CNTs. A magnetic separation study was carried out by Madrakian et al., in which nano-Fe_3O_4 was encapsulated within multi-walled CNTs [56]. The adsorption study found a pH of 7 was the most effective level for removing all of the cationic dyes tested from aqueous solutions. Figure 13.4 presents a representative electron micrograph of magnetically modified multi-walled CNTs used in the study. The Langmuir model was used to analyze the experimental data, which found an adsorption capacity of 250 mg/g for Janus green and 48.1 mg/g for methylene blue.

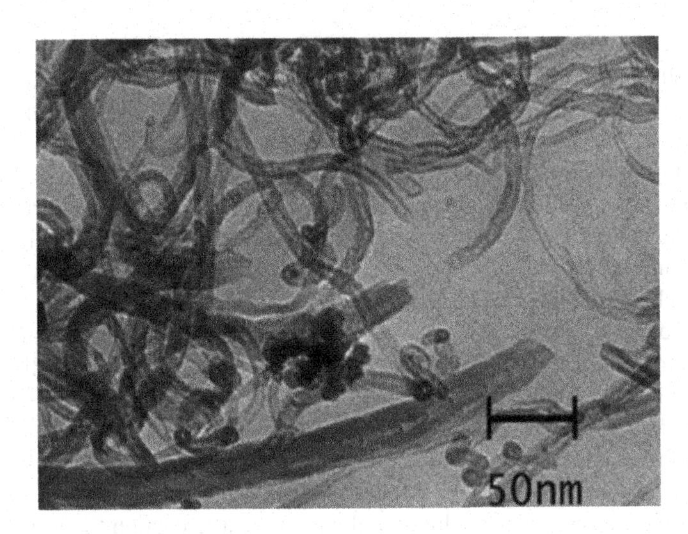

FIGURE 13.4 Electron micrograph image of nano-Fe_3O_4 modified multi-walled CNTs. (Adopted and reproduced with permission from reference [56].)

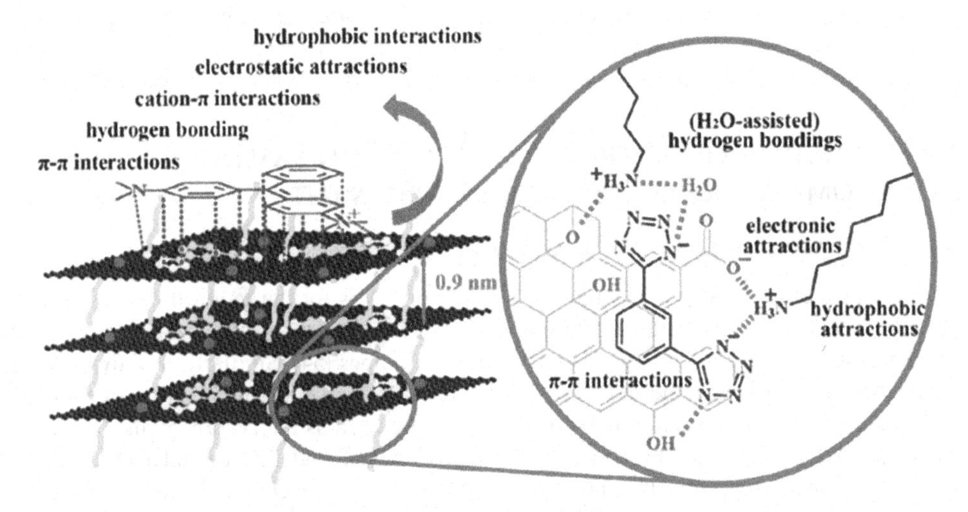

FIGURE 13.5 Schematic representation of the non-covalent interactions between graphene oxide, 1-OA, and malachite green (MG). (Adopted and reproduced with permission from reference [57].)

Another study investigating the removal of dyes from wastewater by Lv et al. found that graphene-based nanomaterials are promising candidates for the removal of inorganic and organic pollutants from an aqueous solution. The high surface area-to-weight ratio and superior chemical stability of the graphene oxide sheets made them a useful absorbent. In addition, Figure 13.5 also shows a schematic illustration of the synergistic effect of the non-covalent interactions occurring between graphene oxide, 1-OA, and malachite green (MG).

13.4.2 Adsorption of Heavy Metals from Wastewater

Graphene has some unique physiochemical properties that indicate it can be used in the removal of heavy metals from wastewater. Generally, there are three conventional techniques for producing graphene nanomaterials. These are 1) carbon nanotube conversion; 2) chemical oxidation-reduction; and 3) mechanical peeling. For instance, a study by Yap et al. synthesized cysteamine-functionalized reduced graphene oxide (r-GO) *via* the thiol-ene click reaction and utilized it for the separation of mercury (Hg (II)) from aqueous solutions [58]. Their study demonstrated the effective removal of trace amounts of heavy metal ions from water. Their study achieved a high elimination rate of 169 19 mg/g, showed high selectivity, and it was possible to improve regeneration capacity [58]. Other studies have also shown that rapid adsorption kinetics, enhanced adsorption capabilities, and increased selectivity is possible for heavy metal ion removal from wastewater samples using chemically modified graphene oxide [59]. It is also known that the solution's pH plays a crucial role in controlling the adhesion of heavy metal ions onto the surface of solid materials. As a result, the surface potential of graphene oxide-based nanomaterials, in particular the protonation and deprotonation activity that occurs during chemical processes, will subsequently be influenced by the pH level of the wastewater [60], thus influencing the number of metal ions that can be adsorbed by the graphene-based nanomaterial [61].

CNTs are a type of adsorbent used in water purification. In most cases, enormous quantities of carbon nanotubes are required for the adsorption of water pollutants, which can be in extremely high concentrations. Therefore, it is essential to comprehend the type of carbon nanotubes used in the specific application, as well as the number of CNTs needed for the purpose.

Studies have also shown that multi-walled CNTs (MWCNTs) can be used for removing metal ions from wastewater. A study by Mubarak et al. that used microwave heating during the synthesis process was able to produce MWCNTs that were capable of removing Zn(II) ions from aqueous solutions. From a starting concentration of 10 mg/L, the MWCNTs were able to remove around 99% of the contaminant [62]. Furthermore, several functionalization techniques, such as free radical polymerization, oxidation, reaction with a diazonium salt, fluorination, and cycloaddition, can be used to increase CNT reactivity and solubility [63, 64]. For instance, a study by Alsohaimi et al. was able to show that between 65 and 85% of the bromate ion, with a starting concentration of 5 mg/L, could be adsorbed by a Fe-CNT nanocomposite in just five minutes [65]. The Fe ions were attached to the walls of the CNT without introducing any flaws into the nanotube structure. The presence of Fe in the nanocomposite played a significant role in the bromate ion's ability to adsorb onto the surface. In particular, there have been a number of studies undertaken to investigate the use of CNTs for the removal of heavy metals from wastewater. These studies showed that pollutant adsorption from wastewater modifies the properties of the CNTs. CNT properties that are changed by the wastewater treatment include 1) their functionalities; 2) hydrophobicity; 3) stability; 4) surface energy; 5) surface charge; and 6) their pore size and pore volume [66]. Thus, there is a great deal of research currently focused on improving the properties and performance of CNTs designed for adsorbing heavy metals from wastewater.

13.4.3 Adsorption of Surfactants from Wastewater Streams

A study by Prediger et al. has examined the adsorption of the non-ionic surfactant TX-100 by several types of graphene. Graphene, also known as graphite oxide (GO), and reduced graphene oxide (r-GO) were among those examined. Their study found that r-GO and GO had the maximum adsorption capabilities for the surfactant TX-100 [67]. Moreover, CNTs were found to be particularly effective at removing aromatic cationic surfactants, as opposed to those that did not contain benzene rings. A study by Cheminski et al. grafted phenyl tetraethyleneglycol (PTEG) units onto graphene oxide using arenediazonium salts with the aim of removing surfactants from aqueous solutions [68]. Their study evaluated the effectiveness of the GO-PTEG-GO to remove cationic (DTAB) and non-ionic (TX-100) surfactants from aqueous solutions. Their study found that factors such as 1) surfactant concentration; 2) pH; 3) sonication and temperature; and 4) duration of treatment were influencing factors that contributed to an optimized adsorption process. Importantly, their study found the PTEG-modified GO enhanced the adsorption process, compared to unmodified GO and other materials.

13.4.4 Adsorption of Active Pharmaceuticals Compounds (APCs)

Carbon-based adsorbents that are effective for the separation of tetracycline include graphene and CNTs. Studies have shown that electrostatic attraction is the reason tetracycline is strongly adsorbed onto graphene nanoparticles [69]. Similarly, a study by Karimi et al. showed that using a magnetic nanocomposite adsorbent composed of r-GO and iron oxide was able to remove phenazopyridine from wastewater samples [70]. Their study revealed that the r-GO and iron oxide nanocomposite was able to remove 91.4% of the phenazopyridine from an aqueous solution. A study by Zhang et al. reported that medications like emodin and physcion could be removed from aqueous solutions using nano zirconium carbide produced using a pre-ceramic polymer method [71].

13.4.5 Adsorption of Phenol and Other Contaminants from Wastewater

A number of studies have demonstrated that CNTs are capable of degrading harmful waterborne pollutants. For instance, studies have found that water molecules and phenolic acid molecules are very weakly connected to the external surface of untreated CNTs. However, when the CNTs are functionalized, both molecules become intensively adsorbed onto their surfaces. The adsorption was enhanced because of the coordinated presence of hydrogen bonds and stacking in the system, which promoted greater phenol binding to CNT–OH when compared to water molecule binding [72]. In addition, the capacity of other types of carbon-based materials to adsorb organic contaminants has also been investigated by researchers [73]. For example, Jiang et al. evaluated the adsorption of two estrogen contaminants (17-ethinyl estradiol and 17-estradiol) by using graphene nanomaterials, CNTs, biochar, and activated carbon [74]. Their study found that graphene-based nanomaterials performed better and the impact of naturally occurring organic matter was lower when compared to biochars, CNTs, and activated carbons. However, CNTs are the preferred absorbent for water filtration [75]. CNTs can be effectively isolated and can remove organic, inorganic, and biological contaminants from wastewater [76]. Moreover, the strong chemical reactivity of CNTs, their enormous surface area, their lower cost, and their lower operational energy requirements have all contributed to their popularity for water purification applications.

13.4.6 Heterogeneous Carbon-Based Nanomaterials in the Photocatalytic Applications

Several studies have shown the importance of carbon-based nanomaterials for photocatalytic applications designed for treating waterborne environmental pollutants [77]. For instance, a study by Lee et al. used a photo-active C60 amino-fullerene immobilized on silica gel to create a visible-light-activated amino-C60-silica photo-catalyst. Their study found the C60 amino-fullerene was effective in degrading pharmaceutical pollutants like cimetidine and ranitidine, and the inactivation of MS-2 bacteriophage [78]. Similar studies have shown that photocatalytic semiconductor nanomaterials, as well as graphene, GO, r-GO, composite semiconductor nanomaterials, and magnetic Fe_3O_4 can be used for treating many organic contaminants [79, 80]. Organic pollutants, with many being organic dyes, are photo-catalytically degraded and adsorbed using hybrid nanocomposite materials based on GO and r-GO, for instance, the use of polyaniline/graphene nanocomposites to break down rose bengal dye when exposed to visible light [81]. The study suggested the presence of graphene sheets in the polyaniline/graphene nanocomposites were able to improve charge separation caused by photo-generated hole-electron pairs and, in turn, improve photocatalytic performance. Thus, further developments into carbon-based nanocomposites as visible-light-activated photo-catalysts for the degradation of waterborne pollutants need to be researched in the future since they have the ability to be an effective wastewater treatment method [82, 83].

13.5 CONCLUSION

The demand for safe drinking and clean water is forever increasing. However, meeting this demand is challenging. The amount of water on the earth is estimated to be around 333 million cubic miles, but only around 2.5% of the total water on earth is regarded to be freshwater while the remaining water is either salty or contaminated. The problem is further complicated by the ineffectiveness of traditional wastewater treatments to eliminate heavy metals, organic contaminants, and certain microorganisms. Therefore, to improve the supply of clean water it is necessary to improve wastewater treatment processes using more modern and effective technologies. Because of their unique optical, electronic, and catalytic properties, carbon-based nanomaterials have demonstrated their ability to be superior absorbers and materials for photocatalysis. Both of these abilities make them ideal methods for developing future wastewater treatment processes. However, a number of challenges still remain to be addressed, such as the optimization and scalability of this material. In addition, with any new technology, there are issues to be considered, such as a lack of modern infrastructure, effective use of resources, and viable economics. Further studies are needed to address the potential toxicity of the new nanomaterials themselves.

REFERENCES

1. Adeleye AS, Conway JR, Garner K, Huang Y, Su Y, Keller AA. Engineered nanomaterials for water treatment and remediation: Costs, benefits, and applicability. *Chem. Eng. J.* 2016; 286: 640–662.
2. Lu H, Wang J, Stoller M, Wang T, Bao Y, Hao H. An overview of nanomaterials for water and wastewater treatment. *Adv. Mater. Sci. Eng.* 2016; 2016.
3. Taghipour S, Hosseini SM, Ataie-Ashtiani B. Engineering nanomaterials for water and wastewater treatment: Review of classifications, properties and applications. *New J. Chem.* 2019; 43: 7902–7927.
4. Santhosh C, Velmurugan V, Jacob G, Jeong SK, Grace AN, Bhatnagar A. Role of nanomaterials in water treatment applications: A review. *Chem. Eng. J.* 2016; 306: 1116–1137.
5. Ghasemzadeh G, Momenpour M, Omidi F, Hosseini MR, Ahani M, Barzegari A. Applications of nanomaterials in water treatment and environmental remediation. *Front. Environ. Sci. Eng.* 2014; 8: 471–482.
6. Tenne R. Inorganic nanotubesandfullerene-like nanoparticles. *Nat. Nanotechnol.* 2006; 1: 103–111.
7. Ajayan PM. Nanotubes from carbon. *Chem. Rev.* 1999; 99: 1787–1800.
8. Falcao EHL, Wudl F. Carbon allotropes: beyond graphite and diamond. *J. Chem. Technol. Biotechnol.* 2007; 82: 524–531.
9. Khan ME. State-of-the-art developments in carbon-based metal nanocomposites as a catalyst: Photocatalysis. *Nanoscale Adv.* 2021; 3: 1887. DOI: 10.1039/d1na00041a.
10. Dresselhaus M, Endo M. Relation of carbon nanotubes to other carbon materials. *Top. Appl. Phys.* 2001; 80: 11–28.
11. Kroto HW. The stability of the fullerenes C-24, C-28, C-32, C-36, C-50, C-60 and C- 70. *Nature.* 1987; 329: 529–531.
12. Campbell EEB, Fowler PW, Mitchell D, Zerbetto F. Increasing cost of pentagon adjacency for larger fullerenes. *Chem. Phys. Lett.* 1996; 250: 544–548.
13. Taylor R, Walton DRM. The chemistry of fullerenes. *Nature.* 1993; 363: 685–693.
14. Johnson RD, Meijer G, Bethune DS. C-60 has icosahedral symmetry. *J. Am. Chem. Soc.* 1990; 112: 8983–8984.
15. Johnson R, Bethune D, Yannoni C. Fullerene structure and dynamics - A magnetic-resonance potpourri. *Acc. Chem. Res.* 1992; 25: 169–175.
16. Diederich F, Thilgen C. Covalent fullerene chemistry. *Science.* 1996; 271: 317–323.
17. Lehn JM, Atwood JL, Davies JED, MacNicol DD, Vogtle F. *Comprehensive Supramolecular Chemistry.* New York: Pergamon, 1996.
18. Bonifazi D, Enger O, Diederich F. Supramolecular [60] fullerene chemistry on surfaces. *Chem. Soc. Rev.* 2007; 36: 390–414.
19. Diederich F, Gomez-Lopez M. Supramolecular fullerene chemistry. *Chem. Soc. Rev.* 1999; 28: 263–277.
20. Moriarty P. Nanostructured materials. *Rep. Prog. Phys.* 2001; 64: 297–381.

21. Shinohara H, Sato H, Saito Y, Takayama M, Izuoka A, Sugawara T. Formation and extraction of very large all-carbon fullerenes. *J. Phys. Chem.* 1991; 95: 8449–8451.
22. Wang BC, Wang HW, Chang JC, Tso HC, Chou YM. More spherical large fullerenes and multi-layer fullerene cages. *J. Mol. Struct. Theochem.* 2001; 540: 171–176.
23. Ugarte D. Curling and closure of graphitic networks under electron-beam irradiation. *Nature.* 1992; 359: 707–709.
24. Bates KR, Scuseria GE. Why are bucky onions round. *Theor. Chem. Acc.* 1998; 99: 29–33.
25. Xia YN, Yang PD, Sun YG, Wu Y, Mayers B, Gates B, Yin Y, Kim F, Yan H. One-dimensional nanostructures: Synthesis, characterization, and applications. *Adv. Mater.* 2003; 15: 353–389.
26. Hu JT, Odom TW, Lieber CM. Chemistry and physics in one dimension: Synthesis and properties of nanowires and nanotubes. *Acc. Chem. Res.* 1999; 32: 435–445.
27. Dresselhaus MS, Ed. *Carbon Nanotubes: Synthesis, Structure, Properties, and Applications.* Berlin, NY: Springer, 2001.
28. Saito R, Fujita M, Dresselhaus G, Dresselhaus MS. Electronic-structure of chiral graphene tubules. *Appl. Phys. Lett.* 1992; 60: 2204–2206.
29. Bandow S, Takizawa M, Hirahara K, Yudasaka M, Iijima S. Raman scattering study of double-wall carbon nanotubes derived from the chains of fullerenes in single-wall carbon nanotubes. *Chem. Phys. Lett.* 2001; 337: 48–54.
30. Thess A, Lee R, Nikolaev P, Dai H, Petit P, Robert J, Xu C, Lee YH, Kim SG, Rinzler AG, Colbert DT. Crystalline ropes of metallic carbon nanotubes. *Science.* 1996; 273: 483–487.
31. Liu YQ, Gao L, Zheng S, Wang Y, Sun J, Kajiura H, Li Y, Noda K. Debundling of single-walled carbon nanotubes by using natural polyelectrolytes. *Nanotechnology.* 2007; 18: 365702.
32. Nish A, Hwang JY, Doig J, Nicholas RJ. Highly selective dispersion of single-walled carbon nanotubes using aromatic polymers. *Nat. Nanotechnol.* 2007; 2: 640–646.
33. Jin H, Jeng ES, Heller DA, Jena PV, Kirmse R, Langowski J, Strano MS. Divalent ion and thermally induced DNA conformational polymorphism on single-walled carbon nanotubes. *Macromolecules.* 2007; 40: 6731–6739.
34. Sun YP, Fu KF, Lin Y, Huang WJ. Functionalized carbon nanotubes: Properties and applications. *Acc. Chem. Res.* 2002; 35: 1096–1104.
35. Holzinger M, Vostrowsky O, Hirsch A, Hennrich F, Kappes M, Weiss R, Jellen F. Sidewall functionalization of carbon nanotubes. *Angew. Chem. Int. Ed.* 2001; 40: 4002–4005.
36. Andreas H. Functionalization of single-walled carbon nanotubes. *Angew. Chem. Int. Ed.* 2002; 41: 1853–1859.
37. Georgakilas V, Kordatos K, Prato M, Guldi DM, Holzinger M, Hirsch A. Organic functionalization of carbon nanotubes. *J. Am. Chem. Soc.* 2002; 124: 760–761.
38. Banerjee S, Hemraj-Benny T, Wong SS. Covalent surface chemistry of single-walled carbon nanotubes. *Adv. Mater.* 2005; 17: 17–29.
39. Niyogi S, Hamon MA, Hu H, Zhao B, Bhowmik P, Sen R, Itkis ME, Haddon RC. Chemistry of single-walled carbon nanotubes. *Acc. Chem. Res.* 2002; 35: 1105–1113.
40. Geim AK, Novoselov KS. *In Nanoscience and Technology: A Collection of Reviews From Nature Journals.* World Scientific, 2010, pp. 11–19.
41. Khan ME, Khan MM, Cho MH. Recent progress of metal–graphene nanostructures in photocatalysis. *Nanoscale.* 2018; 10: 9427–9440.
42. Khan ME, Khan MM, Cho MH. Defected graphene nano-platelets for enhanced hydrophilic nature and visible light-induced photoelectrochemical performances. *J. Phys. Chem. Solids.* 2017; 104: 233–242.
43. Wang X, Blechert S, Antonietti M. Polymeric graphitic carbon nitride for heterogeneous photocatalysis. *ACS Catal.* 2012; 2: 1596–1606.
44. Wang X, Maeda K, Thomas A, Takanabe K, Xin G, Carlsson JM, Domen K, Antonietti M. A metal-free polymeric photocatalyst for hydrogen production from water under visible light. *Nat. Mater.* 2009; 8: 76–80.
45. Barhoum A, Shalan AE, El-Hout SI, Ali GA, Abdelbasir SM, Abu Serea ES, Ibrahim AH, Pal K. A broad family of carbon nanomaterials: Classification, properties, synthesis, and emerging applications. In: *Handbook of Nanofibers* (Eds., Barhoum A, Bechelany M, Makhlouf A). Cham: Springer, 2019.
46. Bandow S. Asaka S, Saito Y, Rao AM, Grigorian L, Richter E, Eklund PC. Effect of the growth temperature on the diameter distribution and chirality of single-wall carbon nanotubes. *Phys. Rev. Lett.* 1998; 80: 3779–3782.
47. Chatterjee A, Deopura BL Carbon nanotubes and nanofibre: An overview. *Fibers Polym.* 2002; 3(4): 134–139.

48. Popov VN. Carbon nanotubes: Properties and application. *Mater. Sci. Eng.* 2004; 43: 61–102.
49. Holt J, Park H, Wang Y, Stadermann M, Artyukhin AB, Grigoropoulos CP, Noy A, Bakajin O. Fast mass transport through sub-2-nanometer carbon nanotubes. *Science.* 2006; 312: 1034–1037.
50. Krishna V, Pumprueg S, Lee S, Zhao J, Sigmund W, Koopman B, Moudgil BM. Photocatalytic disinfection with titanium dioxide coated multi-wall carbon nanotubes. *Process Safe. Environ. Prot.* 2005; 83: 393–397.
51. Khin MM, Nair AS, Babu VJ, Murugan R, Ramakrishna S. A review on nanomaterials for environmental remediation. *Energy Environ. Sci.* 2012; 5: 8075–8109.
52. Hunter R J. *Foundations of Colloid Science*, 2nd ed. Oxford; New York: Oxford University Press, 2001.
53. Lin YH, Cui XL, Bontha J. Electrically controlled anion exchange based on polypyrrole and carbon nanotubes nanocomposite for perchlorate removal. *Environ. Sci. Technol.* 2006; 40: 4004–4009.
54. Fugetsu B, Satoh S, Shiba T, Mizutani T, Lin YB, Terui N, Nodasaka Y, Sasa K, Shimizu K, Akasaka T, Shindoh M. Caged multiwalled carbon nanotubes as the adsorbents for a nity-based elimination of ionic dyes. *Environ. Sci. Technol.* 2004; 38: 6890–6896.
55. Wang S, Ng CW, Wang W, Li Q, Hao Z. Synergistic and competitive adsorption of organic dyes on multiwalled carbon nanotubes. *Chem. Eng. J.* 2012; 197: 34–40.
56. Madrakian T, Afkhami A, Ahmadi M, Bagheri H. Removal of some cationic dyes from aqueous solutions using magnetic-modified multi-walled carbon nanotubes. *J. Hazard. Mater.* 2011; 196: 109–114.
57. Lv M, Yan L, Liu C, Su C, Zhou Q, Zhang X, Lan Y, Zheng Y, Lai L, Liu X, Ye Z. Non-covalent functionalized graphene oxide (GO) adsorbent with an organic gelator for co-adsorption of dye, endocrine-disruptor, pharmaceutical and metal ion. *Chem. Eng. J.* 2018; 349: 791–799.
58. Yap PL, Kabiri S, Tran DN, Losic D. Multifunctional binding chemistry on modified graphene composite for selective and highly efficient adsorption of mercury. *ACS Appl. Mater. Interfaces.* 2018; 11: 6350–6362.
59. Liu X, Ma R, Wang X, Ma Y, Yang Y, Zhuang L, Zhang S, Jehan R, Chen J, Wang X. Graphene oxide-based materials for efficient removal of heavy metal ions from aqueous solution: A review. *Environ. Pollut.* 2019; 252: 62–73.
60. Yu S, Wang X, Liu Y, Chen Z, Wu Y, Liu Y, Pang H, Song G, Chen J, Wang X. Efficient removal of uranium (VI) by layered double hydroxides supported nanoscale zero-valent iron: A combined experimental and spectroscopic studies. *Chem. Eng. J.* 2019; 365: 51–59.
61. Lim JY, Mubarak NM, Khalid M, Abdullah EC, Arshid N. Novel fabrication of functionalized graphene oxide via magnetite and 1-butyl-3-methylimidazolium tetrafluoroborate. *Nano Struct. Nano Obj.* 2018; 16: 403–411.
62. Mubarak NM, Sahu JN, Abdullah EC, Jayakumar NS, Ganesan P. Microwave-assisted synthesis of multi-walled carbon nanotubes for enhanced removal of Zn (II) from wastewater. *Res. Chem. Intermed.* 2016; 42: 3257–3281.
63. Xu J, Cao Z, Zhang Y, Yuan Z, Lou Z, Xu X, Wang X. A review of functionalized carbon nanotubes and graphene for heavy metal adsorption from water: Preparation, application, and mechanism. *Chemosphere.* 2018; 195: 351–364.
64. Lee KM, Wong CP, Tan TL, Lai CW. Functionalized carbon nanotubes for adsorptive removal of water pollutants. *Mater. Sci. Eng. B.* 2018; 236: 61–69.
65. Alsohaimi IH, Khan MA, Alothman ZA, Khan MR, Kumar M. Synthesis, characterization, and application of Fe-CNTs nanocomposite for BrO_3 remediation from water samples. *J. Ind. Eng. Chem.* 2015; 26: 218–225.
66. Das R, Hamid SBA, Ali M, Annuar MS, Samsudin EM, Bagheri S. Covalent functionalization schemes for tailoring solubility of multi-walled carbon nanotubes in water and acetone solvents. *Sci. Adv. Mater.* 2015; 7: 2726–2737.
67. Prediger P, Cheminski T, de Figueiredo Neves T, Nunes WB, Sabino L, Picone CS, Oliveira RL, Correia CR. Graphene oxide nanomaterials for the removal of non-ionic surfactant from water. *J. Environ. Chem. Eng.* 2018; 6: 1536–1545.
68. Cheminski T, de Figueiredo Neves T, Silva PM, Guimaraes CH, Prediger P. Insertion of phenyl ethyleneglycol units on graphene oxide as stabilizers and its application for surfactant removal. *J. Environ. Chem. Eng.* 2019; 7: 102976.
69. Zheng S, Shi J, Zhang J, Yang Y, Hu J, Shao B. Identification of the disinfection byproducts of bisphenol S and the disrupting effect on peroxisome proliferator-activated receptor gamma (PPAR) induced by chlorination. *Water Res.* 2018; 132: 167–176.

70. Karimi-Maleh H, Shafieizadeh M, Taher MA, Opoku F, Kiarii EM, Govender PP, Ranjbari S, Rezapour M, Orooji Y. The role of magnetite/graphene oxide nano-composite as a high-effciency adsorbent for removal of phenazopyridine residues from water samples, an experimental/theoretical investigation. *J. Mol. Liq.* 2020; 298: 112040.
71. Zhang B, Ji J, Liu X, Li C, Yuan M, Yu J, Ma Y. Rapid adsorption and enhanced removal of emodin and physcion by nano zirconium carbide. *Sci. Total Environ.* 2019; 647: 57–65.
72. Moradi F, Ganji MD, Sarrafi Y. Remediation of phenol-contaminated water by pristine and functionalized SWCNTs: Ab initio van derWaals DFT investigation. *Diam. Relat. Mater.* 2018; 82: 7–18.
73. Ersan G, Apul OG, Perreault F, Karanfil T. Adsorption of organic contaminants by graphene nanosheets: A review. *Water Res.* 2017; 126: 385–398.
74. Jiang, L, Liu Y, Liu S, Zeng G, Hu X, Guo Z, Tan X, Wang L, Wu Z. Adsorption of estrogen contaminants by graphene nanomaterials under natural organic matter preloading: Comparison to carbon nanotube, biochar, and activated carbon. *Environ. Sci. Technol.* 2017; 51: 6352–6359.
75. Chen M, Zhou S, Zhu Y, Sun Y, Zeng G, Yang C, Xu P, Yan M, Liu Z, Zhang W. Toxicity of carbon nanomaterials to plants, animals and microbes: Recent progress from 2015-present. *Chemosphere.* 2018; 206: 255–264.
76. Das R, Leo BF, Murphy F. The toxic truth about carbon nanotubes in water purification: A perspective view. *Nanoscale Res. Lett.* 2018; 13: 183.
77. Yi H, Huang D, Qin L, Zeng G, Lai C, Cheng M, Ye S, Song B, Ren X, Guo X. Selective prepared carbon nanomaterials for advanced photocatalytic application in environmental pollutant treatment and hydrogen production. *Appl. Catal. B Environ.* 2018; 239: 408–424.
78. Lee J, Mackeyev Y, Cho M, Wilson LJ, Kim JH, Alvarez PJ. C60 aminofullerene immobilized on silica as a visible-light-activated photocatalyst. *Environ. Sci. Technol.* 2010; 44: 9488–9495.
79. Fu D, Han G, Chang Y, Dong J. The synthesis and properties of ZnO–graphene nano hybrid for photodegradation of organic pollutant in water. *Mater. Chem. Phys.* 2012; 132: 673–681.
80. Liu L, Liu J, Sun DD. Graphene oxide enwrapped Ag_3PO_4 composite: Towards a highly efficient and stable visible-light-induced photocatalyst for water purification. *Catal. Sci. Technol.* 2012; 2: 2525–2532.
81. Ameen S, Seo HK, Akhtar MS, Shin HS. Novel graphene/polyaniline nanocomposites and its photocatalytic activity toward the degradation of rose Bengal dye. *Chem. Eng. J.* 2012; 210: 220–228.
82. Khan ME. State-of-the-art developments in carbon-based metal nanocomposites as a catalyst: Photocatalysis. *Nanoscale Adv.* 2021; 3: 1887.
83. Lv M, Yan L, Liu C, Su C, Zhou Q, Zhang X, Lan Y, Zheng Y, Lai L, Liu X, Ye Z. Non-covalent functionalized graphene oxide (GO) adsorbent with an organic gelator for co-adsorption of dye, endocrine-disruptor, pharmaceutical and metal ion. *Chem. Eng. J.* 2018; 349: 791–799.

14 Microplastic Pollution and Its Detrimental Impact on Coastal Ecosystems and Mid-Ocean Gyres

Derek Fawcett, Cormac Fitzgerald, Jennifer Verduin, and Gérrard Eddy Jai Poinern

CONTENTS

14.1 INTRODUCTION

The production of durable, lightweight, and inexpensive plastic materials has made them a fundamental manufacturing component for a diverse range of products and commodities [1]. Plastics are synthetic polymers composed of long-chain molecules derived from organic and inorganic raw materials like carbon, chloride, hydrogen, oxygen, and silicon [2, 3]. Due to their versatile and desirable properties, annual global production levels have steadily increased to around 322 million tons in 2015 [4]. Plastics are used in products ranging from shopping bags to components in electronics, automobiles, and aircraft. Unfortunately, large quantities of waste plastic materials enter the environment. In particular, large quantities of plastics are used in single-use packaging materials and

DOI: 10.1201/9781003181422-17

consumer goods (39.5%), and household appliances (22.7%) [4]. Importantly, packaging materials, because of their low recovery value (after use), end up going to landfill where they contribute to around 10% of all municipal waste [5]. Consumer goods such as personal care products and cosmetics (anti-aging creams, cleansers and exfoliates, moisturizers, gels, soaps, lotions, and nail polishes), which contain plastic micro-beads, are generally washed down the drain after application [6, 7]. Other sources of waste plastics come from industrial manufacturing, synthetic textiles, and wastewater treatment plant effluent [8]. Four main marine pollutants have been identified: glass, metal, paper, and plastics [9]. Plastics contribute between 60 and 80% to total pollutant levels [10]. In spite of plastics being identified as a major pollutant in the 1970s, it is only in recent years that the problem has been fully appreciated as a global problem [11–13]. Current estimates indicate that between 4.8 and 12.7 million metric tons of plastic materials end up in the marine environment each year [14, 15]. The level of plastic waste entering the marine environment each year *via* rivers is estimated to be between 1.15 and 2.41 million metric tons [16].

The extremely large amounts and widespread distribution of microplastics in the marine environment have attracted considerable global concern. This concern has prompted many countries around the world to undertake plastic monitoring programs. Plastic monitoring programs have primarily focused on coastal surface waters and sediments. However, global distribution studies have also shown that microplastics have spread to the most remote marine environments [17–20]. Plastics are detrimental to marine ecosystems since their resistance to degradation means that they can take many hundreds, and even thousands, of years to completely breakdown [21]. During their time in the marine environment, plastics break up and fragment into small pieces forming microplastics. Microplastics are generally defined as plastic fragments with at least one dimension less than 5 mm [5, 22]. Microplastics can enter marine ecosystems as either primary microplastics (manufactured to have micron-sized dimensions) or secondary microplastics that result from the breakdown of larger plastic materials [23]. Several studies have reported the presence of microplastics of various shapes and sizes (irregular fragments, pellets, spherules, and fibers) in oceans and seas around the world [24–26]. In addition to their widespread pervasiveness, interactions occurring between microplastics and organisms like invertebrates, fish, birds, and mammals have attracted considerable scientific concern [27–31]. For instance, the toxicity of larger plastic materials and their effect on higher trophic-level organisms like seabirds and turtles have been reported for some time [32–35]. However, the widespread distribution of microplastics by ocean currents, exposure to marine organisms, and their harmful effects have only been fully appreciated relatively recently [36, 37]. For instance, monitoring studies have shown that microplastics are highly pervasive and can be found in water columns, coastal sediments, deep sea sediments, and marine organisms [38–41]. Crucially, the availability of microplastics in marine ecosystems and subsequent ingestion of microplastics by marine organisms has far-reaching consequences. Once in the food chain, trophic transfers will result in microplastics ending up in marine species that are subsequently consumed by humans [42, 43]. Furthermore, not only are there health risks associated with the microplastics themselves, but there are also the effects of absorbed toxic compounds like persistent organic pollutants (POPs) and polycyclic aromatic hydrocarbons (PAHs) [44–46]. Once ingested, these compounds are transferred to the tissues of the organism and accumulate over time [47]. Having entered the organism and the food web, subsequent trophic transfer results in the bioaccumulation of these toxic compounds in higher trophic species [48]. Recent studies have also suggested that much smaller nanometer-scale plastics (fragments with at least one dimension that is between one and one hundred nanometers in length), could also pose a problem for aquatic ecosystems. Ultimately, contaminated marine organisms will create a health risk for humans when consumed [49, 50].

The severity of microplastic pollution is increasing. Early plankton studies carried out from the early 1960s to 1990s in North Atlantic surface waters have consistently shown steadily increasing microplastic pollution levels [51], while Belgian beach sediment studies from 1993 through to 2008 have shown an almost tripling in the levels of microplastic pollution [52]. Typically, these studies reveal microplastic pollution is composed of different types of plastic materials. Currently, 80% of the total global demand for plastic materials is for six types and it is usually these types that end up as microplastics. The six types are 1) polyethylene (PE); 2) polypropylene (PP); 3) polyvinyl chloride (PVC); 4) polystyrene (PS) (solid and expandable types); 5) polyethylene terephthalate (PET); and 6) polyurethane (PUR) [4]. Table 14.1 lists plastic materials currently being manufactured, with representative products being produced and typical applications for the products. An inspection of Table 14.1 reveals the wide range of products and applications, and with the large global demand for new products, the demand for plastics will only increase. As a result, larger quantities of microplastics will end up in the marine environment [53]. Other sources of microplastic waste include fishing activities, aquaculture, and a variety of land-based processing facilities. For instance, the fishing industry currently uses a wide range of plastic products like netting, fishing lines, ropes, and traps [54]. Unfortunately, around 18% of the total plastic debris found in the marine environment comes from products lost or discarded by the fishing industry [55]. Typically, fishing gear is made from plastics like polyethylene and polypropylene (both are buoyant in seawater), and nylons (polytetrafluoroethylene), which are non-buoyant [56]. Studies have also reported aquaculture is capable of producing significant amounts of plastic waste [57, 58]. However, the largest contributor to plastic pollution is land-based industrial discharges from processing facilities, textile factory runoffs, and wastewater treatment plant effluents [8, 59, 60]. Unfortunately, the versatile and durable properties that make plastics idea materials for a wide range of products and applications also make them problematic in the environment. For example, their durable nature delays degradation and prevents them from undergoing complete mineralization in the marine environment [61]. During mineralization, the organic compounds that make up plastics break down to produce carbon dioxide, water, and inorganic compounds [62]. Other factors that influence degradation rates in the marine environment include oxygen availability, access to solar (ultraviolet) radiation, and seawater temperatures [63]. Limited access to these factors is the main reason why degradation rates in aquatic environments are slower than in land-based environments. In addition, biological degradation processes resulting from microorganism activity in marine ecosystems are slower than land-based and are influenced by prevailing environmental conditions [64].

TABLE 14.1
Types and sources of common plastics found in the marine environment

Plastic	Name of plastic	Density (Kg m^{-3})	Typical products and usages
CA	Cellulose acetate	1.25–1.50	Insulating coatings, cigarette filters
PP	Polypropylene	0.90	Film for packaging, rope, carpeting, bottle caps, netting
LDPE	Low-density polyethylene	0.92	Plastic bags, bottles, netting, six-pack rings, drinking straws
HDPE	High-density polyethylene	0.94–0.97	Kitchenware, milk and juice cartons
PS	Polystyrene	1.05	Plastic utensils, thermal insulation, food containers, cups
Nylon	Polyamide (nylon)	1.02–1.14	Netting, fishing line, ropes and traps
PVC	Polyvinyl chloride	1.29–1.39	Plastic film, piping, clothing, bottles, cups
PET	Polyethylene terephalicate	1.40	Electrical insulation, piping, kitchenware, beverage bottles
PUR	Polyurethane	1.10–1.50	Foam, sheets, wall panels, piping
PTFE	Polytetrafluoroethylene	2.20	Cooking utensils, electrical insulation, bearings

This chapter focuses on the properties, behavioral mechanisms, and ecological impact of microplastics in the marine environment. The following four sections aim to 1) identify primary and secondary sources of microplastics; 2) examine the various physical and chemical behavioral properties of microplastics in marine ecosystems; 3) identify factors influencing bioavailability and interactions with marine organisms; and 4) provide future perspectives and areas for future research.

14.2 SOURCES OF MICROPLASTICS

The presence of macro-scale plastics in aquatic environments has been an easily recognizable problem for many years. However, in recent years, smaller plastic particles of varying sizes, densities, chemical compositions, and shapes that are present in the marine environment have become collectively known as microplastics [65]. Microplastics are micrometer-scale plastic particles less than 5 mm in size, so many can be seen by the human eye, and numerous other smaller particles can only be seen with the aid of microscopy [66]. They are used in a wide variety of consumer goods, commercial products, and industrial processes (primary microplastics) or result from the disintegration of larger macro-scale plastics by the environment (secondary microplastics) [67]. Because of their proliferation (primary and secondary) at high concentrations in aquatic ecosystems, they have become a serious environmental pollutant in their own right [68, 69]. Their detrimental impact on aquatic ecosystems results from the microplastics being ingested and assimilated into the tissues of organisms [70, 71]. A selection of plastic materials and products commonly found in the marine environment is presented in Table 14.1.

14.2.1 PRIMARY MICROPLASTICS

Microplastics are produced for inclusion in a wide variety of personal care products, commercial products, and industrial applications. In particular, since the incorporation of microplastic beads in personal care products and cosmetics in the 1980s, replacing traditional natural exfoliating agents (like ground almonds, oatmeal, and pumice), their use has dramatically increased [72, 73]. Moreover, microplastics are used in facial peels and cleansers, toothpaste, bath/shower products, shaving creams, deodorants, make-up foundations, nail polishes, eye shadows, blush powders, hairsprays, baby products, lotions, insect repellents, and sunscreens [74, 75]. The quantities of microplastic beads used in these types of products can vary greatly. For instance, personal care products marketed as "micro-beads" or "micro-exfoliates" can contain between 137,000 and 2,800,000 micro-beads per 150 ml bottle [76]. Typically, these products are applied to the body, and after a short period of time are washed off, resulting in the product entering the drainage system. However, in many cases, inadequate waste disposal procedures in the drainage system result in the microplastic beads ending up in the aquatic environment [77, 78]. Microplastic materials have also been included in synthetic clothing and textile printing, and incorporated as abrasives into many commercial and industrial cleaning products. Industrial air-blasting with microplastics is widely used to remove rust and paint from boat hulls in shipyards, and general machinery and engines that are in engineering workshops and spray-painting facilities [79]. Moreover, metallic microplastics are also produced during these air-blasting processes. The generation of metal particles removed from the surfaces being cleaned results in the microplastics being contaminated with the metal particles. If these metal particles are generated from heavy metals like cadmium, chromium, and lead, their combination with microplastic beads leads to greater microplastic toxicity levels in aquatic ecosystems [74, 79].

14.2.2 SECONDARY MICROPLASTICS

Secondary microplastics are produced by the breakdown of larger macro-scale plastic fragments. The breakdown results from the combined effects of physical, biological, and chemical processes present in aquatic environments [80]. These processes change the chemical properties and structural integrity of the plastic fragments, which ultimately leads to fragmentation and the formation of microplastics [81]. Some of these processes include photo-degradation by solar (ultraviolet) radiation, photo-oxidation, abrasion, wave action, and turbulence. Importantly, each of these processes breakdown macro-scale plastics to varying degrees depending upon prevailing exposure and weathering conditions. For instance, prolonged periods of exposure to ultraviolet radiation induce photo-degradation, which causes oxidation of the polymer matrix and results in the bonds between individual atoms within the matrix breaking [82, 83]. Thus, photo-degradation rates will be much higher on beaches due to the higher exposure levels of ultraviolet radiation, increased abrasion resulting from wave turbulence and wind, and increased levels of atmospheric oxygen [67, 84]. This is unlike the breakdown of macro-scale plastics in surface waters or submerged at deeper depths, where the lower levels of ultraviolet radiation, oxygen, and cooler temperatures significantly reduce the breakdown rate [85, 86]. In either case, the ongoing breakdown process continues, and the loss of structural integrity increases fragmentation. As fragmentation progresses, the fragments become smaller over time until they become microplastic in size [87]. Another process that can influence the breakdown rate of macro-scale plastics in surface waters is biological fouling. Floating macro-scale plastics readily attract surface fouling, first with bacteria, followed by an algal mat, and then colonization by invertebrates [88, 89]. Fouling strongly depends on the season of exposure and the prevailing marine environment. Importantly, as fouling continues, the accumulating mass increases until its density exceeds that of seawater and sinks [90]. Thus, fouling ultimately submerges the macroplastics and in the process reduces the levels of oxidation and degradation the macroplastics undergo. The problem is further exacerbated by the lack of natural microbial species in the marine environment that can metabolize plastic materials in common usage, apart from biologically derived polymers such as cellulose and chitin [91]. Thus, current research is investigating the feasibility of developing new biodegradable plastic materials as viable replacements for conventional plastics [92, 93]. However, even the degradation of new biodegradable plastic materials in the relatively cold marine environment will be lengthy and will also expose the floating plastics to fouling. Furthermore, the degradation process of biodegradable plastics will also release microplastics into the marine environment [94]. Ultimately, no matter what type of plastic material enters the marine environment, the larger macroplastics ultimately break down to produce extremely large numbers of microplastics. It is these small microplastics that are easily ingested by marine organisms and that threaten the entire marine ecosystem [95, 96].

14.3 PHYSICAL AND CHEMICAL BEHAVIORAL PROPERTIES OF MICROPLASTICS

14.3.1 Physical Properties

14.3.1.1 Migration

Recent oceanographic studies have shown microplastic pollution is widespread throughout the world's oceans, shorelines, and seabed sediments [97, 98]. Microplastics have even been found in both Arctic and Antarctic regions, as well as many other remote locations around the globe [26, 99–101]. These studies also reveal that microplastics have been entering the marine environment and accumulating for the last four decades [102, 103]. When microplastics enter the marine environment at one location, they begin migrating to other locations. The distribution of microplastics is influenced by their density, location of source, and prevailing environmental conditions like ocean currents, waves, and wind, as seen in Figure 14.1 [104, 105]. Microplastics with densities lower than seawater, such as low-density polyethylene (0.92 kg m^{-3}) and polypropylene (0.90 kg m^{-3}), are buoyant and resistant to degradation and can easily be distributed in marine ecosystems by environmental conditions [106]. For instance, Law et al. found that microplastic pollution from the east coast of the United States of America found its way into the North Atlantic subtropical gyre within

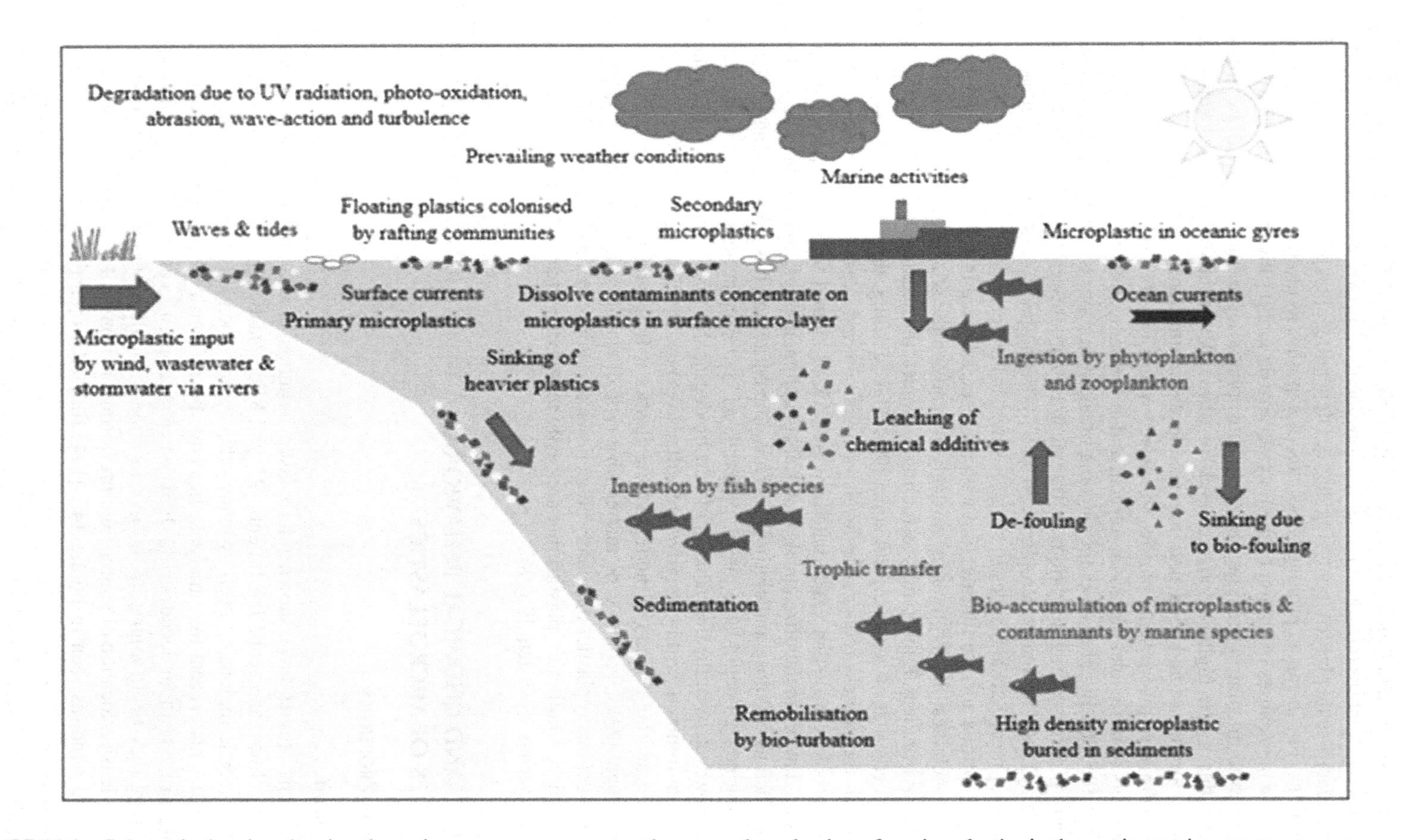

FIGURE 14.1 Schematic drawing showing the main sources, movement pathways, and mechanisms for microplastics in the marine environment.

60 days [107]. Moore et al. estimated that two rivers in Californian delivered around two billion microplastic fragments into coastal waters over a three-day period [108]. In addition, environmental conditions such as strong prevailing winds can promote increased mixing and dispersion of microplastic particles in the upper layers of the water column [109]. Similarly, tides and storm surges can also promote the migration and distribution of surface and sub-surface microplastics [110, 111]. Microplastics denser than seawater like nylon (1.02 – 1.14 kg m^{-3}) and polyvinyl chloride (1.29 – 1.39 kg m^{-3}) can still be transported and distributed by underlying ocean currents [112].

14.3.1.2 Sedimentation in Seabeds

While low-density microplastics float on the surface of the sea, those with densities greater than seawater slowly sink and ultimately accumulate in seafloor sediments [113, 114]. Even the density of floating microplastic debris may increase with longer exposure time in the marine environment [115]. The increase in density results from biological fouling by various organisms, which also makes the debris less hydrophobic [116]. On reaching seawater density, the neutrally buoyant microplastic debris enters and drifts within the water column. Importantly, seawater density increases with depth, so there will be a depth where seawater density is equal to debris density, and the microplastics will be suspended at this depth [117, 118]. However, a study by Ye and Andrady showed that biologically fouled debris at these depths rapidly loses its organism-based coatings and, as a result of decreasing density, microplastics start returning to the surface to repeat the cycle [119]. On the other hand, if the density of the debris is greater than the seawater density, microplastics will accumulate in marine sediments [120]. Studies have not only shown that microplastics accumulate in both beach and shallow marine sediments [113, 121], but they are also found accumulating in the sediments of deep sea areas and submarine canyons [122].

14.3.1.3 Accumulation

Pollution entering the marine environment is a serious problem and global concern. Estimates indicate that between 50 and 80% of all wastes entering the marine environment are composed of plastic materials [123]. Current estimates indicate that around 20 million tons of various plastic materials enter the marine environment each year [124]. Combined with the durability of many plastics (stable for hundreds to thousands of years) and the small size of microplastics, their accumulation in the marine environment is expected to increase [97]. Recent studies indicate that plastic waste will eventually outweigh fish in the world's oceans by the year 2050 [125, 126]. The durability and usefulness of plastics have made them fundamental materials of the modern era, but littering, poor land-based treatment facilities, illegal dumping, and coastal activities by humans have resulted in unacceptable levels of marine pollution [127]. Increasing concentration levels of microplastics and their distribution throughout the global marine environment is worrying since their high concentrations have a significant detrimental impact on aquatic life. The widespread distribution extends from polar regions to equatorial regions, and many remote regions in between, as seen in Table 14.2 [128, 129]. Furthermore, recent studies have identified microplastic distribution patterns that were predicted by oceanic circulation modeling, which confirms the global nature of microplastic pollution [130–132].

14.3.2 Chemical Properties

Plastics are composed of long chains of polymeric molecules produced from a variety of organic and inorganic raw materials such as carbon, chloride, hydrogen, oxygen, and silicon. The source of these raw materials is predominantly derived from fossil fuels, and by varying raw material compositions and processing conditions a wide variety of polymers can be manufactured. The most widely used plastics and their typical uses are shown in Table 14.1. Combined, these plastics equate to around 90% of the current total global production of plastic materials [147]. However, plastics are rarely used in their pure form and are usually combined with a variety of additives to enhance

TABLE 14.2
A selection of microplastic locations and concentrations in various global regions

Location	Sample Type	Concentration	Ref.
Arctic polar waters	Sea	0–1.31 particles m^{-3}	[26, 99]
Belgium, North Sea coast	Sediment, high tide line	125.7 ± 28.9 kg^{-1}	[52]
Chinese, Bohai Sea	Sea	63–201 items kg^{-1}	[133]
China, Yangtze Estuary and East China Sea	Sea	0–144 particles m^{-3}	[134]
Germany, Rugen	Sediment, high tide line	107.3 ± 73.8 kg^{-1}	[135]
India, ship-breaking yard	Sediment	89 mg kg^{-1} *	[136]
North-Western Atlantic	Sea	2500 particles km^{-2}	[107]
New Zealand, Greater Canterbury Region,	High tide line	21.2 ± 16.5 kg^{-1}	[137]
Poland, Gulf of Gdansk,	Middle part of beach	25–53 kg^{-1}	[138]
Portuguese coast	Beaches	332–362 items m^{-2}	[139]
Russia, Kaliningrad region	High tide line	1.3–36.3 kg^{-1}	[140]
Singapore, Strait of Singapore/Strait of Johor	0.5 m above high tide line	0–16 kg^{-1}	[141]
Sweden, harbor	Sediment	50 particles l^{-1}	[142]
United Kingdom, estuary	Sediment	31 particles kg^{-1}	[51, 143, 144]
United States of America, Coastal waters, California	Sea	3 particles m^3	[3, 145]
United States of America, Florida, sub-tidal	Sediment	214 particles l^{-1}	[146]

* Including glass wool.

their material properties and performance [148]. Currently, there are serious concerns regarding the extent to which plastics, including their additives, adversely interact with marine organisms [149]. Importantly, many of these additives can extend plastic degradation times and prolong their presence in the marine environment. Furthermore, the large surface area-to-volume ratio of microplastics makes them prone to contamination by waterborne pollutants, such as persistent organic pollutants (POPs) and aqueous heavy metals [87, 150].

14.3.2.1 Environmental Degradation

Degradation in the marine environment results in changes to the material properties of microplastic materials. Degradation mechanisms such as photo-oxidative degradation, catalytic degradation, ozone-induced degradation, and thermal degradation result in the loss of tensile strength, decolorization, and changes in the structural integrity of the microplastic matrix [151]. Many plastic materials such as LDPE, HDPE, and PP are buoyant in water. Thus, large quantities of debris composed of these plastics are buoyant enough to float in seawater and be exposed to solar ultraviolet radiation. In addition to photo-oxidation, the debris also experiences physical abrasion mechanisms produced by wave action, turbulence, and sand grinding, which ultimately leads to photo-degradation, increased brittleness, and fragmentation [152, 153]. However, due to lower levels of ultraviolet radiation exposure and low water temperatures, degradation rates in the marine environment are much lower than those experienced by microplastics in the terrestrial environment [154]. Importantly, other degradation mechanisms are orders of magnitude lower than ultraviolet radiation-induced photo-oxidative degradation. And as a result, degradation rates in the marine environment are much slower. Thus, based on these slow degradation rates, the current levels of plastic pollution will persist for centuries, even if immediate action is taken to prevent future pollution [97].

14.3.2.2 Leached Additives and Adsorption of Pollutants

Generally, plastics have a structure composed of two phases, namely, crystalline and amorphous. The crystalline phase consists of regularly arranged lattices, while the amorphous phase consists of randomly arranged molecules that give it a loose and flexible structure. The crystalline phase is more condensed and displays higher cohesive forces, whereas the amorphous phase is flexible and tends to behave more like a viscous liquid. Because of their two-phase structure, plastics are rarely used in their pure form. Instead, during the manufacturing process, resins and plasticizers like di-butyl phthalates and bisphenol A (BPA) are added to improve the tensile strength, toughness, and rigidity of the plastic material [155]. In addition, other additives are introduced during manufacture to improve the service life of the plastic. For example, chemicals like nonylphenol are added to reduce oxidative damage, and triclosan is added to reduce microbial degradation [69, 80]. However, these property-enhancing and life-extending additives are also of serious concern because these potentially hazardous materials can leach out and adversely affect marine organisms [149, 156]. Studies have shown that di-butyl phthalates can cause a range of adverse effects in aquatic invertebrates and fish [74, 157]. Other studies have shown that bisphenol A is an endocrine-disrupting agent that can cause permanent morphological changes in a variety of organisms [153, 158]. Microplastics have large surface area-to-volume ratios, which also makes them more susceptible to contamination by a variety of waterborne pollutants. Typically, the highest concentrations of these pollutants are also found in the sea-surface micro-layer, where microplastics with densities less than seawater are present in large numbers. Contaminants such as manufactured polychlorinated biphenyls (PCBs), persistent organic pollutants (POPs) like DDT and DDE, and polycyclic aromatic hydrocarbons (PAHs) are stable and lipophilic chemicals that will readily adhere and accumulate on the hydrophobic surfaces of microplastics [159, 160]. In addition, recent studies have also shown that aqueous metal ions can also be adsorbed onto the surfaces of microplastic materials and accumulate [150, 161, 162]. Thus, microplastics become reservoirs for toxic chemicals and then transport their toxic payload across oceans, polluting otherwise pristine marine environments [100]. Recent studies have also confirmed that microplastics serve as vectors for the delivery of pollutants to marine-based organisms [45, 163]. For instance, Tanaka et al. reported the accumulation of plastic-derived chemicals in the tissues of seabirds feeding on pelagic fish [164]. Other studies have revealed the presence of microplastics in the gut contents of fish and in the gills and digestive organs of mussels [165, 166]. This indicates that the ingestion of microplastics loaded with pollutants is an exposure pathway to a wide variety of marine organisms.

14.4 BIOAVAILABILITY AND BEHAVIORAL PROPERTIES OF MICROPLASTICS

14.4.1 Factors Influencing the Bioavailability of Microplastics

14.4.1.1 Size and Shape

Both microplastic size and shape have been identified as influential parameters when studying ingestion and egestion rates for exposed marine organisms [167]. The small size of microplastics makes them available for a wide variety of organisms that can mistake them for food [70]. Many organisms can exert size and shape preference toward ingested particles [83]. For instance, studies of mussels and lugworms have shown that mussels can ingest microplastics ranging in size from 10 to 30 μm, and lugworms can ingest microplastics ranging in size from 30 to 90 μm, thus indicating that each organism has a preference for a particular size range. However, the reason for this different size preference between mussels and lugworms has not been explained [168]. Similarly, a study found large numbers of 3.0 μm-sized microplastics in the hemolymph of mussels compared to low numbers of 9.6 μm-sized microplastics, indicating the influence of size on ingestion capacity and body residence times for microplastics [169]. Another physical characteristic of microplastics is their shape, which can be non-spherical, fragmented, or as fibers [170]. Importantly, studies have shown that the ingestion of fibers hinders or blocks the passage of food through the intestinal tract of many organisms [171, 172], while non-fiber shapes have been found to increase body residence times in several organisms [173, 174]. However, most laboratory studies have only examined homogeneously shaped and sized microplastics that are typically less than 100 μm. Thus, there is a need for studies to investigate the influence of sizes and shapes on microplastic ingestion.

14.4.1.2 Density

The presence of microplastics in both the sediment and water columns makes it inevitable that they will be ingested by many marine organisms like zooplankton, invertebrates, and fish. The density of the microplastic determines its depth within the water column and its availability to organisms specific to that depth. Microplastics with densities less than seawater, such as polypropylene (0.90 kg m^{-3}) and polyethylene (0.92 to 0.94 kg m^{-3}), will float at the sea surface and encounter suspension-feeding planktivores present in the upper regions of the water column [175]. Higher-density microplastics will sink and accumulate in marine sediments, where they will encounter benthic scavengers and predatory organisms. Furthermore, despite many microplastics being less dense than seawater, colonizing microorganisms and invertebrates create a surface biofilm that results in significant increases in density [176]. Thus, the increasing amount of biomass results in bio-fouled microplastics sinking. In the case of polyethylene shopping bags, neutral buoyancy is achieved after three weeks of biomass growth in the marine environment [116]. Low-density polyethylene (LDPE) microplastics become negatively buoyant after 12 weeks of exposure to the marine environment [176]. Studies indicate that the rate of bio-fouling depends on factors like surface energy and hardness of the plastic material, sunlight levels, seawater conditions, and biomass growth [89, 119]. In addition, subsequent de-fouling of the biofilm by foraging organisms at specific depths in the water column decreases the density, and the microplastic can once again return to the sea surface [67]. This cyclic motion exposes microplastics to a variety of organisms occupying different water column depths at various times. Alternatively, bio-fouled microplastics, as well as high-density microplastics, would continue to sink to the benthos. At the benthos level, the microplastics are once again exposed to a variety of different organisms capable of producing a biofilm, or the microplastics could be ingested by organisms in the surrounding sediments [177].

14.4.1.3 Color Preference

Many species of marine organisms are likely to ingest microplastics due to their resemblance to food. Thus, microplastic colors resembling naturally occurring colors of food types are more likely to be consumed. Results of several studies suggest that many lower-trophic-level organisms, like echinoderms, invertebrates, and zooplankton are susceptible to microplastic ingestion [29, 178, 179], with many planktonic organisms commonly mistaking lightly-colored microplastics for prey [180]. In addition, microplastics may also be consumed as part of a mixture containing phytoplankton and other organic matter [178]. Furthermore, many pelagic invertebrate species are visual raptorial predators and consume microplastics due to their resemblance to food [181]. A study by Carpenter et al. found that fish from the Niantic Bay area of New England only consumed spherical white polystyrene particles when presented with both clear and white particles [102]. Some fish species in the North Pacific Ocean that normally feed on zooplankton may also consume white-, tan-, and yellow-colored microplastics that resemble their food source [180]. Color preferences have also been reported for Carangidae fish located off the coast of Easter Island. These fish consume blue polyethylene micro-particles that resemble blue copepods which are their natural prey [182]. Several studies have also revealed that seabirds selectively ingest lightly-colored microplastics, especially with a brown tint, as opposed to those that are clear or dark in color [183, 184].

14.4.1.4 Abundance in the Marine Environment

Increasing amounts of microplastics in the marine environment amplifies their bioavailability and as a result their uptake by organisms increases. A recent study by Eriksen et al. estimated that there were over five trillion pieces of plastic debris or around 250,000 tons currently polluting the marine environment globally [130], while the Los Angeles River, which flows into the Pacific Ocean, delivers around one billion microplastic fragments daily [185]. With large numbers of microplastics entering the marine environment globally, bioavailability and uptake will be more likely in regions of higher macro-plastic waste abundance. With the progressive fragmentation of macro-plastic debris, the number of microplastics will further increase and escalate the number of microplastics available for uptake by a larger range of organisms. Importantly, ingestion not only depends on the physical and chemical properties of the microplastics, but it also depends on an organism's preferences, anatomy, physiology, and ingestion capacities for certain microplastic sizes, shapes, colors, and surface characteristics [186–188]. Microplastic ingestion by fish species can be significant. For instance, a recent study found that around 10% of marine fishes sampled from the Gulf of Mexico had microplastics in their intestinal tract [189]. Table 14.2 presents some locations and microplastic concentrations in sediments and water columns in various global regions. Some locations have higher levels of microplastics, thus influencing the bioavailability and opportunity for organisms to encounter and ingest microplastics.

14.4.2 Interactions with Marine Organisms

Microplastic pollution is widespread in the global marine environment and has the potential to cause harm to a diverse range of susceptible organisms [190]. The size range of microplastics is comparable to many plankton species which are routinely eaten by many marine invertebrates. Since both microplastics and plankton are present in sediments and the water column, it is expected they will be ingested by a wide variety of predators [191, 192]. For instance, a recent study by Setala et al. reported the ingestion of polystyrene microplastics by mesozooplankton, which were in turn consumed by macrozooplankton [193], demonstrating the trophic transfer of microplastics in the planktonic food chain. Although the amount of microplastics consumed by each zooplankton is small, the ingestion of large numbers of zooplankton by predators will lead to bioaccumulation in larger fish and mammal species [191, 194]. Several *in vitro* and *in vivo* laboratory studies have also reported the ingestion of microplastics by invertebrates and fish. Laboratory-based studies by Hoss and Settle in 1990 were the first to detect and report the ingestion of 100–500 μm microplastic pellets by six different fish species during their early life stages [195], while experimental studies of the mussel *Mytilus edulis* found ingested microplastics accumulated within 12 hours, and large concentrations were present in the hemolymph [166]. A similar study also found that microplastics present in the gills and ciliary movements had transferred (translocated) the microplastics to the digestive tract [166]. A recent study by Wegner et al. reported that nanometer-scale polystyrene fragments (30 nm) reduced the filtering activities of *Mytilus edulis*, and subsequent ingestion of the particles caused the excretion of pseudo-feces [186]. Similarly, ingested microplastics have also been found in the stomach, hepato-pancreas, and gills of crabs *via* the trophic transfer from *Mytilus edulis* [48]. In the North Pacific, around 9% of mesopelagic fish were found to have ingested a variety of microplastic fragments [30]. A study by Lusher et al. found the presence of microplastics in the gastrointestinal tracts of pelagic and demersal fish species in coastal waters around the United Kingdom [143]. The results of similar studies have established the presence of microplastics in the gastrointestinal tracts of fish and mammal species globally and clearly identify microplastic pollution as a widespread problem [33, 168, 196, 197].

14.5 FUTURE PERSPECTIVES FOR FURTHER RESEARCH

The production of durable, lightweight, and inexpensive plastic materials for a wide range of commodities will continue to increase in the future. These versatile materials have many advantageous properties, but microplastic pollution has become a major global environmental problem. The problem arises from the presence of microplastics in marine ecosystems and their subsequent detrimental impact on marine species and ultimately human health [120, 198]. The problem is exacerbated by

TABLE 14.3
Future key research areas for investigating the effects of microplastics and absorbed chemical pollutants in the marine environment

Area	Research field	Ref.
1	***Spatial abundance and accumulation*** Given the occurrence of microplastics in the marine environment, not enough is known about their temporal and spatial concentrations. The literature reports varying trends in accumulating microplastic debris. The variation results from three factors: geographic location, differences in the sampling methods used by different researchers, and anthropogenic activities occurring on beaches. Therefore, future research should address developing internationally recognized standardized methods for collecting temporal and spatial concentration data. This would greatly assist computational modeling of the data in the global marine environment.	[39, 68, 200–202]
2	***Physiochemical properties*** Because microplastics in the marine environment have different sizes, shapes, colors, and compositions. The influence of these properties on various marine organisms needs to be investigated.	[83, 188, 203]
3	***Biodegradation*** A few studies have reported using some microbial strains to degrade plastics. But further studies are needed to fully elucidate the mechanisms involved, the subsequent results would assist in the manufacture of new biodegradable plastic materials.	[151, 92, 149]
4	***Bioavailability*** Further studies are needed to fully investigate the influence of microplastic constituents, their bioavailability, and potential toxicological impacts on various organisms at different trophic levels.	[190, 191, 193, 194]
5	***Ingestion of microplastics*** Laboratory studies are needed to quantify uptake and depuration rates for specific types of microplastics and their impact on various organisms at different trophic levels. In addition, to determine the destination (translocation) of microplastics that can move through gut epithelium in other tissues for different marine organisms at different trophic levels.	[29, 178, 187, 189, 164]
6	***Chemical pollutants and microplastics*** Further studies are needed to investigate adsorption or desorption mechanisms for organic pollutants like POPs and metals in marine environmental conditions. Current data is based on laboratory studies for commonly manufactured plastics and do not take into account the environment or types of marine organisms involved. For example, quantifying uptake and depuration rates of organic pollutants or metals for marine organisms at different trophic levels and possible bioaccumulation of organic pollutants and metals.	[204–208, 159, 143]
7	***Trophic transfers*** Currently, there have been some reports of aquatic contaminants like POPs in the food chain. But the ingestion of microplastics contaminated by chemical pollutants and the nature of microplastics as vectors and trophic transfers through the food chain need further study to determine the scale and level of impact within marine ecosystems.	[169, 173, 209, 210]
8	***Bioaccumulation*** Further laboratory studies are needed to investigate the bioaccumulation and potential bio-magnification of microplastics and absorbed pollutants in marine organisms at higher trophic levels and ultimately in humans.	[211–214]
9	***Nanometer-scale plastics (nanoplastics)*** At present, there is limited information regarding the impact of nanoplastics and their size-specific properties in the marine environment. Thus, there is a need for laboratory studies to determine their impact on marine ecosystems.	[215–217]

ineffective policy enforcement, poor wastewater treatment, and the impractical nature of removing microplastic pollution already present in the marine environment [199].

Current information regarding the impact of microplastics in marine ecosystems and their subsequent uptake and trophic transfer within the food chain is limited. Thus, there are several areas for future research that need to be investigated with regard to the growing global problem of microplastic pollution. Table 14.3 presents a list of future research tasks that will deliver important information needed to address several aspects associated with the presence of microplastics in marine ecosystems and their detrimental effects on marine organisms.

14.6 CONCLUSION

The production of durable, lightweight, and inexpensive plastic materials has made them indispensable products for the manufacture of a wide range of commodities. Due to their durable properties and poor waste plastic management strategies, microplastic pollution is expected to be further exacerbated in the future. Microplastic pollution is found globally, with the largest concentrations typically present in coastal ecosystems and mid-ocean gyres. Both primary and secondary microplastics are found in marine ecosystems. The chapter has discussed their physical and chemical properties, bioavailability, and behavior toward susceptible marine organisms. The physiochemical properties ensure their widespread migration and accumulation in marine environments. Their surface chemistry allows for the accumulation and concentration of toxic substances, thus making them an ideal vector for delivering toxins to a wide variety of marine organisms, while their chemical behavior results in a slow degradation rate that promotes the gradual release of chemical components into the surrounding environment. Moreover, microplastics provide new habitats capable of transporting marine species from one ecosystem to another ecosystem where they threaten local indigenous species. Factors promoting bioavailability and interactions with marine organisms include their size, shape, density, color, and abundance in the ecosystem. The ingestion of microplastics is the most visible and worrying trend because trophic transfer allows microplastics and their surface contaminants to move up the food chain. The chapter also discussed the results of recent studies that reported the ingestion of microplastics by a wide variety of marine organisms. However, few studies have fully evaluated the effects of toxin transfer by microplastics. Nevertheless, the potential for microplastics to move up through the food chain and transfer toxic chemicals to marine organisms is a serious concern. So there is an urgent need to investigate the impact of increasing global microplastic concentrations and their effect on marine environments. A better understanding of the problems associated with microplastics and even smaller nanoplastics particles, the transfer of toxic chemicals, and the ingestion by marine organisms would help both researchers and policymakers develop new marine pollution management strategies.

REFERENCES

1. Thompson RC, Swan SH, Moore CJ, vom Saal FS. Our plastic age. *Philos. Trans. R. Soc. B Biol. Sci.* 2009; 364: 1973–1976.
2. Shah AA, Hasan F, Hameed A, Ahmed S. Biological degradation of plastics: A comprehensive review. *Biotechnol. Adv.* 2008; 26: 246–265.
3. Doyle MJ, Watson W, Bowlin NM, Sheavly SB. Plastic particles in coastal pelagic ecosystems of the North East Pacific ocean. *Mar. Environ. Res.* 2011; 71: 41–52.
4. Plastics Europe. Plastics – The facts 2016: An analysis of European plastics production, demand and waste data, 2016.
5. Barnes DKA, Galgani F, Thompson RC, Barlaz M. Accumulation and fragmentation of plastic debris in global environments. *Philos. Trans. R. Soc. Lond. B. Biol. Sci.* 2009; 364: 1985–1998.
6. Fendall LS, Sewell MA. Contributing to marine pollution by washing your face: Microplastics in facial cleansers. *Mar. Pollut. Bull.* 2009; 58: 1225–1228.
7. Duis K, Coors A. 2016. Microplastics in the aquatic and terrestrial environment: Sources (with a specific focus on personal care products), fate and effects. *Environ. Sci. Eur.* 2016; 28: 2.

8. Sheavly SB, Register KM. Marine debris & plastics: Environmental concerns, sources, impacts and solutions. *J. Polym. Environ.* 2007; 15: 301–305.
9. OSPAR. *OSPAR Pilot Project on Monitoring Marine Beach Litter: Monitoring of Marine Litter on Beaches in the OSPAR Region.* London: OSPAR Commission, 2007.
10. Gregory MR, Ryan PG. *In Marine Debris Springer Series on Environmental Management* (Eds. Coe JM and Rogers DB) Chapter 6. Springer, New York, 1997, pp. 49–66.
11. Carpenter EJ, Smith JKL. Plastics on the Sargasso Sea surface. *Science.* 1972; 175: 1240–1241.
12. Morris AW, Hamilton EI. Polystyrene spherules in the Bristol Channel. *Mar. Pollut. Bull.* 1974; 5(2): 26–27.
13. Wong CS, Green DR, Cretney WJ. Quantitative tar and plastic waste distributions in the Pacific Ocean. *Nature.* 1974; 247: 30–32.
14. Jambeck JR, Geyer R, Wilcox C, Siegler TR, Perryman M, Andrady A, Narayan R, Law KL. Plastic waste inputs from land into the ocean. *Science.* 2015; 347(6223): 768–771.
15. Vannela R. Are we 'Digging our own grave' under the oceans? *Environ. Sci. Technol.* 2012; 46: 7932–7933.
16. Lebreton LCM, van der Zwet J, Damsteeg JW, Slat B, Andrady A, Reisser J. River plastic emissions to the world's oceans. *Nat. Commun.* 2017; 8: 15611.
17. Ivar do Sul JA, Spengler A, Costa MF. Here, there and everywhere: Small plastic fragments and pellets on beaches of Fernando de Noronha (Equatorial Western Atlantic). *Mar. Pollut. Bull.* 2009; 58(8): 1236–1238.
18. Woodall IC, Sanchez-Vidal A, Canals M, Paterson GL, Coppock R, Sleight V, Calafat A, Rogers AD, Narayanaswamy BE, Thompson RC. The deep sea is a major sink for microplastic debris. *R. Soc. Open Sci.* 2014; 1(4): 140371.
19. Lusher AL, Burke A, O'Connor I, Officer R. Microplastic pollution in the Northeast Atlantic Ocean: Validated and opportunistic sampling. *Mar. Pollut. Bull.* 2014; 88: 325–333.
20. Fischer V, Elsner NO, Brenke N, Schwabe E, Brandt A. Plastic pollution of the Kuril–Kamchatka Trench area (NW pacific). *Deep Sea Res. Part II Top. Stud. Oceanogr.* 2015; 111: 399–405.
21. Isobe A, Uchida K, Tokai T, Iwasaki S. East Asian seas: A hot spot of pelagic microplastics. *Mar. Pollut. Bull.* 2015; 101: 618–223.
22. Andrady AL. Microplastics in the marine environment. *Mar. Pollut. Bull.* 2011; 62: 1596–1605.
23. GESAMP. Sources, fate and effects of microplastics in the marine environment: part two of a global assessment Vol. 93, Rep. Stud. GESAMP, 2016.
24. Isobe A, Uchiyama-Matsumoto K, Uchida K, Tokai T. Microplastics in the Southern Ocean. *Mar. Pollut. Bull.* 2017; 114: 623–626.
25. Gajst T, Bizjak T, Palatinus A, Liubartseva S, Krzan A. Sea surface microplastics in Slovenian part of the Northern Adriatic. *Mar. Pollut. Bull.* 2016; 113: 392–399.
26. Lusher AL, Tirelli V, Connor IO, Officer R. Microplastics in Arctic polar waters: The first reported values of particles in surface and sub-surface samples. *Nature.* 2015; 1–9.
27. Ward JE, Kach DJ. Marine aggregates facilitate ingestion of nanoparticles by suspension-feeding bivalves. *Mar. Environ. Res.* 2009; 68(3): 137–142.
28. Zettler E R, Mincer TJ, Amaral-Zettler LA. Life in the 'plastisphere': Microbial communities on plastic marine debris. *Environ. Sci. Technol.* 2013; 47: 7137–7146.
29. Cole M, Lindeque P, Fileman E, Halsband C, Goodhead R, Moger J, Galloway TS. Microplastic ingestion by zooplankton. *Environ. Sci. Technol.* 2013; 47: 6646–6655.
30. Davison P, Asch RG. Plastic ingestion by mesopelagic fishes in the North Pacific subtropical Gyre. *Mar. Ecol. Prog. Ser.* 2011; 432: 173–180.
31. Rezania S, Park J, Md Din MF, Mat Taib S, Talaiekhozani A, Kumar Yadav K, Kamyab H. Microplastics pollution in different aquatic environments and biota: A review of recent studies. *Mar. Pollut. Bull.* 2018; 133: 191–208.
32. Laist DW. Overview of the biological effects of lost and discarded plastic debris in the marine environment. *Mar. Pollut. Bull.* 1987; 18: 319–326.
33. Fry DM, Fefer SI, Sileo L. Ingestion of plastic debris by Laysan albatross and wedge-tailed shearwaters in the Hawaiian Islands. *Mar. Pollut. Bull.* 1987; 18: 339–343.
34. Ryan PG. Effects of ingested plastic on seabird feeding: Evidence from chickens. *Mar. Pollut. Bull.* 1988; 19: 125–128.
35. Bugoni L, Krause L, Petry MV. Marine debris and human impacts on sea turtles in Southern Brazil. *Mar. Pollut. Bull.* 2001; 42: 1330–1334.

36. Wright S, Kelly FJ. Plastic and human health: A micro issue? *Environ. Sci. Technol.* 2017; 51(21): 6634–6647.
37. Anbumani S, Kakkar P. Ecotoxicological effects of microplastics on biota: A review. *Environ. Sci. Pollut. Res.* 2018; 25(15): 14373–14396.
38. Chae DH, Kim IS, Kim SK, Song YK, Shim WJ. Abundance and distribution characteristics of microplastics in surface seawaters of the Incheon/Kyeonggi coastal region. *Arch. Environ. Contam. Toxicol.* 2015; 69: 269–278.
39. Van Cauwenberghe L, Devriese L, Galgani F, Robbens J, Janssen CR. Microplastics in sediments: A review of techniques, occurrence and effects. *Mar. Environ. Res.* 2015; 111: 5–17.
40. Reisser J, Shaw J, Wilcox C, Hardesty BD, Proietti M, Thums M, Pattiaratchi C. Marine plastic pollution in waters around Australia: Characteristics, concentrations, and pathways. *PLoS One*. 2013; 8(11): e80466.
41. Taylor ML, Gwinnett C, Robinson LF, Woodall LC. Plastic microfibre ingestion by deep-sea organisms. *Sci. Rep.* 2016; 6: 33997.
42. Suaria G, Avio CG, Mineo A, Lattin GL, Magaldi MG, Belmonte G, Moore CJ, Regoli F, Aliani S. The Mediterranean plastic soup: Synthetic polymers in Mediterranean surface waters. *Sci. Rep.* 2016; 6: 37551.
43. Rochman CM, Tahir A, Williams SL, Baxa DV, Lam R, Miller JT, Teh FC, Werorilangi S, Teh SJ. Anthrophogenic debris in seafood: Plastic debris and fibers from textiles in fish and bivalves sold for human consumption. *Sci. Rep.* 2015; 5: 1434.
44. Frias JPGL, Sobral P, Ferreira AM. Organic pollutants in microplastics from two beaches of the Portuguese coast. *Mar. Pollut. Bull.* 2010; 60: 1988–1992.
45. Browne MA, Niven SJ, Galloway TS, Rowland SJ, Thompson RC. Microplastic moves pollutants and additives to worms, reducing functions linked to health and biodiversity. *Curr. Biol.* 2013; 23: 2388–2392.
46. Avio CG, Gorbi S, Milan M, Benedetti M, Fattorini D, d'Errico G, Pauletto M, Bargelloni L, Regoli F. Pollutants bioavailability and toxicological risk from microplastics to marine mussels. *Environ. Pollut.* 2015; 198: 211–222.
47. Rochman CM, Hoh E, Kurobe T, Teh SJ. Ingested plastic transfers hazardous chemicals to fish and induces hepatic stress. *Sci. Rep.* 2013; 3: 1–7.
48. Farrell P, Nelson K. Trophic level transfer of microplastic: *Mytilus edulis* (L.) to *Carcinus maenas* (L.). *Environ. Pollut.* 2013; 177: 1–3.
49. Besseling E, Quik JTK, Sun M, Koelmans AA. Fate of nano- and microplastic in freshwater systems: A modeling study. *Environ. Pollut.* 2017; 220: 540–548.
50. Bouwmeester H, Hollman PC, Peters RJB. Potential health impact of environmentally released micro and nano plastics in the human food production chain: Experiences from nanotoxicolgy. *Environ. Sci. Technol.* 2015; 49(15): 8932–8947.
51. Thompson RC, Olsen Y, Mitchell RP, Davis A, Rowland SJ, John AW, McGonigle D, Russell AE. Lost at sea: Where is all the plastic? *Science*. 2004; 304: 838.
52. Claessens M, De Meester S, Van Landuyt L, De Clerck K. Janssen CR. Occurrence and distribution of microplastics in marine sediments along the Belgian coast. *Mar. Pollut. Bull.* 2011; 62: 2199–2204.
53. Ribic CA, Sheavly SB, Rugg DJ, Erdmann ES. Trends and drivers of marine debris on the Atlantic coast of the United States 1997–2007. *Mar. Pollut. Bull.* 2010; 60: 1231–1242.
54. Watson R, Revenga C, Kura Y. Fishing gear associated with global marine catches I. Database development. *Fish. Res.* 2006; 79(1–2): 97–102.
55. Timmers MA, Kistner CA, Donohue MJ. *Marine Debris of the North Western Hawaiian Islands: Ghost Net Identification.* Hawaii Sea Grant Publication, 2005.
56. Klust G. *Netting Materials for Fishing Gear*, Second Ed. Fraham, Surrey: Fishing News Books Ltd., UK, 1982.
57. Hinojosa I, Thiel M. 2009. Floating marine debris in fjords, gulfs and channels of southern Chile. *Mar. Pollut. Bull.* 2009; 58: 341–350.
58. Astudillo JC, Bravo M, Dumont CP, Thiel M. Detached aquaculture buoys in the SE Pacific: Potential dispersal vehicles for associated organisms. *Aquat. Biol.* 2009; 5: 219–223.
59. Comnea-Stancu IR, Wieland K, Ramer G, Schwaighofer A, Lendl B. On the identification of rayon/viscose as a major fraction of microplastics in the marine environment: Discrimination between natural and manmade cellulosic fibers using Frourier Transform Infrared Spectroscopy. *Appl. Spectrosc.* 2017; 71(5): 939–950.

60. Gregory MR, Andrady AL. Plastics in the marine environment. In: *Plastics and the Environment* (Ed. Andrady AL). John Wiley and Sons, 2003.
61. Pospisil J, Nespurek S. Highlights in chemistry and physics of polymer stabilization. *Macromol. Symp.* 1997; 115: 143–163.
62. Andrady AL. Assessment of environmental biodegradation of synthetic polymers: A review. *J. Macromol. Sci.* 1994; 34: 25–75.
63. Tokiwa Y, Calabia BP, Ugwu CU, Aiba S. Biodegradability of plastics. *Int. J. Mol. Sci.* 2009; 10: 3722–3742.
64. Ivar do Sul JA, Costa MF, Barletta M, Cysneiros FJA. Pelagic microplastics around an archipelago of the equatorial Atlantic. *Mar. Pollut. Bull.* 2013; 75: 305–309.
65. Auta HS, Emenike CU, Fauziah SH. Distribution and importance of microplastics in the marine environment: A review of the sources, fate, effects, and potential solutions. *Environ. Int.* 2017; 102: 165–176.
66. Silva AB, Bastos AS, Justino CIL, da Costa JP, Duarte AC, Rocha-Santos TAP. Microplastics in the environment: Challenges in analytical chemistry: A review. *Anal. Chim. Acta.* 2018; 1017: 1–19.
67. Kang JH, Kwon OY, Lee KW, Song YK, Shim WJ. Marine neustonic microplastics around the south eastern coast of Korea. *Mar. Pollut. Bull.* 2015; 96: 304–312.
68. Ryan PG, Moore CJ, van Franeker JA, Moloney CL. Monitoring the abundance of plastic debris in the marine environment. *Philos. Trans. R. Soc. B Biol. Sci.* 2009; 364: 1999–2012.
69. Thompson RC, Moore CJ, vom Saal FS, Swan SH. Plastics, the environment and human health: Current consensus and future trends. *Philos. Trans. R. Soc. B Biol. Sci.* 2009; 364: 2153–2166.
70. Hall NM, Berry KLE, Rintoul L, Hoogenboom MO. Microplastic ingestion by scleractinian corals. *Mar. Biol.* 2015; 162: 725–732.
71. Green DS. 2016. Effects of microplastics on European flat oysters, Ostrea edulis and their associated benthic communities. *Environ. Pollut.* 2016; 216: 95–103.
72. Zitko V, Hanlon M. 1991. Another source of pollution by plastics: Skin cleansers with plastic scrubbers. *Mar. Pollut. Bull.* 1991; 22: 41–42.
73. Betts K. 2008. Why small plastic particles may pose a big problem in the oceans. *Environ. Sci. Technol.* 2008; 42: 8995.
74. Cole M, Lindeque P, Halsband C, Galloway TS. Microplastics as contaminants in the marine environment: A review. *Mar. Pollut. Bull.* 2011; 62: 2588–2597.
75. Eerkes-Medrano D, Thompson RC, Aldridge DC. Microplastics in freshwater systems: A review of the emerging threats, identification of knowledge gaps and prioritization of research needs. *Water Res.* 2015; 75: 63–82.
76. Napper IE, Thompson RC. Characterisation, quantity and sorptive properties of microplastics. *Mar. Pollut. Bull.* 2015; 99: 178–185.
77. Gregory MR. Plastic "scrubbers" in hand cleansers: A further (and minor) source for marine pollution identified. *Mar. Pollut. Bull.* 1996; 32: 867–871.
78. Castaneda RA, Avlijas S, Simard MA, Ricciardi A. Microplastic pollution in St. Lawrence River sediments. *Can. J. Fish. Aquat. Sci.* 2014; 70: 1767–1771.
79. Derraik JGB. 2002. The pollution of the marine environment by plastic debris: A review. *Mar. Pollut. Bull.* 2002; 44: 842–852.
80. Browne MA, Galloway T, Thompson R. Microplastic – An emerging contaminant of potential concern? *Integr. Environ. Assess. Manag.* 2007; 3: 559–561.
81. Yamashita R, Tanimura A. Floating plastic in the Kuroshio current area, western North Pacific Ocean. *Mar. Pollut. Bull.* 2007; 54: 485–488.
82. Mailhot B, Morlat S, Gardette JL. Photo-oxidation of blends of polystyrene and poly (vinyl methyl ether): FTIR and AFM studies. *Polymer.* 2000; 41: 1981–1988.
83. Moore CJ. Synthetic polymers in the marine environment: A rapidly increasing, long-term threat. *Environ. Res.* 2008; 108: 131–139.
84. GESAMP 2014. *Microplastics in the Ocean - A Global Assessment.* Paris, France: GESAMP-IOC IMO/FAO/IOC/WMO/UNIDO/IAEA/UN/UNEP. Joint Group of Experts on the Scientific Aspects of Marine Environmental Protection, 2014.
85. Wang J, Tan Z, Peng J, Qiu Q, Li M. The behaviours of microplastics in the marine environment. *Mar. Environ. Res.* 2016; 133: 7–17.
86. Goldberg ED. Plasticizing the seafloor: An overview. *Environ. Technol.* 1997; 18: 195–202.
87. Rios LM, Moore C, Jones PR. Persistent organic pollutants carried by synthetic polymers in the ocean environment. *Mar. Pollut. Bull.* 2007; 54: 1230–1237.

88. Costerton JW, Cheng KJ. Bacterial biofilms in nature and disease. *Annu. Rev. Microbiol.* 1987; 41: 35–464.
89. Muthukumar T, Aravinthan A, Lakshmi K, Venkatesan R, Vedaprakash L, Doble M. Fouling and stability of polymers and composites in marine environment. *Int. Biodeter. Biodegrad.* 2011; 65(2): 276–284.
90. Railkin AI. *Marine Biofouling: Colonization Processeses and Defenses.* Boca Raton, FL: CRC Press, 2003.
91. Stefatos A, Charalampakis M, Papatheodorou G, Ferentinos G. Marine debris on the seafloor of the Mediterranean Sea: Examples from two enclosed gulfs in Western Greece. *Mar. Pollut. Bull.* 1999; 36: 389–393.
92. Sivan A. New perspectives in plastics biodegradation. *Curr. Opin. Biotechnol.* 2011; 22(3): 422–426.
93. O'Brine T, Thompson RC. Degradation of plastic carrier bags in the marine environment. *Mar. Pollut. Bull.* 2010; 60: 2279–2283.
94. Roy PK, Hakkarainen M, Varma IK, Albertsson AC. Degradable polyethylene: Fantasy or reality. *Environ. Sci. Technol.* 2011; 45: 4217–4227.
95. Shim WJ, Thompson RC. Microplastics in the ocean. *Arch. Environ. Contam. Toxicol.* 2015; 69: 265–268.
96. Law KL, Thompson RC. Microplastics in the seas. *Science.* 2014; 345(6193): 144–145.
97. Dekiff JH, Remy D, Klasmeier J, Fries E. Occurrence and spatial distribution of microplastics in sediments from Norderney. *Environ. Pollut.* 2014; 186: 248–256.
98. International Maritime Organization, IMO. *Plastic Particles in the Ocean May Be as Harmful as Plastic Bags, Report Says.* International Maritime Organization Press Briefing Archives. 27/04/2015, 2015.
99. Liebezeit G, Dubaish F. Microplastics in beaches of the Frisian Islands Spiekeroog and Kachelotplate. *Bull. Environ. Contam. Toxicol.* 2012; 89(1): 213–217.
100. Zarfl C, Matthies M. Are marine plastic particles transport vectors for organic pollutants to the Arctic? *Mar. Pollut. Bull.* 2010; 60: 1810–1814.
101. Lattin GL, Moore CJ, Zellers AF, Moore SL, Weisberg SB. A comparison of neustonic plastic and zooplankton at different depths near the southern California shore. *Mar. Pollut. Bull.* 2004; 49(4): 291–294.
102. Carpenter EJ, Anderson SJ, Harvey GR, Miklas HP, Peck BB. Polystyrene spherules in coastal waters. *Science.* 1972; 178(4062): 749–750.
103. Nel HA, Froneman PW. A quantitative analysis of microplastic pollution along the South-Eastern coastline of South Africa. *Mar. Pollut. Bull.* 2015; 101: 274–275.
104. Kukulka T, Proskurowski G, Moret-Ferguson S, Meyer DW, Law KL. The effect of wind mixing on the vertical distribution of buoyant plastic debris. *Geophys. Res. Lett.* 2012; 39: 7.
105. Magnusson K, Eliasson K, Frane A, Haikonen K, Hulten J, Olshammar M, Stadmark J, Voisin A. Swedish sources and pathways for microplastics to the marine environment: A review of existing data. IVL Swedish Environmental Protection Agency. Report No. C 183, 2016, pp. 65–72.
106. Carvalho D, Baptista Neto JA. Microplastic pollution of the beaches of Guanabara Bay, Southeast Brazil. *Ocean Coast. Manag.* 2016; 128: 10–17.
107. Law KL, Moret-Ferguson S, Maximenko NA, Proskurowski G, Peacock EE, Hafner J, Reddy CM. Plastic accumulation in the North Atlantic subtropical gyre. *Science.* 2010; 329(5996): 1185–1188.
108. Moore, CJ, Lattin GL, Zellers AF. Working our way upstream: A snapshot of land-based contributions of plastic and other trash to coastal waters and beaches of Southern California. In: *Proceedings of the Plastic Debris Rivers to Sea Conference.* Long Beach, CA: Algalita Marine Research Foundation, 2005.
109. Collignon A, Hecq JH, Galgani F, Collard F, Goffart A. Annual variation in neustonic micro- and meso-plastic particles and zooplankton in the Bay of Calvi (Mediterraneane Corsica). *Mar. Pollut. Bull.* 2014; 79(1): 293–298.
110. Sadri SS, Thompson RC. On the quantity and composition of floating plastic debris entering and leaving the Tamar Estuary, Southwest England. *Mar. Pollut. Bull.* 2014; 81(1): 55–60.
111. Desforges JPW, Galbraith M, Dangerfield N, Ross PS. Widespread distribution of microplastics in subsurface seawater in the NE Pacific Ocean. *Mar. Pollut. Bull.* 2014; 79(1): 94–99.
112. Engler RE. The complex interaction between marine debris and toxic chemicals in the ocean. *Environ. Sci. Technol.* 2012; 46(22): 12302–12315.
113. Alomar C, Estarellas F, Deudero S. Microplastics in the Mediterranean Sea: Deposition in coastal shallow sediments, spatial variation and preferential grain size. *Mar. Environ. Res.* 2016; 115: 1–10.
114. Nuelle MT, Dekiff JH, Remy D, Fries E. A new analytical approach for monitoring microplastics in marine sediments. *Environ. Pollut.* 2014; 184: 161–169.

115. Suaria G, Aliani S. Floating debris in the Mediterranean Sea. *Mar. Pollut. Bull.* 2014; 86: 494–504.
116. Lobelle D, Cunliffe M. Early microbial biofilm formation on marine plastic debris. *Mar. Pollut. Bull.* 2011; 62(1): 197–200.
117. Reisser J, Shaw J, Hallegraeff G, Proietti M, Barnes DKA, Thums M, Wilcox C, Hardesty BD, Pattiaratchi C. Millimeter-sized marine plastics: A new pelagic habitat for microorganisms and invertebrates. *PLoS One.* 2014; 9(6): e100289.
118. Browne MA, Crump P, Niven SJ, Teuten EL, Tonkin A, Galloway T, Thompson R. Accumulations of microplastic on shorelines worldwide: Sources and sinks. *Environ. Sci. Technol.* 2011; 45: 9175–9179.
119. Ye S, Andrady AL. Fouling of floating plastic debris under Biscayne Bay exposure conditions. *Mar. Pollut. Bull.* 1991; 22(12): 608–613.
120. Cozar A, Echevarria F, Gonzalez-Gordillo JI, Irigoien X, Ubeda B, Hernandez-Leon S, Palma AT, Navarro S, García-de-Lomas J, Ruiz A, Fernandez-de-Puelles M, Duarte CM. Plastic debris in the open ocean. *Proc. Natl. Acad. Sci.* 2014; 111(28): 10239–10244.
121. Van Cauwenberghe L, Vanreusel A, Mees J, Janssen CR. Microplastic pollution in deep-sea sediments. *Environ. Pollut.* 2013; 182: 495–499.
122. Pham CK, Ramirez-Llodra E, Alt CHS, Amaro T, Bergmann M, Canals M, Company JB, Davies J, Duineveld G, Galgani F, Howell KL. Marine litter distribution and density in European seas, from the shelves to deep basins. *PloS One.* 2014; 9: e95839.
123. Costa MF, Ivar do Sul JA, Santos-Cavalcanti JS, Araujo MCB, Spengler A, Tourinho TS. Small and microplastics on the strandline: Snapshot of a Brazilian beach. *Environ. Monit. Assess.* 2010; 168: 299–304.
124. Avery-Gomm S, O'Hara PD, Kleine L, Bowes V, Wilson LK, Barry KL. Northern fulmars as biological monitors of trends of plastic pollution in the eastern North Pacific. *Mar. Pollut. Bull.* 2012; 64: 1776–1781.
125. World Economic Forum: The Ellen MacArthur Foundation & Company, M. The new plastics economy: Rethinking the future of plastics, 2016.
126. Carson HS, Colbert SL, Kaylor MJ, McDermid KJ. 2011. Small plastic debris changes water movement and heat transfer through beach sediments. *Mar. Pollut. Bull.* 2011; 62: 1708–1713.
127. Hopewell J, Dvorak R, Kosior E. Plastics recycling: Challenges and opportunities. *Phil. Trans. R. Soc. B Biol. Sci.* 2009; 364: 2115–2126.
128. Obbard RW, Sadri S, Wong YQ, Khitun AA, Baker I, Thompson RC. Global warming releases microplastic legacy frozen in Arctic Sea ice. *Earth's Fut.* 2014; 2(6): 315–320.
129. Hidalgo-Ruz V, Thiel M. Distribution and abundance of small plastic debris on beaches in the SE Pacific (Chile): A study supported by a citizen science project. *Mar. Environ. Res.* 2013; 87–88: 12–18.
130. Eriksen M, Lebreton LCM, Carson HS, Thiel M, Moore CJ, Borerro JC, Galgani F, Ryan PG, Reisser J. Plastic pollution in the world's oceans: More than 5 trillion plastic pieces weighing over 250,000 tons afloat at sea. *PLoS One.* 2014; 9(12): e111913.
131. Maximenko N, Hafner J, Niiler P. Pathways of marine debris derived from trajectories of Lagrangian drifters. *Mar. Pollut. Bull.* 2012; 65(1): 51–62.
132. Eriksen M, Maximenko N, Thiel M, Cummins A, Lattin G, Wilson S, Hafner J, Zellers A, Rifman S. Plastic pollution in the South Pacific subtropical gyre. *Mar. Pollut. Bull.* 2013; 68(1): 71–76.
133. Yu X, Peng J, Wang J, Wang K, Bao S. Occurrence of microplastics in the beach sand of the Chinese inner sea: The Bohai Sea. *Environ. Pollut.* 2016; 214: 722–730.
134. Zhao S, Zhu L, Wang T, Li D. Suspended microplastics in the surface water of the Yangtze Estuary System, China: First observations on occurrence, distribution. *Mar. Pollut. Bull.* 2014; 15: 562–568.
135. Hengstmann E, Tamminga M, vom Bruch C, Fischer EK. Microplastic in beach sediments of the Isle of Rugen (Baltic Sea) - Implementing a novel glass elutriation column. *Mar. Pollut. Bull.* 2018; 126: 263–274.
136. Reddy M, Adimurthy S, Ramachandraiah G. Description of the small plastics fragments in marine sediments along the Alang-Sosiya ship-breaking yard, India. *Estuar. Coast. Shelf Sci.* 2006; 68(3–4): 656–660.
137. Clunies-Ross P, Smith G, Gordon K, Gaw S. Synthetic shorelines in New Zealand? Quantification and characterisation of microplastic pollution on Canterbury's coastlines. *N. Z. J. Mar. Freshw. Res.* 2016; 50: 317–325.
138. Graca B, Szewc K, Zakrzewska D, Dołęga A, Szczerbowska-Boruchowska M. Sources and fate of microplastics in marine and beach sediments of the Southern Baltic Sea - A preliminary study. *Environ. Sci. Pollut. Res.* 2017; 24: 7650–7661.

139. Antunes JC, Frias JGL, Micaelo AC, Sobral P. Resin pellets from beaches of the Portuguese coast and adsorbed persistent organic pollutants. *Estuar. Coast. Shelf Sci.* 2013; 130: 62–69.
140. Esiukova E. 2017. Plastic pollution on the Baltic beaches of Kaliningrad region, Russia. *Mar. Pollut. Bull.* 2017; 114: 1072–1080.
141. Ng KL, Obbard JP. Prevalence of microplastics in Singapore's coastal marine environment. *Mar. Pollut. Bull.* 2006; 52: 761–767.
142. Noren F. Small Plastic Particles in Coastal Swedish Waters. N-Research report, commissioned by KIMO Sweden, 2008.
143. Lusher A, McHugh M, Thompson R. Occurrence of microplastics in the gastrointestinal tract of pelagic and demersal fish from the English Channel. *Mar. Pollut. Bull.* 2013; 67: 94–99.
144. Browne MA, Galloway TS, Thompson RC. Spatial patterns of plastic debris along estuarine shorelines. *Environ. Sci. Technol.* 2010; 44: 3404–3409.
145. Moore SL, Gregorio D, Carreon M, Weisberg SB. Composition and distribution of beach debris in orange county, California. *Mar. Pollut. Bull.* 2001; 42(3): 241–245.
146. Graham ER, Thompson JT. Deposit and suspension-feeding sea cucumbers (Echinodermata) ingest plastic fragments. *J. Exp. Mar. Biol. Ecol.* 2009; 368(1): 22–29.
147. Andrady AL, Neal MA. Applications and societal benefits of plastic. *Philos. Trans. R. Soc. B.* 2009; 364: 1977–1984.
148. Teuten EL, Saquing JM, Knappe DRU, Barlaz MA, Jonsson S, Björn A, Rowland SJ, Thompson RC, Galloway TS, Yamashita R, Ochi D. Transport and release of chemicals from plastics to the environment and to wildlife. *Philos. Trans. R. Soc. B.* 2009; 364: 2027–2045.
149. Lithner D, Larsson A, Dave G. Environmental and health hazard ranking and assessment of plastic polymers based on chemical composition. *Sci. Total Environ.* 2011; 409: 3309–3324.
150. Ashton K, Holmes L, Turner A. Association of metals with plastic production pellets in the marine environment. *Mar. Pollut. Bull.* 2010; 60: 2050–2055.
151. Singh B, Sharma N. Mechanistic implications of plastic degradation. *Polym. Degrad. Stabil.* 2008; 93(3): 561–584.
152. Gugumus F. Re-evaluation of the stabilization mechanisms of various light stabilizer classes. *Polym. Degrad. Stabil.* 1993; 39(1): 117–135.
153. Talsness CE, Andrade AJM, Kuriyama SN, Taylor JA, vom Saal FS. Components of plastic: Experimental studies in animals and relevance for human health. *Philos. Trans. R. Soc. B Biol. Sci.* 2009; 364: 2079–2096.
154. Barnes DKA, Milner P. Drifting plastic and its consequences for sessile organism dispersal in the Atlantic Ocean. *Mar. Biol.* 2005; 146: 815–825.
155. Meeker JD, Sathyanarayana S, Swan SH. Phthalates and other additives in plastics: Human exposure and associated health outcomes. *Philos. Trans. R. Soc. Lond. B. Biol. Sci.* 2009; 364: 2097–2113.
156. Lithner D, Damberg J, Dave G, Larsson A. Leachates from plastic consumer productsd screening for toxicity with Daphnia magna. *Chemosphere.* 2009; 74: 1195–1200.
157. Oehlmann JR, Schulte-Oehlmann U, Kloas W, Jagnytsch O, Lutz I, Kusk KO, Wollenberger L, Santos EM, Paull GC, Van Look KJW, Tyler CR. A critical analysis of the biological impacts of plasticizers on wildlife. *Philos. Trans. R. Soc. B Biol. Sci.* 2009; 364: 2047–2062.
158. vom Saal FS, Myers JP. Bisphenol A and risk of metabolic disorders. *J. Am. Med. Assoc.* 2008; 300: 1353–1355.
159. Mato Y, Isobe T, Takada H, Kanehiro H, Ohtake C, Kaminuma T. Plastic resin pellets as a transport medium for toxic chemicals in the marine environment. *Environ. Sci. Technol.* 2001; 35: 318–324.
160. Rochman CM, Lewison RL, Eriksen M, Allen H, Cook AM, Teh SJ. Polybrominated diphenyl ethers (PBDEs) in fish tissue may be an indicator of plastic contamination in marine habitats. *Sci. Total Environ.* 2014; 476–477: 622–633.
161. Holmes LA, Turner A, Thompson RC. Adsorption of trace metals to plastic resin pellets in the marine environment. *Environ. Pollut.* 2012; 160: 42–48.
162. Rochman CM, Hentschel BT, Teh SJ. Long-term sorption of metals is similar among plastic types: Implications for plastic debris in aquatic environments. *PLoS One.* 2014; 9: e85433.
163. Bakir A, Rowland SJ, Thompson RC. Enhanced desorption of persistent organic pollutants from microplastics under simulated physiological conditions. *Environ. Pollut.* 2014; 185: 16–23.
164. Tanaka K, Takada H, Yamashita R, Mizukawa K, Fukuwaka MA, Watanuki Y. Accumulation of plastic-derived chemicals in tissues of seabirds ingesting marine plastics. *Mar. Pollut. Bull.* 2013; 69(1): 219–222.

165. Possatto FE, Barletta M, Costa MF, Ivar do Sul JA, Dantas D. Plastic debris ingestion by marine catfish: An unexpected fisheries impact. *Mar. Pollut. Bull.* 2011; 62(5): 1098–1102.
166. von Moos N, Burkhardt-Holm P, Kohler A. Uptake and effects of microplastics on cells and tissue of the blue mussel *Mytilus edulis* L. after an experimental exposure. *Environ. Sci. Technol.* 2012; 46: 11327–11335.
167. Watts AJR, Lewis C, Goodhead RM, Beckett SJ, Moger J, Tyler CR, Galloway TS. Uptake and retention of microplastics by the shore crab Carcinus maenas. *Environ. Sci. Technol.* 2014; 48: 8823–8830.
168. Van Cauwenberghe L, Claessens M, Vandegehuchte MB, Janssen CR. Microplastics are taken up by mussels (*Mytilus edulis*) and lugworms (*Arenicola marina*) living in natural habitats. *Environ. Pollut.* 2015; 199: 10–17.
169. Browne MA, Dissanayake A, Galloway TS, Lowe DM, Thompson RC. Ingested microscopic plastic translocates to the circulatory system of the mussel, *Mytilus edulis* (L.). *Environ. Sci. Technol.* 2008; 42: 5026–5031.
170. Sutton R, Mason SA, Stanek SK, Willis-Norton E, Wren IF, Box C. Microplastic contamination in the San Francisco Bay, California, USA. *Mar. Pollut. Bull.* 2016; 109: 230–235.
171. Tourinho PS, Ivar do Sul JA, Fillmann G. Is marine debris ingestion still a problem for the coastal marine biota of southern Brazil? *Mar. Pollut. Bull.* 2010; 60(3): 396–401.
172. Rummel CD, Loder MGJ, Fricke NF, Lang T, Griebeler EM, Janke M, Gerdts G. Plastic ingestion by pelagic and demersal fish from the North Sea and Baltic Sea. *Mar. Pollut. Bull.* 2016; 102: 134–141.
173. Murray F, Cowie PR. Plastic contamination in the decapod crustacean *Nephrops norvegicus* (Linnaeus, 1758). *Mar. Pollut. Bull.* 2011; 62: 1207–1217.
174. Au SY, Bruce TF, Bridges WC, Klaine SJ. Responses of *Hyalella azteca* to acute and chronic microplastic exposures. *Environ. Toxicol. Chem.* 2015; 34: 2564–2572.
175. Moret-Ferguson S, Law KL, Proskurowski G, Murphy EK Peacock EE, Reddy CM. The size, mass, and composition of plastic debris in the western North Atlantic Ocean. *Mar. Pollut. Bull.* 2010; 60: 1873–1878.
176. Fazey FMC, Ryan PG. Biofouling on buoyant marine plastics: An experimental study into the effect of size on surface longevity. *Environ. Pollut.* 2016; 210: 354–360.
177. Eich A, Mildenberger T, Laforsch C, Weber M. Biofilm and diatom succession on polyethylene (PE) and biodegradable plastic bags in two marine habitats: Early signs of degradation in the pelagic and benthic zone? *PLoS One.* 2015; 10: e0137201.
178. Long M, Moriceau B, Gallinari M, Lambert C, Huvet A, Raffray J, Soudant P. Interactions between microplastics and phytoplankton aggregates: Impact on their respective fates. *Mar. Chem.* 2015; 175: 39–46.
179. Hart MW. Particle capture and the method of suspension feeding by echinoderm larvae. *Biol. Bull.* 1991; 180(1): 12–27.
180. Shaw DG, Day RH. Colour and form dependent loss of plastic micro-debris from the North Pacific Ocean. *Mar. Pollut. Bull.* 1994; 28(1): 39–43.
181. Greene CH. Planktivore functional groups and patterns of prey selection in pelagic communities. *J. Plankton Res.* 1985; 7(1): 35–40.
182. Ory NC, Sobral P, Ferreira JL, Thiel M. Amberstripe scad *Decapterus muroadsi* (Carangidae) fish ingest blue microplastics resembling their copepod prey along the coast of Rapa Nui (Easter Island) in the South Pacific subtropical gyre. *Sci. Total Environ.* 2017; 586: 430–437.
183. Vlietstra L, Parga JA. Long-term changes in the type, but not amount, of ingested plastic particles in short-tailed shearwaters in the south eastern Bering Sea. *Mar. Pollut. Bull.* 2002; 44: 945–955.
184. van Franeker JA, Blaize C, Danielsen J, Fairclough K, Gollan J, Guse N, Hansen PL, Heubeck M, Jensen JK, Le Guillou G, Olsen B. Monitoring plastic ingestion by the northern fulmar *Fulmarus glacialis* in the North Sea. *Environ. Pollut.* 2011; 159: 2609–2615.
185. Moore CJ, Lattin GL, Zellers AF. Quantity and type of plastic debris flowing from two urban rivers to coastal waters and beaches of Southern California. *J. Integr. Coast Zone Manag.* 2011; 11: 65–73.
186. Wegner A, Besseling E, Foekema EM, Kamermans P, Koelmans AA. Effects of nanopolystyrene on the feeding behavior of the blue mussel (*Mytilus edulis* L.). *Environ. Toxicol. Chem.* 2012; 31: 2490–2497.
187. Sussarellu R, Suquet M, Thomas Y, Lambert C, Fabioux C, Pernet ME, Le Goïc N, Quillien V, Mingant C, Epelboin Y, Corporeau C. Oyster reproduction is affected by exposure to polystyrene microplastics. *Proc. Natl. Acad. Sci. USA.* 2016; 113: 2430–2435.
188. Ogonowski M, Schur C, Jarsen A, Gorokhova E. The effects of natural and anthropogenic microparticles on individual fitness in *Daphnia magna. PLoS ONE.* 2016; 11: 1–20.

189. Phillips MB, Bonner TH. Occurrence and amount of microplastic ingested by fishes in watersheds of the Gulf of Mexico. *Mar. Pollut. Bull.* 2015; 100: 264–269.
190. Laist DW. Impacts of marine debris: Entanglement of marine life in marine debris including a comprehensive list of species with entanglement and ingestion records. In: *Marine Debris-Sources, Impacts and Solutions* (Eds. Coe JM, Rogers DB). New York: Springer Verlag, 1997, pp. 99–139.
191. Desforges JW, Galbraith M, Ross PS. Ingestion of microplastics by zooplankton in the northeast Pacific Ocean. *Arch. Environ. Contam. Toxicol.* 2015; 69: 320–330.
192. Fenchel T. Marine plankton food chains. *Annu. Rev. Ecol. Syst.* 1988; 19–38.
193. Setala O, Fleming-Lehtinen V, Lehtiniemi M. Ingestion and transfer of microplastics in the planktonic food web. *Environ. Pollut.* 2014; 185: 77–83.
194. Jabeen K, Su L, Li J, Yang D, Tong C, Mu J, Shi H. 2017. Microplastics and mesoplastics in fish from coastal and fresh waters of China. *Environ. Pollut.* 2017; 221: 141–149.
195. Hoss DE, Settle L.R. Ingestion of plastic by teleost fishes. In: *Proceedings of the Second International Conference on Marine Debris* (Eds. Shomura RS, Godrey ML), 2–7 April 1989, Honolulu, Hawaii. U.S. Department of Commerce, pp. 693–709. NOAA Tech. Memo. NMFS, NOAA-TM-NMFS-SWFC-154, 1990.
196. Dantas DV, Barletta M, Ferreira da Costa M. The seasonal and spatial patterns of ingestion of polyfilament nylon fragments by estuarine drums (Sciaenidae). *Environ. Sci. Pollut. Res.* 2012; 19(2): 600–606.
197. Rebolledo ELB, Van Franeker JA, Jansen OE, Brasseur SMJM. Plastic ingestion by harbour seals (Phoca vitulina) in The Netherlands. *Mar. Pollut. Bull.* 2013; 67: 200–202.
198. Thevenon F, Carroll C, Sousa J. *Plastic Debris in the Ocean: Characterization of Marine Plastics and Their Environmental Impacts, Situation Analysis Report.* Gland, Switzerland: IUCN, 2014, p. 52.
199. Shaxson L. Structuring policy problems for plastics, the environment and human health: Reflections from the UK. *Philos. Trans. R. Soc. B Biol.* 2009; 364(1526): 2141–2151.
200. Lee J, Lee JS, Jang YC, Hong SY, Shim WJ, Song YK, Hong SH, Jang M, Han GM, Kang D, Hong S. Distribution and size relationships of plastic marine debris on beaches in South Korea. *Arch. Environ. Contam. Toxicol.* 2015; 69: 288–298.
201. Fisner M, Majer AP, Balthazar-Silva D, Gorman D, Turra A. Quantifying microplastic pollution on sandy beaches: The conundrum of large sample variability and spatial heterogeneity. *Environ. Sci. Pollut. Res.* 2017; 24: 13732–13740.
202. Hardesty BD, Lawson T, van der Velde T, Lansdell M, Wilcox C. Estimating quantities and sources of marine debris at a continental scale. *Front. Ecol. Environ.* 2017; 15: 18–25.
203. Fossi MC, Panti C, Gurranti C, Coppola D, Giannetti M, Marsili L, Minutoli R. Are baleen whales exposed to the threat of microplastics? A case study of the Mediterranean fin whale (Balaenoptera physalus). *Mar. Pollut. Bull.* 2012; 64 (11): 2374–2379.
204. Robertson DE. 1968. The adsorption of trace chemicals in sea water on various container surfaces. *Anal. Chim. Acta.* 1968; 42: 533–536.
205. Luis LG, Ferreira P, Fonte E, Oliveira M, Guilhermino L. Does the presence of microplastics influence the acute toxicity of chromium (VI) to early juveniles of the common goby (*Pomatoschistus microps*)? A study with juveniles from two wild estuarine populations. Aquat. Toxicol. 2015; 164: 163–174.
206. Wan Y, Hu J, Yang M, An L, An W, Jin X, Hattori T, Itoh M. Characterization of trophic transfer for polychlorinated dibenzo p dioxins, dibenzofurans, none and mono ortho polychlorinated biphenyls in the marine food web of Bohai Bay, North China. *Environ. Sci. Technol.* 2005; 39(8): 2417–2425.
207. Hu J, Jin F, Wan Y, Yang M, An L, An W, Tao S. Trophodynamic behaviour of 4-nonylphenol and nonylphenol polyethoxylate in a marine food web from Bohai Bay, North China: Comparison to DDTs. *Environ. Sci. Technol.* 2005; 39(13): 4801–4807.
208. Hirai H, Takada H, Ogata Y, Yamashita R, Mizukawa K, Saha M, Kwan C, Moore C, Gray H, Laursen D, Zettler ER. Organic micro-pollutants in marine plastics debris from the open ocean and remote and urban beaches. *Mar. Pollut. Bull.* 2011; 62(8): 1683–1692.
209. Goldstein MC, Rosenberg M, Cheng L. Increased oceanic microplastic debris enhances oviposition in an endemic pelagic insect. *Biol. Lett.* 2012; 8(5): 817–820.
210. Eriksson C, Burton H. Origins and biological accumulation of small plastic particles in fur seals from Macquarie Island. *AMBIO J. Hum. Environ.* 2003; 32(6): 380–384.
211. Besseling E, Wegner A, Foekema EM, van den Heuvel-Greve MJ, Koelmans AA. Effects of microplastic on fitness and PCB bioaccumulation by the lugworm *Arenicola marina* (L.). *Environ. Sci. Technol.* 2012; 47(1): 593–600.
212. Koelmans AA. Modeling the role of microplastics in bioaccumulation of organic chemicals to marine aquatic organisms. A critical review. In: *Marine Anthropogenic Litter*, 2015, pp. 309–324.

213. Paul-Pont I, Lacroix C, Gonzalez FC, Hegaret H, Lambert C, Le Goïc N, Frère L, Cassone AL, Sussarellu R, Fabioux C, Guyomarch J. Exposure of marine mussels Mytilus spp. to polystyrene microplastics: Toxicity and influence on fluoranthene bioaccumulation. *Environ. Pollut.* 2016; 216: 724–737.
214. Ziccardi LM, Edgington A, Hentz K, Kulacki KJ, Kane DS. Microplastics as vectors for bioaccumulation of hydrophobic organic chemicals in the marine environment: A state-of-the-science review. *Environ. Toxicol. Chem.* 2016; 35: 1667–1676.
215. Bhattacharya P, Lin S, Turner JP, Ke PC. Physical adsorption of charged plastic nanoparticles affects algal photosynthetic. *J. Phys. Chem.* 2010; 114: 16556–16561.
216. Cole M, Galloway TS. Ingestion of nanoplastics and microplastics by Pacific oyster larvae. *Environ. Sci. Technol.* 2015; 49: 14625–14632.
217. Besseling E, Wang B, Lurling M, Koelmans AA. Nanoplastic affects growth of *S. obliquus* and reproduction of *D. magna. Environ. Sci. Technol.* 2014; 48: 12336–12343.

15 Quantification of Microplastic Fibers Extracted from South Beach Sediments Located on the South-West Coast of Western Australia

A Preliminary Study

Jennifer Verduin, Julio Sanchez-Nieva, and Derek Fawcett

CONTENTS

15.1 INTRODUCTION

Microplastic contamination of coastal and marine environments is a global problem and is the direct result of human activities. Plastics are composed of long polymer chains made from organic and inorganic precursor materials like carbon, chloride, hydrogen, oxygen, and silicon [1, 2]. Because of their unique and versatile properties, plastics are extensively used in a wide variety of packaging applications, industrial systems, processing facilities, and even cosmetics [3, 4]. The mass production of plastics began in the middle of the twentieth century and since then production levels have steadily increased. For example, between 2004 and 2015, the global production levels of plastics steadily increased from 225 to 322 million tons per year [5]. Unfortunately, because of indiscriminate disposal and poor waste management strategies, plastics have become a major pollutant entering the marine environment [6, 7]. Plastics enter the marine environment in a wide range of sizes and shapes. Once in the environment, they undergo fragmentation and breakdown by microbial

DOI: 10.1201/9781003181422-18

activity [8]. Fragmentation processes include ultraviolet light irradiation, low-temperature breakdown, and mechanical action resulting from wave motion. Regrettably, fragmentation and microbial activity do not completely break down plastic materials. Instead, they are transformed into micrometer-scale plastic particles that are typically less than 5 mm in size [9]. These micrometer-scale plastic particles are commonly referred to as microplastics.

Many of these microplastics are buoyant in water and are easily transported by ocean currents. Because of their buoyancy, microplastics are frequently washed ashore far from their point of entry into the marine environment [10], while non-buoyant microplastics readily settle and accumulate in sediments or are carried by strong underlying currents [11]. Importantly, microplastics are a serious risk to a wide range of organisms in the marine environment. Typical risks include ingestion and accumulation in the gills of fish [12]. A further risk arises from the presence of noxious chemicals, like persistent organic pollutants (POPs), pesticides, and hydrocarbons in the surrounding seawater. These chemicals absorb the exposed surfaces and increase the toxicity of the microplastics [13, 14]. Once absorbed, the microplastics transport their toxic cargo and promote the distribution of toxic chemicals in the marine environment [15]. Importantly, the ingestion of chemically coated microplastics has a detrimental effect on marine organisms. Moreover, once in the food chain, these toxic materials will ultimately end up in marine species that are subsequently consumed by humans, thus presenting a serious risk to human health [16, 17]. In a recent study, researchers reported finding microplastic fibers in human stools [18]. Where the microplastic fibers came from is still unclear. However, participants in the study consumed seafood, food that was wrapped in plastic-based materials and drank water daily from plastic (polyethylene terephthalate, PET) bottles [18]. Current studies in this field suggest that the consumption of microplastics and the presence of harmful surface chemicals could be a serious health risk for humans [19, 20]. From another perspective, coastal environments with attractive beaches have both social and economic value. Plastic pollution reduces the aesthetics and recreational value of beaches, which will in turn have a significant impact on tourism and local fishing activities [21, 22]. Therefore, recent studies have focused on quantifying the number of microplastics present in sandy beach environments in various parts of the world [23, 24].

Recent studies have identified the presence of microplastic fibers in sandy beaches and coastal sediments around the world [25]. Currently, there is no standardized technique for analyzing the quantities of microplastics present in beach sediments. Instead, several methods for both sampling and laboratory analysis are currently being used by several researchers [26, 27]. The solid–liquid separation process, which uses the density differences that exist between different materials has been widely used [9]. In the present work, a new solid–liquid separation process, or elutriation, is used to separate microplastic fibers from beach sands. During the separation process, a continuous upward flow of high-density salt water is used to separate lower-density materials. Microplastic fibers present in the saltwater flow were separated using a filtration unit. The filtrate was subsequently collected and analyzed to determine the quantity and size of microplastic fibers present. The present case study evaluated microplastic fiber accumulation levels present in sediments taken from four small sandy beaches that form South Beach. South Beach, which borders Cockburn Sound, is located on the southwest coast of Western Australia, as seen in Figure 15.1. South Beach is also located nearby population centers, the larger being Fremantle. The popular beaches are frequently used by beach users for a variety of recreational activities.

15.2 MATERIALS AND METHODS

15.2.1 Chemicals and Analytical Considerations

In order to reduce background contamination, glass containers were used where possible. However, during some stages of the experimental procedure, some synthetic materials were used because they could not be replaced by glass, for example, pump fittings and connectors in the elutriation system. To reduce possible contamination sources, all equipment and workplaces were thoroughly cleaned before experimental studies were started. In addition, personnel protection equipment, such

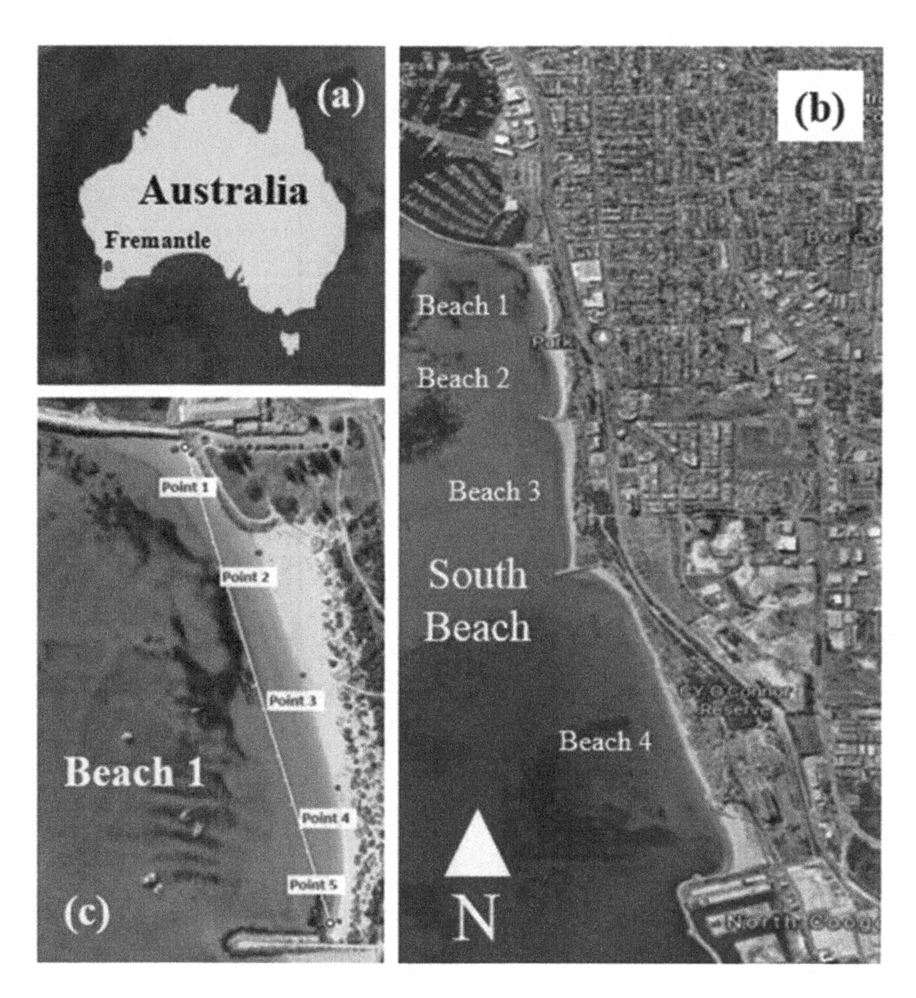

FIGURE 15.1 Microplastic accumulation study (a) orientation map, (b) four small sandy beaches forming South Beach, Cockburn Sound, Western Australia, (c) survey points of Beach 1.

as safety glasses, laboratory coats, and gloves were worn at all times to prevent sample contamination. All materials used in this study were first manually cleaned using water Milli-Q water® and then thoroughly cleaned using analytical grade ethanol (Univar reagent 99.5 %). In both cases, cleaning with each respective solvent was carried out three times before the materials were covered with aluminum foil, as recommended by Nuelle et al. [28]. The hyper-saline solution used in the elutriation system was made from standard cooking salt dissolved in water Milli-Q water®. The hyper-saline solution had a salt saturation of 320 g L^{-1} to achieve a solution density of 1.187 g cm^{-3}. The hyper-saline solution was then filtered using Whatman® paper filters (55 mm/Cat nu-1004-005) to remove any impurities. Furthermore, all chemical treatments, extractions, and filtrations were carried out in a fume hood.

15.2.2 Elutriation System

The elutriation system was designed and in-house fabricated for the purpose of separating microplastic fibers from fluidized beach sediments. Similar elutriation systems have also been developed, tested, and used to extract microplastics from a variety of sediments by floatation and agitation processes [26, 29]. Figure 15.2 schematically presents the experimental setup designed for extracting microplastic fibers from beach sediments. The base of the elutriation system consists of a large two-liter glass container (reservoir) that contains the submersible pump (Aquapro: AP550/Type 01x5,

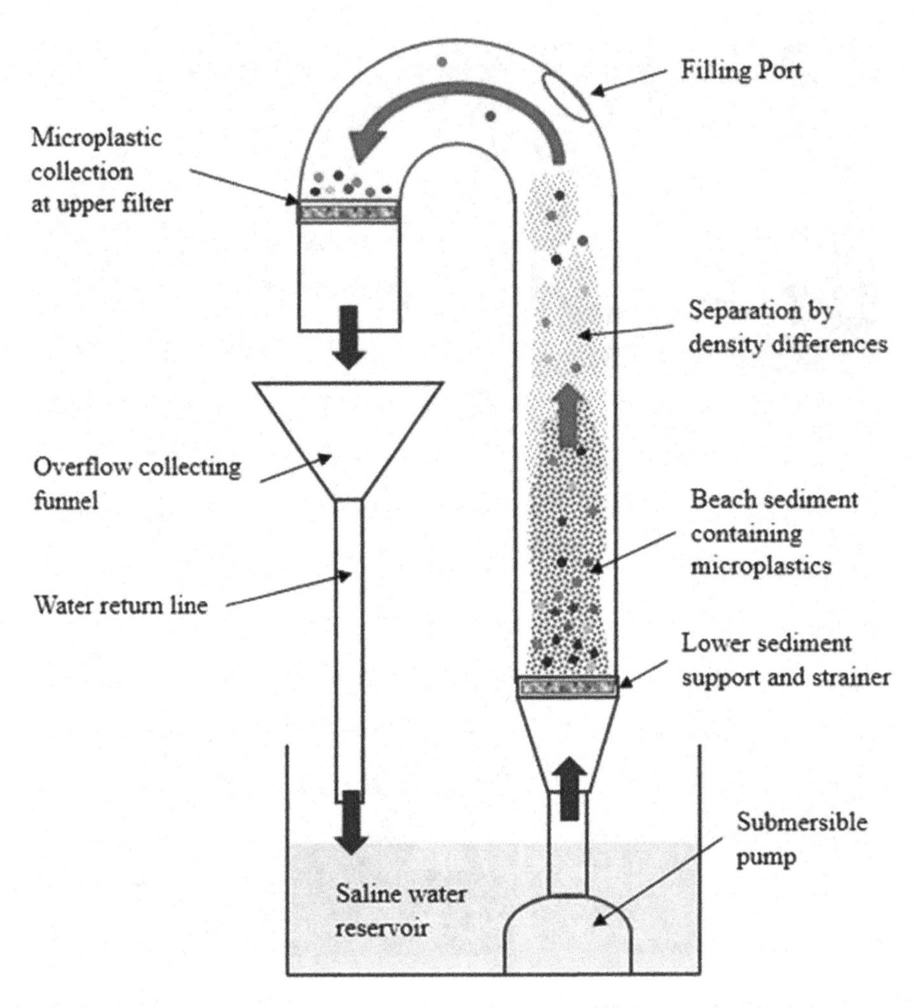

FIGURE 15.2 Schematic of closed loop elutriation system used for extracting microplastic fibers from South Beach sediments.

maximum flow 550 L/h) and the hyper-saline solution. The body of the elutriation system consists of a vertical transparent methacrylate tube 435 mm long with an internal diameter of 40 mm. At the lower end of the tube is a threaded PVC connector (55 mm internal diameter). The connector incorporates the pump-connecting tube and lower wire mesh assembly. The connector contains a vertical funnel, with its small tube end (10 mm diameter) being directly connected to the pump via a 10 mm tube. The larger diameter of the funnel is fitted to the underside of a metal wire mesh sieve (55 mm diameter and 32 μm pore size) designed to support the weight of the sediment. At the upper end of the vertical tube is a fitted and glued PVC U-tube designed to redirect the fluid flow toward the upper filtration connector. At the end of the U-tube is the connector. The connector is 60 mm in length and is easily removed to gain access to the upper wire mesh sieve (55 mm diameter and 32 μm pore size). The upper connector configuration allowed easy access for cleaning and collecting microplastic samples trapped by the sieve. In addition, sediment samples were inserted into the top of the elutriation system through the filling port and hatch assembly. The upward flow of the hyper-saline water from the pump fluidizes the sample sediments at the bottom of the vertical tube. Less dense microplastic fibers are liberated and float upwards. The liberated microplastic fibers are then carried by the water flow around the U-tube to the wire mesh sieve located at the end of the tube. The microplastic fibers are collected in the sieve, while the water continues to flow downwards through the collecting funnel and onto the lower reservoir and recirculated. During the elutriation process, a two-step method was adopted. In the first step, samples were exposed to ten minutes of water flow. During this step, most of the microplastic fibers were recuperated. In the second step, the elutriation system was run for a further one minute so that any remaining microplastic fibers would be flushed through the sieve. The collected microplastic fibers were then transferred from the sieve to filter paper for drying and further investigation. After each sample testing, the elutriation system was thoroughly cleaned using the hyper-saline solution to remove all impurities and prevent cross-contamination between samples.

15.2.3 Reference Microplastic Materials

To evaluate the performance of the elutriation system before examining beach sediments, several types of reference microplastic materials were used to dope clean sand samples. The most common and widespread microplastic pollutants found in the marine environment are polyethylene (PE), polypropylene (PP), polystyrene (PS), polyethylene terephthalate (PET), and polyvinyl chloride (PVC) [30]. In addition, there are also microplastics such as high-density polyethylene (HDPE) and low-density polyethylene (LDPE). HDPE has longer unbranched hydrocarbon chains that make it stronger and more durable than LDPE. Furthermore, both PTFE and nylon have also been detected in aquatic environments [31]. The reference microplastics used in this study included samples of these materials. The reference microplastics were produced from larger plastic materials like blocks, plugs, plastic cups, and bottles. Processing involved using a Dremel® (3000) fitted with a small drum sander attachment (13 mm diameter and 60-grain size abrasive surface) to grind down the bulk plastic materials. After grinding, the respective materials were then mixed with ethanol and underwent sieving to produce microplastics ranging in size from 0.10 to 0.85 mm. However, in the case of nylon, which was sourced from a fishing line, it had to be manually cut to size using scissors. The resultant nylon particle sizes ranged from 0.50 to 1.00 mm. After processing, all microplastic samples were dried at 50 °C for 24 h before being stored in airtight containers.

15.2.4 Elutriation System Evaluation

The effectiveness of the elutriation system was determined by the retrieval of reference microplastics from fine white sand. The sand was initially washed with Milli-Q water® and then subjected to hydrogen peroxide washing for 24 h to remove all organics. The sand was then air-dried in an oven for 48 h at 60 °C. Typically, a total of ten particles of a specific microplastic type were transferred to 100 g of treated sand. The mixture was then thoroughly mixed for 15 minutes. In addition, a further ten particles were selected for comparative microscopy studies. All reference plastics were prepared in this manner, with a total of five samples per microplastic type being prepared. The prepared doped sand samples were then introduced into the elutriation system for processing as discussed above. After processing, the microplastics found in the sieve were inspected, counted, and recorded. Microscopy analysis of samples was performed using an Olympus® SZx12 stereoscopic microscope fitted with an Olympus® DP10 camera and an Olympus® BX51 Compound microscope fitted with an Olympus® DP70 camera.

15.2.5 Beach Sediment Samples

South Beach (see Table 15.2) is made up of four smaller beaches and has a north-south orientation as seen in Figure 15.1. A total of 18 sample points were distributed along the length of South Beach. Both Beaches 1 and 2 are around 400 m in length, with a groyne dividing them. Each beach has five sample points spaced along its length (Beach 1: 1 to 5; Beach 2: 6 to 10). Beach 3 is around 700 m long and has four sample points (11, 12, 13, and 14) and also has groynes at its northerly and southerly ends. The north groyne divides it from Beach 2, and the south groyne divides it from the mostly southerly

Beach 4. Beach 4 is 1500 m long and also has four equally spaced sample points (15, 16, 17, and 18). All four beaches consist of fine quartz white sand containing small shells and fragments of seaweeds and seagrasses. Samples were collected near the breakwater region during calm sea conditions with low wave activity. At each sample point, three one-meter square (1 m^2) plots were selected. Five core (200 mm long and 90 mm in diameter) samples were taken from each square plot. A core sample was taken at each corner and one in the center of the square plot. The core samples from each square were then thoroughly mixed together, and a 100 g sediment sample was taken from the mixture to make the representative sample for this plot. Similarly, a 100 g sediment sample was taken from the other two square plots, thus making the triplicate samples needed for the sample location. Larger organics such as seaweed and seagrass fragments were manually removed from the samples using tweezers. Then the samples were individually introduced into the elutriation system for further processing.

15.3 RESULTS

15.3.1 Recovery Rates of Reference Microplastics and Evaluation of Elutriation System

The recovery rates were experimentally determined and found to range between 84 (± 10) for PTFE and up to almost 100% for other reference microplastics. Microplastics with densities less than the hyper-saline solution density had consistently higher recuperation rates as seen in Table 15.1. Denser microplastics like PET and PTEF had lower recuperation rates of 90 and 84% respectively, thus demonstrating that the elutriation system was very effective in recovering lower-density microplastics.

15.3.2 Quantity of Microplastics Recovered from Beach Sediments

The total amount of beach sediment processed by the elutriation system was 5.4 kilograms. After processing, 234 particles were identified as microplastic fibers by microscopic analysis. This gave an overall mean microplastic density of 43.3 ± 13.3 fibers kg^{-1} for the whole of South Beach. Individual sample points varied between 20 and 80 fibers kg^{-1} as seen in Figure 15.3 (a). For instance, each of the four smaller beaches that made up South Beach displayed varying fiber kg^{-1} numbers. Beach 1 had the highest mean value of 58.00 fibers kg^{-1}, and Beach 3 had the lowest mean value of 35.83

TABLE 15.1
Recovery rates for reference microplastics processed by the elutriation system

Plastic	Name of plastic	Density (kg m^{-3})	Trials	Fibers introduced	Fibers recuperated	% Mean (± std. dev.)
PP	Polypropylene	0.90	5	50	47	94 ± 8
LDPE	Low-density polyethylene	0.92	5	50	48	96 ± 5
HDPE	High-density polyethylene	0.94–0.97	5	50	50	100
PS	Polystyrene	1.05	5	50	47	94 ± 5
Nylon	Polyamide (nylon)	1.02–1.14	5	25	25	100
PVC	Polyvinyl chloride	1.29–1.39	5	50	48	96 ± 8
PET	Polyethylene terephalicate	1.40	5	50	45	90 ± 9
PTFE	Polytetrafluoroethylene	2.20	5	50	42	84 ± 10

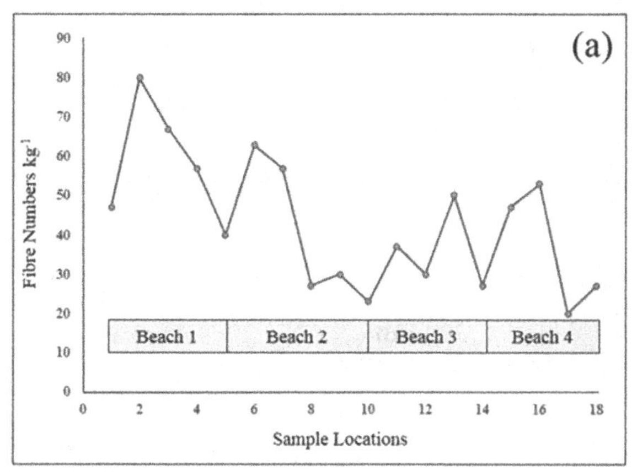

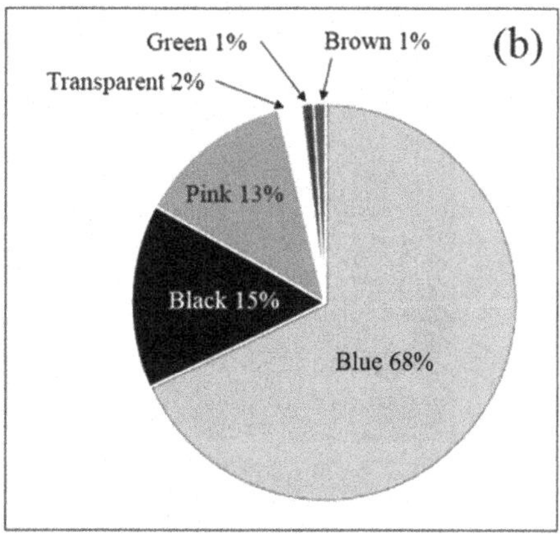

FIGURE 15.3 (a) Microplastic fiber numbers kg^{-1} along the four sandy beaches and (b) the six fiber colors found in beach sediments.

TABLE 15.2
Microplastic fiber size range and fibers kg^{-1} of beach sediments

Site	Location latitude	Longitude	Fiber size range (micron)	Fibers kg^{-1} (range)	Mean beach fibers kg^{-1} (± std. dev.)
Beach 1	32° 04' 20" S	115° 45' 02" E	1039–1753	40–80	58.0 ± 14.2
Beach 2	32° 04' 34" S	115° 45' 05" E	800–1228	23–64	40.0 ± 16.6
Beach 3	32° 04' 58" S	115° 45' 07" E	962–1196	27–50	35.8 ± 8.9
Beach 4	32° 05' 21" S	115° 45' 19" E	948–1824	27–54	36.6 ± 13.7

fibers kg^{-1}. The distribution of fibers kg^{-1} along South Beach is presented in Table 15.2 and reveals larger fiber numbers were found in Beaches 1 and 2.

15.3.3 Optical Microscopy Analysis

Optical microscopy analysis of the samples revealed that the microplastic fibers ranged in size from 800 to 1,824 microns, as seen in Table 15.2. Microplastic fiber size variation was found to be fairly consistent across all four small beaches, and the overall mean value for South Beach was estimated to be 1168 ± 274 microns. Figure 15.4 presents representative microscopic images of fibers found in beach sediments. The fibers have very long lengths compared to their relatively small diameters, which gives them a string-like appearance. Because of their small dimensions, microfibers are normally not seen with the naked eye and not seen in beach sands.

The microscopy analysis not only counted fibers but also identified the various fiber colors present in the sediments. Under the microscope, the fibers are easily distinguishable from sand grains due to their color and characteristic shape. Six colors were clearly identified in the samples, as seen in Figure 15.3 (b). The most dominant color was blue and accounted for around 68% of the total number of fibers counted. There were also significant numbers of black (15%) and pink (13%) fibers. To a lesser extent, there were smaller numbers of transparent (2%), green (1%), and brown (1%) fibers.

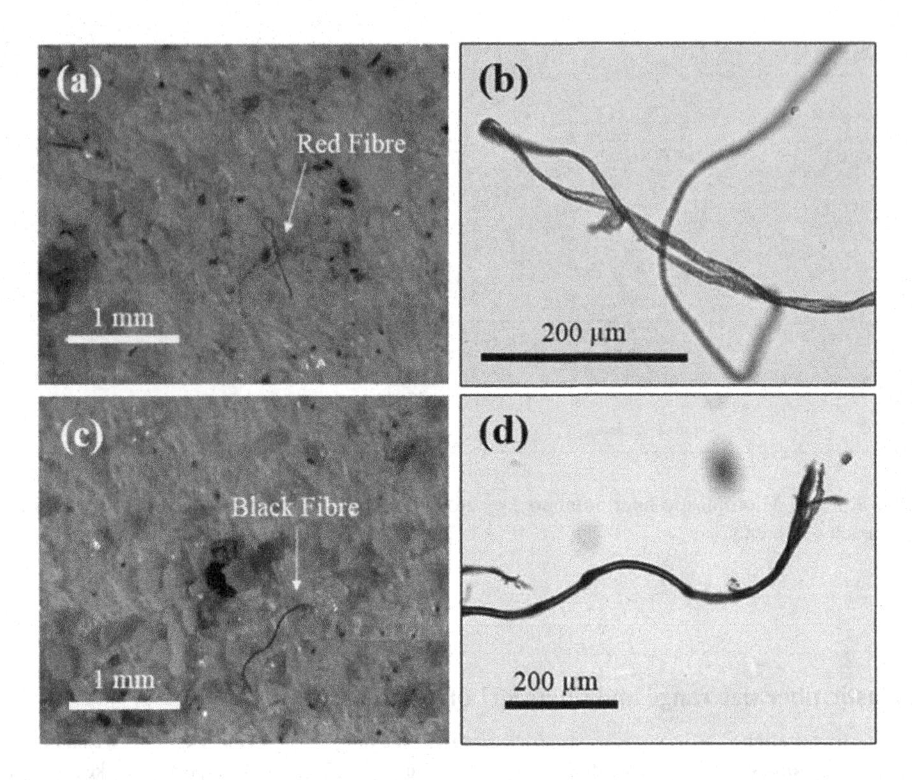

FIGURE 15.4 Representative microscopy images of microplastic fibers seen in beach sediments: (a) red fiber; (b) enlarged view of red fiber; (c) black fiber; and (d) enlarged view of black fiber

15.4 DISCUSSION

In this study, density separation was used to recover microplastic fibers from beach sediments.

The elutriation residues were collected, sorted, and prepared for microscopic analysis. The optical analysis also involved photographing samples to assist in identifying and counting microplastic fibers. Recuperation rates of around 95% were achieved for plastic densities ranging from 0.9 to 1.3 kg m^{-3} and around 84% for densities greater than 1.4. These values are similar to the results reported by other researchers using the elutriation method for collecting microplastic materials. For instance, Kedzierski et al. [26] achieved a recuperation efficiency of around 92.0%, and Claessens et al. [29] achieved an efficiency of around 100% for PVC microplastics. Eight plastic materials were evaluated in this study. The two higher-density plastics, PET and PTFE, had the lowest recuperation efficiencies of 90 ± 9 and 80 ± 10% respectively. The lower-density plastics had much higher recuperation efficiencies that were typically between 94 ± 5 and 100%, thus indicating the lower-density plastics were assisted by buoyancy during density separation.

Studies have shown that the longer that plastic materials are exposed to weathering and disintegration processes, the greater the number of microplastics that are produced [32]. These studies have also revealed that fibers are the most prevalent form of microplastics found in beach sediments [33, 34]. The present study has also confirmed the large number of fibers found in all beach sediments, as seen in Table 15.2. Microplastic fibers are easily transported by wave action and ocean currents. Once deposited on beaches, fibers can also be carried by the wind to accumulate in specific locations [35]. The present study found all four beaches are exposed to the same winds and surface currents that are predominantly from the southwest. Thus, microplastic fibers present in coastal waters are likely to be washed onto all four beaches in similar numbers. However, the study found spatial variability in fiber concentrations along the four small beaches that made up South

Beach. The highest fiber concentration of 58 fibers kg^{-1} found at Beach 1 also coincides with higher levels of recreational activity, as seen in Figure 15.1. Beach 1 is also close to major residential areas and is frequently visited by significant numbers of beach users. Other studies have also reported higher fiber concentrations in beach sediments exposed to high levels of recreational activities [36, 37]. This suggests the most probable cause for differences in fiber concentrations along South Beach is recreational activity. For instance, both Beaches 3 and 4 have lower levels of recreational activity and also have lower fiber concentrations of 35.8 and 36.6 fibers kg^{-1} respectively. The overall mean fiber density for South Beach was found to be 43.3 ± 13.3 fibers kg^{-1}. The fibers ranged in size from 800 to 1,824 microns and had a mean size of 1168 ± 274 microns. These results are similar to those reported in other studies. For instance, Stolte et al. reported fiber concentrations ranging from 17.2 to 73.3 fibers kg^{-1} for beaches on the Isle of Rugen in Germany [38]. Claessens et al. reported a mean fiber concentration of 87.7 ± 25.5 fibers kg^{-1} for beaches along the North Sea coast of Belgium [39]. However, not all reported microplastic fiber concentrations are similar, and there is considerable variation in the literature [38, 40]. The variation in fiber concentration depends on three factors: 1) the geographic location of the study [41, 42]; 2) differences between sampling methods [43, 44]; and 3) levels of anthropogenic activities occurring on beaches [36, 45]. The third factor also had an impact in the present study, with the highest fiber concentration of 58 fibers kg^{-1} occurring for Beach 1, which was also the closest to a major residential area with easy access for beach users. From another perspective, several studies have reported microplastic fiber ingestion by fish, which are then consumed by humans [18, 46, 47]. In addition, microplastics have been detected in commercial table salts, thus highlighting the growing problem of microplastics in the environment [48, 49]. Crucially, the second factor mentioned above highlights the need for a standardized device and procedure for all researchers in this field. A standardized device and procedure would make it easier to carry out routine monitoring programs worldwide and also make it possible to carry out highly repeatable inter-laboratory comparisons. The present study used a novel elutriation system and standard laboratory equipment to recover microplastic fibers from sandy beach sediments. Operation of the elutriation system was straightforward, efficient; and suitable for routine monitoring programs. It could also be used by other researchers to carry out repeatable inter-laboratory comparisons between samples from different global locations

15.5 CONCLUSION

The present study used a purpose-built elutriation system that used hyper-saline solutions to separate microplastic fibers from sandy beach sediments. Evaluation of the elutriation system revealed recuperation rates were typically around 95% for a wide range of microplastic test materials. The high recuperation rates made the elutriation system ideal for recovering microplastic fibers from sandy beach sediments. The present study recovered microplastic fibers from four small sandy beaches that formed South Beach. All four beaches were subjected to the same winds and surface currents that both predominantly come from the southwest. The study found the more northerly beaches had higher concentrations of microplastic fibers in the sediments than the southern beaches. The fiber concentration in the northerly beaches was 58 ± 14.2 fibers kg^{-1} for Beach 1 and 40.0 ± 16.6 fibers kg^{-1} for Beach 2. Both beaches were also subjected to higher levels of recreational activity compared to the southern beaches. The fiber concentrations found in the southern beach were 35.8 ± 8.9 fibers kg^{-1} for Beach 3 and 36.6 ± 13.7 fibers kg^{-1} for Beach 4. These results suggest that recreational activity is an influencing factor in the distribution of microplastic fibers. However, future long-term studies are needed to monitor spatial distributions and concentration levels of microplastic fibers in beach sediments. In addition, future spectroscopic analysis studies are expected to identify individual microplastic materials present in the beach sediments. These longer-term studies will lead to a better understanding of the interaction mechanisms between microplastic fibers and the coastal marine environment.

REFERENCES

1. Shah AA, Hasan F, Hameed A, Ahmed S. Biological degradation of plastics: A comprehensive review. *Biotechnol. Adv.* 2008; 26: 246–265.
2. Doyle MJ, Watson W, Bowlin NM, Sheavly SB. Plastic particles in coastal pelagic ecosystems of the Northeast Pacific ocean. *Mar. Environ. Res.* 2011; 71: 41–52.
3. Derraik JGB. The pollution of the marine environment by plastic debris: A review. *Mar. Pollut. Bull.* 2002; 44: 842–852.
4. Fendall LS, Sewell MA. Contributing to marine pollution by washing your face: Microplastics in facial cleansers. *Mar. Pollut. Bull.* 2009; 58: 1225–1228.
5. Plastics Europe. Plastics: The facts 2016: An analysis of european plastics production, demand and waste data, 2016.
6. Wright SL, Thompson RC, Galloway TS. The physical impacts of microplastics on marine organisms: A review. *Environ. Pollut.* 2013; 178: 483–492.
7. Galgani F, Hanke G, Werner S, De Vrees L. Marine litter within the European marine strategy framework directive. *ICES J. Mar. Sci.* 2013; 70: 1055–1064.
8. Andrady AL. Microplastics in the marine environment. *Mar. Pollut. Bull.* 2011; 62: 1596–1605.
9. Thompson RC, Olsen Y, Mitchell RP, Davis A, Rowland SJ, John AW, McGonigle D, Russell AE. Lost at sea: Where is all the plastic? *Science.* 2004; 304: 838.
10. Moore CJ. Synthetic polymers in the marine environment: A rapidly increasing, long-term threat. *Environ. Res.* 2008; 108: 131–139.
11. Engler RE. The complex interaction between marine debris and toxic chemicals in the ocean. *Environ. Sci. Technol.* 2012; 46: 12302–12315.
12. Gregory MR. Environmental implications of plastic debris in marine settings—Entanglement, ingestion, smothering, hangers-on, hitch-hiking and alien invasions. *Philos. Trans. R. Soc. B.* 2009; 364: 2013–2025.
13. Cole M, Lindeque P, Halsband C, Galloway TS. Microplastics as contaminants in the marine environment: A review. *Mar. Pollut. Bull.* 2011; 62: 2588–2597.
14. Lee YJ, Jang JS, Yang JH. Potential health risks from persistent organic pollutants (POPs) in marine ecosystem. *J. Mar. Biosci. Biotechnol.* 2016; 8: 10–17.
15. Teuten EL, Rowland SJ, Galloway TS, Thompson RC. Potential for plastics to transport hydrophobic contaminants. *Environ. Sci. Technol.* 2007; 41: 7759–7764.
16. Tanaka K, Takada H, Yamashita R, Mizukawa K, Fukuwaka M, Watanuki Y. Accumulation of plastic-derived chemicals in tissues of seabirds ingesting marine plastics. *Mar. Pollut. Bull.* 2013; 69: 219–222.
17. Galloway TS. Micro and nano-plastics and human health. In: Bergmann, M., Gutow, L., Klages, M. (Eds.), *Marine Anthropogenic Litter.* Cham: Springer International Publishing, pp. 343–366, 2015.
18. Parker L. In a first, microplastics found in human poop. http://www.nationalgeographic.com.au/nature/ In a first, microplastics found in human poop.aspx (2018).
19. Bouwmeester H, Hollman PC, Peters R J B. Potential health impact of environmentally released micro- and nanoplastics in the human food production chain: Experiences from nanotoxicolgy. *Environ. Sci. Technol.* 2015; 49: 8932–8947.
20. Wright S, Kelly FJ. Plastic and human health: A micro issue? *Environ. Sci. Technol.* 2017; 51: 6634–6647.
21. Werner S, Budziak A, van Franeker J, Galgani F, Hanke G, Maes T, Matiddi M, Nilsson P, Oosterbaan L, Priestland E, Thompson R, Veiga J, Vlachogianni T. Harm caused by marine Litter. MSFD GES TG Marine Litter - Thematic Report (JRC Technical report; EUR 28317 EN), 2016.
22. Newman S, Watkins E, Farmer A, ten Brink P, Schweitzer JP. The economics of marine litter. In: Bergmann M, Gutow L, Klages M (Eds.), *Marine Anthropogenic Litter.* Cham: Springer International Publishing, pp. 367–394, 2015.
23. Hardesty BD, Lawson T, van der Velde T, Lansdell M, Wilcox C. Estimating quantities and sources of marine debris at a continental scale. *Front. Ecol. Environ.* 2017; 15: 18–25.
24. Browne MA, Crump P, Niven SJ, Teuten E, Tonkin A, Galloway T, Thompson R. Accumulation of microplastic on shorelines worldwide: Sources and sinks. *Environ. Sci. Technol.* 2011; 45: 9175–9179.
25. Ivar do Sul JA, Costa MF. The present and future of microplastic pollution in the marine environment. *Environ. Pollut.* 2014; 185: 352–364.
26. Claessens M, Van Cauwenberghe L, Vandegehuchte MB, Janssen CR. New techniques for the detection of microplastics in sediments and field collected organisms. *Mar. Pollut. Bull.* 2013; 70: 227–233.
27. Sanchez-Nieva J, Perales JA, Gonzalez-Leal JM, Rojo-Nieto E. A new analytical technique for the extraction and quantification of microplastics in marine sediments focused on easy implementation and repeatability. *Anal. Methods.* 2017; 9: 6371–6378.

28. Nuelle MT, Dekiff Jen H, Dominique R, Fries E. A new analytical approach for monitoring microplastics in marine sediments. *Environ. Pollut.* 2014; 184: 161–169.
29. Kedzierski M, Le Tilly V, Bourseau P, Bellegou H, Cesar G, Sire O, Bruzaud S. Microplastics elutriation from sandy sediments: A granulometric approach. *Mar. Pollut. Bull.* 2016; 107: 315–323.
30. Andrady AL. Microplastics in the marine environment. *Mar. Pollut. Bull.* 2011; 62: 1596–1605.
31. Ojeda T, Freitas A, Birck K, Dalmolin E, Jacques R, Bento F, Camargo F. Degradability of linear polyolefins under natural weathering. *Polym. Degrad. Stabil.* 2011; 96: 703–707.
32. Lee J, Hong S, Song YK, Hong SH, Jang YC, Jang M, Heo NW, Han GM, Lee MJ, Kang D, Shim WJ. Relationships among the abundances of plastic debris in different size classes on beaches in South Korea. *Mar. Pollut. Bull.* 2013; 77: 349–354.
33. Hidalgo-Ruz V, Gutow L, Thompson RC, Thiel M. Microplastics in the marine environment: A review of the methods used for identification and quantification. *Environ. Sci. Technol.* 2012; 46: 3060–3075.
34. Esiukova E. Plastic pollution on the Baltic beaches of Kaliningrad region, Russia. *Mar. Pollut. Bull.* 2017; 114: 1072–1080.
35. Liebezeit G, Dubaish F. Microplastics in beaches of the East Frisian Islands Spiekeroog and Kachelotplate. *Bull. Environ. Contam. Toxicol.* 2012; 89: 213–217.
36. Lozoya JP, Teixeira de Mello F, Carrizo D, Weinstein F, Olivera Y, Cedres F, Pereira M, Fossati M. Plastics and microplastics on recreational beaches in Punta del Este (Uruguay): Unseen critical residents? *Environ. Pollut.* 2016; 218: 931–941.
37. Davis W, Murphy AG. Plastic in surface waters of the inside passage and beaches of the Salish Sea in Washington State. *Mar. Pollut. Bull.* 2015; 97: 169–177.
38. Stolte A, Forster S, Gerdts G, Schubert H. Microplastic concentrations in beach sediments along the German Baltic coast. *Mar. Pollut. Bull.* 2015; 99: 216–229.
39. Claessens M, Meester SD, Landuyt LV, Clerck KD, Janssen CR. Occurrence and distribution of microplastics in marine sediments along the Belgian coast. *Mar. Pollut. Bull.* 2011; 62: 2199–2204.
40. Van Cauwenberghe L, Devriese L, Galgani F, Robbens J, Janssen CR. Microplastics in sediments: A review of techniques, occurrence and effects. *Mar. Environ. Res.* 2015; 111: 5–17.
41. Ryan PG, Moore CJ, van Franeker JA, Moloney CL. Monitoring the abundance of plastic debris in the marine environment. *Philos. Trans. R. Soc. B.* 2009; 364: 1999–2012.
42. Lee J, Lee JS, Jang YC, Hong SY, Shim WJ, Song YK, Hong SH, Jang M, Han GM, Kang D, Hong S. Distribution and size relationships of plastic marine debris on beaches in South Korea. *Arch. Environ. Contam. Toxicol.* 2015; 69: 288–298.
43. Fisner M, Majer AP, Balthazar-Silva D, Gorman D, Turra A. Quantifying microplastic pollution on sandy beaches: The conundrum of large sample variability and spatial heterogeneity. *Environ. Sci. Pollut. Res.* 2017; 24: 13732–13740.
44. Hardesty BD, Lawson T, van der Velde T, Lansdell M, Wilcox C. Estimating quantities and sources of marine debris at a continental scale. *Front. Ecol. Environ.* 2017; 15: 18–25.
45. Dekiff JH, Remy D, Klasmeier J, Fries E. Occurrence and spatial distribution of microplastics in sediments from Norderney. *Environ. Pollut.* 2014; 186: 248–256.
46. Lusher A, McHugh M, Thompson R. Occurrence of microplastics in the gastrointestinal tract of pelagic and demersal fish from the English channel. *Mar. Pollut. Bull.* 2013; 67: 94–99.
47. Neves D, Sobral P, Ferreira JL, Pereira T. Ingestion of microplastics by commercial fish off the Portuguese coast. *Mar. Pollut. Bull.* 2015; 101: 119–126.
48. Iniguez ME, Conesa JA, Fullana A. Microplastics in Spanish table salt. *Sci. Rep.* 2017; 7(8620): 1–7.
49. Kim JS, Lee HJ, Kim SK, Kim HJ. Global pattern of microplastics (MPs) in commercial food-grade salts: Sea salt as an indicator of seawater MP pollution. *Environ. Sci. Technol.* 2018; 52: 12819–12828.

16 Genes and Nanogenomics

Gérrard Eddy Jai Poinern, Rupam Sharma, A.F.M. Fahad Halim, Rajeev Kumar Varshney, and Derek Fawcett

CONTENTS

16.1 INTRODUCTION

At present, the world's population is around eight billion people, with a sizable percentage residing in developing and growing economies. Many of these countries are primarily found in Africa and Asia. Many people in these underdeveloped countries contend with daily food shortages that result from environmental damage, socio-economic conditions, or political uncertainty. On the other hand, in industrialized nations, the driving force is to research, develop, and cultivate improved crops that are resistant to droughts, and pests, and ultimately produce high-yielding crops [1, 2]. Achieving zero hunger is one of the United Nations' 17 Sustainable Development Goals, and it cannot be achieved without effective and innovative agricultural systems [3]. Rising population, climate change, environmental contamination, and increasing demands for water and energy all put significant strains on the world's food cultivation and distribution systems. Current agricultural systems consume an enormous quantity of energy. Over two quadrillion British thermal units (BTU) of energy are needed to maintain the world's annual agricultural production of over three billion tons, along with 2.7 trillion cubic meters of water and 187 million tons of fertilizer and insecticides [4]. By 2050, the Food and Agriculture Organization expects the global population to reach 10 billion [5]. This in turn will increase food demand by around 50%, which will have a significant impact on developing nations. There are currently about 815 million undernourished individuals in the world, and it is projected that another two billion will fall into this category by 2050 [3]. Due to the current situation, major adjustments are urgently needed in the way food is produced globally. In the last few decades, scientists have devised a variety of alternative systems to traditional farming methods

DOI: 10.1201/9781003181422-19

in an attempt to increase food production. As our scientific knowledge and capabilities progress with time, there are many advances that can be beneficial to the agricultural industry.

In recent years, nanotechnology (NT) has been proven to hold great promise for enhancing agricultural outputs and addressing issues in the agricultural and environmental spheres that threaten food safety and production. NT had its inception on December 29, 1959, when Noble Prize physicist Richard Feynman gave a talk entitled "There's Plenty of Room at the Bottom, an Invitation to Enter a New Field of Physics" [6]. In his talk, Feynman proposed several concepts and elaborated on the possibility of atomic manipulation. It would take several decades before this could be possible, although in 1972 the topografiner with a resolution of 3 nm in the Z-plane and 400 nm in the X-Y plane [7] came very close. The catalyst for NT development was initiated with the creation of the first scanning tunneling microscope (STM) in 1981 in Zurich and subsequently the atomic force microscope (AFM) in 1985. These scanning probe technologies allowed scientists to image surfaces down to molecular/atomic levels. These microscopes gave an incomparable glimpse into mechanistic processes in real-time as well as imaging and manipulating atoms and molecules in electrolytes without disturbance of the interface. Today, NT is still a developing technology that has the potential to sustainably enhance many fields and modern agricultural research and innovation in a number of ways. These include improving crop productivity through enhanced plant nutrition, precision farming, efficient use of water, protecting crops against vermin and diseases through the evolution of nano-enabled products, and ecological sustainability of contaminated locations through nano-bioremediation [8–10]. Due to their unique physicochemical features, nanoparticles (NPs) have found widespread use in a wide variety of fields, including agriculture, biomedicine, chemistry, optics, pharmaceuticals, food and skincare, and textiles [11]. Many agricultural-based nano-enabled products include nano-fertilizers [12], soil-improving agents [13–15], stimulators that enhance plant growth [16], and nano-sensors [17], which are already on the market and being used in sustainable agricultural practices. In addition, there is substantial evidence that exposure to nanomaterials can improve plant growth, seed germination, and resistance to stress [18]. In addition, NT can be used to monitor a plant's development in real-time, to easily alter its genome, and to express a foreign gene inside its cells (transgene expression) [19, 20].

Transferring biomolecules, such as DNA (deoxyribonucleic acid) and RNA (ribonucleic acid), with the use of nanomaterials, is an exciting area of nano-biotechnology [18]. Importantly, DNA is the set of instructions for controlling a cell's activities, and RNA is the carrier of these instructions from the nucleus. Hence, there is a current widespread interest in genomics, since a genome is an organism's complete set of DNA. Genomics investigates the evolution, functional mechanisms, mapping, and editing of genomes [21]. Crucially, gene silencing [21, 22], genome coding [23], and transgene expression [21–23] all involve the transfer of biomolecules facilitated by nanomaterials. Moreover, plant cell walls have physical and chemical features that make it difficult for biomolecules to be transferred into the cell. For instance, because the cell wall of pollen is chemically inert, it is well-suited to transient gene expression. A study by Lew et al. found that the green fluorescent protein (GFP) encoding of plasmid DNA could be effectively delivered to oil palm pollen with the help of single-wall carbon nanotubes (SWCNTs) coated in imidazolium. Their study found that GFP was well delivered and also very active [22]. Similarly, a study by Kwak et al. showed that a plasmid DNA could be introduced into chloroplasts by employing chitosan-coated SWCNTs. In the study, various plant species were used, and the carriers showed significant temporary expression [19]. Studies involving gene silencing and genome modification using nanomaterials (NMs) have increased in recent years, along with studies involving transgenic expression. Using NMs for the targeted delivery of modified plasmids opens up novel avenues for altering plant genomes [21]. Recent research has demonstrated that DNA-CNT conjugates may be efficiently transmitted to several plant species [24]. For instance, CNT-mediated siRNA delivery has demonstrated great silencing effectiveness in plant cells, while NP-based administration has demonstrated efficient intracellular transferrable capacity [25]. Similarly, a study using polyethyleneimine-coated gold nanoparticle clusters (PEI-AuNPs) delivering siRNA to interact with plant cells with the aim of knocking out particular genes has shown promising results [23]. However, many studies indicate that more improvement is needed when it comes to optimizing NP systems for plant transformation. In addition, more research is needed to determine the long-term effects of genome editing aided by nano-biotechnology techniques [26, 27]. In spite of increased NT research focused on the area of agriculture in recent years, few publications take a comprehensive look at the many forms of NT employed in crop production, especially those designed for new nano-sensors and nano-biotechnology techniques.

1856 represents a very important year, in which two separate scientific works in material sciences and biology would later merge to form the present convergence for the bio-NT field of nanogenomics. In 1856, Michael Faraday, while at the Royal Institute of Great Britain, synthesized the first Au nanoparticles, while trying to study the interaction properties of light with thin films of Au. In that same year, the Augustinian monk Gregor Mendel started his investigations into peas (*Pisum sativum* L.) and noticed that several generational traits in subsequent plants are inherited from the parental plants. These works were published in 1865 and 1866 but remained largely unnoticed until the turn of the 20th century. His work showed that inheritance obeyed exact rules and today, we know that genes are fundamental to the inheritance of traits from parents to offspring and are located in the nucleus of the cells that make up this organism. The genome is the term used to denote the organism's complete set of DNA (deoxyribose nucleic acid). Almost every human cell contains a complete set of about 3 billion DNA base pairs (or letters) that make up the human genome. Other organisms have different numbers of DNA base pairs; for example, the largest DNA virus-Pandoravirus discovered only recently in 2013 by a French team has 2.5 million base pairs whereas a common bacterium such as *E. coli* K-12 has about 4.6 million base pairs that contain in 4288 protein coding genes and the *E. Coli* genomes can vary from 4 to 6 M base pairs dependent on the strains [28]. In the case of plants, their genome can indeed vary greatly, in the case of rice (*Oryza sativa*) one of the most important cereal crops for global food production, it has a size of 430 M base pair and is composed of 12 Chromosomes, Sorghum (*Sorghum bicolor*), also a major food crop, has 750–770 M base pairs and 10 chromosomes and barley 9 (*Hordeum vulgare* L.), one of the earliest domesticated crop plants has 5.1 G bases pairs and 7 chromosomes [29]. Many AFM studies have investigated the use of this nano-device to image conformations of DNA and nucleotides and even manipulate these macromolecules in their native state since its inception [30–31]. These investigations show even the motion of DNA molecules and DNA-enzyme interactions. Bezanilla et al. showed that the AFM could provide images in an aqueous buffer of bare uncoated DNA with about 300 base pairs of 100 nm in length as well as the dynamics and enzymatic degradation of single DNA molecules [32–33].

Innovations and research in genetic engineering and NT can combine together and bring revolution in the field of genome editing. The benefits of having a convergence of nanotech and medicine have already been evident in the field called nanomedicine and the direct ability to move molecular bases such as Thymine-T, Guanine-G or Cytosine-C, and Adenine-A into specific parts of a DNA strand to "write" new sequences, or delete dysfunctional sequences certainly can pave the way to fix genes to create greater crop yields and enhance the ability of the agricultural sector to increase its efficiency for food security. The convergence of nanosciences with genomics is being projected as the new technological paradigm which paves the way for a new concept of "Nanogenomics" (Figure 16.1).

In this chapter, we provide an overview of the current status of the use of NT in plant growth status' monitoring to increase agricultural yield in various important foodcrops, facilitate easy genome modification, and enable technologies for transgene expression in unaltered plant cells for the purpose of creating sustainable agricultural practices. Moreover, the concept of combining NT devices such as Dip-Pen Lithography, to manipulate, control and edit DNA and nucleotides at the molecular level rather than just harnessing the natural molecular machinery of genome editing by agricultural crop scientists is further explored.

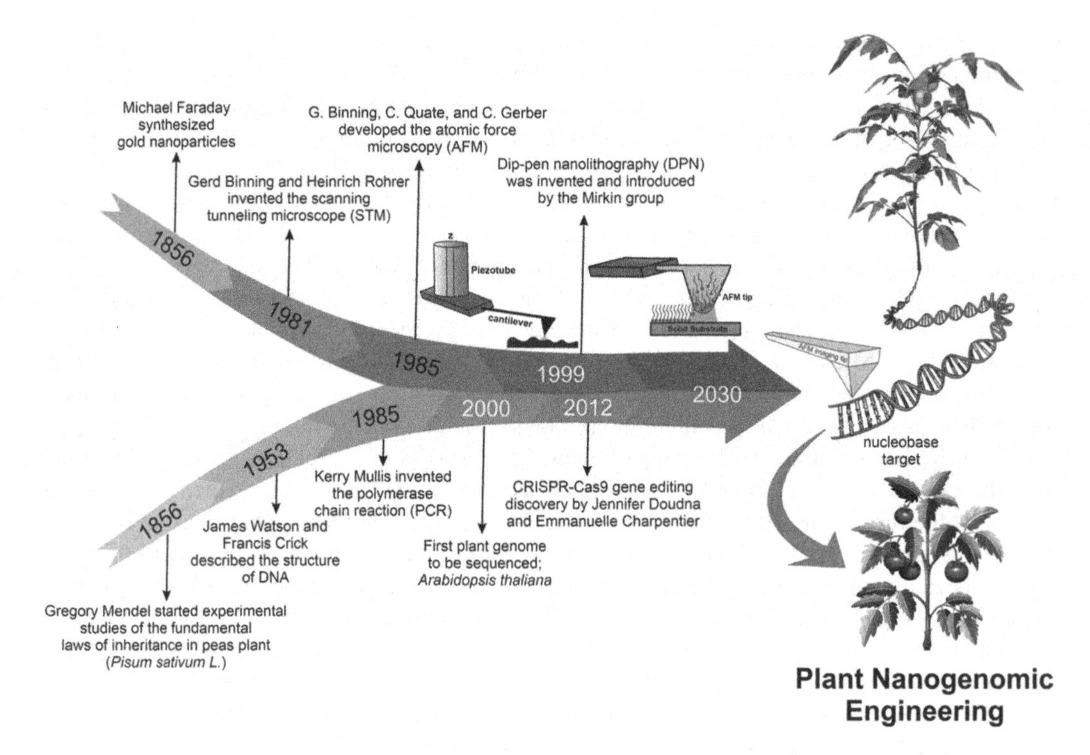

FIGURE 16.1 Nanogenomics: Natural convergence of scanning probe technology in the nanosciences with genomics for potential modification/nanoengineering of plant DNA bases in genes for enhanced crop production.

16.2 SCALE AND NUCLEOTIDES

On average a plant cell (Figure 16.2) is between 10 to 100 microns in size and contains many smaller bodies within a rigid cell wall. It has various organelles that work seamlessly for its life functions and purpose. It has a relatively large vacuole fluid filled organelle that can store food and nutrients, chloroplasts organelles where photosynthesis can take place converting CO_2 and water into simple sugars and give plants their characteristic green color. It also has mitochondrion bodies that are bean shaped organelles within its walls. These mitochondrion bodies are responsible for converting raw molecules into energy that is then accessible to the plant. A Golgi apparatus organelle is also present as well as the nucleus where the genetic blueprint of the plant is stored within its DNA. The chromosomes are in the cell's nucleus and these contain almost all the genes that are the blueprint for that organism. It is tightly packed. For example, a human with 23 pairs of chromosomes, which means 46 chromosomes would unravel into a stretch of DNA that would span about 3 meters long and a width of only 2 nanometers for this double helical structure. This fits in a nucleus of only 5 microns whilst accomodating other bodies as well. It can achieve this feat due to the elaborate, highly multilevel coiling and folding system of packing the DNA into each chromosome. Usually the DNA strand is coiled around bodies called histones like beads and thus a huge number of DNA base pairs can be accommodated into the nucleus. Thus, when packed, the DNA prevents gene expression and when one gene needs to be transcribed, the histones must loosen their attachment to the DNA strand.

Nucleic acids such as DNA are made up of monomers called nucleotides, with each having 3 parts: a 5-carbon sugar molecule, a phosphate group, and a nitrogenous base. In the case of RNA, ribose is the sugar molecule instead of deoxyribose as present in DNA. In DNA, there are nitrogenous bases Adenine (A), Thymine (T), Cytosine (C), and Guanine (G). In the RNA, instead of thymine, it has Uracil (U) present and also A, C, and G. DNA contains the chemical instructions needed to develop and direct the activities of most living organisms and is made up in the shape of two twisting paired

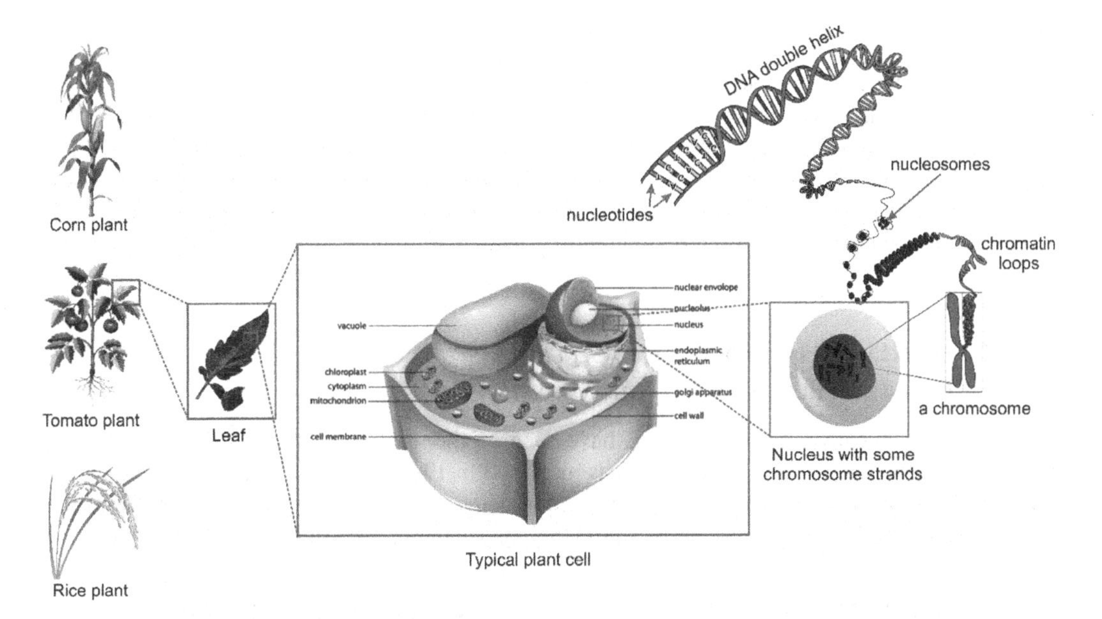

FIGURE 16.2 A plant cell with its various sub-micron components and the location of the chromosomes in the cell's nucleus.

strands of about 2.5 nanometers in width. It is generally referred to as a twisted double helix. The DNA strand is made up of nuclei acids called the nucleotide bases that pair specifically: as A will pair with T, C will pair with G, and it is the order of the As, Cs, Ts, and Gs that determines the meaning of the information that is encoded in a similar fashion as to the order of letters in a word from the alphabet determines the meaning of a word. Since a gene is a stretch of DNA that determines a certain trait of the organism, its expression in that organism is called the phenotype. The gene is a basic unit of hereditary. An organism has different alleles and the set carried by the organism is called the genotype. To enhance crop production, the understanding of genes, their various functions, and the manipulation of these macromolecular structures, have become a fertile area of investigation for many agronomists in their quest for better crops, adaptability to various soils and environments, and pest and pathogenic resistance. Thus genomic-assisted breeding develops much superior cultivars with advanced traits, and these key DNA markers have been identified through developments in various high-throughput technologies of epigenomics, proteomics, transcriptomics, and metabolomics. All these technologies allow us to understand the complex phenotypes of crops of agricultural importance [34–35]. Gene modifications or editing of plants have greatly improved with time and are used today in various instances, with similar labelling terms. This has been the cause of alarm in many corners as food production needs to increase rapidly and at the same time the customer needs to be reassured that their food supply is robust, nutritional, and not detrimental to their overall health.

16.3 KEY GENETICALLY MODIFIED ORGANISM (GMO) QUESTIONS AND ANSWERS

16.3.1 What Is the Difference between Genetic Engineering and Genome Editing?

"Genome editing" (GE) is a relatively new set of technologies that enable scientists to make precise changes to the DNA of a plant, animal, or other living organism. The changes made possible with GE may be as precise as a targeted single base-pair change in the plant's DNA. Current genetic engineering techniques in use do not enable genetic changes to be targeted to specific locations in the plant genome like genome editing techniques do.

16.3.2 What Foods Are Made from Genetically Engineered Plants?

Foods from genetically engineered plants were first introduced into our food supply in the 1990s. The majority of these genetically engineered plants – corn, canola, soybean, and cotton – are typically used to make ingredients that are then used to make other food products. These ingredients include corn starch in soups and sauces, corn syrup as a general-purpose sweetener, and cottonseed oil, canola oil, and soybean oil in mayonnaise, salad dressings, cereals, bread, and snack foods. There are also genetically engineered varieties of potatoes, summer squash, apples, papaya, alfalfa, and sugar beets.

16.3.3 How Long Have Foods from Genetically Engineered Plants Been on the Market?

Foods from genetically engineered plants have been on the market for almost 30 years. The first food from a genetically engineered plant was marketed in 1994. The food was a tomato known as the Flavr Savr. Since then, genetically engineered varieties of corn, cotton, canola, soybeans, sugar beets, apples, potatoes, summer squash, alfalfa, and papaya have been marketed.

16.4 GENETICALLY MODIFIED ORGANISMS (GMO) AND TECHNOLOGY

A GMO is a term used to define any genetically modified organism which can be either plant or animal or any organism whose genome has been modified in the laboratory using genetic engineering tools for the expression of a desired phenotype. This can be achieved in many ways. An organism can be termed as GMO if its genes have been altered, including by nature [36, 37]. However, the Food and Agriculture Organization (FAO) describes a GMO as an organism that has been altered in a way that does not occur naturally by mating or by recombination [38]. GMOs are typically produced by recombinant DNA technology and reproductive cloning [39]. In reproductive cloning, a nucleus is extracted from a cell of the individual to be cloned and is then inserted into the enucleated cytoplasm of a host egg. This results in an offspring that is genetically identical to the individual donor. "Dolly" was the first animal (sheep) produced using this cloning technique in 1996, and thereafter several other animals such as pigs, horses, and dogs, have been created using this cloning technique [40].

In recombinant DNA technology, a gene of interest is collected from a donor cell and inserted into the DNA of another species where the desired trait is to be expressed [41]. There are various steps involved in producing successful GMOs, such as identifying the gene of interest which will express the desired phenotype, copying the gene from the organism, properly inserting that gene into the DNA of another organism, and then successfully growing the new organism with the gene of interest [42]. In recent years, a number of GMOs have been produced in a variety of fields such as agriculture, medicine, research, and environmental management [43]. Although conventional selective breeding programs can also produce organisms with desired traits, it takes a long time, sometimes decades, to produce offspring with desirable traits. Recombinant genetic technologies can genetically modify organisms with desired traits and produce results in much shorter time frames [44]. However, public acceptance of GMOs has been far from widespread. Historically, the first genetically modified organism was a bacterium resistant to the antibiotic kanamycin, which was developed by Herbert Boyer and Stanley Cohen in 1973 [44]. The first genetically modified animal was a mouse which was developed in 1974 by Rudolf Jaenisch at the Salk Institute for Biological Studies in California [45]. The first GMO tobacco plant was produced in 1983 [46]. In terms of crop production, the Flavr Savr tomato was the first commercialized genetically modified food and was released to the market in 1994 by the Californian company Calgene [47]. The first genetically modified animal to be commercialized was the GloFish in 2003 [48], and the first genetically modified livestock animal to be approved for food use was the AquAdvantage salmon in 2015 [49].

From a historical perspective, humans have used traditional modification methods such as selective breeding and cross-breeding to breed plants and animals with more desirable traits since around 8000 BCE. Archaeological sites found in southwest Asia reveal the artificial selection of plants dates back to around 7800 BCE when early farmers developed domestic varieties of wheat [50]. Corn, or maize, which originally grew as a wild grass called teosinte, was selectively bred and resulted in larger ears with more kernels. Thus, the cross-breeding process over many centuries has converted tiny ears with very few kernels to the corn sizes we see today [51]. In 1866, the most revolutionary truth of genetics was uncovered by the Austrian monk Gregor Mendel. His studies revealed that several traits found in pea plants were inherited from their parent plants. Thus, by breeding two different types of peas, Mendel was able to identify the basic process of genetics. However, the first hybrid corn was only produced and sold commercially in 1922. During the 1940s, the use of radiation or chemicals to randomly change an organism's DNA was discovered and used by plant breeders to produce new varieties. However, the basic structure of DNA, the structural unit of life, was discovered in 1953 by James Watson and Francis Crick and was built on the earlier discoveries of chemist Rosalind Franklin. By 1973, biochemists Herbert Boyer and Stanley Cohen had described the transfer process of DNA from one organism to another and termed it "genetic engineering". In 1982, one of the most important milestones in the field was achieved when the US Food and Drug Administration (FDA) approved a GMO-engineered insulin product to treat human diabetes. In 1986, the US federal government established the Coordinated Framework for the Regulation of Biotechnology to regulate the safety of GMOs. And by 1992, the FDA policy also stated that foods from GMO plants must meet the same requirements, including the same safety standards, as foods derived from traditionally bred plants. Following the implementation of the FDA policy, a GMO tomato became the first food available for sale in 1994. Thus, during the 1990s, there was a wave of GMO-produced products such as summer squash, soybeans, cotton, corn, papayas, potatoes, and canola. The arrival of these new GMO products led to the development of international guidelines and standards to determine the safety of GMO foods in 2003 by the World Health Organization (WHO) and the Food and Agriculture Organization (FAO) of the United Nations. By 2005, GMO-produced alfalfa and sugar beets were available for sale in the United States. GMO research crossed another important milestone when the FDA approved genetically engineered salmon as food, made by AquaBounty, in 2015. Following the success of salmon, GMO apples and GMO pink pineapple entered the market in 2017 and 2020, respectively. Recently, an application by Revivicor, Inc., a wholly owned subsidiary of United Therapeutics Corporation, for GalSafe pig was approved in 2020. However, the production of GMO products remains a highly controversial topic in many parts of the world [52].

More recently, scientists developed a new technology known as "genome editing" to create new varieties of crops and animals. These techniques can make changes even more quickly and with greater precision to make current crops more nutritious, drought tolerant, and resistant to insects, pests, and diseases, than traditional breeding techniques. However, there is considerable debate over whether gene editing is considered GMO or not. In 2018, there were rulings issued by two major countries, viz. the US and European Union (EU), which addressed gene editing. The US clarified that gene editing will not be part of the same regulatory oversight as GMOs [53]. Furthermore, the US-Mexico-Canada Agreement (USMCA) included provisions to support gene editing, which was signed into law in 2020. However, the EU announced that any organisms altered by gene-editing techniques would be subjected to the same regulations as GMOs [54].

16.5 CLUSTERED REGULARLY INTERSPACED SHORT PALINDROMIC REPEATS (CRISPR)

Genome editing (GE) or genome engineering is the most recent genetic engineering tool for manipulating the DNA of an organism. It includes editing of the genome by insertion, deletion, or replacement of alleles at particular locations in the genome, which results in predictable and inheritable mutation(s). Genome engineering also ensures the lowest probability of off-target and no integration of exogenous gene sequences [55]. Genome editing technologies that are based on engineered or bacterial nucleases have been developed at a fast pace over the last few years and have demonstrated their usability across several fields including biomedical research. Genome editing can be achieved either *in vitro* or *in vivo* by targeted genomic modifications of DNA by nuclease-induced double-stranded breaks (DSBs). Such DNA DSBs can be repaired either by homology-directed repair (HDR) or non-homologous end-joining (NHEJ), which results in targeted integration or gene disruptions, respectively. There are four enzymatic site-directed nuclease (SDN) families involved in a nucleotide editing mechanism, such as homing endonucleases or meganucleases (HEs) [56], zinc-finger nucleases (ZFNs) [57], transcription activator-like effector nucleases (TALENs) [58], and CRISPR-associated protein (Cas 9) [59].

Meganucleases are homing nucleases found in all microbial genomes with a large recognition site. These enzymes identify and cleave lengthy DNA sequences of 12 to 40 base pairs, which results in double-strand DNA breaks (DSBs). Meganucleases are used to modify all genome types, whether bacterial, plant, or animal. There are many designed meganuclease variants that can cleave unique DNA targets for targeted genomic changes, thus creating important characteristics in various crop species [56]. However, due to problems in the manufacture of these nucleases and designing vectors for their entrance into cells, and off-targeting consequences, meganuclease technology was not successful [60, 61] and is still being developed further.

Zinc-finger nucleases (ZFNs) are artificial endonucleases consisting of a designed zinc-finger protein (ZFP) fused to the cleavage domain of the FokI restriction enzyme, which is designed to cleave a chosen genomic sequence in genome engineering. The cleavage results in double-strand breaks and initiates cellular repair processes which in turn mediates the efficient modification of the targeted locus resulting in remarkably targeted mutagenesis and targeted gene replacement [62]. The zinc-finger domains have DNA-binding domain specificity to 12–18 nucleotide sequences [63].

Transcription activator-like effector nucleases (TALENs) are nonspecific DNA-cleaving nucleases linked to the FokI nuclease domain that can be tailored to target specific sequences for editing the genome. Transcription activator-like effectors (TALEs) can be engineered to bind to practically any desired point in the DNA sequence [64]. However, both ZFN and TALENs technologies are not as popular because of certain disadvantages, such as complex design, lower engineering feasibility, lower specificity, low efficacy, and an inability for gene knockout and RNA editing [65]. Therefore, to overcome the disadvantages of HEs, ZFNs, and TALENs, a new technology called CRISPR-Cas technology was pioneered and developed. Since its inception, it has gained popularity among genetic engineers globally. Clustered regularly interspaced short palindromic repeats (CRISPR) are currently the most extensively used genome editing technique worldwide. This is due to its simple design, cost-effectiveness, high efficiency, good reproducibility, high engineering feasibility, ability to create gene knockout, RNA editing, and quick cycle operation [66].

CRISPR-Cas is an adaptive defense system of bacteria and archaea that protects them from phages, viruses, and other foreign genetic material being inserted into their genetic inherent [67]. Using CRISPR, bacteria can snip out parts of virus DNA or other foreign DNA and keep a bit of it behind to help them recognize and defend against invaders the next time they attack. This immune defense system has been studied and is now used for genome editing. The CRISPR-Cas9 system consists of two key molecules: 1) an RNA-guided Cas9 endonuclease and 2) a single-guide RNA (sgRNA). The Cas9 endonuclease acts as a pair of "molecular scissors" that can cut the two strands of DNA at a specific location in the genome so that bits of DNA can be removed or added. The guide RNA (gRNA), which consists of a small piece of pre-designed RNA sequence (about 20 bases long) located within a longer RNA scaffold and is complementary to those of the target DNA sequence in the genome, is used to find and bind to a specific sequence in the DNA. Once the scaffold part binds to DNA, the pre-designed sequence of the RNA "guides" Cas9 to the same location in the target DNA sequence and then cuts across both strands of the DNA. Although Cas9 is the enzyme that is most often used, other enzymes like Cpf1 can also be used. Once the DNA is cut, researchers use the cell's own DNA repair machinery to add or delete pieces of genetic material, or to make changes to the DNA by replacing an existing segment with a customized DNA sequence [68].

Cas9 endonucleases' HNH domain cuts one strand of sgRNA, while the RuvC-like domain cuts the opposite strand of dsDNA, resulting in double-strand breaks (DSBs). As a result, the endogenous repair system automatically repairs DSBs *in vivo*, utilizing error-prone non-homologous end-joining (NHEJ) or homology-directed repair (HDR), resulting in massive insertion or fragment replacement [69]. The CRISPR/Cas systems are more efficient and straightforward to use for genome editing because their specificity toward editing is dictated by nucleotide complementarity of the guide RNA to a specific sequence without complex protein engineering. Therefore, many researchers have applied CRISPR/Cas tools to gene functional analysis [70].

CRISPR/Cas9-based gene editing has been carried out in plants and involves the following steps: 1) selecting genes of interest for editing and designing spacers for selected genes; 2) preparing transformation carrier or ribonucleoprotein (RNP); 3) delivering foreign nucleotides or proteins into plant cells; 4) identifying edited lines in T0 generation by next generation sequencing (NGS); 5) selecting null plants with the gene edited in T1 and confirming them by NGS in T2 generation; 6) obtaining homozygous edited lines and the evaluation of expression of the target gene; and 7) using the null lines for breeding new variety.

However, there are still many ethical concerns involved with genome editing. At present, genome editing is generally limited to somatic cells, and the genetic changes are not passed from one generation to the next. Germline cell and embryo genome editing bring up a number of ethical challenges since these procedures can result in genetic materials being passed hereditarily from one generation to the next. To avoid these concerns, germline cell and embryo genome editing have been made illegal in many countries.

16.6 CURRENT NANOTECHNOLOGY USAGE IN MAJOR FOOD CROPS

16.6.1 Nanoagriculture

As nanoscience/NT encompasses several fields of physical, chemical, and biological sciences, this interdisciplinary field has emerged as a growing industry with many engineered products as well as creating tools to advance research and innovation in the agricultural sector and related sectors of crop production. While crop production is well established in many parts of the world, the global agri-food production system is inefficient due to poor uptake of nutrients by plants, inefficient use of fertilizers and biocides as well as food wastage after production and harvest [71]. Other challenges of pests and pathogens add to this burden and decrease our global food production goals and security.

At the nano-scale, the inherent properties of nanomaterials allow for unique interactions and these can be leveraged and optimized for better pathogen protection, water use in plants, agrochemical uptake as well as plant sensor platforms and animal health surveillance and control all leading to making agri-food systems more sustainable [72]. In parallel to nanomedicine, novel nanomaterials in agriculture can be used for targeted delivery of appropriate agrochemicals and the payload as well. For example, it is estimated that 1–2.5 million tons of pesticide active ingredients are used globally, mainly in the agricultural sector, with 40% being herbicides and the rest as pesticides and fungicides [73]. While most of these chemical compounds are to have distinct modes of action and targets, these can only be used if proven not to persist in the environment beyond a set soil half-lives limit of days or weeks. Nonetheless, due to the widespread use of pesticides and these chemical compounds being extensively applied over vast areas in agriculture, pesticides are a global chemical pollution threat, and hard to control in the biosphere. Thus, there is a need for nano-pesticide technologies as well as nanosynthetic technologies reliant on green non-toxic biogenic methods [74]. Nano-fertilizer is another area where the fertilizer payload to the crop can be used to increase production and minimize the waste of chemicals. This technology allows the use of delivery templates for the general bio-fortification of crops to enhance the nutritional value of foods.

While nanomaterials play an integral part in the delivery platform technology to enhance plant growth in agricultural R&I, there is however a largely untapped area of NT devices that could be used effectively for genomic modification or engineering of genes. While most agricultural scientists have focused their attention on genome-based molecular technologies to improve crop yield and efficiencies in agricultural production, a nanotechnological tool was developed by AFM researchers in 1999, whereby molecular structures of alkanethiols entities were delivered with a cantilever onto a surface in a technique called "Dip-Pen" lithography [75]. In this case an AFM tip was used to write in a positive mode using alkanethiol molecules directly onto a surface with a linewidth of 30 nm. This direct write method was further demonstrated by Demer et al. whereby oligonucleotides were written onto gold substrates. Moreover, in 2011, Liang et al. showed that AFM nanolithography could be used to nanograft thiolated DNA structures on gold surfaces [76]. Pyne at UCL demonstrated that the AFM was well-suited for *in situ* DNA investigations and the results of oligonucleotides are comparable to similar diffraction studies made with X-rays [77]. Capitalizing on these parallel developments in NT devices, there is an opportunity to use these AFM techniques to a greater extent in engineering and manipulating molecular bases compared to utilizing the molecular machinery at the biological level.

Current farming practices are proving to be largely unsustainable methods for delivering food security due to serious factors like rapidly increasing population, limited supplies of freshwater, and global warming [78]. NT-based Dip-Pen Nanolithograpy plant technologies offer new solutions for improving crop production. For instance, nanomaterials are capable of delivering biomolecules directly into plants [79]. In particular, nanomaterial-delivered biomolecules can be directed toward plant cells and organelles to optimize delivery [80]. Importantly, nanomaterials have been evaluated for diverse functions such as 1) nano-fertilizers; 2) nano-pesticides; and 3) nano-biosensors [81, 82]. The following section discusses and summarizes current NT application studies in five major food crops: 1) rice; 2) wheat; 3) maize; 4) potato; and 5) chickpeas.

16.6.2 Rice

Rice (*Oryza sativa spp.*) is a major cash crop that is produced around the globe. It is grown in a wide range of local environments. Importantly, rice is one of the leading food crops for over half of the world's population [83]. However, to meet the increasing global demand for rice and other agricultural produce, novel technologies like NT are being investigated as methods for improving crop productivity. There are several NT applications currently being investigated for use in agriculture in general and rice in particular. The following discusses NT applications being explored for rice.

Nanoparticles have attracted considerable interest for their potential use as fertilizers and pesticides since their use has been shown to enhance food crop productivity. The large surface area and small size of the nanomaterials promote increased interactions and enhance the uptake of nutrients for crop fertilization applications. Similar interactions and crop improvements have also been seen for pesticide applications [84, 85]. Hence, nano-fertilizers and nano-pesticides are receiving increased attention worldwide owing to their positive effect on crop production [86].

In plants, zinc is essential for photosynthesis, carbohydrate and phosphorus metabolism, and for promoting grain development [87]. Rice is an essential source of energy, vitamins, minerals, and rare amino acids to the global population, many of whom consume it as a daily staple food [88]. Zinc is also an essential element for the human body, but the levels of zinc found in rice are very low [89]. Hence, zinc-based fertilizers are being used in different ways, but their effectiveness is limited due to factors like the fertilizer dripping rate and leaching by rain [90]. In addition, iron, aluminum oxides, clay minerals, and humus present in the soil can also adsorb and fix Zn^{2+} ions, thus reducing the effectiveness of the Zn fertilizer in the soil and causing adverse effects on the local ecosystem [91]. To avoid such problems, zinc oxide (ZnO) nano-fertilizers are being used and have been found to increase the zinc content in grains and promote greater crop yields [92]. Factors such as soil texture, structure, and colloidal content have a smaller impact on nano-zinc fertilizers when compared to conventional fertilizers. In addition, their smaller particle size and higher surface area, combined with the controlled release of zinc, facilitate greater uptake and utilization by rice plants [93]. Similarly, other studies have reported higher growth rates and higher yields when rice crops were treated with biogenic silver nanoparticles [94]. Furthermore, a study by Valoji et al. found that higher growth rates and better rice grain quality resulted when nano-NPK fertilizers were used instead of conventional fertilizers [95].

Food crop bio-fortification is an approach currently being used to improve grain quality. In rice, bio-fortification was carried out using nano-chelated iron fertilizers. The use of chelated iron fertilizers was found to increase plant height, panicle length, grain weight, and paddy yield. The bio-fortification treatment also resulted in significant enrichment levels of nitrogen, phosphorus, and potassium in the rice grains [96]. Importantly, rice bio-fortification *via* the application of nano-chelated iron fertilizer is possible without any genetic manipulation of the plants.

Nano-sensors for detecting a variety of elements at extremely low concentrations *via* a physiochemical transducer have also been used in crop cultivation practices, food quality monitoring, and food packing systems [97, 98]. Importantly, nano-sensors enable direct communication between plants and farmers, thus promoting crop productivity by supplying water, fertilizer, and pesticide management when needed by the plants [99]. A number of different types of nano-sensors have been developed and used with plants; these include carbon-based electrochemical, fluorescence resonance energy, nanowire, and plasmonic nano-sensors. They have been used to detect small concentrations of urea, fertilizers, and pesticides, and detect various microorganisms [100, 101].

Nanomaterials have also been used to improve crop quality and protection. For instance, carbon nano-fibers (CNFs) containing foreign DNA were delivered, through microinjection, to create genetically modified golden rice. The result of this treatment produced an increase in the concentration of vitamin A in the rice grains [102]. Recently, scientists from the Indian Institute of Technology (IIT) Kanpur, India, developed a novel nanoparticle-based biodegradable carbonoid metabolite (BioDCM) that was designed to protect agricultural crops, especially rice crops, from fungal and bacterial infections. BioDCM is a biodegradable nanoparticle system with a metabolite extracted from a naturally occurring common soil fungus, viz. *Trichoderma asperellum* strain, TALK1, which can be used as an effective organic antimicrobial agent. The nanoparticle-based system is also combined with a carbonaceous degradable covering to provide protection against crop diseases and enrichment of the surrounding soil. The nanoparticle-based system also helps enhance more effective plant defense mechanisms, thereby promoting better crop productivity. The invention will also assist farmers to overcome the pre-mature degradation of rice crops [103].

16.6.3 Wheat

Wheat is an important staple food for the people of South Asia, America, and Europe. To meet the growing demand for wheat, modern technologies like NT are needed to improve crop production. Compared to rice, NT applications directed toward wheat production are still new and in their early stages. However, there are tremendous prospects for using nanomaterials to enhance production, improve surveillance, and advance pest and disease management strategies. In particular, the development of a new generation of nano-pesticides and nano-fertilizers has the potential to revolutionize wheat production.

Polymeric, ceramic, and metallic nanomaterials are currently being used in bio-fortification applications. Their usage ensures the controlled release of nutrients to the soil environment. Importantly, this form of application also increases nutrient uptake and prevents the leaching of potentially harmful inorganic fertilizers into fragile ecosystems. Because of their small size and large surface area, nanoparticles can significantly increase the bioavailability of micronutrients and hence reduce the amount of fertilizer needed. Nano-fertilizers contain nutrients in nano-scale quantities. Typically, nano-scale coating-based fertilizers, nano-scale additives fertilizers, and nano-scale fertilizers are being used for bio-fortification. The commonly incorporated micronutrients found in nano-fertilizers are Zn and Fe [104].

Studies have shown the application of nano-fertilizers improves the solubility and dispersion of insoluble nutrients in soils, reduces nutrient immobilization (soil fixation), and increases the bioavailability of the nutrients [105]. Moreover, nano-fertilizers are easily absorbed by plants and provide a nutrient supply in the soil or on the plant for longer durations [106]. For instance, when chitosan nanoparticles loaded with nitrogen, phosphorus, and potassium (NPK) were delivered to wheat plants for foliar uptake, significant increases in harvest index, crop index, and mobilization index were reported [107]. Other studies have also reported that both soil and foliar fertilization methods delivered increased nutrient uptake and crop yield when nanoparticle-based fertilizers were used [104].

Seed priming with nanoparticles has also been found to improve metabolism rates and deliver nutrients more effectively than conventional methods. Currently, ZnO NPs are being used in foliar sprays to promote the growth of wheat. The ZnO NPs were found to accumulate in apoplasts *via* the stomata by crossing the epidermis and were also absorbed by mesophyll cells [108]. Several other metallic nanoparticles, such as copper (Cu) and silver (Ag), have also been used to increase drought resistance in wheat plants [109–113]. Carbon nanotubes have been used to enhance wheat seed germination and promote plant growth by improving the uptake of water, calcium (Ca), and Fe [114, 115]. Hydrated graphene ribbons (HGRs), which are novel and biocompatible, can help promote the germination of aged wheat seeds and also enhance plant resistance to oxidative stress [116]. Titanium dioxide (TiO_2) NPs, which can regulate enzyme activity involved in nitrogen metabolism, have also been found to promote root growth and shoot development in wheat plants [117, 118]. Similarly, nanometer-scale aluminum oxide (nAl_2O_3), ZnO, and cerium oxide (CeO_2) have been found to promote seed germination and promote better crop production levels [119, 120]. Ag NPs can be used to alleviate salt stress in wheat by decreasing oxidative stress through the modification of antioxidant enzyme activities. The NPs were also found to reduce sodium (Na) translocation from the plant roots to their shoots, which ultimately led to an increase in plant growth [121].

Enhancing crop yields through effective pathogen detection and disease management is of utmost importance. In recent years, researchers have developed biosensors and electronic noses for the detection of pathogens in wheat. For instance, gold (Au) nanoparticle-based immune sensors have been developed to detect Karnal bunt (*Tilletia indica*) disease in wheat [122]. Electronic noses (EN) fitted with metal-oxide-semiconductor sensors have been used to assess deoxynivalenol contamination levels [123], insect infestation [124], and fungal infections in wheat grains [125]. Ag NPs were used in controlling *B. sorokiniana* infection which causes spot blotch disease in wheat [118], and Savi et al. used ZnO NPs (diameter 30 nm) to efficiently reduce *F. graminearum* levels in wheat grains [126]. Seeds treated with metal nanoparticles such as Zn, Ag, Fe, Mn, and Cu can promote a defensive reaction in wheat seedlings infected with *Pseudocercosporella herpotrichoides* [127]. Chitosan nanoparticles (CS/NPs) can be used to control *Fusarium* growth in wheat [128]. Furthermore, a study by Teodoro et al. found that nanostructured alumina had insecticidal activity toward *Sitophilus oryzae* and *Rhyzopertha dominica* [129]. Interestingly, inorganic nanostructured alumina is a low-cost and reliable alternative for controlling insect pests, thus highlighting the need for further studies into nanoparticle-based technologies for developing effective pest control management strategies for wheat.

16.6.4 Maize

Maize (*Zea mays*) is another important food crop having global importance. It also has importance as a feedstock for biofuels. Maize originated in Mexico around 8,700 years ago and since then it has been an important food crop. Currently, the United States is the largest maize producer and contributes around 30% to the global maize market [130]. Recent studies have revealed that the application of nanoparticles to maize can significantly improve plant growth, yield, and crop quality [131, 132].

Several types of nanoparticles are being used to improve both the growth rates and production levels of maize. For instance, zinc is one such supplement since maize is often found to be deficient in zinc. To overcome this problem, studies by Umar et al. found the application of nanometer-scale ZnO significantly improved maize growth [133]. The study also found bio-fortification not only improved overall Zn content but also enhanced crop yields. A study by El-Naggar et al. found the application of nano-silica, in the form of a nano-fertilizer, increased plant height, improved chlorophyll content, and enhanced protein content [134].

However, various abiotic factors can influence the growth of maize, and drought is a major stressor [135]. Therefore, developing drought-tolerant maize varieties and sustainable NT-based strategies is considered the most economical and effective approach for successful maize production [136]. Under drought conditions, plants should have enough copper (Cu) content to carry out photosynthesis. Cu is also essential for plant respiration and for the metabolism of carbohydrates and proteins [137]. Therefore, applying nano-CuO priming could enhance drought tolerance in maize. Since studies have shown that nano-CuO priming can promote greater leaf water status and increase both chlorophyll and carotenoid content. Importantly, nano-CuO priming also helps in maintaining photosynthesis and plant protective mechanisms. This in turn results in an effective balance between growth and drought stress response. Overall it increases plant biomass and grain productivity under drought stress conditions [136].

Late wilt (LWD) is one of the diseases caused by soil and seed-borne fungi, and it affects maize during or after flowering. The disease causes significant economic losses in several parts of the world. One effective method to treat the disease is to coat maize seeds with silica nanoparticles [138]. The treatment also improves the germination rate of maize seeds. In addition, nano-silica has also been used as a pesticide against four common pests that infect stored maize [134].

16.6.5 Potato

Potato (*Solanumtuberosum* L.) is the fourth globally important food crop after rice, wheat, and maize. Potato belongs to the Solanaceae family, and the plant tubers are consumed by both humans and livestock for nutrition. Potatoes are also used as a source of starch for alcohol production [139]. The growth and production of potatoes are seriously affected by a number of abiotic and biotic stresses, which will be further exacerbated by global warming in the future [140]. One consequence of climate change is the salinization of land and water, which has a direct impact on agricultural crop production. In particular, potatoes are seriously affected by salinization. In an attempt to mitigate the effects of climate change, scientists have investigated using different types of nanoparticles like biochar, nano-potassium, and nano-silicon to promote greater water content, improve leaf area and chlorophyll content, increase free proline content, enhance antioxidant enzyme activity, and promote greater nutrient uptake. Thus, by improving these parameters, scientists believe that crop yields and the quality of potato tubers will increase [141]. This is very important because the morphological, physiological, and biochemical processes of potato plants are seriously hampered by increasing salinity [141]. Thus, nano-silica and nano-potassium can be used to promote plant tolerance toward increasing salinity by improving water status, increasing the photosynthetic rate, decreasing oxidative injuries, and modulating some osmolytes and phytohormones [142, 143].

As mentioned above, potatoes are sensitive to heat, cold, and drought. Therefore, climate change is expected to have a negative impact on many potato-cultivating areas around the world. Another measure to mitigate the effects of climate change is to develop and use nano-fertilizers [144]. Studies have reported higher potato production rates when nano-fertilizers are used instead of conventional chemical fertilizers [145, 146]. One common problem found in potato crops is potato wilt, which is caused by the bacterium *Ralstonia solanacearum*. Since bacterial disease is difficult to control, there is increasing interest in developing new alternative plant disease management strategies to reduce the dependence on synthetic chemicals. Using NT-based strategies is a new method for dealing with fungal diseases in plants [147]. For instance, Khairy et al. reported the use of nano-chitosan particles to inhibit bacterial growth in potatoes [148].

Another feature that has appeared in agricultural crops in recent years is iron (Fe) fortification. Fe fortification is designed to mitigate the iron deficiency experienced by humans and results in anemia. Since Fe uptake *via* foods is more effective than consuming supplements, research has focused on developing nano-rust particles to iron-fortify potatoes. This method is designed to help reduce anemia in humans, and studies also indicate that nano-Fe-treated potatoes grow faster and bigger than non-treated ones. Another major problem associated with potatoes is sprouting during storage. When sprouting occurs, it is very fast and also releases significant amounts of α-solanine. Importantly, α-solanine is a toxic substance and causes human health problems. Sprouting also decreases the commercial value of the potato. Because of these problems, scientists have developed a new nanomaterial called hydrophobic nano-silica that is designed to inhibit the sprouting process in potatoes [149]. The process involves the potato being immersed in the hydrophobic solution which deposits a coating. Once removed from the solution and dried, the coating prevents sprouting, thereby preventing α-solanine production. Importantly, the particle-based coating does not penetrate through the potato skin and can be readily washed off before cooking. This eco-friendly technology ensures food safety and improves the storage life of the potato.

16.6.6 Chickpeas

Chickpea, *Cicer arientinum L.* (family: Fabaceae; sub-family: Faboidae) is considered a king of pulses and occupies a predominant position among the pulses family. There are different types such as gram, Bengal gram, garbanzo bean, as well as Egyptian pea. In recent years, a variety of metal nanoparticles such as aluminum (Al), silica (Si), zinc (Zn), and metal oxide nanoparticles like ZnO and titanium dioxide (TiO_2) have been considered for agricultural uses [150–152]. For chickpeas, nanoparticles have been utilized as nano-fertilizers for enhancing plant production and growth. For instance, ZnO nanoparticles are widely used for treating agricultural crops, including chickpeas. Foliar sprays of aqueous solutions containing ZnO nanoparticles improve the overall biomass accumulation in the treated chickpea seedlings [153]. However, ZnO nanoparticle concentrations of 10 ppm have been found to exert adverse effects on root growth. Hence, it is very important to use the particles in the appropriate size range and concentration to avoid such adverse effects. Some researchers have treated chickpea seeds with Zn- and Fe-based polymers to improve germination rates. Chitosan-conjugated Ag nanoparticles have also been used to enhance the seed germination rates of chickpeas. Some studies have reported high germination percentages (93%) for chickpea seeds treated with Ag-CS nanoparticles [154]. Studies by Pandey et al. [155] and Tripathi et al. [156] investigated the effects of nanoparticles on chickpeas and found that the nanoparticles promoted enhanced shoot and root length. Also, being nano, the particle size promoted the absorption of inorganic nutrients and accelerated the breakdown of organic substances during the photosynthetic process, thereby increasing the photosynthetic rate. A study by Saharan et al. showed that chitosan, which is a cationic polysaccharide with β-(1-4)-linked d-glucosamine (deacetylated unit) and N-acetyl-d-glucosamine (acetylated unit), possesses properties such as biocompatibility, biodegradability, non-toxicity, and antimicrobial activity, and is a good material for use in agricultural NT applications [157]. Chitosan has been used to deliver various biomolecules for a variety of different purposes. For example, thiamine was conjugated with chitosan nanoparticles (TCNP) and used against stress caused by the wilt pathogen; *Fusarium oxysporum f. sp. ciceri* in chickpeas [153]. Recently, a study by Das and Dutta used ZnO NPs for nano-priming chickpea seeds. Their study found the treatment was able to enhance the storage period and suppress infection by *F. oxysporum* [159]. Similarly, a study by Kaur et al. found that Ag nanoparticles (Ag NPs) were effective against chickpea fungal diseases 160]. In the area of genomic studies with respect to chickpeas varieties, the genome sequence of the *kabuli* chickpea variety was made by the team of Varshney et al. in 2013, and it contains an estimated 28,269 genes [161].

16.7 NANO-DELIVERY PLATFORM TECHNOLOGIES

When a biomolecule is delivered to an organism for biomedical purposes, it must also withstand the effects of an unfavorable physiological environment before reaching its target. On reaching its target, the desired amount of active compound must be delivered at a suitable rate over time to achieve the desired therapeutic effect. Therefore, to achieve this goal, the drug delivery system must overcome several challenges before it can deliver drug agents in a safe and reproducible manner at the desired target site. In addition, the prolonged availability at the intended site of action defines the potency and efficacy of the exogenously administrated bioactive molecule. Apart from the intrinsic factors related to the nature of the molecule itself, certain extrinsic factors, such as the physiological state of the receptor organism, enzymatic machinery, and external pH in the surrounding environment, reduce the effectiveness of the drug's performance [162].

A drug delivery system is defined as an engineered device that is used to transport a biomedical or pharmaceutical compound throughout the body and to release its active ingredient in a controlled manner. Possible physical, chemical, or enzymatic disruptions of the active ingredients are reduced by conjugating the active ingredients with another material or by encapsulating the active ingredients within a protective shell-like structure. These types of modifications also help in releasing the active ingredient in a controlled manner and increase its bioavailability. Further, possible undesirable side effects resulting from the unspecific systemic distribution of the active ingredient are also minimized since it is released at a specific site [163].

On another biomedical front, advanced methods like gene therapy were developed because of the Human Genome Project, and the progress in elucidating the molecular basis of genetic diseases has accelerated the technology further. Gene therapy involves the use of nucleic acids as functional molecules to treat a wide range of diseases [164]. Along with recent discoveries like RNA interference (RNAi) and CRISPR-based genome editing, nucleic acids have attracted significant interest aimed at understanding the human genome [165, 166]. Gene therapy uses genetic engineering tools to segment DNA which alters the expression of the target gene. It is also used to modify the biological properties of living cells for therapeutic purposes. There are many gene therapy products that have been approved by several regulatory agencies globally. The most recent is the authorization of mRNA vaccines to fight the COVID-19 virus [167].

Targeted gene delivery is a complex process and involves crossing several biological barriers. Hence, both scientific and technical issues must be resolved before the benefits of gene therapy can be harnessed [168]. For instance, the injection of naked plasmid DNA has poor transfection efficiency and is not very effective. To overcome such difficulties, the DNA needs to be delivered using a combination of various types of vectors. For instance, there are several different viral and non-viral vectors designed for the delivery of genetic material across cell membranes and other biological barriers to reach the cell nucleus. These types of vectors are contributing to the success of gene therapy techniques [169]. There are many viral vectors available for gene therapy systems, such as retrovirus, adenovirus (types 2 and 5), adeno-associated viruses, herpes virus, pox virus, human foamy virus (HFV), and lentivirus [170]. However, before using viral vectors for gene therapy, all viral vector genomes are modified by deleting some areas of their genomes. This deleting process prevents their replication and makes them safe to use. However, the use of viral vectors for gene therapy has a number of problems. For instance, viral vectors' immunogenicity triggers the inflammatory system and leads to the degeneration of transferred tissue, toxin production, mortality, insertion mutagenesis, and limits their transgenic capacity [171]. One solution to this problem is to use a non-viral mode for delivering DNA, and this alternative approach has gained in popularity with many researchers [172]. The non-viral mode for delivering DNA generally involves chemical methods like cationic liposomes and polymers, or physical methods like gene guns, electroporation, particle bombardment, and ultrasound utilization. The efficiency of these types of delivery systems is generally less than viral systems in gene transduction. However, their cost-effectiveness, availability, and more importantly their lower immune system response have made them more effective for gene delivery [173].

Progress in NT has resulted in the development of nanoparticle-based delivery systems. Roco et al. were the first to note that nanomaterials could be used for drug delivery applications [174]. Since then, several nano-scale DNA carriers have been developed. Typical nanoparticle carriers include 1) polymeric [175]; 2) silica-based hybrids [176]; 3) gold nano-particle-based hybrids [177]; 4) two-dimensional nanomaterials [178]; and lipid-based systems [179]. Nano-delivery systems designed for drug therapy have several advantages, such as the controlled release of active ingredients and targeted delivery in a time-dependent manner *via* passive or active targeting methods. In passive targeting methods, the drug nano-carrier design is based on pathophysiological features that allow the accumulation of the nano-sized drug on the targeted tissue. Active targeting involves the coupling or assembly of surface-active ligands onto the surface of the drug carrier system. The ligands can recognize and interact with a receptor on the target cell. Because of the interaction between ligands and receptors, the drug delivery specificity and nanoparticle uptake are greatly enhanced [163]. Different types of ligands such as engineered antibodies, growth factors [180], vitamins [181], and aptamers [182] have been successfully tested *in vitro* by several researchers.

For gene editing, the CRISPR/Cas9 complex needs to operate on the nuclear genome; and its components need to be transferred into the targeted nucleus. Like DNA delivery systems, efficient delivery of CRISPR/Cas9 to target tissues or cells also must overcome considerable challenges. The most important challenge is to overcome tissue barriers and cell membranes. Presently, CRISPR/Cas9 is being delivered by various methods such as non-viral vectors, viral vectors, and physical methods. Among these methods, the virus-mediated gene delivery system is the most widely used method. The method involves integrating CRISPR/Cas9-encoding sequences into the viral genome and then releasing the CRISPR/Cas9 gene complex into targeted cells. However, during this process, the viral vectors may integrate into host cells and cause problems such as mutations and carcinogenesis, and activate an immune response [183, 184]. Additionally, the virus has a limited load-carrying capacity to deliver CRISPR/Cas9 DNA [185]. Hence, the efficiency of viral vectors to deliver the CRISPR/Cas9 system safely still remains to be determined.

Out of all the physical methods, electroporation is the most commonly used method for transferring CRISPR/Cas systems into cells. Single-cell microinjection is another physical delivery tool for delivering CRISPR/Cas9 and has been widely used in embryo gene editing and transgenic animal production. The transfer of Cas9 DNA or protein components using microinjection has shown high transduction efficiency and low cytotoxicity. However, it is time-consuming and labor-intensive, which limits its application to a small number of species [186]. In recent years, studies have shown that nanomaterials can be used for gene delivery. Several nano-carrier-based systems have been developed for delivering CRISPR/Cas9. These include cationic liposomes, lipid nanoparticles (LNPs), cationic polymers, vesicles, and gold nanoparticles [187, 188]. Furthermore, exosomes are also another promising non-viral nano-carrier method for delivering the CRISPR/Cas9 system in both *in vitro* and *in vivo* applications. This type of nano-carrier serves as a unique biomolecular tool for expanding the use of gene editing technologies [188].

16.8 FUTURE PERSPECTIVES AND CONCLUDING REMARKS

The use of nanometer-scale devices has enabled researchers to detect and treat infections, nutrient deficiencies, and many other health problems long before the symptoms are evident at the macro-scale level. Nanomaterials used in the medical field have significant advantages such as increased penetration, more surface area and greater uptake of active ingredients, and specific targeting and controlled release of payloads. These advantageous features make NT-based technologies and materials ideal for many applications in agriculture [189]. Like biomedical delivery systems, delivery systems designed for agriculture are also quite important for the delivery of nano-pesticides and nano-fertilizers. In addition, NT-based systems can be used for the transfer of genetic material for plant improvement, thus revolutionizing traditional agriculture systems. To date, nanomaterials are used in agricultural applications such as nano-pesticide formulations, nano-fertilizers, and plant growth regulators (PGRs). Nanomaterials are also being developed for bio-pesticides including nucleic acid-based pesticides and pheromones [190–192]. Importantly, in 2019, IUPAC selected ten chemical technologies most likely to impact human society in the future, and nano-pesticides were ranked first for their potential low impact on the environment and human health. [193]. Thus, nano-agrochemical delivery systems *via* nanomaterials have the greatest potential for improving the efficacy of agricultural inputs, alleviating environmental pollution, saving labor costs, and reducing health issues for humans. This in turn significantly contributes to the maintenance of sustainable agricultural systems and improves food security. Furthermore, nanomaterials will have an important role to play in the future development of more sustainable agricultural practices [194]. In addition, the use of nanoparticle-assisted delivery of genetic material is being currently being advanced to assist in the development of insect-resistant plant varieties. For instance, DNA-coated gold nanoparticles are being used as bullets in a "gene gun" system, to achieve gene transfer into plant cells and tissues [195]. Recently, a study by Zhang et al. showed that when different gold nanoparticle shapes and sizes were used to carry RNAi and then injected into GFP-overexpressing of *N. benthamiana* leaves, the bar-shaped Au nanoparticles had the highest transfer efficiency [196]. Other studies have shown several other types of nanoparticles, such as mesoporous silica nanoparticles, and single and multiple wall nanotubes (SWNTs and MWNTs), have the potential to be used as nano-vectors for gene delivery in plant genome engineering procedures [197].

Since its inception in the 2000s alongside other AFM probes, the technology behind "Dip-Pen" nanolithography (DPN) has greatly evolved and can today achieve nano-fabrication and material discovery [198]. Currently, there are various methods of DPN such as electrochemical, thermal, and other forms based on single cantilever, 1D multipen, and 2D 55,000 cantilever technologies. In 2002, the Northwestern group pioneered the use of DPN to deliver directly oligonucleotides onto metallic and insulator substrates. In this case, the DNA was modified with a hexanethiol group and features ranging from micrometre structures to less than 100 nm were created and the sequence specific binding properties of the DNA were not lost even after transfered and patterned onto the surface. The important step was to silanize the tip prior to the DNA dipping and the team found that the positively hydrophilic tip was readily wetted by the DNA ink so that it could be used for several hours between recoating procedures [199]. Capitalizing on the need to modify genetic material and specific DNA bases, and the technological advances made in DPN, there is a scope that direct, *in situ* modification of specific strands of DNA is a possibility as shown in Figure 16.3.

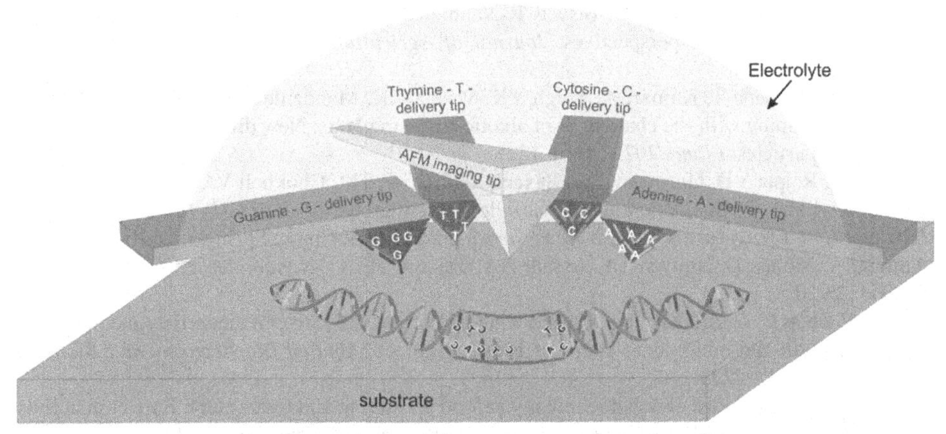

FIGURE 16.3 Concept of imaging and molecular in situ modification of DNA bases in its native electrolyte by Dip-Pen Nanolithography.

In principle, these stylus technologies can place or replace specific DNA bases such as Thymine, Guanine, Cytosine, and Adenine, to repair and modify dysfunctional genes and even synthesize novel combinations of DNA bases without relying solely on the DNA inherent molecular machinery, which allows the next generation of agroscientists to explore new and robust varieties of plants and enhance crop production. This will be in a sense the accomplishment of Feyman's dream of the new field of NT that he envisaged in 1959.

REFERENCES

1. Chokheli V, Rajput V, Dmitriev P, Varduny T, Minkina T, Singh RK, Singh A. Status and policies of GM crops in Russia. In *Policy Issues in Genetically Modified Crops*, pp. 57–74. Academic Press, 2021.
2. Rizwan M, Ali S, Qayyum MF, Ok YS, Adrees M, Ibrahim M, Zia-ur-Rehman M, Farid M, Abbas F. Effect of metal and metal oxide nanoparticles on growth and physiology of globally important food crops: A critical review. *Journal of Hazardous Materials* 2017; 322: 2–16.
3. Usman M, Farooq M, Wakeel A, Nawaz A, Cheema SA, ur Rehman H, Ashraf I, Sanaullah M. Nanotechnology in agriculture: Current status, challenges and future opportunities. *Science of the Total Environment* 2020; 721: 137778.
4. Kah M, Tufenkji N, White JC. Nano-enabled strategies to enhance crop nutrition and protection. *Nature Nanotechnology* 2019; 14(6): 532–540.
5. Food and Agriculture Organization (FAO). The future of food and agriculture–Trends and challenges. *Annual Report* 2017; 296: 1–180.
6. Feynman RP. There's plenty of room at the bottom. *Engineering and Science* 1960; 23(5): 22–36.
7. Young R, Ward J, Scire F. The topografiner: An instrument for measuring surface microtopography. *Review of Scientific Instruments* 1972; 43(7): 999–1011.
8. Chauhan R, Yadav HO, Sehrawat N. Nanobioremediation: A new and a versatile tool for sustainable environmental clean-up: Overview. *Journal of Materials and Environmental Science* 2020; 11(4): 564–573.
9. Cecchin I, Reddy KR, Thome A, Tessaro EF, Schnaid F. Nanobioremediation: Integration of nanoparticles and bioremediation for sustainable remediation of chlorinated organic contaminants in soils. *International Biodeterioration & Biodegradation* 2017; 119: 419–428.
10. Rizwan M, Singh M, Mitra CK, Morve RK. Ecofriendly application of nanomaterials: Nanobioremediation. *Journal of Nanoparticles* 2014; 2014: 1–7.
11. Shende S, Rajput VD, Gade A, Minkina T, Fedorov Y, Sushkova S, Mandzhieva S, Burachevskaya M, Boldyreva V. Metal-based green synthesized nanoparticles: Boon for sustainable agriculture and food security. *IEEE Transactions on NanoBioscience* 2021; 21(1): 44–54.
12. Raliya R, Saharan V, Dimkpa C, Biswas P. Nanofertilizer for precision and sustainable agriculture: Current state and future perspectives. *Journal of Agricultural and Food Chemistry* 2017; 66(26): 6487–6503.
13. Rajput VD, Minkina T, Kumari A, Singh VK, Verma KK, Mandzhieva S, Sushkova S, Srivastava S, Keswani C. Coping with the challenges of abiotic stress in plants: New dimensions in the field application of nanoparticles. *Plants* 2021; 10(6): 1221.
14. Shende SS, Rajput VD, Gorovtsov AV, Saxena P, Minkina TM, Chokheli VA, Jatav HS, Sushkova SN, Kaur P, Kizilkaya R. Interaction of nanoparticles with microbes. In *Plant-Microbes-Engineered Nano-Particles (PM-ENPs) Nexus in Agro-Ecosystems*, pp. 175–188. Cham: Springer, 2021.
15. Amrane A, Mohan D, Nguyen TA, Assadi AA, Yasin G (Eds.). *Nanomaterials for Soil Remediation*. Elsevier, 2020.
16. Sharma P, Bhatt D, Zaidi MGH, Saradhi PP, Khanna PK, Arora S. Silver nanoparticle-mediated enhancement in growth and antioxidant status of Brassica juncea. *Applied Biochemistry and Biotechnology* 2012; 167(8): 2225–2233.
17. Duncan TV. Applications of nanotechnology in food packaging and food safety: Barrier materials, antimicrobials and sensors. *Journal of Colloid and Interface Science* 2011; 363(1): 1–24.
18. Shang Y, Hasan MK, Ahammed GJ, Li M, Yin H, Zhou J. Applications of nanotechnology in plant growth and crop protection: A review. *Molecules* 2019; 24(14): 2558.
19. Kwak SY, Lew TTS, Sweeney CJ, Koman VB, Wong MH, Bohmert-Tatarev K, Snell KD, Seo JS, Chua NH, Strano MS. Chloroplast-selective gene delivery and expression in planta using chitosan-complexed single-walled carbon nanotube carriers. *Nature Nanotechnology* 2019; 14(5): 447–455.

20. Kwak SY, Wong MH, Lew TTS, Bisker G, Lee MA, Kaplan A, Dong J, Liu AT, Koman VB, Sinclair R, Hamann C. Nanosensor technology applied to living plant systems. *Annual Review of Analytical Chemistry* 2017; 10: 113–140.
21. Demirer GS, Silva TN, Jackson CT, Thomas JB, Ehrhardt D, Rhee SY, Mortimer JC, Landry MP. Nanotechnology to advance CRISPR–Cas genetic engineering of plants. *Nature Nanotechnology* 2021; 16(3): 243–250.
22. Lew TTS, Park M, Wang Y, Gordiichuk P, Yeap WC, Mohd Rais SK, Kulaveerasingam H, Strano MS. Nanocarriers for transgene expression in pollen as a plant biotechnology tool. *ACS Materials Letters* 2020; 2(9): 1057–1066.
23. Zhang H, Cao Y, Xu D, Goh NS, Demirer GS, Cestellos-Blanco S, Chen Y, Landry MP, Yang P. Gold-nanocluster-mediated delivery of siRNA to intact plant cells for efficient gene knockdown. *Nano Letters* 2021; 21(13): 5859–5866.
24. Demirer GS, Zhang H, Matos JL, Goh NS, Cunningham FJ, Sung Y, Chang R, Aditham AJ, Chio L, Cho MJ, Staskawicz B. High aspect ratio nanomaterials enable delivery of functional genetic material without DNA integration in mature plants. *Nature Nanotechnology* 2019; 14(5): 456–464.
25. Demirer GS, Zhang H, Goh NS, Pinals RL, Chang R, Landry MP. Carbon nanocarriers deliver siRNA to intact plant cells for efficient gene knockdown. *Science Advances* 2020; 6(26): eaaz0495.
26. Zhang H, Demirer GS, Zhang H, Ye T, Goh NS, Aditham AJ, Cunningham FJ, Fan C, Landry MP. DNA nanostructures coordinate gene silencing in mature plants. *Proceedings of the National Academy of Sciences of the United States of America* 2019; 116: 7543–7548.
27. Cunningham FJ, Goh NS, Demirer GS, Matos JL. Nanoparticle-mediated delivery towards advancing plant genetic engineering. *Trends Biotechnology* 2015; 8(36): 882–897.
28. Blattner FR, Plunkett III G, Bloch CA, Perna NT, Burland V, Riley M, Collado-Vides J, Glasner JD, Rode CK, Mayhew GF, Gregor J. The complete genome sequence of Escherichia coli K-12. *Science* 1997; 277(5331): 1453–1462.
29. Mayer KF, Taudien S, Martis M, Šimková H, Suchánková P, Gundlach H, ... Stein N. Gene content and virtual gene order of barley chromosome 1H. *Plant Physiology* 2009; 151(2): 496–505.
30. Samori B, Siligardi G, Quagliariello C, Weisenhorn AL, Vesenka J, Bustamante CJ. Chirality of DNA supercoiling assigned by scanning force microscopy. *Proceedings of the National Academy of Sciences* 1993; 90(8): 3598–3601.
31. Hansma HG, Browne KA, Bezanilla M, Bruice TC. Bending and straightening of DNA induced by the same ligand: Characterization with the atomic force microscope. *Biochemistry* 1994; 33(28): 8436–8441.
32. Bezanilla M, Drake B, Nudler E, Kashlev M, Hansma PK, Hansma HG. Motion and enzymatic degradation of DNA in the atomic force microscope. *Biophysical Journal* 1994; 67(6): 2454–2459.
33. Hansma HG, Laney DE, Bezanilla M, Sinsheimer RL, Hansma PK. Applications for atomic force microscopy of DNA. *Biophysical Journal* 1995; 68(5): 1672–1677.
34. Varshney RK, Bohra A, Roorkiwal M, Barmukh R, Cowling WA, Chitikineni A, Lam HM, Hickey LT, Croser JS, Bayer PE, Edwards D. Fast-forward breeding for a food-secure world. *Trends in Genetics* 2021; 37(12): 1124–1136.
35. Samantara K, Bohra A, Mohapatra SR, Prihatini R, Asibe F, Singh L, Reyes VP, Tiwari A, Maurya AK, Croser JS, Wani SH. Breeding more crops in less time: A perspective on speed breeding. Biology 2022; 11(2): 275.
36. Chilton MD. Nature, the first creator of GMOs. *Forbes*. Retrieved 24 November 2022.
37. Blakemore E. The first GMO is 8,000 years old. *Smithsonian*. Retrieved 24 November 2022.
38. Food and Agriculture Organization (FAO). Chapter 2: Description and definitions. Retrieved 15 November 2022.
39. Fridovich-Keil JL, Diaz JM. Genetically modified organism. *Encyclopaedia Britannica*. Retrieved 19 October 2022.
40. Stocum D. Somatic cell nuclear transfer. *Encyclopaedia Britannica*. https://www.britannica.com/science/somatic-cell-nuclear-transfer.
41. Khan S, Ullah MW, Siddique R, Nabi G, Manan S, Yousaf M, Hou H. Role of recombinant DNA technology to improve life. *International Journal of Genomics* 2016; 2016: 2405954.
42. Phillips T. Genetically modified organisms (GMOs): Transgenic crops and recombinant DNA technology. *Nature Education* 2008; 1(1): 213.
43. Buiatti M, Christou P, Pastore G. The application of GMOs in agriculture and in food production for a better nutrition: Two different scientific points of view. *Genes Nutrition* 2013; 8(3): 255–270.
44. Zhang C, Wohlhueter R, Zhang H. Genetically modified foods: A critical review of their promise and problems. *Food Science and Human Wellness* 2016; 5(3): 116–123.

45. Jaenisch R, Mintz B. Simian virus 40 DNA sequences in DNA of healthy adult mice derived from preimplantation blastocysts injected with viral DNA. *Proceedings of the National Academy of Sciences of the United States of America* 1974; 71(4): 1250–1254.
46. Bevan MW, Flavell RB, Chilton MD. A chimaeric antibiotic resistance gene as a selectable marker for plant cell transformation. *Nature* 1983; 304(5922): 184.
47. Bruening G, Lyons JM. The case of the FLAVR SAVR tomato. *California Agriculture* 2000; 54(4): 6–7.
48. Vazquez-Salat N, Salter B, Smets G, Houdebine LM. The current state of GMO governance: Are we ready for GM animals? *Biotechnology Advances* 2012; 30(6): 1336–1343.
49. Pollack A. Genetically engineered salmon approved for consumption (19 November 2015). *The New York Times*. ISSN 0362-4331. Retrieved 27 October 2022.
50. Balter M. Farming was so nice, it was invented at least twice. http://news.sciencemag.org/archaeology/2013/07/farming-was-so-nice-it-was-invented-least-twice.
51. The Evolution of Corn. Genetics learning center. *University of Utah*, July 2015. http://learn.genetics.utah.edu/content/selection/corn/.
52. Smyth SJ. The human health benefits from GM crops. *Plant Biotechnology Journal* 2020; 18(4): 887–888.
53. Van Eenennaam AL, Wells KD, Murray JD. Proposed U.S. regulation of gene-edited food animals is not fit for purpose. *NPJ Science Food* 2019; 20: 3.
54. Dederer HG, Hamburger D. Are genome-edited micro-organisms covered by Directive 2009/41/EC?-Implications of the CJEU's judgment in the case C-528/16 for the contained use of genome-edited micro-organisms. *Journal of Law and Biosciences* 2022; 9(1): lsab033.
55. Bhattacharya A, Parkhi V, Char B. Genome editing for crop improvement: A perspective from India. *In Vitro Cellular & Developmental Biology – Plant* 2021; 57: 565–573.
56. Daboussi F, Stoddard TJ, Zhang F. Engineering meganuclease for precise plant genome modification. In: *Advances in New Technology for Targeted Modification of Plant Genomes*, pp. 21–38. New York, NY: Springer, 2015.
57. Urnov FD, Rebar EJ, Holmes MC, Zhang HS, Gregory PD. Genome editing with engineered zinc finger nucleases. *Nature Reviews Genetics* 2010; 11(9): 636–646.
58. Joung JK, Sander JD. TALENs: A widely applicable technology for targeted genome editing. *Nature Reviews Molecular Cell Biology* 2013; 14(1): 49–55.
59. Chin JS, Chooi WH, Wang H, Ong W, Leong KW, Chew SY. Scaffold-mediated non-viral delivery platform for CRISPR/Cas9-based genome editing. *Acta Biomaterials* 2019; 90: 60–70.
60. Jin L, Lange W, Kempmann A, Maybeck V, Günther A, Gruteser N, Baumann A, Offenhäusser A. High-efficiency transduction and specific expression of ChR2opt for optogenetic manipulation of primary cortical neurons mediated by recombinant adeno-associated viruses. *Journal of Biotechnology* 2016; 233: 171–180.
61. Rey-Rico A, Cucchiarini M. Controlled release strategies for rAAV mediated gene delivery. *Acta Biomaterials* 2016; 29: 1–10.
62. Carroll D. Genome engineering with zinc-finger nucleases. *Genetics* 2011; 188(4): 773–782.
63. Davies JP, Kumar S, Sastry-Dent L. Use of zinc-finger nucleases for crop improvement. *Progress in Molecular Biology and Translational Science* 2017; 149: 47–63.
64. Boch J. TALEs of genome targeting. *Nature Biotechnology* 2011; 29(2): 135–136.
65. Sun B, Yang J, Yang S, Ye RD, Chen D, Jiang Y. A CRISPR-Cpf1-assisted non-homologous end joining genome editing system of mycobacterium smegmatis. *Biotechnology Journal* 2018; 13(9): e1700588.
66. Asmamaw M, Zawdie B. Mechanism and applications of CRISPR/Cas-9-mediated genome editing. *Biologics: Targets and Therapy* 2021; 15: 353–361.
67. Marraffini LA, Sontheimer EJ. CRISPR interference: RNA directed adaptive immunity in bacteria and archaea. *Nature Reviews Genetics* 2010; 11(3): 181–190.
68. Jiang F, Doudna JA. CRISPR-Cas9 structures and mechanisms. *Annual Review of Biophysics* 2017; 46: 505–529.
69. Liu J, Chen J, Zheng X, Wu F, Lin Q, Heng Y, Tian P, Cheng Z, Yu X, Zhou K, Zhang X. GW5 acts in the brassinosteroid signalling pathway to regulate grain width and weight in rice. *Nature Plants* 2017; 3: 17043.
70. Lino CA, Harper JC, Carney JP, Timlin JA. Delivering CRISPR: A review of the challenges and approaches. *Drug Delivery* 2018; 25: 1234–1257.
71. Poinern GE, Fawcett D. Food waste valorization: New manufacturing processes for long-term sustainability. Elsevier.

72. Rodrigues SM, Demokritou P, Dokoozlian N, Hendren CO, Karn B, Mauter MS, Lowry GV. Nanotechnology for sustainable food production: Promising opportunities and scientific challenges. *Environmental Science: Nano* 2017; 4(4): 767–781.
73. Fenner K, Canonica S, Wackett LP, Elsner M. Evaluating pesticide degradation in the environment: Blind spots and emerging opportunities. *Science* 2013; 341(6147): 752–758.
74. Shah M, Fawcett D, Sharma S, Tripathy SK, Poinern GE. Green synthesis of metallic nanoparticles via biological entities. *Materials* 2015; 8(11): 7278–7308.
75. Piner RD, Zhu J, Xu F, Hong S, Mirkin CA. "Dip-pen" nanolithography. *Science* 1999; 283(5402): 661–663.
76. Liang J, Castronovo M, Scoles G. DNA as invisible ink for AFM nanolithography. *Journal of the American Chemical Society* 2012; 134(1): 39–42.
77. Pyne A, Thompson R, Leung C, Roy D, Hoogenboom BW. Single-molecule reconstruction of oligonucleotide secondary structure by atomic force microscopy. *Small* 2014; 10(16): 3257–3261.
78. Lowry GV, Avellan A, Gilbertson LM. Opportunities and challenges for nanotechnology in the agritech revolution. *Nature Nanotechnology* 2019; 14: 517–522.
79. Hofmann T, Lowry GV, Ghoshal S, Tufenkji N, Brambilla D, Dutcher JR, Gilbertson LM, Giraldo JP, Kinsella JM, Landry MP, Lovell W. Technology readiness and overcoming barriers to sustainably implement nanotechnology-enabled plant agriculture. *Nature Food* 2019; 1: 416–425.
80. Demirer GS, Chang R, Zhang H, Chio L, Landry MP. Nanoparticle-guided biomolecule delivery for transgene expression and gene silencing in mature plants. *Biophysics Journal* 2018; 114: 217a.
81. Kah M, Tufenkji N, White JC. Nano-enabled strategies to enhance crop nutrition and protection. *Nature Nanotechnology* 2019; 14: 532–540.
82. Giraldo JP, Wu H, Newkirk GM, Kruss S. Nanobiotechnology approaches for engineering smart plant sensors. *Nature Nanotechnology* 2019; 14: 541–553.
83. FAO. *A Regional Rice Strategy for Sustainable Food Security in Asia and the Pacific*. Bangkok: FAO Regional Office for Asia and the Pacific, 2003.
84. Derosa M, Monreal C, Schnitzer M, Walsh R, Sultan Y. Nanotechnology in fertilizers. *Nature Nanotechnology* 2010; 5: 91.
85. Meena RS, Yadav RS. Yield and profitability of groundnut (Arachis hypogaea L) as influenced by sowing dates and nutrient levels with different varieties. *Legume Research* 2015; 38(6): 791–797.
86. Wang Y, Deng C, Rawat S, Cota-Ruiz K, Medina-Velo I, Gardea-Torresdey JL. Evaluation of the effects of nanomaterials on rice (*Oryza sativa* L.) responses: Underlining the benefits of nanotechnology for agricultural applications. *ACS Agricultural Science & Technology* 2021; 1(2): 44–54.
87. Cakmak I. Enrichment of cereal grains with zinc: Agronomic or genetic biofortification? *Plant Soil* 2008; 302: 1–17.
88. Yan S, Wu F, Zhou S, Yang J, Tang X, Ye W. Zinc oxide nanoparticles alleviate the arsenic toxicity and decrease the accumulation of arsenic in rice (*Oryza sativa* L.). *BMC Plant Biology* 2021; 21: 150.
89. Lu L, Tian S, Liao H, Zhang J, Yang X, Labavitch JM, Chen W. Analysis of metal element distributions in rice (*Oryza sativa* L.) seeds and relocation during germination based on X-ray fluorescence imaging of Zn, Fe, K, Ca, and Mn. *PLoS ONE* 2013; 8(2): e57360.
90. Wu F, Fang Q, Yan SW, Pan L, Tang X, Ye W. Effects of zinc oxide nanoparticles on arsenic stress in rice (*Oryza sativa* L.): Germination, early growth, and arsenic uptake. *Environmental Science and Pollution Research* 2020; 27: 26974–26981.
91. Elemike EE, Uzoh IM, Onwudiwe DC, Babalola OO. The role of nanotechnology in the fortification of plant nutrients and improvement of crop production. *Applied Sciences* 2019; 9: 499.
92. Zhang H, Wang R, Chen Z, Cui P, Lu H, Yang Y, Zhang H. The effect of zinc oxide nanoparticles for enhancing rice (*Oryza sativa* L.) yield and quality. *Agriculture* 2021; 11: 1247.
93. Samart S, Chutipaijit S. Growth of pigmented rice (*Oryza sativa* L. cv. Rice berry) exposed to ZnO nanoparticles. *Materials Today: Proceedings* 2019; 17: 1987–1997.
94. Ikhajiagbe B, Igiebor FA, Ogwu MC. Growth and yield performances of rice (*Oryza sativa* var. nerica) after exposure to biosynthesized nanoparticles. *Bulletin of the National Research Centre* 2021; 45: 62.
95. Valojai STS, Niknejad Y, Amoli HF, Tari DB. Response of rice yield and quality to nano-fertilizers in comparison with conventional fertilizers. *Journal of Plant Nutrition* 2021; 44(13): 1971–1981.
96. Fakharzadeh S, Hafizi M, Baghaei MA, Etesami M, Khayamzadeh M, Kalanaky S, Akbari ME, Nazaran MH. Using nanochelating technology for biofortification and yield increase in rice. *Science Reports* 2020; 10: 4351.

97. Younas A, Yousaf Z, Riaz N, Rashid M, Razzaq Z, Tanveer M, Huang S. Role of Nanotechnology for enhanced rice production. In *Nutrient Dynamics for Sustainable Crop Production* (Ed. Meena RS). Springer Nature Singapore Pte Ltd., 2020.
98. Sanzari I, Leone A, Ambrosone A. Nanotechnology in plant science: To make a long story short. *Frontiers in Bioengineering and Biotechnology* 2019; 7: 120.
99. Wang P, Lombi E, Zhao FJ, Kopittke PM. Nanotechnology: A new opportunity in plant sciences. *Trends Plant Science* 2016; 21: 699–712.
100. Giraldo JP, Landry MP, Faltermeier SM, McNicholas TP, Iverson NM, Boghossian AA, Reuel NF, Hilmer AJ, Sen F, Brew JA, Strano MS. Plant nanobionics approach to augment photosynthesis and biochemical sensing. *Nature Materials* 2014; 13: 400–408.
101. Rai V, Acharya S, Dey N. Implications of nanobiosensors in agriculture. *Journal of Biomaterials and Nanobiotechnology* 2012; 3: 315–324.
102. Abd-Elsalam KA, Alghuthaymi MA. Nanodiagnostic tools in plant breeding. *Journal of Plant Pathology and Microbiology* 2014; 5: e107.
103. Mishra S.K, Kumar P, Balamurugan R, Mandal M, Kannan C. Bio-degradable- carbonoid-metabolite nanoparticles for crop protection. *IPA No 202111061106,* 2022.
104. Kahn MK, Pandey A, Hamurcu M, Gezgin S, Athar T, Rajput VD, Gupta OP, Minkina T. Insight into the prospects for nanotechnology in wheat biofortification. *Biology* 2021; 10(11): 1123.
105. Naderi MR, Danesh-Shahraki A. Nanofertilizers and their roles in sustainable agriculture. *International Journal of Agriculture and Crop Sciences* 2013; 5(19): 2229.
106. Rameshaiah GN, Pallavi J, Shabnam S. Nanofertilizers and nano sensors – An attempt for developing smart agriculture. *International Journal of Engineering Research and General Science* 2015; 3(1): 314–320.
107. Abdel-Aziz HM, Hasaneen MN, Omer AM. Nano chitosan-NPK fertilizer enhances the growth and productivity of wheat plants grown in sandy soil. *Spanish Journal of Agricultural Research* 2016; 14(1): 0902.
108. Zhu J, Li J, Shen Y, Liu S, Zeng N, Zhan X, White JC, Gardea-Torresdey J, Xing B. Mechanism of zinc oxide nanoparticle entry into wheat seedling leaves. *Environmental Science: Nano* 2020; 7: 3901–3913.
109. Ahmed F, Javed B, Razzaq A, Mashwani ZU. Applications of copper and silver nanoparticles on wheat plants to induce drought tolerance and increase yield. *IET Nanobiotechnology* 2021; 15(1): 68–78.
110. Taran N, Storozhenko V, Svietlova N, Batsmanova L, Shvartau V, Kovalenko M. Effect of zinc and copper nanoparticles on drought resistance of wheat seedlings. *Nanoscale Research Letters* 2017; 12(60): 1–6.
111. Almutairi ZM, Alharbi A. Effect of silver nanoparticles on seed germination of crop plants. *Journal of Advanced Agricultural Technologies* 2015; 10(1): 283–288.
112. Hafeez A, Razzaq A, Mahmood T, Jhanzab HM. Potential of copper nanoparticles to increase growth and yield of wheat. *Journal of Nanoscience With Advanced Technology* 2015; 1(1): 6–11.
113. Yasmeen F, Razzaq A, Iqbal MN, Jhanzab HM. Effect of silver, copper and iron nanoparticles on wheat germination. *International Journal of Biological Sciences* 2015; 6(4): 112–117.
114. Villagarcia H, Dervishi E, de Silva K, Biris AS, Khodakovskaya MV. Surface chemistry of carbon nanotubes impacts the growth and expression of water channel protein in tomato plants. *Small* 2012; 8(15): 2328–2334.
115. Miralles P, Church TL, Harris AT. Toxicity, uptake, and translocation of engineered nanomaterials in vascular plants. *Environmental Science and Technology* 2012; 46(17): 9224–9239.
116. Hu X, Zhou Q. Novel hydrated graphene ribbon unexpectedly promotes aged seed germination and root differentiation. *Scientific Reports* 2014; 4: 3782.
117. Mahmoodzadeh H, Aghili R. Effect on germination and early growth characteristics in wheat plants (*Triticum aestivum* L.) seeds exposed to TiO_2 nanoparticles. *Journal of Chemical Health Risks* 2014; 4(1): 29–36.
118. Mishra S, Singh BR, Singh A, Keswani C, Naqvi AH, Singh HB. Biofabricated silver nanoparticles act as a strong fungicide against Bipolaris sorokiniana causing spot blotch disease in wheat. *PLoS ONE* 2014; 9(5): e97881.
119. Riahi-Madvar A, Rezaee F, Jalali V. 2012. Effects of alumina nanoparticles on morphological properties and antioxidant system of Triticum aestivum. *Iranian Journal of Plant Physiology* 2012; 3: 595–603.
120. Ramesh M, Palanisamy K, Babu K, Sharma NK. Effects of bulk & nano-titanium dioxide and zinc oxide on physio-morphological changes in Triticum aestivum Linn. *Journal of Global Bioscience* 2014; 3: 415–422.

121. Mohamed AKS, Qayyum MF, Abdel-Hadi AM, Rehman RA, Ali S, Rizwan M. Interactive effect of salinity and silver nanoparticles on photosynthetic and biochemical parameters of wheat. *Archives of Agronomy and Soil Science* 2017; 63(12): 1736–1747.
122. Singh S, Singh M, Agrawal VV, Kumar A. An attempt to develop surface plasmon resonance based immunosensor for Karnal bunt (*Tilletia indica*) diagnosis based on the experience of nano-gold based lateral flow immune-dipstick test. *Thin Solid Films* 2010; 519: 1156–1159.
123. Campagnoli A, Cheli F, Polidori C, Zaninelli M, Zecca O, Savoini G, Pinotti L, Dell'Orto V. Use of the electronic nose as a screening tool for the recognition of durum wheat naturally contaminated by deoxynivalenol: A preliminary approach. *Sensors* 2011; 11: 4899–4916.
124. Wu J, Jayas DS, Zhang Q, White ND, York RK. Feasibility of the application of electronic nose technology to detect insect infestation in wheat. *Canadian Biosystems Engineering* 2013; 55: 3.1–3.9.
125. Eifler J, Martinelli E, Santonico M, Capuano R, Schild D, Di Natale C. Differential detection of potentially hazardous fusarium Species in wheat grains by an electronic nose. *PLoS ONE* 2011; 6(6): e21026.
126. Savi GD, Piacentini KC, de Souza SR, Costa ME, Santos CM, Scussel VM. Efficacy of zinc compounds in controlling Fusarium head blight and deoxynivalenol formation in wheat (*Triticum aestivum* L.). *International Journal of Food Microbiology* 2015; 205: 98–104.
127. Panyuta O, Belava V, Fomaidi S, Kalinichenko O, Volkogon M, Taran N. The effect of presowing seed treatment with metal nanoparticles on the formation of the defensive reaction of wheat seedlings infected with the eyespot causal agent. *Nanoscale Research Letters* 2016; 11(1): 92.
128. Kheiri A, Jorf SM, Malihipour A, Saremi H, Nikkhah M. Application of chitosan and chitosan nanoparticles for the control of Fusarium head blight of wheat (*Fusarium graminearum*) in vitro and greenhouse. *International Journal of Biological Macromolecules* 2016; 93: 1261–1272.
129. Teodoro S, Micaela B, David KW. Novel use of nano-structured alumina as an insecticide. *Pest Management Science* 2010; 66(6): 577–579.
130. Piperno DR, Ranere AJ, Holst I, Iriarte J, Dickau R. Starch grain and phytolith evidence for early ninth millennium B.P. maize from the Central Balsas River Valley, Mexico. *Proceedings of the National Academy of Sciences of the United States of America* 2009; 106: 5019–5024.
131. Burke DJ, Pietrasiak N, Situ SF, Abenojar EC, Porche M, Kraj P, Lakliang Y, Samia AC. Iron oxide and titanium dioxide nanoparticle effects on plant performance and root associated microbes. *International Journal of Molecular Sciences* 2015; 16: 23630–23650.
132. Ngo BQ, Dao TT, Nguyen CH, Tran XT, Nguyen TV, Khuu TD, Huynh TH. Effects of nanocrystalline powders (Fe, Co and Cu) on the germination, growth, crop yield and product quality of soybean (Vietnamese species DT-51). *Advances in Natural Sciences: Nanoscience and Nanotechnology* 2014; 5: 015016.
133. Umar W, Hameed MK, Aziz T, Maqsood MA, Bilal HM, Rasheed N. Synthesis, characterization and application of ZnO nanoparticles for improved growth and Zn biofortification in maize. *Archives of Agronomy and Soil Science* 2021; 67(9): 1164–1176.
134. El-Naggar ME, Abdelsalam NR, Fouda MMG, Mackled MI, Al-Jaddadi MA, Ali HM, Siddiqui MH, Kandil EE. Soil application of nano silica on maize yield and its insecticidal activity against some stored insects after the post-harvest. *Nanomaterials (Basel)* 2020; 10(4): 739.
135. Webber H, Ewert F, Olesen JE, Müller C, Fronzek S, Ruane AC, Bourgault M, Martre P, Ababaei B, Bindi M, Ferrise R. Diverging importance of drought stress for maize and winter wheat in Europe. *Nature Communications* 2018; 9: 4249.
136. Van Nguyen D, Nguyen HM, Le NT, Nguyen KH, Nguyen HT, Le HM, Nguyen AT, Dinh NT, Hoang SA, Van Ha C. Copper nanoparticle application enhances plant growth and grain yield in maize under drought stress conditions. *Journal of Plant Growth Regulations* 2022; 41: 364–375.
137. Ambrosini VG, Rosa DJ, de Melo GW, Zalamena J, Cella C, Simão DG, da Silva LS, Dos Santos HP, Toselli M, Tiecher TL, Brunetto G. High copper content in vineyard soils promotes modifications in photosynthetic parameters and morphological changes in the root system of 'Red Niagara' plantlets. *Plant Physiology and Biochemistry* 2018; 128: 89–98.
138. El-Shabrawy EM. Use silica nanoparticles in controlling late wilt disease in maize caused by harpophora maydis. *The Egyptian Journal of Applied Sciences* 2021; 36(3): 1–19.
139. Abdallah IS, Atia MA, Nasrallah AK, El-Beltagi HS, Kabil FF, El-Mogy MM, Abdeldaym EA. Effect of new preemergence herbicides on quality and yield of potato and its associated weeds. *Sustainability* 2021; 13: 9796.

140. Cocozza C, Abdeldaym EA, Brunetti G, Nigro F, Traversa A. Synergistic effect of organic and inorganic fertilization on the soil inoculum density of the soil borne pathogens Verticillium dahliae and Phytophthora spp. under open-field conditions. *Chemical and Biological Technologies in Agriculture* 2021; 8: 24.
141. Mahmoud AWM, Abdeldaym EA, Abdelaziz SM, El-Sawy MB, Mottaleb SA. Synergetic effects of zinc, boron, silicon, and zeolite nanoparticles on confer tolerance in potato plants subjected to salinity. *Agronomy* 2019; 10: 19.
142. Kafi M, Nabati J, Saadatian B, Oskoueian A, Shabahang, J. Potato response to silicone compounds (micro and nanoparticles) and potassium as affected by salinity stress. *Italian Journal of Agronomy* 2019; 14(3): 162–169.
143. Abdelaziz ME, Atia MA, Abdelsattar M, Abdelaziz SM, Ibrahim TA, Abdeldaym EA. Unravelling the role of *Piriformospora indica* in combating water deficiency by modulating physiological performance and chlorophyll metabolism-related genes in *Cucumis sativus*. *Horticulturae* 2021; 7: 399.
144. Mijweil K, Abboud AK. Growth and yield of potato (*Solanum tuberosum* L.) as influenced by nano-fertilizers and different planting dates. *Research on Crops* 2018; 19(4): 649–654.
145. Abd El-Azeim MM, Sherif MA, Hussien MS, Tantawy IA, Bashandy SO. Impacts of nano- and non-nanofertilizers on potato quality and productivity. *Acta Ecologica Sinica* 2020; 40: 388–397.
146. Thul ST, Sarangi BK, Pandey RA. Nanotechnology in agroecosystem: Implications on plant productivity and its soil environment. *The Journal of Expert Opinion on Environmental Biology* 2013; 2: 1.
147. El-Mohamedy RS, El-Aziz MEA, Kamel S. Antifungal activity of chitosan nanoparticles against some plant pathogenic fungi in vitro. *Agricultural Engineering International* 2019; 21(4): 201–209.
148. Khairy AM, Tohamy MR, Zayed MA, Mahmoud SF, El-Tahan AM, El-Saadony MT, Mesiha PK. Eco-friendly application of nano-chitosan for controlling potato and tomato bacterial wilt. *Saudi Journal of Biological Sciences* 2022; 29(4): 2199–2209.
149. Zhang L, Li M, Zhang G, Wu L, Cai D, Wu Z. Inhibiting sprouting and decreasing α-solanine amount of stored potatoes using hydrophobic nanosilica. *ACS Sustainable Chemical Engineering* 2018; 6(8): 10517–10525.
150. Barik TK, Sahu B, Swain V. Nanosilica-from medicine to pest control. *Parasitology Research* 2008; 103: 253–258.
151. Nair R, Varghese SH, Nair BG, Maekawa T, Yoshida Y, Kumar DS. Nanoparticulate material delivery to plants. *Plant Science* 2010; 179: 154–163.
152. Khot LR, Sankaran S, Mari Maja J, Ehsani R, Schuster EW. Applications of nanomaterials in agricultural production and crop protection: A review. *Crop Protection* 2012; 35: 64–70.
153. Burman U, Saini M, Kumar P. Effect of zinc oxide nanoparticles on growth and antioxidant system of chickpea seedlings. *Toxicological & Environmental Chemistry* 2013; 95(4): 605–612.
154. Anusuyan S, Nibiya Banu K. Silver-chitosan nanoparticles induced biochemical variations of chickpea (*Cicer arietinum* L.). *Biocatalysis and Agricultural Biotechnology* 2016; 8: 39–44.
155. Pandey AC, Sanjay SS, Yadav RS. Application of ZnO nanoparticles in influencing the growth rate of *Cicer arietinum*. *Journal of Experimental Nanoscience* 2010; 5: 488–497.
156. Tripathi S, Sonkar SK, Sarkar S. Growth stimulation of gram (*Cicer arietinum*) plant by water soluble carbon nanotubes. *Nanoscale* 2011; 3: 1176–1181.
157. Saharan V, Mehrotra A, Khatik R, Rawal P, Sharma SS, Pal A. Synthesis of chitosan based nanoparticles and their in vitro evaluation against phytopathogenic fungi. *International Journal of Biological Macromolecules* 2013; 62: 677–683.
158. Sathiyabama M, Indhumathi M. Chitosan thiamine nanoparticles intervene innate immunomodulation during Chickpea-Fusarium interaction. *International Journal of Biological Macromolecules* 2022; 15(198): 11–17.
159. Das G, Dutta P. Effect of nanopriming with zinc oxide and silver nanoparticles on storage of chickpea seeds and management of wilt disease. *Journal of Agricultural Science and Technology* 2022; 24(1): 213–226.
160. Kaur P, Thakur R, Duhan JS, Chaudhury A. Management of wilt disease of chickpea in vivo by silver nanoparticles biosynthesized by rhizospheric microflora of chickpea (*Cicer arietinum*). *Journal of Chemical Tenology and Biotechnology* 2018; 93(11): 3233–3243.
161. Varshney RK, Song C, Saxena RK, Azam S, Yu S, Sharpe AG, Cannon S, Baek J, Rosen BD, Tar'an B, Millan T. Draft genome sequence of chickpea (Cicer arietinum) provides a resource for trait improvement. *Nature Biotechnology* 2013; 31(3): 240–246.
162. Vega-Vasquez P, Mosier NS, Irudayaraj J. Nanoscale drug delivery systems: From medicine to agriculture. Frontiers in Bioengineering and Biotechnology 2020; 8: 79.

163. Felice B, Prabhakaran MP, Rodriguez AP, Ramakrishna S. Drug delivery vehicles on a nano-engineering perspective. *Materials Science and Engineering C* 2014; 41: 178–195.
164. Guijarro-Munoz I, Compte M, Alvarez-Vallina L, Sanz L. Antibody gene therapy: Getting closer to clinical application? *Current Gene Therapy* 2013; 13(4): 282–290.
165. Watson DJ. The human genome project: Past, present, and future. *Science* 1990; 248: 44–49.
166. Fire A, Xu S, Montgomery MK, Kostas SA, Driver SE, Mello CC. Potent and specific genetic interference by double-stranded RNA in *Caenorhabditis elegans. Nature* 1998; 391: 806–811.
167. Polack FP, Thomas SJ, Kitchin N, Absalon J, Gurtman A, Lockhart S, Perez JL, Pérez Marc G, Moreira ED, Zerbini C, Bailey R. Safety and efficacy of the BNT162b2 mRNA COVID-19 vaccine. *New England Journal of Medicine* 2020; 383: 2603–2615.
168. High KA. Gene therapy: The moving finger. *Nature* 2005; 435: 577–578.
169. Nayerossadat N, Maedeh T, Ali PA. Viral and nonviral delivery systems for gene delivery. *Advances in Biomedical Research* 2012; 1: 27.
170. Huang Y, Liu X, Dong L, Liu Z, He X, Liu W. Development of viral vectors for gene therapy for chronic pain. *Pain Research Treatment* 2011; 2011: 968218.
171. Katare DP, Aeri V. Progress in gene therapy: A review. *IJTPR* 2010; 1: 33–41.
172. Shi J, Kantoff PW, Wooster R, Farokhzad OC. Cancer nanomedicine: Progress, challenges and opportunities. *Nature Review Cancer* 2017; 17: 20–37.
173. Hirai H, Satoh E, Osawa M, Inaba T, Shimazaki C, Kinoshita S, Nakagawa M, Mazda O, Imanishi J. Use of EBV-based vector/HVJ-liposome complex vector for targeted gene therapy of EBV-associated neoplasms. *Biochemical and Biophysical Research Communications* 1997; 241: 112–118.
174. Roco MC, Williams RS, Alivisatos P. *Nanotechnology Research Directions: IWGN Workshop Report: Vision for Nanotechnology in the Next Decade.* Berlin: Springer, 2000.
175. Jawahar N, Meyyanathan SN. Polymeric nanoparticles for drug delivery and targeting: A comprehensive review. *International Journal of Health and Applied Sciences* 2012; 1(4): 217–223.
176. Carvalho AM, Cordeiro RA, Faneca H. Silica-based gene delivery systems: From design to therapeutic applications. *Pharmaceutics* 2020; 12(7): 649.
177. Shan Y, Luo T, Peng C, Sheng R, Cao A, Cao X, Shen M, Guo R, Tomás H, Shi X. Gene delivery using dendrimer-entrapped gold nanoparticles as nonviral vectors. *Biomaterials* 2012; 33: 3025–3035.
178. Kim J, Kim H, Kim WJ. Single-layered MoS2-PEI-PEG nanocomposite-mediated gene delivery controlled by photo and redox stimuli. *Small* 2016; 12: 1184–1192.
179. Li L, Zahner D, Su Y, Gruen C, Davidson G, Levkin PA. A biomimetic lipid library for gene delivery through thiol-yne click chemistry. *Biomaterials* 2012; 33: 8160–8166.
180. Lee K, Silva EA, Mooney DJ. Growth factor delivery-based tissue engineering: General approaches and a review of recent developments. *Journal of the Royal Society Interface* 2010; 8: 153–170.
181. Chen S, Zhao X, Chen J, Kuznetsova L, Wong SS, Ojima I. Mechanism-based tumor-targeting drug delivery system: Validation of efficient vitamin receptor-mediated endocytosis and drug release. *Bioconjugate Chemistry* 2010; 21: 979–987.
182. Colombo M, Mizzotti C, Masiero S, Kater MM, Pesaresi P. Peptide aptamers: The versatile role of specific protein function inhibitors in plant biotechnology. *Journal of Integrative Plant Biology* 2015; 57: 892–901.
183. Xu CL, Ruan MZC, Mahajan VB, Tsang SH. Viral delivery systems for CRISPR. *Viruses* 2019; 11: 28.
184. Yip BH. Recent advances in CRISPR/Cas9 delivery strategies. *Biomolecules Therapy* 2020; 10: 839.
185. Chew WL, Tabebordbar M, Cheng JK, Mali P, Wu EY, Ng AH, Zhu K, Wagers AJ, Church GM. A multifunctional AAV-CRISPR-Cas9 and its host response. *Nature Methods* 2016; 13: 868–874.
186. Duan L, Ouyang K, Xu X, Xu L, Wen C, Zhou X, Qin Z, Xu Z, Sun W, Liang Y. Nanoparticle delivery of CRISPR/Cas9 for genome editing. *Frontiers in Genetics* 2021; 12: 673286.
187. Chin JS, Chooi WH, Wang H, Ong W, Leong KW, Chew SY. Scaffold-mediated non-viral delivery platform for CRISPR/Cas9-based genome editing. *Acta Biomaterialia* 2019; 90: 60–70.
188. Chen F, Alphonse M, Liu, Q. Strategies for nonviral nanoparticle based delivery of CRISPR/Cas9 therapeutics. *Wiley Interdisciplinary Reviews Nanomedicine and Nanobiotechnology* 2020; 12: e1609.
189. Li P, Huang Y, Fu C, Jiang SX, Peng W, Jia Y, Peng H, Zhang P, Manzie N, Mitter N, Xu ZP. Eco-friendly biomolecule-nanomaterial hybrids as next generation agrochemicals for topical delivery. *EcoMat* 2021; 3(5): e12132.
190. Huang B, Chen F, Shen Y, Qian K, Wang Y, Sun C, Zhao X, Cui B, Gao F, Zeng Z, Cui H. Advances in targeted pesticides with environmentally responsive controlled release by nanotechnology. *Nanomaterials* 2018; 8(2): 102.

191. Adams CB, Erickson JE, Bunderson L. A mesoporous silica nanoparticle technology applied in dilute nutrient solution accelerated establishment of zoysiagrass. *Agrosystems, Geosciences & Environment* 2020; 3(1): e20006.
192. Fincheira P, Tortella G, Seabra AB, Quiroz A, Diez MC, Rubilar O. Nanotechnology advances for sustainable agriculture: Current knowledge and prospects in plant growth modulation and nutrition. *Planta* 2021; 254(4): 1–25.
193. Ten GB. Chemical innovations that will change our world: IUPAC identifies emerging technologies in chemistry with potential to make our planet more sustainable. *Chemistry International* 2019; 41(2): 12–17.
194. Pulizzi F. Nano in the future of crops. *Nature Nanotechnology* 2019; 14(6): 507.
195. Vijayakumar PS, Abhilash OU, Khan BM, Prasad BLV. Nanogold-loaded sharp-edged carbon bullets as plantgene carriers. *Advances in Functional Materials* 2010; 20: 2416–2423.
196. Zhang H, Goh NS, Wang JW, Pinals RL, González-Grandío E, Demirer GS, Butrus S, Fakra SC, Del Rio Flores A, Zhai R, Zhao B. Nanoparticle cellular internalization is not required for RNA delivery to mature plant leaves. *Nature Nanotechnology* 2022; 17: 197–205.
197. Zhi H, Zhou S, Pan W, Shang Y, Zeng Z, Zhang H. The promising nanovectors for gene delivery in plant genome engineering. *International Journal of Molecular Sciences* 2022; 23: 8501.
198. Liu G, Petrosko SH, Zheng Z, Mirkin CA. Evolution of dip-pen nanolithography (dpn): From molecular patterning to materials discovery. *Chemical Reviews* 2020; 120(13): 6009–6047.
199. Demers LM, Ginger DS, Park SJ, Li Z, Chung SW, Mirkin CA. Direct patterning of modified oligonucleotides on metals and insulators by dip-pen nanolithography. *Science* 2002; 296(5574): 1836–1838.

Index

D

E

F

G

H

N

O

P

Printed in the United States
by Baker & Taylor Publisher Services